ESSENCE OF

STATISTICS

Geoffrey R. Loftus & Elizabeth F. Loftus

UNIVERSITY OF WASHINGTON

Brooks/Cole Publishing Company

MONTEREY, CALIFORNIA

Brooks/Cole Publishing Company
A Division of Wadsworth, Inc.

Printed in the United States of America

10 9 8 7 6 5 4 3 2 1

Library of Congress Cataloging in Publication Data

Loftus, Geoffrey R.
    Essence of statistics.

    Bibliography: p.
    Includes index.
    1. Social sciences—Statistical methods.
2. Statistics. I. Loftus, Elizabeth F.,
1944-   . II. Title.
HA29.L8357   519.5   81-12312
ISBN 0-8185-0475-7   AACR2

Subject Editor: C. Deborah Laughton
Production Service: Cobb/Dunlop Publisher Services, Inc.,
                    New York
                    Maureen P. Conway, Project Editor
Production Coordinator: Cece Munson
Manuscript Editor: Rhoda Blecker
Interior Design: Mina Greenstein
Cover Photo: Geoffrey Loftus
Typesetting: Science Press, Ephrata, Pennsylvania

This book is dedicated to
Edith L. Loftus and Russell Loftus,
without whom the book would
have consisted only of Chapter 17.

ESSENCE OF
STATISTICS

Contents

Preface

We have written this book for three reasons. First, we wanted to present the mathematical rationale underlying various statistical concepts in a way that didn't inflict terror on those who were uncomfortable with mathematics to begin with. Second, we wanted to range over sufficient territory that various commonly used statistical techniques such as repeated-measures analysis of variance and planned comparisons would be included within our scope. Finally, we have found in our courses that a somewhat unusual organization—involving an initial heavy emphasis on basic probability and probability distributions—is highly useful in terms of providing an underlying unity for the material that follows. This organization is reflected in the book.

Statistics is fundamentally a mathematical discipline. For the social science student who enjoys working with mathematics, the task of learning statistics can be a source of pleasure and fulfillment. But for the other student—the one who is frightened by mathematics—the acquisition of statistical knowledge is often accompanied by frustration and pain. We try in this book to cater to both these students. Our strategy is to initially present a particular statistical concept in its most intuitive form, and within the context of a clear, concrete example. Once the intuitions are firmly implanted, then we unveil the general mathematical machinery that corresponds to them, and the relationships between the intuitions and the mathematics are emphasized and reemphasized. Finally, we come to the high-level mathematical underpinnings that are enlightening for the mathematically sophisticated, but at the same time difficult for the uninitiated and generally nonessential for understanding subsequent concepts. These are presented as "digressions." They can be read for pleasure and for deeper understanding or they can be skipped without breaking the flow of the reasoning.

This book also stems from our observations over the years of the sorts of experimental designs commonly used in the social sciences. Most of these designs—and their associated statistical analyses—are somewhat complex, involving multiple factors and repeated measures, post-hoc tests, planned comparisons, and assorted other statistical paraphernalia. Although one could teach these topics by having the student read a short introductory statistics textbook to acquire the fundamentals and then directing the student to a higher-level book, this strategy causes the student to confront changes in notation and loss of continuity. We have found that such disruptions tend to elicit some degree of statistical disorientation, and we thus prefer our own strategy in which the student travels the entire route using the same text.

We believe that statistics, like any other branch of mathematics, is best viewed, and taught, as a cumulative discipline—that is, with any given topic building on the cumulation of the previous topics. That is how this book is organized. We start with the elements of probability theory (Chapter 1) and move on to the general concepts of random variables and probability distributions (Chapter 2). Probability distributions are then used as a means of unifying most of the material in the rest of the book. Most important is a continual emphasis that the process of hypothesis testing does not change throughout a variety of tests (sign test, z-test, t-test, F-test, χ^2 test). All that changes is the distribution of the test statistic under consideration.

The "cumulative philosophy" also dictates a fairly natural order of progression through various of the topics we have chosen to include. As we have just suggested, understanding hypothesis testing rests on the understanding of probability distributions, which in turn requires knowledge of basic probability. Following chapters on these topics, and a chapter on descriptive statistics, we shift to inferential statistics, which constitutes the bulk of the book.

The various experimental situations, along with their associated tests, progress from simple to complex. The first test described is the sign test (based on the binomial distribution), followed by the z-test (based on the normal, which is a "continuous version" of the binomial). Next comes a t–test, which is similar to a z-test except that it is used when the variance of the underlying probability distribution is unknown rather than known. On the heels of the t-test comes a one-way analysis of variance (ANOVA) situation, which is like a t-test situation, but incorporates more than two conditions. Along with the ANOVA situation we describe the F-test, which is like a z- and t-test in the sense that the test statistic in all cases consists of some measure of variance between groups divided by some measure of variance within groups. After one-way ANOVA comes multifactor ANOVA, which is characterized as being like a single-factor ANOVA; but with more than one independent variable. The major new concept introduced here is that of an interaction. Repeated-measures designs are then described as being like a multifactor ANOVA design with one of the factors being "subjects." Here we clearly describe the rationale for using the subject-by-condition interaction as the error term. Following a chapter on the rudiments of regression and correlation, we cover post-hoc tests, planned comparisons, and measuring the strength of a relationship.

By its very nature, the cumulative philosophy embodied in the book limits flexibility with respect to the order in which topics can be covered. This inevitability of progression benefits the student who can more easily see how complex concepts are built on a foundation of simpler ones. The student thereby ends up viewing statistics as a coherent, unified structure rather than a collection of unrelated subtopics.

But the book is not structured so as to make the topic order completely immutable. For example, Chapter 7—covering sampling distributions—is placed within the context of hypothesis testing with normal distributions. Although the information in this chapter is critical, the placement of the chapter within the book is somewhat arbitrary. Likewise, the topics of parameter estimation (Chapter 9), power (Chapter 8), chi-square (Chapter 16), and nonparametric techniques (Chapter 17) are placed where they seem to fit best, but they can all be taken up earlier in the book should the individual instructor desire.

So much for why we organized the book the way we did. Now a few words on how we feel about the way statistical technique is currently applied in social science research. We emphasize hypothesis testing in this book. This emphasis reflects the prevalence of hypothesis testing in social science journals, and the inescapable requirement that students learn to interpret what they read. However, this emphasis should *not* be taken to mean that we believe hypothesis testing is the only, or even the best, statistical vehicle for eliciting interesting information from data. In fact we believe there are a number of analysis techniques that are better for this purpose. These include:

1. A clear initial portrayal of the data via descriptive measures, emphasizing sample statistics, and their associated confidence intervals.
2. A clear assessment of the statistical power in an experiment.
3. Clear alternative hypotheses, which are tested via planned comparisons.
4. Delineation of the strength of association between variables via measures such as r^2 and ω^2.

The first technique is stressed throughout the book. The second is dealt with in a separate chapter on power (Chapter 8) along with an increasing emphasis on confidence intervals, beginning in Chapter 10. We describe the third and fourth techniques, along with the reasons why we believe them to be important, in Chapter 15.

This book has been a long time in the making, and a variety of individuals have provided help, advice, and criticism along the way. To all we extend our heartfelt thanks. Students in our classes who read various portions of the manuscript and caught innumerable errors were Ann Bockman, Tim Cahn, Harold Carter, Marisa Chu, Bill Erdly, Steve Gillispie, Frank Iachelli, Lynn Maclin, Liz Moore, Angie Mylar, Karen Nichols, Steve Okiyama, Davis Pitt, Connie Sroufe, Leo Stefurak, John Takami, Kelly Tissell, and Shelly White. Tim Cahn and Bill Erdly struggled to provide exercise answers. Various reviewers read the manuscript in painstaking detail; their advice has been exceedingly valuable. Those who have chosen to identify themselves are Jeffrey O. Miller, James Thissen, and Brian Wandell, Elizabeth Lynn, and Thomas E.

Malloy. Various typists have labored uncomplainingly to make yet more revisions and meet yet more deadlines. They are Mary Kolpacoff, Beverly Rengert, Kathy Sullivan, Jackie Taghon, and Gail Workman. The production crew, including Cece Munson, Maureen Conway, Rhoda Blecker, and Lavon Armstrong, did a great job transforming a tattered manuscript into a finished product. And finally a special word of thanks is due to Deborah Laughton, our Brooks/Cole editor, who, despite having to put up with endless torment and abuse from us, managed to do a magnificent job of pulling it all together.

Geoffrey R. Loftus
Elizabeth F. Loftus

Introduction

We live in a world riddled with uncertainty, and we seem to be constantly grappling with questions whose answers are elusive and indefinite. Will the stock market be up or down tomorrow? What will the professor pull out of her bag of tricks for the final exam? Will Autumn Mist manage to show in the seventh race? Does our bridge partner hold the ace of clubs or doesn't he?

The twin disciplines of *probability* and *statistics* have evolved out of attempts to cope with this state of affairs.[1] Probability and statistics may be viewed as two sides of the same coin. Probability theory deals with the question: Given a particular state of the world, what consequences are likely to follow? Statistical theory deals with the converse question: Given a particular set of consequences, what state of the world may have given rise to them?

A simple example will illustrate. Suppose that in the United States a newborn baby has an equal chance of being a boy or a girl. In probabilistic terms, this state of the world is expressed by the assertion that "the probability that a newborn baby will be a boy is equal to 0.5." Furthermore, an elementary application of probability theory allows us to compute a specific consequence of this state of the world: Out of any given sample of newborns, about half should be boys. (What is meant by "about" is an interesting, more sophisticated question dealt with by probability theory and will be spelled out in more detail in later chapters.)

Now consider the following situation: An anthropologist studying a primitive culture notices that over a year's time only 35% of newborn infants are boys. What is the anthropologist to make of this set of consequences? One possibility is that the culture is quite different from U.S. culture in terms of what determines an infant's sex. Another possibility, however, is that it has simply been an

---

[1]Probability theory made its debut in the 17th century courtesy of the French philosopher and mathematician Blaise Pascal (of whom we shall hear more later). Pascal was approached by a friend, the Chevalier de Méré, who was evidently possessed with a passion for games of chance. It was Pascal's subsequent effort to work out gambling systems that eventually led to the concepts embodied in modern probability theory. Thus, probability, like so many other endeavors, was born of practical necessity.

1

off-year for baby boys; that is, the apparent discrepancy is due merely to chance. As it turns out, the anthropologist would be able to use statistical theory to make an educated guess as to which of these two states of the world is most likely correct.

We shall have more to say about the relationship between probability and statistics later in this chapter. In the meantime we turn to a discussion of the role of probability and statistics in science.

UNCERTAINTY IN SCIENCE

As indicated by the dilemma of our roving anthropologist, scientific inquiry is not exempt from uncertainty. Some sciences, notably genetics, are constructed almost entirely on a foundation of probability and statistics. Even physics, once thought to represent the quintessence of deterministic endeavor, now finds itself trying to cope with basic particles of matter that, astonishingly, can only be viewed as probability distributions. It is therefore the case that a knowledge of probability and statistics constitutes an indispensable tool for any scientist.

Theories and Experiments

A major goal of science is to construct *theories* which, ideally, are simple, parsimonious descriptions of some aspect of reality. Any specific theory should lead to *predictions* about consequences that should ensue from some particular set of circumstances. An *experiment* consists of setting up such circumstances and observing whether the consequences predicted by the theory do indeed follow. If they do, then the theory receives support. If they do not, the theory (or part of the theory) is disconfirmed.

Determining whether an experiment has yielded the consequences predicted by a theory may seem simple. But often it's not, because the annoying uncertainty to which we keep alluding may creep in to obscure the picture. And this leads to the key use of statistical theory in science: **Statistical theory is used to determine whether or not an obtained set of results may reasonably have occurred given that a particular theory is correct.** This reasoning may sound somewhat confusing, and we will offer an example that we hope will provide some illumination. This example is not meant to present formal statistical methodology (that's the task of the rest of the book), but to provide an *intuitive flavor* for statistical methodology and the means by which it is applied to experimental data.

Strategies for Influencing Attitudes

Imagine that a social psychologist has a theory involving the circumstances under which a person's attitudes toward some topic are altered. Stemming from notions about "speaker authority," the theory makes the prediction that a speech by a well-dressed person will be more effective in changing listeners' attitudes than a speech by a not-so-well-dressed person. How would we design an experiment to test this prediction?

Variables.   Before plunging into this topic, we must digress briefly to discuss a critical concept in scientific experimentation—that of a *variable*. A variable is just what you probably think it is—anything that varies—or, more precisely, anything that can be in more than one state. Table I-1 gives some examples of variables. The amount of gasoline in your car's tank is a variable inasmuch as it can be in any one of a number of states—it can be zero gallons, or 1 gallon, or 3.17 gallons or whatever, up to the capacity of the tank. Likewise, the outfit a person wears on a given day is a variable. On one day the outfit might consist of a blue pinstripe suit, whereas on another day the outfit might consist of a pair of blue jeans and a sweatshirt (each a different "state" of outfit). A third example of a variable is a person's attitude to whether or not marijuana should be legalized. A given person may believe at one time that marijuana should not be legalized and then believe at a subsequent time that it should.

In an experiment we are generally concerned with answering the question: Does a change in some variable lead to a change in some other variable? Recall that the question we have posed in our present example is: Does a speech by a well-dressed person change attitudes (say, about marijuana legalization) more than the same speech by a not-so-well-dressed person? In terms of variables this question becomes: Does a change in the "mode of dress" variable (from well-dressed to not-so-well-dressed) cause a change in the "attitude toward marijuana legalization" variable?

TABLE I-1

Variable	Possible States of the Variable	
I. Amount of gasoline in a car's tank	State 1:	0 gallons
	State 2:	1 gallon etc.
II. Mode of dress	State 1:	blue pinstripe suit; white button-down collar; regimental tie; black shoes
	State 2:	Faded, patched blue jeans; sweatshirt; sandles etc.
III. Attitude toward legalization of marijuana	State 1:	Definitely opposed to legalization (1 on rating scale)
	State 2:	Somewhat opposed to legalization (2 on rating scale)
		.
		.
		.
	State 7:	Definitely in favor of legalization (7 on rating scale)

1   Manipulate state of the goodness-of-dress (independent) variable

<table>
<tr><td>Condition 1:
blue pinstripe suit state</td><td>Condition 2:
blue jeans and sweatshirt state</td></tr>
</table>

2   Measure state of the attitude
toward marijuana legalization (dependent)
variable

<table>
<tr><td>Subject 1: Attitude = 5
Subject 2: Attitude = 6
Subject 3: Attitude = 4
Subject 4: Attitude = 7
Subject 5: Attitude = 3</td><td>Subject 1: Attitude = 3
Subject 2: Attitude = 6
Subject 3: Attitude = 4
Subject 4: Attitude = 3
Subject 5: Attitude = 4</td></tr>
</table>

$$\text{Mean} = \frac{5 + 6 + 4 + 7 + 3}{5} \qquad \text{Mean} = \frac{3 + 6 + 4 + 3 + 4}{5}$$

$$= 5 \qquad\qquad\qquad\qquad = 4$$

FIGURE I-1

To investigate whether a change in one variable causes a change in another, we *manipulate* or *experimentally control* the state of the variable we think is doing the causing, and we *measure* the state of the other variable. The variable whose state we manipulate in an experiment is referred to as an *independent variable,* whereas the variable whose state we measure is referred to as a *dependent variable.* (These terms make sense, since we think of the state of the dependent variable as being dependent on the state of the independent variable.) Figure I-1 illustrates how this strategy would apply to the example at hand. Since we believe that mode of dress is doing the causing, we manipulate the state of that variable. That is, we will have a person give a speech about marijuana legalization to two groups of subjects. While speaking to the first group of subjects, the speaker wears his pinstripe suit. He then dons sweatshirt and blue jeans and proceeds to give the identical speech to a second group of subjects. These two different states of the independent variable correspond to what we refer to as *conditions* of the experiment. Since in our example experiment we have opted to look at two different states of the independent variable, the experiment has two conditions.

After the speech has been given to each group of subjects, we measure the attitude toward marijuana legalization held by each person in the two groups. We can perform this measurement simply by asking each person to *rate* his or her attitude on a scale from 1 to 7, where 1 means "definitely think marijuana should not be legalized" and 7 means "definitely think marijuana should be legalized." Let us imagine that this experiment yields the attitude ratings that are listed in the lower part of Figure I-1.

Data analysis.   We have now collected our data. In this hypothetical experiment, as is often the case, the data consist of *numbers*. Our job is now to take these numbers and use them to answer our original questions—that is, to make conclusions about whether or not a difference in the "mode of dress" variable causes a difference in the "attitude toward marijuana" variable.

How do we get from data to conclusions? Well, to the extent that people's attitude ratings in Condition 1 (the pinstripe condition) tend to be higher than the ratings in Condition 2 (the blue jeans condition), we would tend to conclude that mode of dress *does* have an effect on attitude change. Thus, we would tend to think the prediction held true and our original theory would receive support. Conversely, to the extent that the attitude ratings were about the same in the two conditions, we would tend to conclude that "mode of dress" does not have an effect on attitude change and that our theory was then in trouble.

Just looking at all the numbers, it is difficult to ascertain whether the attitude ratings differ in the two conditions. Some subjects seem to have high ratings in the pinstripe condition, but other subjects in the pinstripe condition have fairly low ratings. Although the scores in the blue jean condition seem in general to be somewhat lower than the scores in the pinstripe condition, there are also both high and low scores within the blue jean condition.

Thus, interpretation of these data does not appear to be altogether straight-forward. As a start we would like to have *one* number from each condition that is indicative or representative of the scores in that condition. Comparison of the two conditions would then consist merely of comparing the two representative scores. A possible candidate for a representative number is the arithmetic average, or *mean,* of the scores in a given condition. The mean is a *descriptive statistic* of a set of scores and will be dealt with in detail in a subsequent chapter. For now suffice it to say that the mean is simply the sum of all the scores divided by the total number of scores. At the bottom of Figure I-1 we have computed the mean for each of the two conditions. The mean attitude rating turns out to be 5 in the pinstripe condition and 4 in the blue jean condition. Hence, we discover that for the subjects in our experiment the mean attitude rating is higher in the pinstripe than in the blue jean condition.

Are we now to conclude that this difference in mean score is caused by the difference in the speaker's outfit? Although this *may* be what's going on, it may also be the case that the speaker's outfit had no effect whatsoever on the subjects' attitudes. Instead, the difference in the means may be due to the fact that, by chance, some people who were favorable toward marijuana happened to be chosen as subjects in the pinstripe condition and/or by chance some people who didn't like marijuana happened to be chosen as subjects in the blue jean condition. Another possible explanation is that one of the subjects in the pinstripe condition got the scale confused and wrote down a 7 when he meant to write down a 1. Or maybe the research assistant who was recording the data erred in copying some of the scores. In short we can't really be certain whether the difference in the dependent variable was really due to a change in the independent variable or whether the difference was merely due to other factors not under our control (which we shall refer to as *chance factors*).

What Governs Our Belief
About What's Going On?

Let us refer to the hypothetical outcome we have just described as "Outcome 1." Suppose that instead of coming out as depicted by Outcome 1, the data in our experiment had come out in the way designated in Table I-2 as "Outcome 2." Although in Outcome 2 the means of the pinstripe and blue jean conditions are still 5 and 4, respectively, we are intuitively a little more confident that this difference in the means is due to "mode of dress" as opposed to chance factors. This is because the data are in a sense more stable—*all* the people in the pinstripe condition have a score of 5, whereas all the people in the blue jean condition have a score of 4. It doesn't make much sense to attribute the difference between conditions in Outcome 2 to chance factors, because if chance factors *were* causing the difference between conditions, we would expect the same chance factors to cause some variation of the scores *within* a given condition.

Table I-2

Outcome 2

Pinstripe Condition	*Blue Jean Condition*
Subject 1: Attitude = 5	Subject 1: Attitude = 4
Subject 2: Attitude = 5	Subject 2: Attitude = 4
Subject 3: Attitude = 5	Subject 3: Attitude = 4
Subject 4: Attitude = 5	Subject 4: Attitude = 4
Subject 5: Attitude = 5	Subject 5: Attitude = 4

Mean = $\dfrac{5 + 5 + 5 + 5 + 5}{5}$ = 5

Mean = $\dfrac{4 + 4 + 4 + 4 + 4}{5}$ = 4

Outcome 3

Subject 1: Attitude = 7	Subject 1: Attitude = 1
Subject 2: Attitude = 7	Subject 2: Attitude = 1
Subject 3: Attitude = 7	Subject 3: Attitude = 1
Subject 4: Attitude = 7	Subject 4: Attitude = 1
Subject 5: Attitude = 7	Subject 5: Attitude = 1

Mean = $\dfrac{7 + 7 + 7 + 7 + 7}{5}$ = 7

Mean = $\dfrac{1 + 1 + 1 + 1 + 1}{5}$ = 1

Outcome 3 (also given in Table I-2) shows an even more extreme outcome. Here all individual scores in the pinstripe condition are 7, whereas all individual scores in the blue jean condition are 1. Here, as in Outcome 2, the fact that there is no variation within a condition leads us to suspect that effects of chance factors are negligible and that any difference between conditions must be due to the speaker's mode of dress. Additionally, however, we note that the mean difference between the two conditions is very large in Outcome 3: the mean attitude score is 7 in the pinstripe condition, but only 1 in the blue jean condition. This large difference strengthens our belief that "mode of dress" must be rather important in terms of influencing attitudes.

These examples should convince us that there are two major aspects of the data on which we base our conclusions: (1) the extent to which individual scores within a condition bounce around, and (2) the extent to which the *mean* scores of the various conditions differ from one another.

PROBABILITY REVISITED

The above example and its various outcomes are meant to illustrate a point we made earlier: uncertainty typically exists in scientific endeavor. No matter *how* the data come out, we cannot (unfortunately) draw completely unequivocal conclusions in terms of answering the question at hand.

If we had observed Outcome 3, we would have been quite certain that the independent variable, "mode of dress," was influencing the dependent variable, "attitude toward marijuana legalization." If we had observed Outcome 2, we would still have been fairly sure that the independent variable was influencing the dependent variable, but not as sure as we would have been had we observed Outcome 3. Finally, had we observed Outcome 1, we would not have been at all sure whether the independent variable was being influenced by the dependent variable or whether the difference between the means of the two conditions was due merely to chance.

All this probably seems very vague, imprecise, and unscientific. How are we ever to construct theories and carry out scientific research in general when we apparently always have to hedge on the answers to our experimental questions? There are two replies—one pessimistic and the other optimistic. The pessimistic reply is that there will probably always be chance factors lurking within our experiments. Although (as will be discussed in subsequent chapters) we usually try to reduce these chance factors, we can't eliminate them. We're stuck with them, and we have to take their existence into account when interpreting data from experiments.

The optimistic reply is that we can be much more precise about our answers using probability theory. That is, a statement such as "the probability that such-and-such is true is 92%" is considerably more useful than a statement such as "I'm pretty sure that such-and-such is true." This is because the former statement is made within the context of a precise mathematical framework and thus has a precise meaning and precise implications. The latter statement, in contrast, involves such vague factors as the speaker's definition of "pretty sure" and the listener's interpretation of the phrase.

The methods by which we get from data to a conclusion based on probability are the major topics dealt with in the rest of this book. In various chapters we will be examining different types of experimental situations and seeing how the data from such situations can be made to yield precise probabilistic conclusions. In order to have a background for this material it is, of course, necessary to have a knowledge of probability theory itself. Hence, in the next few chapters, we will be dealing with the rudiments of probability. This material will appear abstract and somewhat removed from social science, but everything you learn about probability will bear exquisite fruit in the subsequent chapters. (Additionally, you will find a knowledge of probability exceedingly useful if you enjoy dabbling in games of chance like poker, horse racing, lotteries, or backgammon.)

MEASUREMENT SCALES

It is appropriate to complete this chapter with a discussion of *measurement scales*. A measurement scale is essentially a device whereby some physical entity is represented by a number, this number being the *scale value* corresponding to the physical entity. The issue of scales is actually exceedingly complex and extensively discussed within the topic of measurement theory. What follow are the merest rudiments of what the issue is all about.

Four Types of Scales

Classical measurement theory recognizes four basic types of scales: nominal scales, ordinal scales, interval scales, and ratio scales. The types of scales are distinguished in terms of the degree to which arithmetic operations ordinarily performed with numbers (for example, declaring two numbers to be equal; declaring one number to be larger than another; adding numbers together) make sense in terms of the physical things that the numbers are designed to represent. Let us clarify this remark with some examples of the four types of scales.

Nominal scales.   A nominal scale assigns numbers to objects only to distinguish the objects from one another. Suppose, for example, that we want to classify people by the color of their hair. We might do so via the following nominal scale:

Blonds are assigned a 1.
Brunettes are assigned a 2.
Redheads are assigned a 3.
Black-haired people are assigned a 4.

Now what arithmetic operations make sense in terms of the 1s, 2s, 3s, and 4s assigned by our scale? Many operations would clearly *not* make sense. For example it would not make sense to *add* numbers; that is, blond hair (1) plus red hair (3) would not add up to black hair (4). In fact with a nominal scale the *only* operation that makes sense is the equality operation: 3 = 3, in that two redheads are equal to one another within our hair color classification. Likewise, 1 ≠ 2 is meaningful, in that a blond is not equal to a brunette.

Ordinal scales.   An ordinal scale is applied to any set of things *ordered* in some way. A good example of a set of ordered objects is a set of solids ordered with respect to hardness; that is, if solid A will scratch solid B, then solid A is defined as harder than solid B. Suppose, as an example, we had four solids: steel, mahogany, balsa wood, and diamond. Using our scratching test, we would discover them to be ordered, from softest to hardest: balsa, mahogany, steel, diamond.

Now suppose we wanted to assign numbers to reflect this hardness. A possible such set might be:

    balsa: 1
    mahogany: 2
    steel: 3
    diamond.   4

What arithmetic operations would be applicable to these numbers? Certainly as with a nominal scale, equality/inequality would make sense in that such statements as 1 = 1 and 3 ≠ 4 are true characterizations of the things (in this case, the solids) that these numbers represent. In addition we can make use of the notion of less than (<) and more than (>). Thus, for example, the statements

    4 > 2

and

    1 < 3

both make sense in that they represent that diamond (4) is *harder* (>) than mahogany (2), and that balsa (1) is not as hard (<) as steel (3).

Interval scales.   With an ordinal scale it would *not* make sense to consider *differences* among values. For instance the difference between the scale values corresponding to steel and balsa (3 − 1 = 2) is the same numerically as the difference between diamond and mahogany (4 − 2 = 2). However, the difference in the corresponding hardnesses are not necessarily equal.

An interval scale uses numbers to represent things that are ordered in a special way such that the differences between the scale values of the things are meaningful. A good example is the common Fahrenheit scale of temperature. Consider four temperatures: 30°, 40°, 80°, and 90°. Here in addition to the equality operation (30° = 30°; 40° ≠ 80°) and the ordering operation (40° > 30°, 40° < 80°), it also makes sense to consider differences between values. For instance the difference between 90° and 80° is in a real sense equal to the difference between 30° and 40°. Likewise the difference between 80° and 40° can be reasonably viewed as greater than the difference between 30° and 40°.

Ratio scales.   A ratio scale uses numbers to represent things that are representable by an interval scale, but in addition have a meaningful zero point. Height is a good example of such a scale, in the sense that a height of zero is a meaningful entity.

All arithmetic operations that are meaningful in terms of an interval scale (equality, ordering, subtraction) are also meaningful in terms of a ratio scale. In addition, one can add numbers (for example, 5 ft + 6 ft = 11 ft) and divide numbers (for example, 5 ft is 5/20 = 1/4 of 20 ft).

Permissible Transformations

These four scales are also distinguished from one another in terms of what is referred to as the transformations that are permissible on the scale values. Suppose we have a particular scale whose values are $V_1, V_2, V_3 \ldots$ and so on. (In our hair color example the scale contained the four values $V_1 = 1$, $V_2 = 2$, $V_3 = 3$, and $V_4 = 4$). Suppose we wished to select a set of new scale values V_i' such that the new values accomplished the same things in their representational abilities as did the old values. What new values can we choose?

A moment's reflection should convince us that as we go from nominal to ordinal to interval to ratio scales, we are increasingly restricted in terms of the types of transformations that are permissible. Let us discuss the restrictions for the four types of scales in turn.

Nominal scales.   Here we are permitted any transformation we like, with the one restriction that for any two values V_i and V_j and their transformed values V_i' and V_j',

$$V_i' \neq V_j' \quad \text{if and only if} \quad V_i \neq V_j$$

That is, if two original scale values (such as $V_1 = 1$ and $V_2 = 2$ representing blond and brown hair) are different, their transformed values in the new scale must also be different. So, for example, we could let the new scale values be $V_1' = 10$ and $V_2' = 7$ for blond and brown hair. But we could not let $V_1' = V_2' = 10$ represent both blond and brown hair.

Ordinal scales.   Here we have the further restriction that

$$V_i' > V_j' \quad \text{if and only if} \quad V_i > V_j$$

Thus, suppose in the hardness example we used we wanted new scale values to represent the hardness of our four solids. It would be permissible to use 102 for steel and 99 for mahogany, but it would not be permissible to use 102 for steel and 110 for mahogany.

Interval scales. Here only *linear* transformations may be used—that is, transformations of the form

$$V'_i = aV_i + b$$

where *a* and *b* are constants.

An example of such a transformation is that which transforms temperature in Fahrenheit to temperature in Celsius. For any value (V_i) in Fahrenheit, the transformation that produces the corresponding value (V'_i) in Celsius is:

$$V'_i = 0.56V_i - 18$$

Thus, a temperature of 50° F would be (0.56) (50) − 18 = 10° C.

Ratio scales. Finally, the only transformation permissible on a ratio scale is multiplication by a constant:

$$V'_i = cV_i$$

An example of such a transformation changes length in feet (V_i) to length in meters (V'_i):

$$V'_i = 0.305V_i$$

For example, 6 ft would be (6) (0.305) = 1.83 m.

Scales and Statistical Applications

The data we gather in an experiment generally consist of numbers which, in turn, are scale values corresponding to the measurement of some physical entity. The way we treat our data is determined by the type of scale we use. We can roughly dichotomize data into those measured on nominal and ordinal scales versus those measured on interval and ratio scales. In the latter case we are able to carry out what are referred to as *parametric procedures,* which constitute the vast bulk of this book's subject matter. In the former case we must make use of nonparametric procedures, which are the topic of Chapter 17.

The preceding remarks probably seem to imply that any set of data can be unambiguously classified as being measured on one type of scale or another. Unfortunately, this is by no means always the case. If we are measuring height, we can be quite sure we have an interval scale; if we are measuring hardness of solids, we have an ordinal scale. But what about measuring attitudes with a rating scale as in the example described above? This is certainly at least an ordinal scale (since attitudes are ordered from "disagree" to "agree"), but probably not a ratio scale (since there is not an obvious meaningful zero point).

Whether it is an ordinal or an interval scale is somewhat of a value judgment. We have implicitly treated it as an interval scale, since computation of a mean turns out to be a parametric operation. But this is not a self-evident truth, and the ambiguity we suffer in the determination of the scale dictates concomitant caution in interpretation of the results.

Probability Theory 1

In this chapter we will cover some very fundamental aspects of mathematical probability theory. As it turns out, a convenient and relatively simple-to-understand background for probability theory is provided by the framework of *set theory*. The pages to follow therefore assume a familiarity with the basic concepts of set theory. If you have any doubts about the degree of your familiarity with set theory, you should read Appendix A before proceeding.

PROBABILITY: BASIC RULES AND TOOLS

Everybody has an intuitive idea of what probability is. *Probability* has to do with the likelihood that some event in which you are interested will happen. When you flip a coin, you know that it has some probability of coming up heads. On any given day there is some probability that it will snow in the Sahara Desert, but this probability, we would intuit, is smaller than the probability that a flipped coin will turn up heads.

The probability that some event will occur is represented by assigning to that event a number ranging from zero (signifying that the event definitely will *not* occur) to one (signifying that the event definitely *will* occur). The probability that a flipped coin will turn up heads is 0.5, or halfway between the two extremes. The fact that this probability is halfway between the two extremes indicates that the coin is equally likely to turn up heads or tails. On the other hand the probability that it will snow in the Sahara Desert tomorrow is close to zero, inasmuch as the event will almost definitely not occur.

Situations and Outcomes of Situations

When discussing probability, we are typically concerned with some *situation* and with the events that constitute *outcomes* of that situation. For example, a situation might consist of going on a picnic, with the outcomes consisting either of rain or of sunshine. Or the situation might consist of inviting someone on a

Digression 1.1

Probability and Everyday Life

People base a good deal of their behavior on their feelings about probabilities. Sometimes this is done explicitly—for example, most professional gamblers have both an intuitive feeling and a mathematically solid knowledge of the probabilities of various outcomes involving cards, dice, roulette wheels, or whatever. Most people, however, base behavior on only a subjective feeling of what various probabilities are. Given the choice of planning to go either on a picnic tomorrow or to the movies, most people would opt for the movies if the weatherman said that the probability of rain tomorrow were 95%. Person X's decision about whether or not to ask Person Y for a date can depend on the probability that Person Y will answer yes. A skier's decision about whether to ski down a very steep chute will depend upon what the skier thinks the probability is that she will crash and get hurt, and so on.

As noted in the previous chapter, social scientists tend to be professionally interested in probability because conclusions based on experiments can typically be stated only probabilistically. This is the principal reason that probability is discussed in books such as this one. However, since so many everyday events are based on probabilistic happenings, it is to everyone's advantage (whether interested in social science or not) to have a fairly solid idea of what probability is all about.

date with the outcomes being that the person either will or will not accept the invitation.

Table 1-1 lists some situations that provide convenient examples to use when discussing probability theory. For each situation we have stated some hypothetical outcome in which we imagine ourselves to be interested. For

TABLE 1-1.   Situations and Outcomes of Interest

Situation	*Outcome of Interest*
Flip a coin	Coin comes up heads
Throw a die	Die comes up 2
Draw a card	Card is a heart
Select a random person from a class	Person has passed the first hour exam
Pick a random day in Seattle	It doesn't rain that day
Flip a coin three times	Two of the three flips come up heads

instance, we may flip a coin and call heads to decide who will pay for dinner. Thus, the occurrence of a head would be the outcome of interest. Throwing a die during a Monopoly game, we may need a two in order to land on Boardwalk. If we draw a card during a poker game, we may need a heart to complete a flush.

Let's focus on one of these situations—rolling a die—and consider the *set* of all possible outcomes. This set, generally referred to as *S*, is called the *sample space* and is the universal set with which we are concerned in any probabilistic situation. When we roll a die, of course, $S = \{1, 2, 3, 4, 5, 6\}$.

Each possible outcome (member of the sample space) is referred to as an *elementary event*. Of interest is the *number* of elementary events in the sample space. We shall refer to this number as $n(S)$. In this case, $n(S)$ is equal to 6 since there are 6 possible outcomes. Now we consider the set of elementary events that make up the outcome in which we are interested. We shall refer to this set as *A*. In this case the outcome we are interested in is the die coming up a two. Therefore,

$$A = \{2\}$$

How many elementary events are there in *A*? Continuing with the notation introduced above, we refer to the number of elementary events in *A* as $n(A)$; and in our example we can easily see that $n(A)$ is equal to one.

We are now ready to provide our first formal definition of probability.[1] The probability of the event that we are interested in, which we will denote $p(A)$, is obtained as follows: **$p(A)$ equals the number of elementary events in the outcome set divided by the number of elementary events in the sample space,**

or

$$p(A) = \frac{n(A)}{n(S)} \qquad (1.1)$$

In this case, $p(A) = 1/6 = 0.167$.

Let us consider another situation—drawing a card from a deck. Suppose we are interested in drawing a heart. In terms of the sample space, $S = \{$all cards in standard deck$\}$. Notice that there are 52 members of this set, or 52 elementary events; thus, $n(S) = 52$. Let's denote *H* as the set of all elementary events corresponding to the outcome "drawing a heart." In this case $H = \{$all hearts in the deck$\}$. Since there are 13 hearts, this means that there are 13 elementary

---

[1]The issue of how probability should be defined has been under some debate for several centuries, and several different definitions have evolved. The definition we offer here is quite restricted and we will loosen it somewhat in a subsequent section. Although the definitions of probability provided in this book will suffice for the purposes at hand, the reader is warned that there are other definitions we will not cover. Discussion of these can be found in books concerned specifically with probability theory.

events corresponding to this outcome, or $n(H) = 13$. Therefore, using Equation 1.1,

$$p(H) = \frac{n(H)}{n(S)} = \frac{13}{52} = 0.250$$

In these examples it was fairly simple to calculate the number of elementary events in the sample space and the number of elementary events in the outcome of interest. Often it is somewhat less simple. Suppose, for example, that the situation consists of flipping a coin three times, and the outcome of interest consists of getting exactly two heads and one tail (in any order). What is the probability of this outcome?

First let us determine the sample space in this situation. To deal with this issue, we'll initially consider what might constitute an elementary event. It seems reasonable to suppose that one elementary event might consist of getting a head on all three flips, or "head, head, head." Another elementary event might consist of getting tails on the first and third flip and a head on the second flip, or "tail, head, tail," and so on. If H refers to getting a head, and T refers to getting a tail, all elementary events can be listed systematically:

$$S = \{(HHH), (HHT), (HTH), (HTT), (THH), (THT), (TTH), (TTT)\}$$

By counting up the number of elementary events in S, we determine that $n(S) = 8$. Remember that our outcome of interest consists of getting two heads and a tail in any order. Let us designate this outcome as A. Looking at all possible elementary events, it is easy to list the elementary events corresponding to A:

$$A = \{(HHT), (HTH), (THH)\}$$

Again, simply by counting, we determine that $n(A) = 3$.

Now our task of computing the probability of getting two heads and one tail in any order becomes very simple. We simply plug $n(A)$ and $n(S)$ into Equation 1.1 and find that

$$p(A) = \frac{n(A)}{n(S)} = \frac{3}{8} = 0.375$$

Notice that by having listed all the elementary events in S, we are in a position to compute any probability we wish that involves this particular situation. For example suppose that we are interested in the probability of getting a head, a head, and a tail *in that order*. We see by inspecting our listing of S that there is one elementary event—namely, HHT, corresponding to that outcome. Referring to this outcome as B, we see that $n(B)$ is equal to one. Therefore,

$$p(B) = \frac{n(B)}{n(S)} = \frac{1}{8} = 0.125$$

More General Laws of Probability

In all examples that we have discussed so far, we have been making the important, implicit assumption that all elementary events have an equal probability of occurring. In terms of the examples the assumption is that we have a fair die (equal probability of coming up any given number between one and six), a fair coin (equal probability of coming up heads or tails), a well-shuffled deck, and so on. Now we would like to describe some more general laws about probability that don't require this assumption.

The addition law for mutually exclusive events. Consider again some situation that has a sample space S. Suppose there are a number of possible outcomes of that situation (events), and let the sets corresponding to these events be A, B, ..., N. For convenience of notation we will also refer to the events themselves as A, B, ..., N. Now the addition rule may be stated as follows. If all the sets A, B, ..., N are *mutually exclusive*, then

$$p(A \cup B \cup \cdots \cup N) = p(A) + p(B) + \cdots + p(N). \quad (1.2)$$

That is, to get the probability of the *union* of all these events, we simply compute the probability of each individual event and sum these probabilities.

As an example suppose that the situation consists of picking a card from a standard deck. In this case we will let

A = heart
B = spade
C = diamond

We know that $p(A) = p(B) = p(C) = 1/4$ if we are drawing from a fair deck. For the purposes of illustrating the addition law, it is important to note that these three events would be mutually exclusive. That is, if you drew a heart, you couldn't draw a spade or a diamond; if you drew a spade, you couldn't draw a diamond or a heart; and so on.

Suppose now that we were interested in the probability of drawing a heart *or* a spade *or* a diamond. This, of course, would correspond to the probability of $A \cup B \cup C$. By Equation 1.2, this probability would be equal to

$$p(A \cup B \cup C) = p(A) + p(B) + p(C) = \frac{1}{4} + \frac{1}{4} + \frac{1}{4} = \frac{3}{4}$$

The addition rule applied to mutually exclusive and exhaustive events. One consequence of this rule is often very useful. Suppose once again that we have a sample space, S, and we have several possible outcomes, A, B, ..., N. Suppose furthermore that the sets corresponding to these events are mutually exclusive and exhaustive. What kinds of things do we then know about the probabilities of these events? First, since the events are mutually *exhaustive*, we know that

$$A \cup B \cup \cdots \cup N = S$$

Second, since S is the sample space of all possible outcomes of our situation, we know it must be true that $p(S) = 1.0$; that is, *some* outcome in S has to occur. **Finally, since A, B, ..., N have been defined to be mutually exclusive as well as mutually exhaustive, we can apply the addition rule to get**

$$p(A) + p(B) + \cdots + p(N) = 1.0.$$

Suppose, for example, that I know that the probability of hitting a red traffic light at the corner of Grapevine and Gopher is 0.56. What is the probability that I will hit a yellow or a green light at this particular intersection?

Well, we know that

$$p(\text{red} \cup \text{green} \cup \text{yellow}) = p(\text{red}) + p(\text{green}) + p(\text{yellow})$$

since red, green, and yellow are mutually exclusive. Furthermore,

$$p(\text{red} \cup \text{green} \cup \text{yellow}) = 1.0$$

since red, green, and yellow are mutually exhaustive. Hence,

$$p(\text{red}) + p(\text{green}) + p(\text{yellow}) = 1.0.$$

Since $p(\text{red}) = 0.56$,

$$p(\text{green}) + p(\text{yellow}) = 1.0 - 0.56 = 0.44$$

Thus, the probability is 0.44 that I will run into a yellow or a green light.

Complements.   The notion of dealing with probabilities of two events whose corresponding sets are complements of one another follows directly from the above discussion. To be precise, suppose that we have a sample space S and two possible outcomes A and B. Suppose further that the sets corresponding to A and B are mutually exclusive and exhaustive (with respect to S, which, remember, we are counting as the universal set). Then, by definition, B is the complement of A, or $B = \overline{A}$. Since A and B are mutually exclusive and exhaustive, we know that

$$p(A \cup B) = p(S) = p(A) + p(B) = 1.0$$

Therefore, we can state the general rule that, since $p(B) = p(\overline{A})$,

$$p(A) + p(\overline{A}) = 1.0$$

or

$$p(A) = 1.0 - p(\overline{A})$$

Likewise,

$$p(\overline{A}) = 1.0 - p(A)$$

We therefore come up with the complement rule: the probability of some event is one minus the probability of the complement of that event.

To illustrate this rule, suppose that the situation of interest consists of picking a random person and determining that person's age in years. Suppose we know that the probability that the person is 21 or over is 0.8. What's the probability that the person's age is less than 21? Well, since "21 or over" and "less than 21" are complements of one another, we know that the probability that the person is less than 21 is 1.0 minus 0.8, or 0.2.

1.1 When outcomes A and B are mutually exclusive and exhaustive, then

$$p(A) = 1.0 - p(B)$$

and

$$p(B) = 1.0 - p(A)$$

Events that are not mutually exclusive.   In preceding sections we have discussed how one computes probabilities involving events that are mutually exclusive. Suppose we have two events that are *not* mutually exclusive. How do we compute the probability of the unions of such events? **When we are concerned with two such events, the general formula is:**

$$p(A \cup B) = p(A) + p(B) - p(A \cap B) \qquad (1.3)$$

That is, we still add the two individual probabilities, but we then subtract the probability of the intersection. (You may wonder about the case when there are *more* than two events. In this case the situation becomes more complicated. Although we won't discuss it, it would be a useful exercise for you to try to work out an example of it.)

To get an intuitive feel for this law, consider a situation in which we are drawing a card from a standard deck. Suppose that

J = drawing a jack

and

H = drawing a heart

It should be fairly evident that

$$p(J) = 4/52$$

since there are four jacks in a 52-card deck, and

$$p(H) = 13/52$$

since there are 13 hearts in the deck. What is $p(J \cap H)$—that is, the probability of drawing the jack of hearts? Well, since there is only one jack of hearts in the deck (that is, only one elementary event in the set $J \cap H$), this probability must be $1/52$. Using Equation 1.3, $p(J \cup H)$, the probability of drawing a jack or a heart or both is

$$p(J \cup H) = p(J) + p(H) - p(J \cap H)$$
$$= 4/52 + 13/52 - 1/52 = 16/52 = 4/13$$

In the context of this example it should be easy to see why we have to subtract the intersection—in counting up the 13 hearts and the 4 jacks, we counted the jack of hearts twice, so we must subtract it.

Let us note parenthetically that Equation 1.3 is a *general* formula. To see why this is so, we note that if A and B are mutually exclusive, then

$$A \cap B = \phi$$

or the intersection of A and B is the empty set. Since there are by definition zero events in the empty set, we know by Equation 1.1 that

$$p(A \cap B) = p(\phi) = \frac{n(\phi)}{n(S)} = \frac{0}{n(S)} = 0$$

Thus,

$$p(A \cup B) = p(A) + p(B) - p(\phi) = p(A) + p(B).$$

So we see that when A and B are mutually exclusive, Equation 1.3 reduces to Equation 1.2.

Use of Venn Diagrams to Get a Feeling for Probabilities

Venn diagrams are convenient pictorial representations of sets and probabilities. In Appendix A we illustrate how Venn diagrams are useful for showing the relations of sets to one another. They are similarly useful for showing how probabilities are calculated. To demonstrate this usefulness, consider the following example: Let S be a sample space that consists of the set of 12 people in a sauna. Note that $n(S) = 12$. Let sets M and L be characterized as follows:

M = all males in the sauna

Suppose there are four males, so $n(M) = 4$.

L = all left-handed people in the sauna

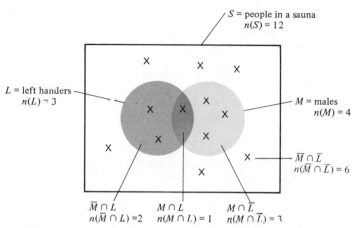

S = people in a sauna
n(S) = 12

L = left handers
n(L) = 3

M = males
n(M) = 4

$\overline{M} \cap \overline{L}$
$n(\overline{M} \cap \overline{L}) = 6$

$\overline{M} \cap L$
$n(\overline{M} \cap L) = 2$

$M \cap L$
$n(M \cap L) = 1$

$M \cap \overline{L}$
$n(M \cap \overline{L}) = 3$

FIGURE 1-1. Representation of two sets, "left-handers" and "males," in a Venn diagram. The universal set consists of 12 people in a sauna. Each member of the universal set is represented by an x.

Suppose there are three left-handers, so $n(L) = 3$. Finally, suppose that there is one left-handed male in the sauna, or

$$n(M \cap L) = 1$$

Figure 1-1 shows how these sets are represented in a Venn diagram. Using the frequencies of various sets shown in the Venn diagram, it is easy to compute the corresponding probabilities. For example,

$$p(M) = n(M)/n(S) = 4/12 = 0.333$$
$$p(L) = n(L)/n(S) = 3/12 = 0.250$$
$$p(\overline{M}) = n(\overline{M})/n(S) = 8/12 = 0.667$$
$$p(\overline{L}) = n(\overline{L})/n(S) = 9/12 = 0.750$$
$$p(M \cap L) = n(M \cap L)/n(S) = 1/12 = 0.083$$
$$p(\overline{M} \cap L) = n(\overline{M} \cap L)/n(S) = 2/12 = 0.167$$

and so on.

Frequencies to probabilities. Notice now that the probability of event X, $p(X)$, is obtained from the corresponding number of elementary events or *frequency*, $n(X)$, simply by dividing by $n(S)$ (which is equal to 12 in the present example). Since we are typically more interested in the probabilities than we are in the frequencies, we may as well just do the following: We divide all the various frequencies from the Venn diagram in Figure 1-2, by $n(S) = 12$ to get probabilities, and then make a new Venn diagram that has probabilities rather than frequencies as its components. This new Venn diagram is illustrated in Figure 1-2.

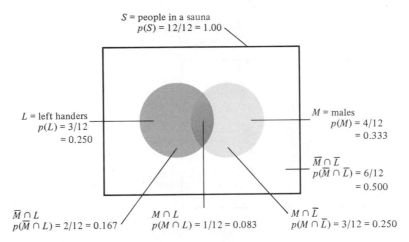

S = people in a sauna
$p(S) = 12/12 = 1.00$

L = left handers
$p(L) = 3/12$
$= 0.250$

M = males
$p(M) = 4/12$
$= 0.333$

$\overline{M} \cap \overline{L}$
$p(\overline{M} \cap \overline{L}) = 6/12$
$= 0.500$

$\overline{M} \cap L$
$p(\overline{M} \cap L) = 2/12 = 0.167$

$M \cap L$
$p(M \cap L) = 1/12 = 0.083$

$M \cap \overline{L}$
$p(M \cap \overline{L}) = 3/12 = 0.250$

FIGURE 1-2.   The numbers of Figure 1-1 have now all been divided by $n(S)$ to represent the probabilities instead of the frequencies (numbers).

A Different Sort of Venn Diagram: The Contingency Table

As we have noted, Venn diagrams are convenient pictorial representations of sets and probabilities. As such they are widely used and useful things to know about. However, let us look with a slightly more jaundiced eye at a Venn diagram—say, at the Venn diagrams shown in Figures 1-1 and 1-2. There is something unappealingly lopsided about this representation. For instance, why should the representation of $M \cap L$ (a little sliver) look so different from the representation of $\overline{M} \cap \overline{L}$ (a big rectangle with an oddly shaped hole in it)? Also, it is somewhat awkward to label the various things. For instance, the arrows from $p(M)$ and $p(M \cap \overline{L})$ might, to the uninitiated eye, appear to be pointing to the same piece of the Venn diagram. Finally each of the two sets of interest—males and left-handers—has a convenient complement. But suppose that in addition to left- and right-handers there were a few ambidextrous people, and we wanted to take them into account. How would we represent this additional information in the Venn diagram? It would not be straightforward.

A different representation of a Venn diagram for solving these problems is called a *contingency table*. Table 1-2 shows how the Venn diagrams of Figures 1-1 and 1-2 become a contingency table. First the sample space, *S*, is represented by a rectangle, just as before. Now to represent the partition of the sample space into *M* and $\overline{M}$, we simply divide the rectangle in half horizontally, creating two rows. The top half represents *M* (males) and the bottom half represents $\overline{M}$ (females). Likewise, to partition the sample space into *L* and $\overline{L}$, we divide the rectangle in half vertically, creating two columns. The left half of the figure represents *L* (left-handers) whereas the right half represents $\overline{L}$ (right-handers). The four boxes or *cells* formed by these two divisions very neatly represent $M \cap L$, $M \cap \overline{L}$, $\overline{M} \cap L$, and $\overline{M} \cap \overline{L}$. (These intersections are often called *joint*

TABLE 1-2.   A 2 × 2 Contingency Table Representation of Various Probabilities

	L (left-handers)	$\overline{L}$ (right-handers)	
M (males)	$n(M \cap L) = 1$ $p(M \cap L) = 1/12 = 0.083$	$n(M \cap \overline{L}) = 3$ $p(M \cap \overline{L}) = 3/12 = 0.250$	$n(M) = 4$ $p(M) = 4/12 = 0.333$
$\overline{M}$ (females)	$n(\overline{M} \cap L) = 2$ $p(\overline{M} \cap L) = 2/12 = 0.167$	$n(\overline{M} \cap \overline{L}) = 6$ $p(\overline{M} \cap \overline{L}) = 6/12 = 0.500$	$n(\overline{M}) = 8$ $p(\overline{M}) = 8/12 = 0.667$
	$n(L) = 3$ $p(L) = 3/12 = 0.250$	$n(\overline{L}) = 9$ $p(\overline{L}) = 9/12 = 0.750$	$n(S) = 12$ $p(S) = 12/12 = 1.0$

events when we are talking about probabilities. We will henceforth use the terms "intersection" and "joint event" interchangeably.) All relevant frequencies and probabilities are very easy to depict. Notice that the frequencies and probabilities of M, $\overline{M}$, L, and $\overline{L}$ are represented in the far right and lower margins of the table.

Notice also that the probabilities and frequencies of the two left-hand column cells add up to corresponding frequencies and probabilities in the margin at the bottom left. The same holds true for the two cells in the right column, the two cells in the top row, and the two cells in the bottom row. Typically, at the bottom right-hand corner, we put $n(S)$, or the total number of elementary events, in the entire sample space. Notice that $n(S)$ is (a) the sum of the two row totals, (b) the sum of the two column totals, and (c) the sum of the four cell frequencies.

This particular kind of contingency table is referred to as a 2 × 2 contingency table because it has 2 columns × 2 rows. But what about the problem of partitioning the sample space into more than simply an event and its complement? Suppose that in addition to the three left-handers, there were also two ambidextrous people, one male and one female, in our sauna. Now, it would not really be appropriate to claim that there were three left-handed people and nine right-handed people, as has been done in the contingency table in Table 1-2. Rather, we would want to somehow represent the information that, in terms of handedness, there are three left-handers, two ambidextrous people, and seven right-handers. Using a contingency table, this is very easy to do; we simply increase the number of columns in the table from two to three (thereby making it a 3 × 2 contingency table). This has been done in Table 1-3, where we have also depicted the information that one of the two ambidextrous people was male and the other was female.

The notion of a contingency table is a very important one. We reiterate here (because it can't be reiterated too much) the fact that a contingency table is a very powerful, convenient, and clear way of representing information. Contingency tables will crop up in many places throughout this book under many

TABLE 1-3.   A 3 × 2 Contingency Table

	L (left-handers)	A (ambidextrous)	R (right-handers)	
M (males)	$n(M \cap L) = 1$ $p(M \cap L) = 1/12 = 0.083$	$n(A \cap M) = 1$ $p(A \cap M) = 1/12 = 0.083$	$n(R \cap M) = 2$ $p(R \cap M) = 2/12 = 0.167$	$n(M) = 4$ $p(M) = 4/12 = 0.333$
$\overline{M}$ (females)	$n(\overline{M} \cap L) = 2$ $p(\overline{M} \cap L) = 2/12 = 0.167$	$n(A \cap \overline{M}) = 1$ $p(A \cap \overline{M}) = 1/12 = 0.083$	$n(R \cap \overline{M}) = 5$ $p(R \cap \overline{M}) = 5/12 = 0.417$	$n(\overline{M}) = 8$ $p(\overline{M}) = 8/12 = 0.667$
	$n(L) = 3$ $p(L) = 3/12 = 0.250$	$n(A) = 2$ $p(A) = 2/12 = 0.167$	$n(R) = 7$ $p(R) = 7/12 = 0.583$	$n(S) = 12$ $p(S) = 12/12 = 1.00$

different guises. If you don't completely understand all the concepts involved in the notion of a contingency table, we strongly urge you to go back and reread this section.

Relationship Between Probabilities and Frequencies: In the Long Run

You may have noticed that we have been switching back and forth in a rather cavalier fashion between frequencies and probabilities. We do this because probabilities and frequencies are very closely tied to one another. One way of thinking about the probability of some event is to consider probability as equivalent to the proportion of times (relative frequency) that that event will occur in the long run. ("In the long run" refers to the hypothetical circumstance in which an event has virtually an infinite number of opportunities to occur.) For example, you throw a die and you are interested in the outcome of getting a four. We know that the probability of getting a four is one in six. Now on any given throw of the die you either will come up with a four or you won't. However, over a very large number of throws of the die—thousands and thousands of throws— we expect the proportion of times the die comes up four to be about one in six. Thus, given a very large number of possible occurrences, we would expect results close to what probability predicts.

Bernoulli's theorem. It is mathematically demonstrable that the more chances some event has to occur (in the example above, the more tosses of the die), the closer we can expect the relative frequency of the event to be to the probability of the event. To illustrate, suppose that we throw a die 12 times. We might expect the relative frequency of rolling a four to be the probability of a four, or one in six. Therefore, of the 12 throws, we might expect $\frac{1}{6}$ or two of the throws to come up fours. But we wouldn't be terribly surprised if the frequency of fours that we actually got turned out to be as low as zero (a relative frequency of zero). We likewise wouldn't be shocked if we actually got as many as, say, four fours—a relative frequency of $\frac{1}{3}$.

Now suppose that we flip the coin 1200 instead of 12 times. The probability in this case would be extremely low that we would actually obtain a proportion as low as zero (not a single four in 1200 rolls) or a frequency as high as one in three (400 fours in 1200 throws). Mathematically, this notion is known as Bernoulli's theorem, and its formulation follows. Suppose that we take any number, ϵ, which we can make as small as we like as long as it is greater than zero. Let us further designate the number of rolls of the die (or, in general, the number of opportunities some event has to occur) as n and the frequency with which a four actually shows up in those n opportunities as $f(4)$. It is then the case that

$$P\left\{\left|\frac{f(4)}{n} - \frac{1}{6}\right| \geq \epsilon\right\}$$

approaches zero as n approaches infinity.

CONDITIONAL PROBABILITY AND INDEPENDENCE

In the last section we outlined some very general notions of probability and gave some general rules and tools for thinking about and calculating probabilities of various events. In this section, we will move along to some more complex (and very useful) topics of probability—*conditional probability* and *independence*.

Conditional Probability

The notion of conditional probability is easiest to describe using an example. Suppose we have a sample space consisting of a 120-person college class [hence, $n(S) = 120$]. Let us define a few sets within this sample space.

R = {all people who have done the reading for the midterm}.

Suppose that $n(R) = 90$. [Thus, $n(\overline{R}) = 120 - 90 = 30$.]

P = {all people who passed the midterm}

Suppose that $n(P) = 75$. [Thus, $n(\overline{P}) = 120 - 75 = 45$.] Finally, suppose that 70 people have both done the reading *and* passed the midterm; thus $n(P \cap R) = 70$.

Filling in a contingency table.   We now have enough information to represent this situation in a contingency table, which is shown in Table 1-4. First the frequency of 70 corresponding to $n(P \cap R)$ goes in the upper left-hand cell of the table. We can fill in the other three cells of the tables as follows: We know that of the 90 people who did the reading, 70 passed the exam. Therefore, the other 20 people who did the reading must not have passed and must be contained

TABLE 1-4.   A 2 × 2 Contingency Table

	R (done reading)	$\overline{R}$ (not done reading)	
P (passed exam)	$n(P \cap R) = 70$ $p(P \cap R) = 70/120 = 0.583$	$n(P \cap \overline{R}) = 5$ $p(P \cap \overline{R}) = 5/120 = 0.042$	$n(P) = 75$ $p(P) = 75/120 = 0.625$
$\overline{P}$ (did not pass exam)	$n(\overline{P} \cap R) = 20$ $p(\overline{P} \cap R) = 20/120 = 0.167$	$n(\overline{P} \cap \overline{R}) = 25$ $p(\overline{P} \cap \overline{R}) = 25/120 = 0.208$	$n(\overline{P}) = 45$ $p(\overline{P}) = 45/120 = 0.375$
	$n(R) = 90$ $p(R) = 90/120 = 0.750$	$n(\overline{R}) = 30$ $p(\overline{R}) = 30/120 = 0.250$	$n(S) = 120$ $p(S) = 120/120 = 1.00$

in $R \cap \overline{P}$, which is represented by the lower left-hand cell. Likewise, of the 75 people who passed the exam, 70 did the reading; therefore, the other 5 must not have done the reading. So these 5 people are in $\overline{R} \cap P$, and a 5 goes in the upper right-hand cell. Finally we can get the frequency for the lower right-hand cell (representing $\overline{R} \cap \overline{P}$) in any one of three ways. First we know that of the 30 people who didn't do the reading, 5 passed the exam; therefore, the other 25 must not have passed. Or we can use the fact that of the 45 people who did not pass, 20 did the reading. Therefore, the other 25 must not have done the reading. Or finally, of the 120 total people in the class, $70 + 20 + 5 = 95$ of them are already accounted for in three of the cells; therefore, the other 25 must be in the fourth cell.

Unconditional probabilities.   Let us for the moment concern ourselves with the probabilities that are shown in the margin which, appropriately enough, are sometimes called *marginal probabilities*. For example, the probability that a person in the class has done the reading, $p(R)$, is $90/120 = 0.750$; and the probability that a person in the class has not done the reading, $p(\overline{R})$, is $30/120 = 0.250$. [Or we can compute it as $p(\overline{R}) = 1 - p(R) = 1 - 0.750 = 0.250$.] Likewise, the probability that the person has passed the exam is 0.625, and the probability that the person has not passed the exam is 0.375. These marginal probabilities are also referred to as *unconditional probabilities*. Thus, for example, we would say that the unconditional probability that a random person in the class has done the reading is 0.750.

Conditional probabilities: Restricting the sample space.   Imagine that we pick a random person from the class and *we ask the person if he or she has done the reading*. If the person answers "yes," we know that he or she is one of those 90 people in the left column of Table 1-4. With this information in mind let us now consider the probability that the person has passed the exam. To compute this probability, we note that we have reduced our sample space to those 90 people who have done the reading. Furthermore, we know that of the 90 people who did the reading, 70 passed the exam. Therefore, we conclude that the probability of this particular person having passed the exam is

$$\frac{n(P \cap R)}{n(R)} = \frac{70}{90} = 0.778$$

To reiterate, we see that if we *know* a person is one of those who has done the reading, then that person's *conditional probability* of having passed the exam is 0.778—which is different from the *unconditional probability* of 0.625 that a randomly selected person from the class has passed the exam. We designate this conditional probability as

$$p(P|R) - 0.778$$

The vertical bar is a shorthand way of expressing the phrase "given that." Therefore, the above statement would be read as: the conditional probability that a person has passed the exam *given that* the person has done the reading is 0.778.

> 1.2 The conditional probability of A given B, $p(A|B)$, is the probability that A occurs *assuming that* B is true.

The formula for conditional probability. We will now derive a general formula for computing conditional probabilities. Since general formulas are often easier to understand if they are derived in conjunction with an example, let's continue using the above example.

To derive our general formula, we first make use of the fact that our hypothetical person is known to be in set R. (That is what the phrase "given that the person has done the reading" means.) We therefore know that this person is one of $n(R)$ people, who have not done the reading, and that we don't have to worry about the $n(\overline{R})$ people, who have not done the reading. How many of the $n(R)$, or 90, people who have done the reading have also passed the exam? We see instantly from Table 1-4 that $n(P \cap R)$ or 70 of the total of 90 people have done the reading. Therefore, the formula for computing the conditional probability, $p(P|R)$ in our example, can be represented as

$$p(P|R) = \frac{n(R \cap P)}{n(R)}$$

Suppose now that we go through and divide the numerator and the denominator of the above equation by $n(S)$ (in this case, 120). We then end up with the formula

$$p(P|R) = \frac{n(R \cap P)/n(S)}{n(R)/n(S)} = \frac{p(R \cap P)}{p(R)}$$

This is the formula that is typically used—it is expressed in terms of probabilities rather than frequencies, and it allows us to eliminate the step of using frequences when we have probabilities to start with (which, as it turns out, is often the case). **The general formula for the conditional probability of** A **given** B **is therefore:**

$$p(A \mid B) = \frac{p(A \cap B)}{p(B)} \qquad (1.4)$$

That is, to find any conditional probability $p(A \mid B)$, we simply divide the probability of the joint event, $p(A \cap B)$, by the unconditional probability of B, $p(B)$.

More illustrations of how to use Equation 1.4. Continuing with our example, let's compute some other conditional probabilities using our 2×2 contingency table. Suppose that we are interested in the probability that a person has passed the exam, given that the person has *not* done the reading. This would be $p(P \mid \overline{R})$ which, according to Equation 1.4, we simply compute by

$$p(P \mid \overline{R}) = \frac{p(P \cap \overline{R})}{p(\overline{R})} = \frac{0.042}{0.250} = 0.168$$

Hence, we see that the probability that a person passed the exam given that the person did not do the reading is only 0.168. Below we will go through and compute all other pertinent conditional probabilities from this table. We suggest that you go through these conditional probabilities very carefully and make sure that you understand where they all come from.

$$p(\overline{P} \mid R) = \frac{p(\overline{P} \cap R)}{p(R)} = \frac{0.167}{0.750} = 0.223$$

$$p(\overline{P} \mid \overline{R}) = \frac{p(\overline{P} \cap \overline{R})}{p(\overline{R})} = \frac{0.208}{0.250} = 0.832$$

$$p(R \mid P) = \frac{p(R \cap P)}{p(P)} = \frac{0.583}{0.625} = 0.933$$

$$p(R \mid \overline{P}) = \frac{p(R \cap \overline{P})}{p(\overline{P})} = \frac{0.167}{0.375} = 0.445$$

$$p(\overline{R} \mid P) = \frac{p(\overline{R} \cap P)}{p(P)} = \frac{0.042}{0.625} = 0.067$$

$$p(\overline{R} \mid \overline{P}) = \frac{p(\overline{R} \cap \overline{P})}{p(\overline{P})} = \frac{0.208}{0.375} = 0.555$$

Things that have to add to 1 versus things that don't have to add to 1. Referring back to the various conditional probabilities we have just been computing, suppose we add the conditional probabilities that (1) a person passes the exam given that he or she has done the reading, and (2) a person does not pass the exam given that he or she has done the reading. Thus,

$$p(P|R) = 0.778$$
$$+ p(\overline{P}|R) = 0.222$$
$$\overline{p(P|R) + p(\overline{P}|R) = 1.000}$$

This example is meant to provide an intuitive illustration of the following rule: For any two events A and B,

$$p(A|B) + p(\overline{A}|B) = 1.0$$

What this means is that once you have established that an outcome is in state B, then it must either be in state A or in state $\overline{A}$; that is, $p(A|B)$ and $p(\overline{A}|B)$ are complements of each other relative to B, and thus have to add to 1. Referring to the example, once you establish that a person has done the reading for the exam then (since all people who did the reading took the exam) that person must either have passed or not passed the exam.

On the other hand, there is no such simple relationship between the probabilities $p(A|B)$ and $p(A|\overline{B})$. They may just happen to add to 1, but usually they will add to something else. We spell all this out because it often confuses people.

Relationship Between the Conditional Probability $p(A|B)$ and the Conditional Probability $p(B|A)$

Confusion also tends to arise about the relationship (or lack thereof) between the conditional probabilities $p(A|B)$ and $p(B|A)$. One real-life instance of where confusion and consequently misleading conclusions have arisen has to do with the relationship between marijuana smokers and heroin users. In particular the following sort of argument is occasionally used to demonstrate that marijuana is bad. First some data are offered: "Of all heroin users interviewed in the last 10 years, 80% of them started off smoking marijuana." These data are then purported to support the conclusion that if one is a marijuana smoker, one's chances of ultimately using heroin are very high.

However, can these data really be used to support this conclusion? Let's look at this situation systematically. In particular, we will consider the sample space of all people and designate the following two events:

M = {all people who smoke marijuana}
H = {all people who use heroin}

Now what do the data say? The data say that 80% of heroin users smoked marijuana, or

$$p(M|H) = 0.8$$

But this fact is then used inappropriately to support the conclusion that the probability of using heroin given that one is a marijuana smoker is very high, or that

$p(H|M)$ is very high

The fallacy in this argument, of course, is that while the data involve $p(M|H)$, the conclusion involves $p(H|M)$, and there is no necessary relationship between the two.

Just to make this concrete, let's suppose that our universe consists of only 100 people [$n(S) = 100$]. Suppose further that, of these 100 people, 50 smoke marijuana [$n(M) = 50$]; 10 take heroin [$n(H) = 10$]; and 8 both smoke marijuana and take heroin [$n(H \cap M) = 8$]. We can easily see then that $p(M) = 0.5, p(H) = 0.1$, and $p(H \cap M) = 0.08$. Therefore,

$$p(M|H) = \frac{p(M \cap H)}{p(H)} = \frac{0.08}{0.10} = 0.80$$

as stated. However, it is also true that

$$p(H|M) = \frac{p(M \cap H)}{p(M)} = \frac{0.08}{0.50} = 0.16$$

Thus, the probability of taking heroin given that one is a marijuana smoker is in this example actually quite low.

To really debunk this argument, it is sufficient to point out that although many heroin users started off using marijuana, *all* heroin users started off drinking water. But no one claims that drinking water leads to heroin usage.

Bayes' theorem.   Just what is the relationship between $p(A|B)$ and $p(B|A)$? These two entities are related by a fairly complex formula known as Bayes' theorem which we will first present. We will then explain where this formula came from. **Bayes' theorem is**

$$p(A|B) = \frac{p(B|A)p(A)}{p(B|A)p(A) + p(B|\overline{A})p(\overline{A})} \tag{1.5}$$

Bayes' theorem originated as Proposition 9 in Thomas Bayes' work. "An Essay Towards Solving a Problem in the Doctrine of Chances." The work was published posthumously in 1764 and refined by later mathematicians and scientists. To see how this formula arose, we first note that the numerator can be changed to

$p(B|A)p(A) = p(B \cap A)$

Similarly, the two terms in the denominator can be changed to

$p(B|A)p(A) = p(B \cap A)$

and

$$p(B|\overline{A})\, p(\overline{A}) = p(B \cap \overline{A})$$

Now the denominator reduces to

$$p(B \cap A) + p(B \cap \overline{A}) = p(B)$$

And the whole equation becomes

$$p(A|B) = \frac{p(B \cap A)}{p(B)}$$

which we recognize as our now-familiar equation for conditional probability.

Illustration of the use of Bayes' theroem.   Suppose you are a baseball fan, and your favorite player on the hometown team is Joe Shablotnik. You know a couple of things about the team and about Joe. First of all you know that Joe's batting average is 100—that is, every time Joe comes up to bat, his probability of getting a hit is 0.1. Second, you know that the rest of the team has a combined batting average of 300—that is, on the average, when anyone on the team other than Joe comes up to bat, his probability of getting a hit is 0.3.

Knowing all this, you turn on the radio one day. You know Joe's team is at bat, and you hear, "It's a hit!" Now you want to compute the probability that it was Joe who had gotten the hit. To do this, let's define a couple of different events. First of all,

J – {Joe at bat}

$\overline{J}$ = {Joe not at bat}

Note that since Joe is one of nine men on the team, and only one person is at bat at any given time,

$$p(J) = 1/9$$

and

$$p(\overline{J}) = 8/9$$

Second,

H – {batter gets a hit}

Since Joe's batting average is 100, the probability that a hit is obtained given Joe is at bat is 0.1, or

$$p(H|J) = 0.1$$

Likewise, since the rest of the team has an average of 300, the probability that a hit is obtained given Joe is not at bat is 0.3, or

$$p(H|\overline{J}) = 0.3$$

Now we want to know the probability $p(J|H)$—that is, given that a hit was made, we want to know the probability that it was Joe who made the hit. To do this, we simply plug everything we have into Equation 1.5 and come out with the following.

$$p(J|H) = \frac{p(H|J)p(J)}{p(H|J)p(J) + p(H|\bar{J})p(\bar{J})}$$

$$= \frac{(0.1)(1/9)}{(0.1)(1/9) + (0.3)(8/9)} = 0.04$$

Contingency table representation.   We could also have solved this problem by putting the information that we had been given into a contingency table. Table 1-5 shows such a contingency table. Here, the sample space is partitioned in two ways: Joe versus not Joe and hit versus not hit. We know that the probability that Joe is at bat is 1/9; therefore, the probability that Joe is not at bat is 8/9. How do we calculate the probability of a hit and not a hit? We know that the probability of a hit given Joe is 0.1. Therefore,

$$p(H|J) = 0.1 = \frac{p(H \cap J)}{1/9}$$

and so

$$p(H \cap J) = (0.1)(1/9) = 0.011$$

Hence, we can fill in the upper left-hand cell—the probability of a joint event, $H \cap J$. Likewise, we know that the probability of a hit given not Joe is 0.3. Therefore,

$$p(H|\bar{J}) = \frac{p(H \cap \bar{J})}{p(\bar{J})}$$

TABLE 1-5.   Contingency Table Representation of the Shablotnik Example

	J *(Joe at bat)*	$\bar{J}$ *(not Joe at bat)*	
H (hit)	$p(H \cap J) = p(H|J)p(J)$ $= (0.100)(0.111)$ $p(H \cap J) = 0.011$	$p(H \cap \bar{J}) = p(H|\bar{J})p(\bar{J})$ $= (0.300)(0.889)$ $p(H \cap \bar{J}) = 0.267$	$p(H) = 0.278$
$\bar{H}$ (not a hit)	$p(\bar{H} \cap J) = 0.100$	$p(\bar{H} \cap \bar{J}) = 0.622$	$p(\bar{H}) = 0.722$
	$p(J) = 1/9 = 0.111$	$p(\bar{J}) = 8/9 = 0.889$	$p(S) = 1.00$

or

$$0.3 = \frac{p(H \cap \bar{J})}{8/9}$$

and so

$$(0.3)(8/9) = 0.267 = p(H \cap \bar{J})$$

Hence, we can also fill in the upper right-hand cell. Knowing what should go into these two cells, we are able to fill in the other two cells, thereby filling up the entire contingency table. Having a filled contingency table puts us in a position to compute any conditional probabilities we want. In particular we are able to compute the conditional probability we want, $p(J|H)$, by simply using our familiar equation,

$$p(J|H) = \frac{p(J \cap H)}{p(H)} = \frac{0.011}{0.278} = 0.04$$

So we see, once again, the value of representing our information in a contingency table, if possible.

Independence

Let's go back to our 120-person class (Table 1-4) and look at a couple of interesting conditional probabilities from a slightly different perspective.

Two dependent events.   First, given that a person has done the reading, we have seen that his or her conditional probability of passing the exam is

$$p(P|R) = \frac{p(P \cap R)}{p(R)} = \frac{0.583}{0.750} = 0.778$$

However, given that the person has *not* done the reading, the conditional probability of passing is only

$$p(P|\bar{R}) = \frac{p(P \cap \bar{R})}{p(\bar{R})} = \frac{0.042}{0.250} = 0.168$$

What this indicates is that there seems to be some *connection* or *dependence* between doing the reading and passing the exam—namely, doing the reading seems to increase a person's probability of passing the exam (just as we might expect).

These two conditional probabilities can also be viewed in terms of their relationship to the unconditional probability that a random person selected from the class has passed the exam, which is 0.625. We note that a person who has done the reading has a probability, 0.778, of passing which is greater than this unconditional probability. Conversely, a person who has *not* done the reading has a conditional probability of passing, 0.168, which is *lower* than the unconditional probability.

Two independent events.   Now let us change our example somewhat. We will keep the same 120-person class and we will still assume that 75 of the 120 people have passed the exam and that the other 45 people have not passed the exam. However, instead of then partitioning the class according to whether or not people have done the reading, let's partition the class according to whether the people are females or not females (that is, females or males). Let's assume that of the 120 people in the class, 24 of them are female and the other 96 are not female. Furthermore, let's suppose that 15 people are both female and have passed the exam. We now can get a new 2 × 2 contingency table which we have shown in Table 1-6. We see, of course, that the unconditional probability that a random person in the class has passed the exam is still 0.625. Let's compute the conditional probabilities of passing the exam given that a person is a female or that a person is not a female.

$$p(P|F) = \frac{p(P \cap F)}{p(F)} = \frac{0.125}{0.200} = 0.625$$

$$p(P|\overline{F}) = \frac{p(P \cap \overline{F})}{p(\overline{F})} = \frac{0.500}{0.800} = 0.625$$

As we can see, whether a person is a female, not a female, or just a random person in the class, that person's probability of having passed the exam is 0.625. In other words, unlike the case of doing versus not doing the reading, passing the exam seems to have nothing whatsoever to do with whether a person is a female or not a female. Another way of stating this is that the two events—passing the exam and being a female—are *independent* of one another.

Using these examples, we have tried to convey an intuition notion of what independence means. Now we shall use these intuitions and examples to arrive at

TABLE 1-6.   A 2 × 2 Contingency Table Illustrating Independent Events

	F (female)	$\overline{F}$ (male)	
P (passed exam)	$n(F \cap P) = 15$ $p(F \cap P) = 15/120 = 0.125$	$n(\overline{F} \cap P) = 60$ $p(\overline{F} \cap P) = 60/120 = 0.500$	$n(P) = 75$ $p(P) = 75/120 = 0.625$
P (did not pass exam)	$n(F \cap \overline{P}) = 9$ $p(F \cap \overline{P}) = 9/120 = 0.075$	$n(\overline{F} \cap \overline{P}) = 36$ $p(\overline{F} \cap \overline{P}) = 36/120 = 0.300$	$n(\overline{P}) = 45$ $p(\overline{P}) = 45/120 = 0.375$
	$n(F) = 24$ $p(F) = 24/120 = 0.20$	$n(\overline{F}) = 96$ $p(\overline{F}) = 96/120 = 0.80$	$n(S) = 120$ $p(S) = 120/120 = 1.00$

a formal mathematical rule for what is meant by statistical independence. **This rule is: two events, A and B, are independent of one another if, and only if,**

$$p(A \mid B) = p(A \mid \overline{B}) = p(A)$$

Actually, it is enough to say that two events are independent if and only if $p(A \mid B)$ is equal to $\underline{p(A)}$. We can show mathematically that if $p(A \mid B)$ is equal to $p(A)$ then $p(A \mid \overline{B})$ must also be equal to $p(A)$. Additionally, we can show that if $p(A \mid B)$ is greater than $p(A)$, then $p(A \mid \overline{B})$ has to be less than $p(A)$. This should make intuitive sense. Going back to our initial example, if the probability of passing given that you have done the reading is *higher* than the unconditional probability of passing the exam, then we would intuit that the conditional probability of passing the exam given that you have not done the reading would be lower than the unconditional probability— just as, indeed, it is.

Computing Probabilities of Joint Events

In our discussions of conditional probability, there has been an implicit assumption that it's in some sense more fundamental to compute the probability of joint event, $A \cap B$, than it is to compute a conditional probability like $p(A \mid B)$ or $p(B \mid A)$. That is, since the formula for conditional probability includes as one of its components the probability of a joint event, the suggestion is that we always know what the probability of the joint event is before we know what some related conditional probability is. However, things do not always turn out this way. In many instances when we are trying to compute probabilities of things, it turns out that conditional probabilities are much more available than are probabilities of joint events. Therefore, in many instances we take our conditional probability formula and jiggle it around in order to go in the direction which allows us to compute the probability of a joint event from the conditional and unconditional probabilities that we presumably know beforehand.

The general rule.   To do this, we begin with our earlier formula for conditional probabilities,

$$p(A \mid B) = \frac{p(A \cap B)}{p(B)}$$

and simply multiply each side of the equation by $p(B)$ to obtain

$$p(A \mid B)p(B) = p(A \cap B)$$

Notice that since it is also true that

$$p(B \mid A) = \frac{p(A \cap B)}{p(A)}$$

and therefore that

$$p(B|A)p(A) = p(A \cap B)$$

we can come up with the following general rule:

the probability of the joint event, $p(A \cap B)$, can always be computed as

$$p(A \cap B) = p(A|B)p(B) = p(B|A)p(A) \qquad (1.6)$$

Under what conditions would we have a conditional probability, like $p(A|B)$ or $p(B|A)$, together with unconditional probabilities such as $p(A)$ and $p(B)$ and want to compute the probability of the intersection, $p(A \cap B)$? Let's consider the following problem.

Suppose that there are six people in a seminar that meets once per week. Of these six people, four are males and two are females. Let's imagine that during the first six weeks the seminar meets, each person in the seminar will present one paper. The order in which the six people give their presentations is determined as follows: The first week the presenter will be chosen randomly from the six people. The next week, the presenter will be chosen randomly from the remaining five people, and so on. (Such a procedure is referred to as "sampling without replacement," a designation indicating that once a person has been randomly selected from the total pool of people, that person is not replaced in the pool and thus does not have the opportunity to be selected again.)

Now suppose we want to know what the probability is that the two females in the class will give the first two presentations. How do we compute it? For ease of discourse let's first create some designations. We will designate F_1 to mean "female presents on the first week," F_2 to mean "female presents on the second week," M_1 to mean "male presents on the first week," and so forth. What we really want to compute then is the joint probability of the two events $(F_1 \cap F_2)$; stated in English, we want to know the probability of the joint event, "one female presents on the first week and the other female presents on the second week."

At first glance it is not obvious what this joint probability is. However, we can use Equation 1.6 to express the joint probability in terms of a conditional probability and an unconditional probability.

$$p(F_1 \cap F_2) = p(F_1)p(F_2|F_1).$$

Let's first compute $p(F_1)$. On the first week there are six people who can potentially present, and two of them are females. Therefore, the probability that a female will be chosen to present on the first week is 2/6 or 1/3. Now what about the conditional probability $p(F_2|F_1)$—that is, the probability of selecting a female on the second week given that a female has also been selected on the first week? Remember that if a female has been selected on the first week, then of the five people left in the pool out of which the speaker will be selected on the second week, four are male and only one is female. Therefore, given that one female has been selected on the first week, the probability that the other female will be selected on the second week is 1/5. We now have both ingredients necessary to use Equation 1.6 and compute the joint probability we want. Specifically, the probability that a female is selected on both the first and the second weeks is

$$p(F_1 \cap F_2) = p(F_1)p(F_2|F_1) = (1/3)(1/5) = 1/15 = 0.067$$

The multiplication rule for joint events when the two events are independent. We have noted that Equation 1.6 is a general formula; that is, it will work in any situation. However, suppose, as often turns out to be the case, that two events A and B in which we are interested are *independent* of one another. What does this imply regarding our equation for finding the probability of a joint event? We first notice that by definition, if A and B are independent, then

$$p(B|A) = p(B)$$

It is also of course true that

$$p(A|B) = p(A)$$

Therefore, referring to Equation 1.6, we can either substitute $p(B)$ for $p(B|A)$ or $p(A)$ for $p(A|B)$. In either case we conclude that when A and B are independent events

$$p(A \cap B) = p(A)p(B) \tag{1.7}$$

Equation 1.7 simplifies life considerably when we are dealing with independent events. It says that when we have two independent events the probability of the joint event—that is, the probability of both events occurring—is simply the product of their two individual, unconditional probabilities. We will show, using a couple of examples, how to make use of this fact. Suppose first that we are playing some odd game that involves flipping a coin and throwing a die. What is the probability of the joint event that the die will come up a five and also that the coin will turn up a head?

We first observe that these two events are independent of one another, because the coin has no knowledge of and will not be affected by how the die comes up, and vice versa. Therefore, to get the probability of this joint event, we simply use Equation 1.7 and multiply the two unconditional probabilities. We know that when we throw a die, $p(5) = 1/6$; when we toss a coin, $p(H) = 1/2$. Therefore, the probability of *both* those things happening is

$$p(H \cap 5) = p(H)p(5) = (1/2)(1/6) = 1/12 = 0.083$$

In games of chance that involve throwing two dice, the multiplication rule comes in very handy, because we are often interested in the probability that the first die comes up something and the second die comes up something else. For instance, suppose that we are interested in the probability that both dice come up sixes. We know that the probability of the first die coming up a six is one in six, and the probability of the second die coming up a six is also one in six. These two events are independent of one another; that is, neither die is affected by what the other one does. Therefore, to get the probability that they both come up a six, we once again multiply the two unconditional probabilities together:

$$p(6 \cap 6) = (1/6)(1/6) = 1/36 = 0.028$$

1.3 The probability $p(A \cap B)$ can *always* be computed by the formula(s)

$$p(A \cap B) = p(A)p(B|A) = p(B)p(A|B)$$

When A and B are independent, then the simpler formula,

$$p(A \cap B) = p(A)p(B)$$

can be used.

Extension of the multiplication rule to more than two independent events. So far we have discussed finding the intersections of two independent events. Suppose, however, that we had more than two events, all mutually independent, which we shall refer to as events $A, B, \ldots, N$. To compute the probability of the intersection of all these events (in other words, the probability that they all will occur) we can simply extend our multiplication rule to the following.

$$p(A \cap B \cap \ldots \cap N) = p(A)p(B) \ldots p(N) \tag{1.8}$$

For example, suppose we were flipping a coin, drawing a card from a deck, and throwing a die. What is the probability of the joint occurrence of the following three events: (a) The coin comes up heads (H). (b) The ace of spades (AS) is drawn from the deck. (c) The die comes up a 3. We can simply use the extension of our multiplication rule, Equation 1.8, as follows:

$$p(H \cap AS \cap 3) = p(H)p(AS)p(3) = (1/2)(1/52)(1/6) = 0.0016$$

Or suppose that we were simply flipping the coin three times and wanted to know the probability of getting a head on all three flips. We know that the probability of getting a head on the first flip is $1/2$, the probability of getting a head on the second flip is $1/2$, and the probability of getting a head on the third flip is also $1/2$. Since the three flips are all independent of one another, we can get the probability of this joint event by simply multiplying the unconditional probabilities, or

$$p(H \cap H \cap H) = (1/2)(1/2)(1/2) = 1/8 = 0.125$$

Use of the Multiplication Rule in Conjunction with the Addition Rule

We have talked about two major rules in this chapter: the addition rule for finding probabilities of unions of events (Equation 1.2), and the multiplication rules for finding probabilities of intersection of events (Equations 1.6 and 1.7). In our examples so far we have, for simplicity, talked about situations in which the use of only one of these rules is required. Often, however, we have to compute probabilities involving use of both rules together. Let's consider one fairly simple example of such a situation. Suppose once again that we throw two dice and are interested in the probability that the sum of these two dice is 11. For example, suppose we need a sum of 11 to be able to land on and buy Boardwalk in a Monopoly game. Let's compute this probability in two ways: first in the most straightforward (but tedious) way possible—simply by counting elementary events. Then we'll do it again, making use of the rules we have learned for computing probabilities of unions and joint events. We hope you will agree that the second way of doing it is easier than the first way.

Elementary events revisited.   First let's go back to our notion of a sample space. If we are to throw two dice, what would be the sample space of all possible outcomes of this situation? To answer this question, we first have to consider what might constitute an elementary event. In this case an elementary event would consist of a pair of numbers, each number ranging from one to six. For example (1, 1) would signify that the first die came up a one and the second die also came up a one. Likewise, (3, 5) would signify that the first die came up a three and the second die came up a five; (5, 3) would signify that the first die came up a five and the second die a 3; and so on. Now let's systematically list all the elementary events in the sample space, as has been done in Table 1-7. Notice that there are 36 elementary events in the sample space—this is reasonable; the first die can come up any one of six ways and for each of those six ways the

TABLE 1-7.   All Ways in Which a Throw of Two Dice Might Come Out

Second Die Is a	First Die Is a					
	1	2	3	4	5	6
1	(1, 1)	(1, 2)	(1, 3)	(1, 4)	(1, 5)	(1, 6)
2	(2, 1)	(2, 2)	(2, 3)	(2, 4)	(2, 5)	(2, 6)
3	(3, 1)	(3, 2)	(3, 3)	(3, 4)	(3, 5)	(3, 6)
4	(4, 1)	(4, 2)	(4, 3)	(4, 4)	(4, 5)	(4, 6)
5	(5, 1)	(5, 2)	(5, 3)	(5, 4)	(5, 5)	(5, 6)
6	(6, 1)	(6, 2)	(6, 3)	(6, 4)	(6, 5)	(6, 6)

second die can also come up any one of six ways. So the total number of ways in which the dice can come up is six times six or 36; and using our notation, $n(S) = 36$. How many of these 36 elementary events involve the dice summing to 11? Going through Table 1-7, we discover that only two—(6, 5) and (5, 6)—will do the trick. Referring to the outcome "sums to 11" as A, we therefore see that

$$n(A) = 2$$

Going back to our Equation 1.1, we can therefore compute the probability that the dice will sum to 11 as follows:

$$p(A) = \frac{n(A)}{n(S)} = 2/36 = 1/18 = 0.056$$

Using the addition and multiplication rules. As we noted, the former method of computing a probability is very tedious, involving a constructed list of all the elementary events, a lot of searching and counting, and so on. Suppose we look at the problem from a different perspective. We first ask ourselves under what circumstances the dice total to 11. It is not very difficult to figure out that there are only two possible ways in which this could happen: the first die comes up a six and the second die comes up a five, or its reverse—the first die comes up a five and the second die comes up a six.[2] Let's denote these two possible outcomes as B and C, respectively. Now, $p(A)$—the probability that we are interested in—is equal to the probability of the union of events B and C. Also, since events B and C are mutually exclusive (if we get a six–five, we cannot get a five–six, and vice versa), we can use the addition rule, or

$$p(A) = p(B) + p(C)$$

Now how do we get the probabilities B and C? We note that both these events consist of intersections of two independent events. Event B consists of getting a five on the first die (which happens with probability 1/6) and getting a six on the second die (which also happens with probability 1/6). Since these two events are independent, the probability of their intersection may be computed simply by multiplying the two individual probabilities, or

$$p(B) = (1/6)(1/6) = 1/36 = 0.028$$

Event C can be computed exactly analogously; its probability is also 0.028. Therefore, to get the probability of the event we are interested in,

$$p(A) = p(B) + p(C) = 0.028 + 0.028 = 0.056$$

We have gone through this second way of computing the probability in a painstakingly slow, step-by-step fashion so that you will be absolutely clear

---

[2]When working with these situations involving two dice, it is often convenient to imagine the dice being of different colors—for example the first die being red and the second die being green. You may find it easier to keep them apart that way.

about where each step came from. As you can probably see, however, with a little practice you can zip through the steps—certainly more quickly than it takes to list all the possible elementary events, count the relevant ones, and so on.

We've covered a fairly substantial amount of material in this chapter. Our basic goal was to provide you with some rules and some facility in computing probabilities of various events.

These are the rules you ought to remember:

1. If we can characterize some situation as having some number of equally probable outcomes which we call elementary events, then we can similarly characterize some event of interest as constituting a subset of those elementary events. The probability of that event is then computed by the equation

$$p(A) = \frac{n(A)}{n(S)}$$

2. In *any* situation, we can compute the probability of the union of two events A and B (that is, the probability that one event or the other or both will happen), using the general formula

$$p(A \cup B) = p(A) + p(B) - p(A \cap B)$$

3. A special case of computing probability of unions occurs when we have a number of mutually exclusive events. We can then use our addition rule or

$$p(A \cup B \cup \cdots \cup N) = p(A) + p(B) + \cdots + p(N)$$

4. The conditional probability of some event A given that we know the outcome of some other event B is computed as follows:

$$p(A|B) = \frac{p(A \cap B)}{p(B)}$$

5. Likewise, if we want to compute the probability of the intersection of two events (that is, the probability that both events will occur, or the probability of a joint event), we can use the general formula

$$p(A \cap B) = p(A)\,p(B|A) = p(B)\,p(A|B)$$

6. A special case of computing probabilities of intersections occurs when a number of events, A, B, ..., N are independent. We can use the multiplication rule, or

$$p(A \cap B \cap \cdots \cap N) = p(A)\,p(B)\ldots p(N)$$

PROBLEMS

1. Suppose that we have a universal set, W, consisting of all people. Verbally describe the complements of the following sets.

   1. All people who are left-handed.
   2. All people who are left-handed and color-blind.
   3. All people who are left-handed or color-blind.
   4. All left-handed, color-blind females.
   5. All people who are left-handed or male but not color-blind.

2. Classify the following sets as finite, countably infinite, or uncountably infinite.

   1. The set of all four-letter strings.
   2. The set of all letter strings.
   3. The set of all amounts that a human being could weigh.
   4. The set of exact weights of 100 human beings.
   5. The set of all grains of sand in the Sahara Desert.
   6. The number of times a pair of dice could be thrown before double sixes turn up.

3. Consider W = all people. Given this universal set, make up examples of pairs of sets that are

   1. Mutually exclusive and exhaustive.
   2. Mutually exclusive but not mutually exhaustive.
   3. Not mutually exclusive but mutually exhaustive.
   4. Neither mutually exclusive nor mutually exhaustive.

4. Let S be the set of all restaurants in Walla Walla, Washington. Assume S contains 96 elementary events. Define F as the set of all French restaurants and assume that F contains 32 elementary events. Define T as the set of restaurants which require men to wear ties and jackets, and assume that T contains 48 elementary events. There are 32 French restaurants which require ties and jackets.

   1. Represent the situation as a Venn diagram—fill in the appropriate numbers of elementary events.
   2. Represent the situation as a contingency table. Now compute the following.

(1) $p(F)$	(6) $p(F \cap \overline{T})$	(11) $p(\overline{F}\|\overline{T})$
(2) $p(T)$	(7) $p(\overline{F} \cup \overline{T})$	(12) $p(T\|F)$
(3) $p(F \cap T)$	(8) $p(F\|T)$	(13) $p(T\|\overline{F})$
(4) $p(F \cup T)$	(9) $p(F\|\overline{T})$	(14) $p(\overline{T}\|F)$
(5) $p(\overline{F} \cap T)$	(10) $p(\overline{F}\|T)$	(15) $p(\overline{T}\|\overline{F})$

5. There are 84 passengers on an airliner. Define the following two events:

   F: Person is traveling first class.
   M: The person is male.

   Suppose that there are 32 first-class passengers and 30 males. Suppose further that $p(F \cup M) = 0.61905$.

   1. Construct a contingency table depicting the preceding information. Compute the following, based on your contingency table:

   2. $p(F\|M)$

   3. $p(M\|F)$

   4. $p(\overline{F}\|\overline{M})$

   5. $p(\overline{M}\|\overline{F})$

   6. $p(\overline{M} \cap \overline{F})$

   7. $p(\overline{F}\|M)$

   8. $p(\overline{M}\|\overline{F})$

   9. $p(\overline{M}\|F)$

6. Consider the set, S, of all gas stations in the country. Set G is the set of stations open and selling gas, and set E is the set of all Exxon stations.
   Suppose that:

   $$p(G) = 0.4$$
   $$p(G \cup E) = 0.5$$
   $$p(E) = 0.2$$

   Represent this situation as a contingency table and then calculate

   1. $p(G \cap E)$
   2. $p(G\|E)$
   3. $p(G\|\overline{E})$
   4. $p(\overline{E}\|G)$
   5. $p(E\|\overline{G})$

7. Assume that there are 100 cars in East Elbow, Idaho. Of these cars 25 are Fords, 25 are Jaguars, and the remaining 50 are Volkswagens. Twenty of the 100 cars are blue. The two events "type of car" and "being blue" are independent. Suppose now that a random car is picked from East Elbow.

1. Make up a contingency table that represents this situation.
2. What is the probability that the car will be a blue Ford?
3. What is the probability that the car will be a Jaguar that is not blue?
4. What is the probability that the car will be blue and either a Jaguar or a Volkswagen?
5. What is the probability that the car will be neither blue nor a Volkswagen?

8. Assume that there is a set, S, of 96 people traveling on an airplane. Of these 96 people, 24 are in the first-class section (set F) and 32 are flying youth-fare (set Y). Due to a slip-up, four youth-fare travelers are in the first-class section.

    Find the following probabilities:

    1. $p(Y \cup F)$
    2. $p(F | \overline{Y})$
    3. $p(\overline{Y} | F)$
    4. $p(\overline{Y} | \overline{F})$
    5. $p(\overline{F} | Y)$

9. In the town of Glimp, Alaska, there are three types of restaurants: American (A), French (F), and Greek (G). There are equal numbers of the three types of restaurants.

    Additionally some restaurants have a liquor license (L) whereas others do not ($\overline{L}$). As it happens, half the restaurants in town have a liquor license whereas the other half do not. We know the following:

    $$p(L \cap A) = 1/3$$
    $$p(\overline{L} | G) = 1$$

    Now compute:

    1. $p(F | L)$
    2. $p(L | A)$
    3. $p(L | \overline{F})$
    4. $p(F | \overline{L})$
    5. $p(\overline{A} | L)$

10. Suppose two dice, a black die and a white die, are thrown. Two outcomes are defined as follows:

    $A = \{$white die comes up a five or a six$\}$

    $B = \{$black die comes up a six$\}$

    Compute the following.
    (*Hint:* Make a contingency table)

    1. $p(A)$
    2. $p(B)$

3. $p(A \cap B)$
4. $p(A \cup B)$
5. $p(A|B)$
6. $p(B|A)$
7. $p(B|\overline{A})$
8. $p(\overline{A}|B)$
9. $p(\overline{A}|\overline{B})$

Suppose we define a third event in addition to the first two:

   $C = \{$black die comes up a 1$\}$

Compute the following:

10. $p(A \cap C)$
11. $p(B \cap C)$
12. $p[(B \cup C)|A]$

11. Suppose that we have two dice, die A and die B, and we throw them both together. Compute the following probabilities:

   1. At least one six is obtained.
   2. A double six is obtained.
   3. A two and a one are obtained.
   4. Die A turns up six and die B turns up one.
   5. A "double" is obtained (double 1, double 2, and so on).
   6. The number on die A and the number of die B differ by one.
   7. The number on die A is one greater than the number on die B.
   8. No ones or twos are obtained.

12. Suppose that I flip a coin and draw a card from a deck.

   1. What is the probability that the coin comes up heads *and* that I get a diamond?
   2. What is the probability that the coin comes up heads *or* I get a diamond, but not both?
   3. What is the probability that the coin comes up heads or tails or that I get a spade (or both)?
   4. What is the probability that the coin comes up heads or tails and that I get a spade?

13. Joe Shablotnik is drinking beer after a baseball game. To decide how many beers to drink, he plans to use the following procedure.

   To decide on the first beer, he throws a die. If the die comes up one to five, he drinks a beer; otherwise he goes home. If he gets to drink his first beer, he decides on the second beer by throwing the die again when he finishes. This time if the die comes up one to four, he drinks a second beer; otherwise he goes home. To decide on the third beer: If he gets to drink his second beer, he decides on a third beer by throwing the die again when he finishes. This time, if the die comes up one to three he drinks a third beer; otherwise he goes home. He will definitely go home after the third beer.

What is the probability that Joe drinks

1. No beers—he just goes home?
2. One beer and then he goes home?
3. Two beers and then he goes home?
4. Three beers?

14. Suppose that I teach a seminar with five students, three females, and two males. I decide to randomly pick one student to lead this week's seminar and one student to lead next week's seminar. I do this by putting the five names into a hat, drawing out one name for this week and then, not replacing the name in the hat, drawing another name for next week. (This technique is called "sampling without replacement.") What are the following probabilities?

1. This week's seminar is led by a female.
2. This week's seminar is led by a male and next week's seminar is led by a female.
3. This week's and next week's seminars are led by members of the same sex.
4. At least one of the seminars is led by a female.

15. Consider two dice, die C and die D. When thrown together, the two dice are completely normal except for one strange thing. If die C comes up "4" (which, of course, happens with probability $1/6$) then it casts a cosmic force field over die D which in turns causes the numbers 1 through 6 on die D to have the following probability of occurrence.

$p(1) = 1/2$
$p(2) = 1/8$
$p(3) = 1/8$
$p(4) = 1/8$
$p(5) = 1/16$
$p(6) = 1/16$

Compute the probabilities of the following outcomes of throwing two dice:

1. Die $C = 1$ and die $D = 5$.
2. Die $C = 1$ or 2 or 3, and die $D = 1$ or 2.
3. Die $C = 3$ or 4, and die $D = 1$.
4. Die $C = 4$ and die $D = 3$.

16. A peculiar deck of cards has only red jacks. (The black suits have elevens instead of jacks.)

Define:

H: as "draw a heart"

J: as "draw a jack"

Note that

$$p(H) = 1/4$$

$$p(J) = 2/52 = 1/26$$

I claim that drawing the jack of hearts is equivalent to the event $H \cap J$, and the probability of drawing the jack of hearts is

$$p(H \cap J) = p(H)\, p(J) = (1/4)(1/26) = 1/104$$

My friend, however, claims that since there are 52 cards and one jack of hearts, the probability of drawing the jack of hearts is $1/52$. Who is right? What error has been committed?

17. Joe Shablotnik is interested in the appearance of unidentified flying objects (UFOs). Checking the records, he discovers that there have been many more reported UFO sightings between the hours of noon and 6 p.m. than between the hours of midnight and 6 a.m. He therefore concludes that UFO pilots prefer flying around in daylight rather than during the night.

    What is wrong with Joe's reasoning?

18. Joe Shablotnik is trying to decide on a career and is choosing between being a lion-tamer or a lumberjack. Joe discovers that more lumberjacks than lion-tamers are killed every year. Therefore, Joe decides to be a lion-tamer.

    Is this a reasonable thing for Joe to do? Prove (by making up an example) that even though the fact noted above is correct, Joe's probability of being killed can be *greater* if he is a lion-tamer than if he is a lumberjack.

    (*Hint:* consider the universal set made up *only* of people who are lion-tamers or lumberjacks. You may assume that nobody is both a lion-tamer and a lumberjack).

19. Show that if A and B are independent then $\overline{A}$ and $\overline{B}$ are also independent.

    (*Hint:* Remember that $(A - B)(C - D) = AC - BC - AD + BD$.)

20. Suppose that you have a sample space, S, and two events A and B. Suppose further that

$$p(A|B) > p(A)$$

Prove that $p(A|\overline{B}) < p(A)$.

(*Hint:* This should make sense in intuitive terms. Using a familiar example, suppose that the unconditional probability of passing an exam is 0.75, but the conditional probability of passing given that you have done the reading is 0.85. It should then make sense that the probability of passing given that you *didn't* do the reading would be less than 0.75.)

21. Suppose that there are two events A and B and

$$p(A|B) = p(A)$$

Prove $p(A|\overline{B}) = p(A)$.

(*Hint:* Do *not* try to prove this just by making up an example. Your proof must be for the general case.)

2

Random Variables and Distributions

The principal goal of this chapter is to provide a firm understanding of and feeling for what is called a *probability distribution*. The notion of a probability distribution is a central concept in this book, and discussions of statistical issues that will appear in subsequent chapters will all revolve around probability distributions. Thus, an understanding of probability distributions is a crucial prerequisite for understanding things to come.

Before plunging into probability distributions, however, we will first provide a brief introduction to *frequency distributions*. We do this here because the relation between frequency and probability distributions is similar to the relation between sets and probability; in each case the former is useful for understanding the latter. We should note briefly here that frequency distributions are also highly useful in their own right as a means of describing data, but we will defer a detailed discussion of this aspect of frequency distributions until the next chapter.

AN INTRODUCTION TO FREQUENCY DISTRIBUTIONS

Suppose that you and 19 of your friends get together and collect the following sort of data. Each of you flips a coin four times and counts the number of heads obtained out of the four flips. Table 2-1 lists some hypothetical data that might result from such a hypothetical exercise.

Random Variables

In order to organize the relationship between these data and distributions, it is useful to introduce the notion of a random variable. To see what a random variable is, consider some sample space—for example, the sample space consisting of you and your 19 fellow coin-flippers. A random variable is nothing more than a function which assigns a *number* to each member of the sample space. Corresponding to any given sample space, there is an infinite number of possible random variables. Table 2-2 lists several different sample spaces and for each

TABLE 2-1. Hypothetical Data: Number of Heads in Four Flips of a Coin from 20 Four-Flip Sequences

Person	Number of Heads	Person	Number of Heads
1	3	11	1
2	1	12	2
3	1	13	2
4	3	14	1
5	1	15	2
6	2	16	1
7	0	17	2
8	2	18	3
9	4	19	3
10	4	20	3

TABLE 2-3. Different Sample Spaces and Examples of Random Variables Applying to Them

Sample Space	Random Variables
1. All people in a classroom	a. Age of person in years
	b. Weight of person in pounds
	c. Number of hairs on person's head
	d. Number of ears person has
2. All books on a bookshelf	a. Number of pages in book
	b. Copyright date of book
	c. Weight of book in grams
3. All cars with California license plates	a. Year car was built
	b. Horsepower of car's engine
	c. Number of gears in car's transmission
	d. Number of clutch pedals car has
	e. Largest digit on car's license plate
4. All possible five-card poker hands a person could get	a. Number of aces in the hand
	b. Number of pairs in the hand
	c. Number of spades in the hand

sample space gives some examples of random variables that could reasonably apply.

Construction of a Frequency Distribution

Let's return to our 20 coin-flippers. The random variable here assigns to each person the number of heads that person obtained in four flips of the coin. Recall that the 20 resulting numbers are listed in Table 2-1. Briefly, a frequency distribution simply lists with each possible value that the random variable can assign (the numbers 0–4 in this case) the

TABLE 2-3.   Frequency
Distribution of the Data shown in
Table 2-1. With Each Value the
Frequency with Which that
Value Occurs

Number of Heads	Frequency
0	1
1	6
2	6
3	5
4	2

frequency with which that value occurred. More formally, the process works by two steps.

Determination of V.   To construct a frequency distribution for a particular random variable operating on a particular sample space, we first determine *the set of all possible values that the random variable could assign.* Let us refer to this set as V. In this case,

$$V = \{0, 1, 2, 3, 4\}$$

Frequencies.   Having determined V, we simply go through our data, count the number of instances of each member of V, and list the result as has been done in Table 2-3. Thus, Table 2-3 constitutes the frequency distribution of the Table 2-1 data.

> 2.1 A frequency distribution simply lists with every possible value the frequency with which that value occurred in the data.

Histograms

Since pictures are typically easier to understand than numbers, frequency distributions are often represented in pictorial form. The frequency distribution in Table 2-3 has accordingly been transformed into the picture in Figure 2-1. This sort of picture is referred to as a *frequency histogram,* another name for a bar graph. On the abscissa (horizontal axis) of the graph we have listed the members of V (which, recall, are the values that the random variable can assign, in this case the numbers 0 through 4). Above each of these five values, we have drawn a bar ascending to the ordinate[1] (vertical axis) value that corresponds to

---

[1]Many people have difficulty remembering which axis is the ordinate and which is the abscissa. A convenient mnemonic to help solve this problem is the following. When you say the word "abscissa" note that your mouth makes a horizontal line, like the horizontal axis. Likewise, when you say the word "ordinate," your mouth makes a (more or less) vertical line.

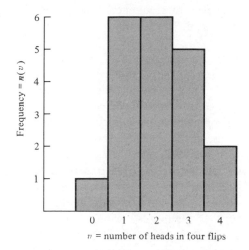

FIGURE 0-1. Frequency distribution of the number of heads in four flips of a coin. The bar over 0 reflects the fact that a frequency of one person got zero heads; the bar over 1 reflects the fact that a frequency of six people got one head, and so on.

the frequency with which that particular number occurred. Thus, the bar over the value 0 goes up to 1 to represent the fact that we obtained a single score of 0. Likewise, the bar over the value 1 rises to a frequency of 6 to represent the fact that there were 6 1s, and so on. Basically that's all a frequency histogram is. Notice that if we sum all the ordinate values on the histogram, this sum must be 20—the total number of scores we had at the outset.

PROBABILITY DISTRIBUTIONS

We will now discuss how a frequency distribution may be changed into an equivalent *probability distribution* and what the relationship is between the two. This discussion will be very similar to that in the last chapter, where we discussed how Venn diagrams (and contingency tables) could be used to represent either frequencies or probabilities.

Frequencies to Probabilities and Vice Versa

Recall that when talking about sets, we began by considering numbers (that is, frequencies) of elementary events. For example we talked about a sample space consisting of $n(S) = 12$ people, the set of $n(L) = 2$ people who were left-handed, and so on. We saw, however, that we could divide all relevant frequencies such as $n(L)$ by the total number of elementary events in the sample space $n(S)$ which would produce for us the *probabilities* of various events instead of the frequencies of elementary events corresponding to the events. We have thereby seen that the information conveyed by frequencies and probabilities is much the same, but they have somewhat different interpretations, which we discussed in the last chapter.

When we are concerned with distributions, we can go back and forth from frequencies to probabilities in exactly the same way. Whenever we plot a frequency distribution, we have some number, which we will henceforth refer to as n, that corresponds to the total number of cases. (Thus, n corresponds to what we were calling $n(S)$ before.) In the coin-flipping example described above, $n = 20$ since there were 20 coin-flip sequences, each of which resulted in some number of heads. **Suppose now that we divide each frequency in a frequency distribution by** n**.** The result would be the *probability* of occurrence (as opposed to the frequency of occurrence) of each member of V. Carrying out this procedure for the coin-flipping produces the probability distribution shown in Table 2-4. Figure 2-2 which depicts this probability distribution graphically, as a probability histogram. Notice that on the ordinate of Figure 2-2 distribution is now probability, which can range from 0 to 1, as opposed to frequency, which can range from 0 to n. As we mentioned earlier, the interpretation of the data changes somewhat when we translate from frequency to probability. Frequencies are descriptive of a set of data. In terms of probability, however, the question of interest is the following. Assuming we select some random member of our sample space—one particular coin-flipper in the example we have been using—what is the *probability* that this member will

---

TABLE 2-4. All Possible Four-Flip Sequences of Heads (H) and Tails (T) Along with the Number of Heads in Each Sequence

Sequence	Number of Heads
HHHH	4
HHHT	3
HHTH	3
HHTT	2
HTHH	3
HTHT	2
HTTH	2
HTTT	1
THHH	3
THHT	2
THTH	2
THTT	1
TTHH	2
TTHT	1
TTTH	1
TTTT	0

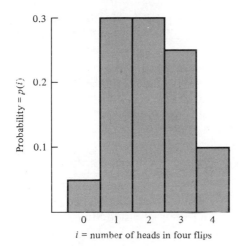

FIGURE 2-2   An empirical probability distribution. This probability distribution was obtained by taking the frequency distribution of Figure 2-1 and dividing each frequency (ordinate) value by n, the total number of scores in the situation. The height of each bar now represents the probability of occurrence of each of the abscissa values.

have been assigned any particular value by the random variable? For example, Figure 2-2 indicates that the probability of a randomly selected coin-flipper obtaining two heads is 0.3.

> 2.2 Probability and frequency distributions provide the same information in different ways. A frequency distribution lists the frequency with which each value occurred and a probability distribution lists the probability that each value will occur.

Empirical Versus Theoretical Distributions

We will refer to the sorts of probability distributions that we have been talking about so far as *empirical probability distributions*. Empirical means "based on data," and to reiterate we construct an empirical probability distribution by collecting some data (for example, the number of heads obtained in coin-flip sequences), using these data to construct a frequency distribution, and then dividing each frequency by n to obtain a probability distribution.

We will refer to a somewhat different sort of probability distribution as a *theoretical probability distribution*. To construct a theoretical probability distribution, we do not collect any data—instead, we begin with various considerations that allow us to figure out what the probability distribution *ought to* look like if we were to go out and collect some data. To provide a feeling for what is meant by this, we will describe the construction of a theoretical probability distribution for the number-of-heads-in-four-flips random variable that we discussed above.

Members of *V*.   As was the case when we constructed an empirical frequency distribution, our first step is to determine the set *V* of all values that the random variable can assign. If the random variable under consideration assigns some number of heads to a four-flip sequence, then the resulting value must be an integer between 0 and 4. It's still the case that

$$V = \{0, 1, 2, 3, 4\}$$

Probability of each member of *V*.   Here is where things change when we are dealing with a theoretical as opposed to an empirical distribution. When we had data, we simply computed the frequency of each member of *V* (by counting) then transformed that frequency into an empirical probability (by dividing by *n*).

Now we don't have any data. Nonetheless, it's still possible to compute the probability of each member of *V* using the concepts we developed in the last chapter. Table 2-4 lists all possible sequences of heads and tails that we could get in four flips of the coin. The first sequence consists of four heads; the second sequence consists of heads on the first three flips and tails on the fourth; and so on. As you can see, there are 16 possible sequences, each equally probable. How many of these sequences involve getting no heads? The answer is only one of them—the sequence that consists of all tails. Likewise, we can see that four of the sequences involve getting exactly one head, six involve getting exactly two heads, four involve getting exactly three heads, and one of the sequences involves getting all four heads. Since, as noted, each of the 16 sequences is equally probable, we determine that the probability of getting zero heads is 1/16, the probability of getting one head is 4/16, the probability of getting two heads is 6/16, the probability of getting three heads is 4/16, and the probability of getting four heads is 1/16. This provides us with all the information we need in order to construct the theoretical probability distribution we originally wanted. This distribution is shown in Figure 2-3.

The distribution shown in Figure 2-3 is an instance of a particular type of theoretical probability distribution called a *binomial distribution*. We will provide a detailed discussion of a binomial distribution in Chapter 4. For our present purposes we just use it as an example of a reasonably easy-to-construct theoretical distribution.

> 2.3 A theoretical probability distribution uses a priori considerations rather than data to compute the probability of each member of *V*.

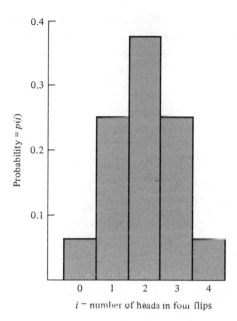

FIGURE 2-3. A theoretical probability distribution of the number of heads that ought to be obtained in four flips of a coin. Based on a priori considerations we determined that the probability of obtaining zero heads is 0.0625; the probability of obtaining one head is 0.250, and so on.

Theoretical Frequency Distributions

Recall that when we collect data, we can construct an empirical frequency distribution, which in turn can be used to construct an empirical probability distribution. When on the other hand we start with a theoretical probability distribution, we can use it to generate a theoretical frequency distribution. We do this simply by multiplying each probability by n where n is the total number of cases we plan to have. An example of this process is shown in Table 2-5 where

TABLE 2-5.   A Theoretical Frequency Distributor of Number of Heads in Four Flips of a Coin (the number in the sample space, n, is assumed to be 128)

Value (v)	Probability	Frequency = Probability × 128
0	0.0625	8
1	0.250	32
2	0.375	48
3	0.250	32
4	0.0625	8

we use our theoretical number-of-heads-in-four-flips *probability* distribution to generate a theoretical *frequency* distribution assuming $n = 128$. Thus if we dispatched 128 individuals to flip a coin four times apiece, our best guess is that eight of them would obtain zero heads, 32 of them would obtain one head, and so on.

Digression 2.1

Comparing Theoretical and Empirical Distributions: A Preview of Things to Come

Let's return to the binomial distribution show in Figure 2-3. As we have discussed, this is the probability distribution that we *expect* to get if we perform the hypothetical experiment of having a number of people each flip a coin four times. However, this expectation involves a number of assumptions about the world. First of all it assumes that the coins being flipped are fair coins—that they have equal probability of coming up heads or tails. Second, it assumes that the people flipping the coins are flipping them "normally"—that is, not in any strange way that will bias the coin to come up heads or tails. It also assumes that the people are reporting their results accurately. If all these assumptions are true, then Figure 2-3 depicts the probability distribution.

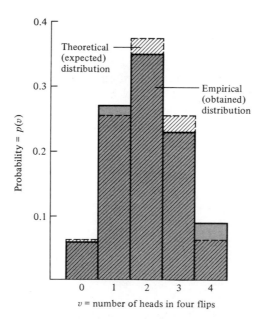

FIGURE D-1.   One way in which an empirical probability distribution might correspond to a theoretical probability distribution. In this case the correspondence is quite good, and we conclude that all is well with the world.

Suppose now that we actually perform this experiment; that is, we take a group of people and ask them each to flip a coin four times. We then get their data and use the data to generate an empirical frequency distribution and then an empirical probability distribution using the methods that we discussed before. Suppose that the empirical probability distribution came out as depicted in Figure D-1. In this case the empirical probability distribution seems to correspond pretty well to the theoretical probability distribution that we expect. It would therefore seem reasonable to conclude that all is well with the world. The coin is probably fair, the people are probably flipping the coins correctly, reporting their data accurately, and so forth.

However, suppose that the obtained probability distribution came out looking like the one in Figure D-2. In this case the empirical distribution that is obtained does not look at all like what we expect. In particular there seems to be an overabundance of sequences containing four heads and three heads, and not nearly enough sequences containing zero, one, and two heads. What would we do in such a situation? Probably we would question some of the assumptions that are listed above. We might conclude either that the coin is not fair or perhaps that the people weren't flipping the coins correctly. In any case the point of this digression is to show how theoretical and empirical probability

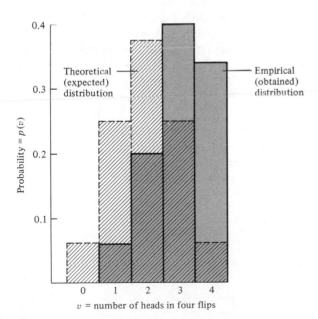

FIGURE D-2.   Another way in which the empirical probability distribution might correspond to the theoretical probability distribution. In this case the correspondence is very poor. Hence, we would conclude that something is probably wrong with the assumptions on which the theoretical probability distribution was based.

distributions are used to evaluate data. In general before doing an experiment, we determine what some theoretical probability distribution will look like if our assumptions are true. We then *do* the experiment in order to get an empirical probability distribution from the data. If the obtained and theoretical distributions coincide fairly well, we conclude that the assumptions in which we were interested are probably true. On the other hand if the obtained and theoretical probability distributions do not seem to coincide, we conclude that one or more of the assumptions is probably false. All this will be discussed much more in later chapters; we have mentioned it briefly now just to give some indication of the ultimate use of all this information about probability distributions.

Different Types of Probability Distributions

In Appendix A we note that a fundamental characteristic of a set is the number of members contained by that set. In particular a set could be *finite, countably infinite,* or *uncountably infinite.* Let's consider this characteristic with respect of V, the set of all possible values a random variable can assign. When V is finite or countably infinite, we are dealing with what is called a *discrete* probability distribution. When, in contrast, V is uncountably infinite, we are dealing with a *continuous* probability distribution. Let's elaborate on this distinction by considering a sample space consisting of people. As we have seen, there are numerous random variables that could be applied to such a sample space. We have already described the random variable that assigns some number of heads in four flips of a coin. As we have seen, for this random variable,

$$V = \{0, 1, 2, 3, 4\}$$

Because the set contains a finite number of elements, we can say that V is finite. Now imagine a different situation, one in which each person in the sample space rolls a die until a six is obtained. The random variable assigns to each person the *number of rolls prior to obtaining a six*. Here V would include all integers from one to infinity. That is,

$$V = \{1, 2, \ldots, \infty\}$$

Thus, we can call V countably infinite.

Digression 2.2

A Geometric Distribution

The number-of-rolls-prior-to-the-first-six probability distribution is known as a geometric distribution. To construct this distribution, we proceed as follows: the probability that the random variable assigns the value one, that is, that a six will occur on the first throw, is one in six (1/6). What about the probability that the random variable assigns a two? For this to be the case two events would have to occur. First we would have to get something other than a six (which we will call a not-six) on the first throw, which happens with probability (5/6). Second, we would have to get a six on the second throw, which happens with probability (1/6). These two events—getting a not-six on the first throw and a six on the second throw—are independent of one another, since during any throw the die is uninfluenced by what happened on any other throw. Therefore, the probability that both events happen—the probability of their intersection—is simply the product of the two probabilities, or

$$p(\text{not-6} \cap 6) = (5/6)(1/6) = 5/36 = 0.139$$

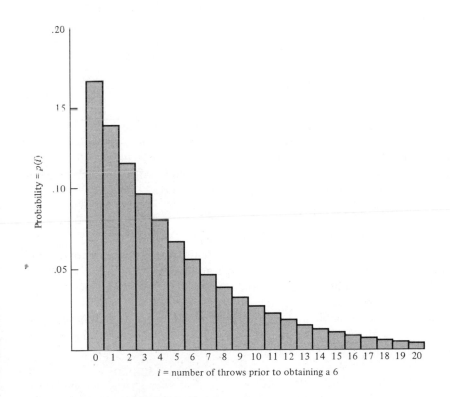

i = number of throws prior to obtaining a 6

FIGURE D-3.   A theoretical probability distribution representing the number of throws of a die prior to obtaining a 6. We could potentially plot probabilities of an infinite number of abscissa values. However, due to practical considerations we truncate the distribution when the corresponding probabilities become vanishingly small. In this case, the probability that the number of throws will be greater than 20 is less than 1%; so we arbitrarily truncate the distribution at this point.

What is the probability that V is equal to three—that is, the probability that three throws occur prior to the first six? This is simply the probability that a not-six occurs on each of the first two throws and a six occurs on the third throw. Again, these three events are all independent of one another, so the probability is

$$p(\text{not-6} \cap /6) = (5/6)(5/6)(1/6) = 25/216 = 0.116$$

In general, the probability that the value is equal to any integer i may be obtained by the following equation.

$$p(i) = p(\text{throw not-6 } i - 1 \text{ times, followed by 6})$$

$$= (5/6)^{i-1} (1/6)$$

Naturally it is not possible to actually construct the probability distribution for all the infinite values that the random variable can potentially assign. Notice, however, that for very large values (for instance, 20) the probabilities associated with these values start to get so tiny that we're not really concerned with them anymore. Therefore, to depict our probability distribution, we can arbitrarily truncate the distribution at some value— say, 20—as depicted in Figure D-3.

Both these random variables are, however, discrete in the sense that only certain, specific values are assigned—for the number-of-rolls-prior-to-a-six random variable, a two or a three could be assigned, but not a 2.6 or 3.24. Contrast these random variables to a third that would assign a *weight* to each person. In this case, V would be uncountably infinite, or

$$V = \{\text{any positive real number}\}$$

and the resulting distribution would be continuous.

To be somewhat more formal about the distinction between a continuous and a discrete random variable, let's consider any two values, v_1 and v_3, that a particular random variable can assign, and suppose that the only restriction on v_1 and v_3 is that v_3 is greater than v_1, that is, $(v_1 < v_3)$.

Now a random variable under consideration is defined as *continuous* if no matter what v_1 and v_3 are chosen to be, we can find another v_2 that is in between them—that is, there exists a v_2 such that

$$v_1 < v_2 < v_3$$

So, for example, if the random variable we were dealing with assigned a weight to every person, no matter what two weights you chose (say, $v_1 = 150.000$ pounds and $v_3 = 150.001$ pounds), you could always find another potential weight between the first two which the random variable could also assign (for example, $v_2 = 150.0005$ pounds).

On the other hand a random variable is defined as discrete if two values v_1 and v_3 can be found such that the random variable cannot assign any value between them. So, for example, if we were talking about number of heads in four flips of a coin, we might choose v_1 to be 2 and v_3 to be 3. We then could never find a v_2 which was equal to 2.5 or 2.72 or anything in between 2 and 3. This random variable would be discrete by our definition.

2.4 When V, the set of all possible values that a random variable can assign, is finite or countably infinite, the resulting probability distribution is discrete. When V is uncountably infinite, the resulting probability distribution is continuous.

Some Important Characteristics of Probability Distributions

So far we have constrained a detailed discussion of probability distributions to those that are discrete. We have avoided talking about probability distributions of continuous random variables because we will have to think about such distributions in a slightly different way from the way we think about distributions of discrete random variables. As a background for talking about continuous probability distributions, however, it will be helpful to enumerate a few important characteristics of the discrete probability distributions we've already discussed.

Ordinate values must sum to 1.0 for discrete distributions. Let's go back to Figure 2-3, which, you will recall, depicted a theoretical probability distribution for each possible number of heads in four flips of a coin. Notice that the five possible outcomes—0, 1, 2, 3, or 4 heads—are mutually exclusive and exhaustive events. That is to say, one of them has to happen, and if one happens, then none of the others can happen. Recall from our discussion in the previous chapter that the probabilities of mutually exclusive and exhaustive events must sum to 1.0. Since in Figure 2-3 the probability of each of these five events (values) is represented by the corresponding ordinate value of the probability distribution, this means that the sum of all the ordinate values must be 1.0. Looking at the figure, we can easily ascertain that this is the case.

This particular fact—that the ordinate values on the probability distribution must sum to 1.0—is true for *any* discrete probability distribution, either theoretical or empirical. In general, a probability distribution lists probabilities for a number of mutually exclusive and exhaustive events—and so the probabilities of all these events must sum to 1.0.

The area under the curve must also sum to 1.0.   The distribution shown in Figure 2-3 may be viewed somewhat differently.[2] How would we compute the *area* under the curve? This is not difficult to do—the curve is, after all, made up of a series of bars, or rectangles—and we can easily compute the area of any rectangle since the area of a rectangle is equal to its height times its base. In this case the height of each rectangle is equal to the corresponding probability and the width of each rectangle is equal to 1.0. So the area of each rectangle over each abscissa value *v* may be characterized as

$$p(v) \times 1 = p(v)$$

Therefore, the area contained in each rectangle is equal to the ordinate value of the rectangle. The sum of all the areas, then, must be equal to the sum of all the ordinate values, which in turn is equal to 1.0. This characteristic, true of *all* probability distributions—both discrete and continuous—is, to reiterate, that the area under the curve representing the probability distribution is equal to 1.0.

Probabilities of intervals.   Again let's refer back to Figure 2-3. Suppose we inquire about the probability that the random variable assigns a value contained within some *interval*. Suppose for example that we are concerned with the probability that the number of heads fell in the interval from 2 to 4 (inclusive). We could view this probability in either of two ways. First we could simply take the sum of the probabilities of 2, 3, and 4 heads, or

$$p(2 \le v \le 4) = \sum_{v=2}^{4} p(v) = p(2) + p(3) + p(4)$$

$$= 6/16 + 4/16 + 1/16 = 11/16 = 0.687$$

Alternatively, as discussed above, we could look at this probability as being equal to the area under the curve—that is, the combined area of the three rectangles corresponding to values of 2, 3, and 4. Figure 2-4 shows how this probability is represented by the equation

$$p(2 \le v \le 4) = \sum_{i=2}^{4} p(v) (1) \tag{2.1}$$

Notice that Equation 2.1 simply says: sum the areas of the three rectangles to get the probability.

The idea that probability can be viewed as an area under a curve will be very important in the next section when we discuss continuous probability distributions.

---

[2]The fact that a histogram is referred to as a "curve" may clash with the commonplace notion that a curve is a smooth, willowy, arc-like entity. Mathematically, however, any function, be it arc-like or lumpy, is referred to as a curve. The area under the curve is simply the area between the curve and the abscissa. Also, in the absence of any qualification, the expression "area under the curve" refers to the area under the curve between the abscissa values of $-\infty$ and $+\infty$.

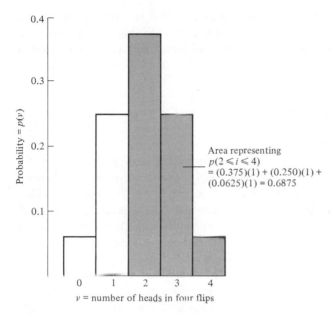

FIGURE 2-4. A theoretical probability distribution with a representation of the probability of some interval. This probability is represented by the area under the curve within the appropriate abscissa boundaries. Specifically the shaded area represents the probability that the number of heads in four flips of a coin will be between 2 and 4, inclusive.

2.5 Probability can always be characterized as the area under some portion of a curve.

Probability Distributions of Continuous Random Variables

As we have noted, continuous probability distributions involve somewhat different concepts than do discrete probability distributions. We shall begin this section by describing the nature of these differences and providing a general definition of what a continuous probability distribution is. Then we will try to elucidate the relationship between discrete and continuous distributions.

As usual we proceed on the assumption that abstract concepts are easier to understand if they are described within the context of an example. So let's suppose that our sample space consists of all male adults in the city of Seattle and that the random variable of interest is the one that assigns a *height* (in inches) to each member of the sample space. For the sake of argument even though some men may in actuality be shorter or taller, let's characterize V in this situation as

$V = \{$any real number between 65 and 75$\}$

We should make a comment about this characterization of *V* before we proceed. "Any real number" means that we are automatically dealing with a continuous random variable, because if we choose any two distinct members of *V* (say, 70.000 in. and 70.001 in.), we could easily find a third member (for instance 70.0005 in.) that is between the first two.

The probability of an exact value.   Recall that when we were concerned with a discrete probability distribution (such as the one depicted in Figure 2-3), we talked about the probability of exact values—for example, we asserted that the probability of getting *exactly* two heads in four flips of a coin is 0.25. But now we state a very important fact—namely, that *the probability that a continuous random variable will assign any exact value is 0.* To see why this is so, imagine that we are interested in finding some male who is *exactly* 72 in. tall. We argue (correctly) that if we find such an individual, then the fact stated above must be false. So we march out into the population, tape measure in hand, and we come upon a person who, according to our tape measure, seems to be exactly 72 in. tall. However, this apparent exactness will invariably turn out to be illusory and will stem from the fact that our measuring instrument (the tape, in this case) is not perfect. That is, if we could find a better measuring tape, accurate to, say, six decimal places, we might find that the man was 72.000001 or 71.999999 in. tall. But conceivably, you argue, even this very precise measuring tape could measure the person to be 72.000000 in. in height. This may be quite true, but if we were then to find an even better tape, which could measure out to enough decimal places, we would eventually find a discrepancy— the person will *not* be 72.000000—on out to an infinite number of zeros—in. tall.[3]

Probabilities of intervals.   We therefore see that when we are concerned with a continuous distribution, it no longer makes sense to talk about probabilities of exact values. It *does,* however, make sense to talk about the probability that a value will fall within some *interval.* For example in our population a certain proportion of males will have heights greater than or equal to 68 in. but less than 72 in. Notationally we can express this as

$$p(68 \le h < 72) = ?$$

This is the second important aspect of continuous probability distributions with which we must be concerned: we always talk in terms of probabilities of *intervals* rather than probabilities of exact values.

---

[3]It is sometimes hard for people to admit that the probability of an exact value is zero. However, nearly everyone will admit that the probability of an exact value is very, very tiny. If you wish, you can think of this probability of being "very, very tiny" rather than "zero." For all practical purposes the ensuing arguments will be the same.

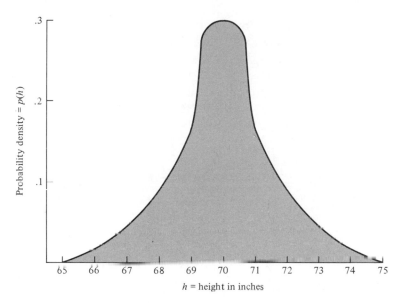

FIGURE 2-5.   An example of a typical continuous probability distribution.

Areas under curves.   To see *how* we talk about probabilities of intervals, let us first see what a typical continuous probability distribution looks like. Figure 2-5 shows the distribution corresponding to our height example. Notice that this distribution is smooth, in contrast with the curves that represent discrete probability distributions (which were made up of discrete bars and therefore were lumpy). Notice also that the abscissa of the curve still represents all the values the random variable can assign. However, the ordinate no longer represents probability. It now represents something new which is termed *probability density*. Odd though it may seem, we will never be very concerned with probability density. Probabilities are represented by areas under curves, not by densities.

The next two facts that we are about to state are crucial to your knowledge of continuous probability distributions. For the moment just accept these facts. In the next section we will try to illustrate why they are so.

Fact 1: The area under the curve is equal to 1.0 (just as was the case for discrete distributions).

Fact 2: The probability that a value falls within some interval *is represented by the area under the curve contained within that interval* (just as was the case with discrete distributions). So for example, Figure 2-6 shows the same curve as Figure 2-5, along with a depiction of the probability that the height of a random male will be between 68 and 72

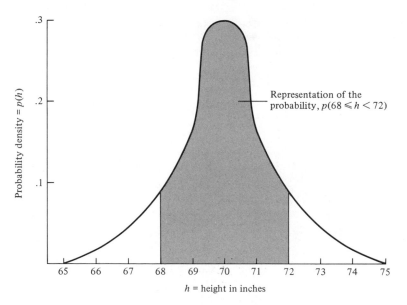

FIGURE 2-6.   A continuous probability distribution along with a representation of the probability that a score falls within some interval. This representation is the area under the curve within that interval.

in.[4] This probability is depicted to be 0.78, or 78% of the total area under the curve.

Digression 2.3

Calculation of Areas

You may feel somewhat uncomfortable with this state of affairs. With discrete distributions everything was relatively easy. To depict or compute some probability, we had to be concerned only with ordinate values. Even using the (seemingly perverse) alternate technique of

---

[4]Confusion sometimes arises with regard to the term "between," as in "between 68 and 72 in." Should the limits—68 and 72 in., in this case—be included? For all practical purposes it doesn't really make any difference whether they are included or not, since we have argued that the probability of an exact value like 68 or 72 is zero anyway. For logical consistency, however, we generally include one limit and exclude the other in our "between." So one interval might be

$$68 \le h < 72$$

and an adjacent interval might then be

$$72 \le h < 73$$

Using this scheme, we make all our intervals mutually exclusive (and, if we like, exhaustive).

computing areas under curves was relatively easy, since calculating areas of rectangles is relatively straightforward. Now it appears we have to be concerned with calculating very oddly shaped areas. How do we do it?

There are two answers to this question. First the relevant areas of all probability distributions with which we will be concerned have already been computed and put into tables by various industrious individuals. (Some of these tables are reproduced in Appendix F of this book.) The second answer has to do with the existence of a branch of mathematics known as *integral calculus,* specifically concerned with computation of areas under curves. It was by using integral calculus that the aforementioned industrious individuals managed to construct the various tables of relevant areas. Since we have the tables, we don't need to know about integral calculus. For convenience of discourse, however, we will borrow a piece of notation from integral calculus called an *integral sign.* An integral sign is used to denote the area under some curve, $p(x)$, within the interval whose limits are a and b. In general

$$\text{area under } p(x) \text{ between } a \text{ and } b = \int_a^b p(x)dx$$

Thus, to represent the fact that the shaded-in area of Figure 2.6 is equal to 0.78, we would write

$$\int_{68}^{72} p(h)dh = 0.78 \tag{D.1}$$

Notice the similarity between Equation D.1 and Equation 2.1, which gave the probability that a value of a discrete distribution fell within some interval. In fact there is a great deal of similarity between a summation sign and an integral sign. The expression

$$\sum_{i=2}^{4} p(i)(1)$$

means we should sum the areas of a finite number of bars whose heights are $p(i)$ and whose widths are 1. Likewise, the expression

$$\int_{68}^{72} p(h)dh$$

means we should sum an infinite number of "bars" whose heights are $p(h)$ and whose widths (dh) are infinitely skinny.

Relationships between discrete and continuous probability distributions. The preceding sections have, as noted, provided the crucial facts that you need to know about continuous probability distributions. We now will attempt to give you an understanding of why these facts are true by relating them to what you already know about discrete probability distributions.

Let's stick with our example of a probability distribution of height, and, for the moment, let's view height as a discrete, not a continuous, random variable. To do this, we suppose that anyone whose height is greater than or equal to x, but less than $x + 1$ inches, is assigned a height of x inches (for example, anyone with a height between 67 and 68 in. would be assigned 67). Now we can characterize V as

$$V = \{65, 66, 67, 68, 69, 70, 71, 72, 73, 74\}$$

Figure 2-7 shows the distribution that we might construct with this particular V. We note two interesting things about this distribution: First, the area under the curve is 1.0 (as we have already argued that it must be for any discrete distribution). Second, the probability that a person's height will fall between 68 and 72 in. is

$$p(68) + p(69) + p(70) + p(71)$$
$$= 0.12 + 0.27 + 0.27 + 0.12$$
$$= 0.78$$

just as it was before.

Now let's concentrate on one of these values—say, the value 68. We know that the probability that a person is between 68 and 69 in. tall is 0.12. Suppose, however, that at this point we decide that we'd like a little more accuracy—we'll measure everyone's height to an accuracy of 0.1 in. as opposed to just 1.0 in. Thus, everyone who is between 68.0 and 68.1 in. will be designated 68.0 in. tall; everyone between 68.1 and 68.2 in. will be designated as 68.1 in. tall, and so on.

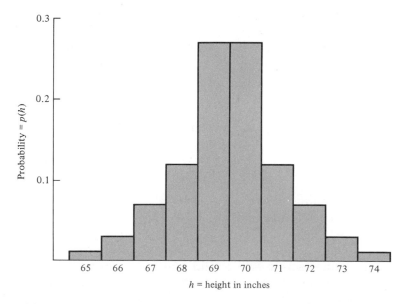

FIGURE 2-7.   Height as a random variable represented as a discrete probability distribution. The bar over each abscissa value corresponds to the probability that a random person falls in the interval between that value and the next highest integer value. (For example, the bar over 68 would represent the probability that a random person falls between 68 and 69 inches.)

Now let's consider the bar in Figure 2-7 that corresponds to 68 in. As we know, the fact that this bar goes up to 0.12 indicates that 12% of the sample space range between 68 and 69 in. tall. But with our new accuracy campaign, we are now able to distinguish people *within* this range. In particular suppose that all the people within this range are distributed as indicated in Table 2-6—that is, 5.8% of these people are between 70 and 70.1 in., 6.7% of them are between 70.1 and 70.2 in. and so on. Figure 2-8 depicts the way in which the 68-in. histogram bar has been "adjusted" to reflect the fine-grained data of Table 2-6. Over the original bar going up to 0.12 is superimposed a series of narrower bars. Notice that the average height of the narrow bars is still 0.12, the average of the original bar. This reflects the fact that "on the average" the probability that a person will fall within the 68–69-in. range is still 0.12. However, the fact that for example there are more people in the small interval from 68.9–69 than in the small interval from 68.0–68.1 is reflected in that the narrow bar representing the former small interval is taller than the narrow bar representing the latter small interval

Several aspects of Figure 2-8 should be attracting your attention. First the bars of Figure 2-11 look like those in a familiar old probability histogram (like the one shown in Figure 2-7). However, they are different in a fundamental way: The ordinate value of a particular bar no longer represents the probability of the abscissa value of the bar. For example the fact that the bar over 68.0 goes up to a value of 0.070 does *not* mean that the probability is 0.070 that a person is between 68.0 and 68.1 in. tall—rather, the bar has been judiciously constructed so that the *area under the curve* represents this probability instead. This area, of course, is simply the base times the height of the bar, or

$$p(68.0 \leq h < 68.1) = (0.1)(0.070) = 0.0070$$

Second, however, the area under the curve between 68 and 69 in. has not changed. This area (which may be obtained by adding the areas of all the

TABLE 2-6.   Data Concerning All Males in the Height Range of 68–69 in. and Distribution of Heights within this Range

Height (h) Range	Males in this Range
$68.0 \leq h < 68.1$	5.8%
$68.1 \leq h < 68.2$	6.7%
$68.2 \leq h < 68.3$	7.5%
$68.3 \leq h < 68.4$	8.3%
$68.4 \leq h < 68.5$	9.2%
$68.5 \leq h < 68.6$	10.0%
$68.6 \leq h < 68.7$	11.2%
$68.7 \leq h < 68.8$	12.3%
$68.8 \leq h < 68.9$	13.6%
$68.9 \leq h < 69.0$	15.4%

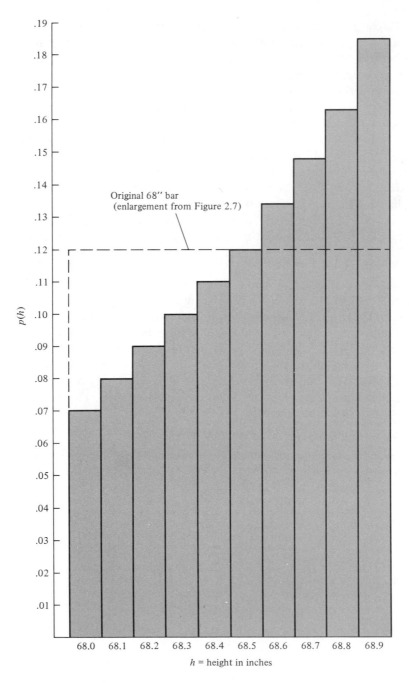

FIGURE 2-8.   One of the bars from Figure 2-7 has been modified to represent the distribution of heights within the interval. Tall bars represent the fact that there are relatively many people in the corresponding subintervals; short bars reflect the fact that there are relatively few people in the corresponding subintervals.

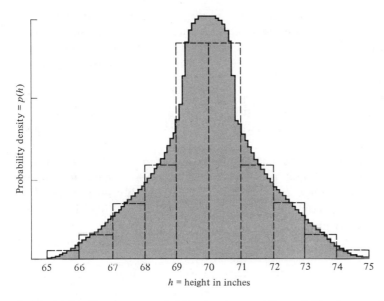

FIGURE 2-9. The discrete probability distribution of Figure 2-7 has been modified to represent the fact that finer-grained data have been obtained. The dashed bars represent the original histogram of Figure 2-7. The lumpy curve represents the subdividing of the intervals as was done in Figure 2-8.

individual narrow bars) remains 0.12. This makes sense, since the area between 68 and 69 in. still represents the probability that a person is between 68 and 69 in. tall. We would not expect this probability to change merely because we have gathered finer-grained information.

Figure 2-9 now shows the entire distribution after the technique described above has been applied to each of the original bars of Figure 2-7. This histogram looks considerably smoother and more akin to a continuous distribution than does Figure 2-7.

This same procedure of subdividing the bars may now be reapplied to the new histogram in Figure 2-9. That is, we could opt for yet more accuracy and measure heights to 0.01 inch, in which case the histogram would become even smoother. Continuing to reapply the procedure—making our measurements more and more refined—would ultimately result in the continuous distribution that is shown in Figure 2-5.

SUMMARY

The principal goal of this chapter was to familiarize you with various types of probability distributions. To do this, we first described *frequency* distributions, since frequency distributions are easy to understand and provide a convenient background against which probability distributions could be viewed.

There are several important facts about probability distributions which we discussed.

1. Probability distributions stem from the application of some random *variable* to some *sample space*. A random variable assigns some number to each member of the sample space; a probability distribution simply lists the probability of occurrence of each value assignable by the random variable.

2. Probability distributions may be classed as *empirical* or *theoretical*. An empirical probability distribution is derived from data. A theoretical probability distribution is derived from a priori considerations and does not depend on data.

3. Probability distributions may be classed as *discrete* or *continuous*. If a distribution is discrete, then the set of all values that the random variable can assign is either a finite or a countably infinite set. If a distribution is continuous, then the set of all values that the random variable can assign is an uncountably infinite set.

4. For discrete probability distributions the sum of all ordinate values must equal 1.0.

5. For all probability distributions the area under the curve must equal 1.0.

6. For all probability distributions the probability of a value falling within some interval can be represented by the area under the curve within the bounds of that interval.

7. For a continuous probability distribution the probability of an exact value is zero.

8. For a continuous probability distribution all probabilities that are considered are probabilities of intervals.

PROBLEMS

1. A political incumbent has a survey taken on how well she is doing in office. Each participant in the survey rates the incumbent on a scale from 1–5 where 1 signifies "abominable" and 5 signifies "fantastic."

   The results of the 70 respondents produce the following ratings:

   ```
   2  1  3  3  2  1  3  4  2  1  4
   1  4  1  5  3  4  1  1  2  1  2
   2  3  1  1  1  2  1  3  4  4  5
   1  4  1  4  4  4  2  4  2  3  5
   3  1  1  1  5  5  3  2  5  5  3
   4  1  3  4  4  3  3  4  3  3  1
   4  5  2  3  5  5  4  5  3  4  4
   ```

   1. Represent these numbers as a frequency distribution.
   2. Represent them as probability distributions.

2. Perusing the want ads in a newspaper, we find 50 houses with the following prices (in thousands of dollars):

250	260	130	260	810	480	630	950	110	980
540	130	290	820	750	680	290	940	920	590
530	480	370	960	350	910	160	850	430	110
140	370	960	510	260	190	730	600	110	90
930	820	100	770	650	120	60	550	760	410

1. Plot a frequency distribution of house prices.
2. Convert this distribution to a probability distribution.

3. A die is thrown and its value is noted. Denote the value a, which can range from 1–6.

   A random variable, X, is created as follows:

   $$X = 3a^2 + 2a - 1$$

   1. What are the members of V where V consists of all values that X can take on?
   2. Compute the probability distribution of V.

4. A family plans to have three children. Suppose that each child has a probability 0.5 of being a boy, and suppose the sex of each child is independent of the sex of every other child.

   1. What are the members of V where V is the set of possible numbers of boys the family could have?
   2. Compute the probability distribution over the members of V.

5. Suppose a player bats twice in a baseball game. On the first at-bat his probability, P, of getting a hit is 0.25. On the second at-bat, his probability of getting a hit is determined as follows:

   If he got a hit on the first at-bat, $P = 0.35$.
   If he did not get a hit on the first at-bat, $P = 0.25$.

   1. What are the members of V where V is the set of possible numbers of hits the player could get during his two at-bats?
   2. Compute the probability distribution of the members of V.

6. Joe Shablotnik has gone out for beer to celebrate his safe return from a scuba diving expedition. *After drinking his first beer,* Joe decides that the number of subsequent beers he will drink that night will be determined in the following way. Before each potential round of beer, Joe will throw a die. If the die comes up 1, 2, 3, or 4 Joe will drink that round of beer. If the die comes up 5 or 6, Joe will not drink the round of beer but will go home instead.

   1. What are the members of V, where V is the set of possible numbers of beers Joe could drink that night before going home?
   2. Is V a finite, countably infinite, or uncountably infinite set?
   3. Fill in the probability distribution below of the number of beers Joe will drink before going home.

4. What is the probability that Joe will drink *more than* four beers before going home?

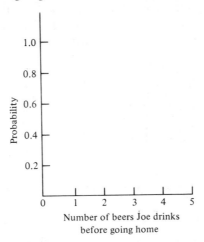

Number of beers Joe drinks
before going home

7. A professor decides to grade papers according to the following scheme. For any given paper
   —she rolls a die. If the die comes up 6, the paper receives an A. Otherwise
   —she rolls the die again. If the die comes up 5 or 6, the paper receives a B. Otherwise
   —she rolls the die again. If the die comes up 4, 5, or 6, the paper receives a C. Otherwise
   —she rolls the die again. If the die comes up 3, 4, 5, or 6, the paper receives a D. Otherwise the paper receives an F.
   1. Plot the probability distribution of paper grades.
   2. Suppose that 1296 students turn in papers. Plot the expected frequency distribution for grades.

8. Antoine the accountant sits down one night to fill out three tax returns. He decides that the number of tax returns he will fill out before quitting that night will be determined in the following way: before each tax return, he will draw a card from a well-shuffled deck. If the card comes up a spade, he will quit; otherwise, he'll fill out the tax return. (Naturally, if he fills out all three tax returns, he'll quit.)
   1. What are the members of V where V is the set of possible number of tax returns Antoine could fill out that night?
   2. What is the probability of each member of V?

9. Two dice are thrown and the sum, s, is observed.
   1. Another random variable, x, is created using the equation

      $$x = s$$

      Plot the probability distribution of this random variable.
   2. Another random variable is created as follows:

$y = 2s^2$ if at least one 6 shows up
$y = s^2$ if no 6 shows up

Plot the probability distribution of this random variable.

10. Below is a probability distribution for weight of salmon caught in the state of Washington. (Assume that weight is a continuous random variable).

   1. Calculate the probability that the weight of a salmon caught in Washington will be greater than or equal to 1 point, but less than or equal to 3 points.
   2. Calculate the probability that a Washington salmon will be less than 2 pounds in weight.
   3. Calculate the probability that the weight of a salmon caught in the state of Washington will be greater than or equal to 2.5 points.

   (*Hint:* You shouldn't have to compute any strange areas.)

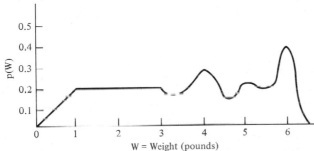

11. A probability distribution is represented by the following function:

   $p(x) = 0$ for $x < 0$

   $p(x) = x$ for $0 \leq x < 0.50$

   $p(x) = 0.5$ for $0.5 \leq x < 2.00$

   $p(x) = 2.50 - x$ for $2.00 \leq x < 2.50$

   $p(x) = 0$ for $x \geq 2.50$

   1. Plot $p(x)$ as a function of x.
   2. What is $p(0.1 \leq x \leq 0.25)$?
   3. What is $p(0 \leq x < 0.5)$?
   4. What is $p(x < 1.0)$?
   5. What is $p(x > 1.0)$?
   6. What is $p(0 \leq x \leq 1.25)$?
   7. What is $p(x > 0)$?

12. In a strange version of basketball, a shooter shoots foul shots (each successful shot worth 1 point). A shooter keeps shooting until he or she fails, up to 5 shots.

   Plot the probability distribution for the number of points a shooter will get if the shooter's probability of making any given shot is

1. 9/10   3. 1/4
2. 1/2   4. 1/8

13. Joe Shablotnik is playing a game of "Dazzle." Each trial of a Dazzle game involves throwing a black die, then a white die, and finally flipping a coin. The score, x, obtained on the trial is then obtained by the following formula:

x = number on white die plus number on black die plus 1 if coin turns up heads

x = number on white die plus number on black die if coin turns up tails

1. List the members of V where V is the set of scores that Joe could potentially get on each trial.
2. Compute the probability of all members of V.

In "Superdazzle" the rules are the same, but the dice and coin used are such that if the white dies comes up a 1, it exerts a magic forcefield which causes the coin to always turn up heads.

1. List the members of V for Superdazzle.
2. Compute the probability of all members of V.

14. George the pollster plans to interview two individuals and inquire as to their attitudes toward birth control. George gives each individual a score of 0 if he or she is unalterably opposed to birth control, a score of 2 if he or she firmly believes in birth control, and a score of 1 if he or she doesn't care. George will then sum the scores from the two individuals to obtain a "sample opinion" score.

1. What are the members of V where V is the set of possible sample opinions?
2. Suppose that each individual has probabilities of 0.10, 0.50, and 0.40 of producing scores of 0, 1; and 2. What is the probability distribution over the members of V?

15. Suppose a coin is flipped. Let $a = 4$ if the coin comes up heads and $a = 2$ if the coin comes up tails. Now a die is thrown. Let b equal the number showing on the die; thus b can range from 1–6.

1. Letting a and b be random variables, compute their probability distributions.
2. Suppose a new random variable, X, is created such that $X = b - a$. Determine the probability distribution of X.
3. Suppose a new random variable, y, is created such that $y = b + X$. Determine the probability distribution of y.

16. Consider a continuous probability density function, $P(x)$. Is it possible for $P(x)$ to be greater than 1.0? If so make up an example of such a situation. If not explain why not.

Descriptive Measures 3

In the previous chapter we began to deal with the difference between data on the one hand and theory on the other. This difference was embodied principally in the distinction between empirical and theoretical distributions. In this chapter we will maintain that distinction, and we will cover several topics. Initially we will return to the topic of empirical frequency distributions. In the previous chapter our focus was on frequency distributions as an aid to understanding probability distributions. Here, in contrast, we will be interested in frequency distributions as a means of organizing and elucidating a set of data. Our second topic of this chapter will be that of *descriptive statistics,* which are measures designed to crisply summarize the critical information in a collection of numbers. Finally, we will return to theoretical probability distributions and discuss how analogous descriptive measures may be used to summarize them. You will need the information in Appendix B to understand the notation in this chapter.

FREQUENCY DISTRIBUTIONS REVISITED

Consider a set of data consisting of exam scores obtained by 50 students. Table 3-1 lists these scores. What are we to make of them?

Just looking at these scores, we can't make much of them. The human mind is not designed to assimilate 50 separate numbers, and the present case is no exception. What we need are methods for summarizing or boiling down the interesting information from this set of numbers. Construction of a frequency distribution is the first step in this process. The random variable of interest here assigns to each of $n = 50$ students an exam score. The set V of all possible assignable values is:

$$V = \{0, 1, 2, \ldots, 100\}$$

Figure 3-1 shows the frequency distribution that results (for simplicity we have not included the area of abscissa corresponding to scores below 40 since there were no scores in that range).

TABLE 3-1.   Hypothetical Data: Exam Scores of Each Person in a 50-Person Class

Person	Score	Person	Score	Person	Score	Person	Score	Person	Score
1	66	11	47	21	79	31	86	41	59
2	50	12	55	22	89	32	91	42	65
3	72	13	64	23	73	33	53	43	73
4	78	14	79	24	94	34	70	44	74
5	75	15	67	25	88	35	78	45	91
6	41	16	77	26	71	36	77	46	99
7	84	17	76	27	69	37	87	47	49
8	82	18	55	28	80	38	48	48	97
9	62	19	97	29	63	39	81	49	92
10	60	20	57	30	48	40	95	50	79

The picture in Figure 3-1 provides a start in terms of acquiring a feeling for the data in Table 3-1. We can immediately tell, for example, that all scores fall within the range 40–100 and that scores tend to be bunched in the region 60–90. There are, however, two ways of reconstructing this frequency distribution so as to make it yet more meaningful. These two ways involve a *cumulative* distribution and the use of *class intervals*.

Cumulative Distributions and Percentiles

A cumulative frequency distribution associates with each value v the number of scores that were *of size v or smaller*. Table 3-2 and Figure 3-2 depict the construction of a cumulative frequency distribution from the data in Table 3.1. So for example a score of 55 has a cumulative frequency of 9 since there were 9 scores of 55 or less. Note that cumulative frequencies must always start at 0 and end at n. In interpreting a cumulative distribution, we pay attention to the *slope* of the graph. In intervals of high frequency (such as in the 70s) the

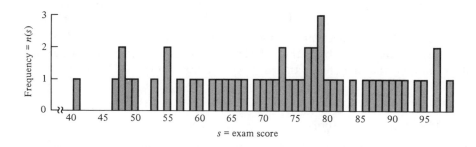

FIGURE 3-1.   An empirical frequency distribution of the scores from Table 3-1.

TABLE 3-2.   Construction of a Cumulative Frequency Distribution from the Data in Table 3-1

Value	Freq.	Cum. Freq.	Value	Freq.	Cum. Freq.	Value	Freq.	Cum Freq.
40	0	0	60	1	12	80	1	35
41	1	1	61	0	12	81	1	36
42	0	1	62	1	13	82	1	37
43	0	1	63	1	14	83	0	37
44	0	1	64	1	15	84	1	38
45	0	1	65	1	16	85	0	38
46	0	1	66	1	17	86	1	39
47	1	2	67	1	18	87	1	40
48	2	4	68	0	18	88	1	41
49	1	5	69	1	19	89	1	42
50	1	6	70	1	20	90	0	42
51	0	6	71	1	21	91	2	44
52	0	6	72	1	22	92	1	45
53	1	7	73	2	24	93	0	45
54	0	7	74	1	25	94	1	46
55	2	9	75	1	26	95	1	47
56	0	9	76	1	27	96	0	47
57	1	10	77	2	29	97	2	49
58	0	10	78	2	31	98	0	50
59	1	11	79	3	34	99	1	50

slope is steep. In intervals of low frequency (as in the 40s and in the 90s) the slope is shallow.

Cumulative probability distributions.   Recall that an empirical frequency distribution can be changed into an empirical probability distribution by dividing each frequency by n. Exactly the same thing can be done with a cumulative distribution. Figure 3-3 shows the same data as does Figure 3-2, but the ordinate has been divided by n (50 in this example) to provide cumulative probability.

Percentiles.   The cumulative probability of any given score multiplied by 100 is referred to as that score's *percentile*. A percentile of some score is, in some sense, the measure of "goodness" of that score. For instance a score of 60 on this exam would have a percentile of $(0.24)(100) = 24$, which would mean that only 24% of the students performed as poorly as 60. A score of 90, in contrast, has a percentile of $(0.84)(100) = 84$, meaning that 84% of the class got a score of 90 or below. Another way of looking at a score of 90 is that only $100 - 84 = 16\%$ of the class performed better. Percentiles are often used in conveying the results of standardized tests such as the Scholastic Aptitude Test (SAT) or the Graduate Record Examination (GRE). So if you have a percentile rating of 89 on your verbal GRE, you should feel good because this means that only 11% of those taking the exam did better than you.

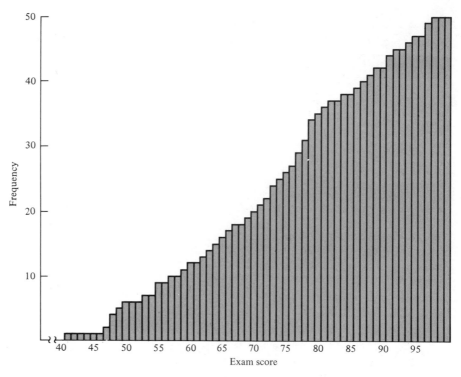

FIGURE 3-2.   The frequency distribution of Figure 3-1 has been transformed into a cumulative frequency distribution.

> 3.1 The percentile of a given score, x, is the percentage of scores in the distribution that are as low as or lower than x.

Class Intervals

The second modification of the original frequency distribution shown in Figure 3-1 involves the division of our set of values into intervals called class intervals. So, for example, the first interval might consist of values from 40 to 49; the second interval might consist of values from 50 to 59; and so on until the last interval consisted of values from 90 to 100. These intervals, rather than the individual values, can then be listed along the abscissa, as has been done in Figure 3-4. We then simply proceed to get the frequency of the number of scores within each of these intervals. For example, there are five scores in the interval 40 to 49, six scores in the interval 50 to 59, and so on. This class interval distribution allows us to get somewhat more of the "big picture." We can see, for example, that there are a large number of scores in the 70s, relatively few in the 40s, and so on. The disadvantage of class intervals, of course, is that we lose

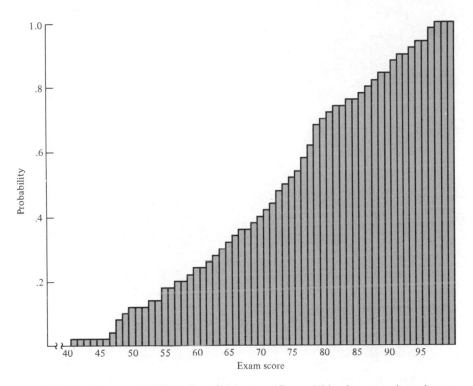

FIGURE 3-3.   The cumulative frequency distribution of Figure 3-2 has been transformed into a cumulative probability distribution.

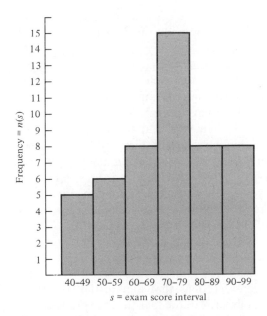

FIGURE 3-4.   The frequency distribution of Figure 3-1 has been reconstructed using class intervals.

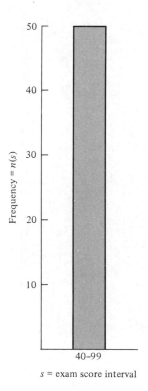

FIGURE 3-5.   An example of a preposterously large class interval.

information, specifically information corresponding to the distribution of scores within each interval.

Interval size.   An obvious question that arises in such a situation is: How big should the intervals be? In our example we chose to make the interval of size 10—but there is no reason why it could not have been 5, 8, 7, or anything else. The answer to this question is: the size of the interval is up to you. You can make it whatever you want. Several rules of thumb are useful. First the interval should be big enough that it doesn't run into the problem of having annoyingly small frequencies occurring at each interval. On the other hand it is unwise to make the interval too big because then you lose too much information. Figure 3-5 exemplifies this latter extreme—here the same data have been plotted with only one interval. The size of the interval is very big, ranging from 40 to 100, and therefore includes the entire range of the data. As you can see, Figure 3-5 is virtually useless—it merely tells us that there were 50 scores in all, and they ranged from 40 to 100.

Finally there might be other considerations that dictate the size of the intervals. In the situation we have just described, the fact that the grade of "A" typically corresponds to the interval from 90 to 100, the grade of "B" typically corresponds to the interval 80 to 90, and so on, suggests that these intervals might be appropriate ones to use.

> 3.2 When there are many possible values that a score can be, it is useful to construct class intervals. The size of each interval is arbitrary and is typically dictated by practical considerations.

DESCRIPTIVE STATISTICS TO DESCRIBE DATA

The preceding section was designed to show how a set of data may be summarized via transformation into some sort of frequency distribution. Often, however, we wish to summarize a set of data even more concisely. To do so, it is customary to extract two numbers from the data—one reflecting the *average* or *central tendency* of the data and the other reflecting the *dispersion* of scores in the data. To get a flavor for central tendency and dispersion, imagine that you have just gotten an exam back and have discovered that your score was 80 out of a possible 100 points. You are naturally interested in evaluating how your score stacked up relative to those of others in the class. What kinds of information would you like to have in order to make such an evaluation? The first such piece of information might be the class average. That is, you might want a single score that represented or typified the scores received by everyone in the class. Then you could see how your score compared to that average score. If the class average turned out to be 75, then your score would be above the average and you would probably feel happy. On the other hand, if the class average turned out to be 85, then you would be below the average and you probably wouldn't be so happy.

Let's suppose that in fact the typical or average score turned out to be 75. You now know that you did better than the average, but you might still want more information. In particular, you might want to know how closely the exam scores clustered around the average of 75. Suppose for example that the majority of the scores in the class took on values like 73, 76, 78, 75, and so on. In this case your score of 80 would be in a sense "far above the average" and would therefore be spectacular! On the other hand, suppose that there were a lot of scores in the class with values of 80, 85, 90, and 95 and also a lot of very low scores such as 60, 50, 55, and so on—in other words suppose that instead of being tightly clustered around the average of 75, the scores were, in general, very spread out. In this case, although your score of 80 is above the average, it would still be sort of typical. That is, there would still be quite a number of scores in the class that were better than your score.

As noted above and illustrated in this example, we will be concerned with two major types of descriptive statistics—measures involving the average (often

TABLE 3-3.   Exam Scores from an 11-Person Quiz Section

$x_1 = 8$		
$x_2 = 5$		
$x_3 = 7$		
$x_4 = 9$		
$x_5 = 8$	$\text{Mean} = \dfrac{\sum_{i=1}^{11} x_i}{n}$	
$x_6 = 1$		
$x_7 = 3$		
$x_8 = 4$	$= 66/11 = 6$	
$x_9 = 7$		
$x_{10} = 7$		
$x_{11} = 7$		

referred to as the central tendency) of the scores, and measures involving dispersion (often referred to as *variation*). Let's take up these two types of measures in turn. As a means of illustrating them, let's consider quiz scores of a hypothetical 11-person class. These hypothetical scores are shown in Table 3-3.

Measures of Central Tendency

Measures of central tendency, or "the average," are probably familiar concepts to most people. In everyday life we hear things about the "average family income," average number of divorces per year," "earned run average" of a baseball pitcher, and so on. In this section we shall systematize the notion of an average by talking about three different types of averages: the mean, the median, and the mode.

The mean.   The *mean* of a set of scores is typically the measure of central tendency that a person has in mind when he or she thinks of "the average." **The mean is the sum of all of our scores we have divided by n (the number of scores we have). Hence, the mean, which we will denote by the letter M, is obtained by the following formula:**

$$M = \frac{\sum_{i=1}^{n} x_i}{n}$$

As shown in Table 3-3, the mean of the scores in our example is simply

$$M = \frac{\sum_{i=1}^{11} x_i}{11} = \frac{66}{11} = 6$$

TABLE 3-4.   The Scores from Table 3-3 Have Been Ranked from Highest to Lowest, Allowing Easy Computation of the Median and the Mode

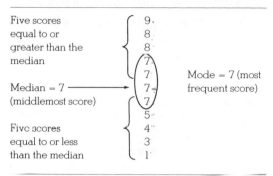

Five scores equal to or greater than the median

Median = 7 (middlemost score)

Five scores equal to or less than the median

9
8
8
7
7
7
7
5
4
3
1

Mode = 7 (most frequent score)

The median.   Another measure of central tendency is known as the *median*. The median is obtained in the following way. First one takes the set of scores and *ranks* them; that is, the scores are rearranged such that they are listed in order from the highest down to the lowest (or from the lowest to the highest if you prefer). Table 3-4 shows the data from Table 3-3 in ranked form. **The median is then the middlemost of the ranked scores—that is, the score that has the same number of scores equal to or greater than and equal to or less than it.** Referring to the data shown in Tables 3-3 and 3-4, we can easily see that the median is 7—there are 5 scores equal to or greater than 7, and there are also 5 scores equal to or less than 7.

Notice that in making up an example to use in computing the median, we have conveniently provided ourselves with an odd number of scores (specifically, 11 scores). This means that we have one and only one middlemost score. Suppose, however, that we wanted to compute the median of the scores shown in Table 3-5. There, we have eight, or an even number of scores, and there is no

TABLE 3-5.   Example of the Median of an Even Number of Scores

7
6
5

Three scores equal to or greater than middlemost scores

5
4

Median = mean of two middlemost scores = (5 + 4)/2 = 9/2 = 4.5

2
1
1

Three scores equal to or less than middlemost scores

single middlemost score. However, we can see that there are *two* scores that are middlemost. In this case if we consider the 5 and the 4 as the two middlemost ranked scores, then there are three scores (the 7, the 6, and the 5) equal to or greater than these two middlemost scores and three other scores (the 2 and the two 1s) equal to or less than these two middlemost scores. In a situation involving an even number of scores, the median is defined as the mean of the two middlemost scores, or in this case, $(4 + 5)/2 = 4.5$.

The mode.   Our final measure of central tendency is the mode. **The mode is the score that occurs with the greatest frequency.** In Table 3-4, there are one 9, two 8s, four 7s, one 5, one 4, one 3, and one 1. The most frequently occurring score is the 7, and so the mode of these scores is 7.

As you have probably figured out, there may arise a situation in which there is more than one mode. Such a situation occurs when two or more scores are tied for most frequent occurrence. Table 3-6 gives an example of such a situation. Here there are two 6s and two 3s, and only one of each of the other scores. Therefore, this set of scores would contain two modes—the 6 and the 3. This distribution would be called *bimodal,* referring to the fact that it has two modes. It would be perfectly possible for a set of scores to be trimodal (three modes), quatramodal (four modes), and so on.

The fact that a distribution may have multiple modes actually renders the mode a very poor measure of central tendency. Some would argue that it is not really a measure of central tendency at all. But it is occasionally used that way, so you should be aware of it.

Common errors made when computing the median and the mode.   As we have seen, computation of the three measures of central tendency—the mean, the median, and the mode—are all relatively simple and straightforward. Probably the most familiar type of measure of central tendency is the mean, and people rarely make errors (other than arithmetic errors) when they are computing the mean. However, there are several common errors when computing the median and the mode. We shall just go through examples of these types of errors in order to provide some defense against them.

TABLE 3-6.   An Example of a Bimodal Set of Scores

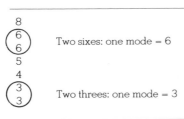

Let's imagine that Pacific Northwest Bell has decided to study the number of telephone calls made in a set of seven adjacent phone booths in Seattle–Tacoma Airport. To do this, the phone company monitors each of the seven phone booths for a period of one hour and comes up with the data shown in Table 3-7 the number of which simply consist of calls made from each booth. As can be seen, eight calls were made from Booth A, six calls from Booth B, and so on. Computation of the median and the mean are shown in the lower part of Table 3-7. Here we have ranked the data, and it's easy to see that the median or middlemost score is 5, and the mode or most frequent score is 1. However, when this example was given on an examination to a set of bright University of Washington undergraduates, a very large percentage of them made the errors shown at the bottom of Table 3-7, claiming that the mode was 8 and the median was 7.

There are two apparent sources of these errors. The first involves the misconception that the mode is equal not to the most frequent but to the largest score. The second source of error is taking the middlemost score without first ranking the data. In this case the middlemost score would appear to be 7, since it shows up in the middle phone booth.

We're not sure why, but these errors, as noted, appear to occur fairly frequently. Be careful to avoid them.

Which measure of central tendency is appropriate?   We have now discussed three measures of central tendency. An obvious question one might ask is: Which measure is best as a descriptive statistic for a set of data? **The rule of thumb is: use the mean unless there is some reason not to use the mean.** There are two reasons for this rule. First the mean is relatively easy to compute. When computing the median or the mode, one is forced to endure the

---

TABLE 3-7.   Number of Phone Calls made from Seven Telephone Booths: Computation of Median and Mode

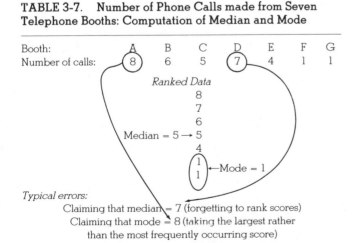

Booth:	A	B	C	D	E	F	G
Number of calls:	8	6	5	7	4	1	1

Ranked Data

8
7
6
Median = 5 → 5
4
1
1   ←Mode = 1

Typical errors:
    Claiming that median = 7 (forgetting to rank scores)
    Claiming that mode = 8 (taking the largest rather
    than the most frequently occurring score)

tedium of listing scores, ranking scores, counting scores, and so on. But none of this is necessary to compute the mean. One merely adds up and divides, which is quite easy to do, particularly with a pocket calculator. The second reason for using the mean is that the mean possesses some nice mathematical properties that make it useful for performing the sorts of statistical tests we will be discussing in subsequent chapters.

You now may ask: What would be reasons for *not* using the mean? Unfortunately there are no set rules for times you should not use the mean; a general rule is that it is a bad idea to use the mean when you have "extreme scores" lurking within your data. As an example consider the following experiment, which was informally performed by one of the authors (GL) using an experimental paradigm invented by the other author (EL). In this paradigm a subject is given a category name (for example, fruit) along with a letter (such as P). The subject's task is then to produce as quickly as possible a member of the category beginning with the letter. (Hence a reasonable response to fruit–P would be "pear.") The dependent variable of interest is the amount of time required to produce the answer. Table 3-8 shows some of the questions along with the answers and the reaction times. As you can see, the typical reaction time seemed to be about half a second—except for one question (body–W) that produced the very long reaction time of about 10 seconds.

Let's compute the mean, median, and mode for these data. The mean, shown in Table 3-8, is 2.46 sec. Notice that this number, which is supposed to be

---

TABLE 3-8. Reaction-Time Data Obtained Informally by G. Loftus: Illustration of a Situation in Which the Mean Would Not Be a Good Statistic To Use as a Measure of Central Tendency

Question	Answer	Reaction time
Fruit—p	"pear"	$x_1 = 0.6$ sec.
Country—f	"France"	$x_2 = 0.4$ sec.
Boy's name—j	"John"	$x_3 = 0.6$ sec.
Season—f	"fall"	$x_4 = 0.7$ sec.
Body part—w	"waist"	$x_5 = 10.0$ sec.

$$\text{Mean} = \frac{\sum_{i=1}^{5} x_i}{5} = \frac{0.6 + 0.4 + 0.6 + 0.7 + 10.0}{5.0} = 2.46 \text{ sec.}$$

Ranked Scores

$$\text{Median} = 0.6 \rightarrow \begin{matrix} 10.0 \\ 0.7 \\ 0.6 \\ 0.6 \\ 0.4 \end{matrix}$$

Mode = 0.6

representative or typical of the set of scores, is not really representative or typical of any of the scores. It certainly isn't representative of the four scores around a half-second, and it is also not at all representative of the 10-second score. Therefore it is not a very good measure of central tendency to use in this particular case. On the other hand we see that the median and the mode both turn out to be 0.6 sec. This score is not representative of the one long score (10 sec.), but it is at least representative of *most* of the scores. Therefore the median or the mode would probably be a better measure of central tendency to use in this example.

When faced with a choice between the median and the mode, the experimenter typically uses the median. The reason for this is that, as noted above, the mode often turns out to be a somewhat awkward statistic. Suppose, for example, there were no scores occurring more than once in the data. In this case there would be no mode (or there would be as many modes as there were scores). Likewise, as we have seen, a set of scores may be bimodal or trimodal. If this is the case, we tend to feel uncomfortable using the mode because we are seeking one unique score to use as a measure of central tendency.

3.3 Three measures of central tendency are the mean, the median, and the mode. The mean is generally used unless a few extreme scores make the median more appropriate. The mode is rarely used.

Measures of Variation

We now come to the second type of descriptive statistic in which we are interested—the measure of variation or dispersion. Recall that a measure of dispersion should represent the extent to which the scores are clustered together versus the extent to which they are spread out.

Deviation scores. It would be convenient to have a measure that is small to the extent that the scores are tightly clustered together and is large to the extent that the scores are spread apart. Therefore, what seems to be called for is some kind of measure of the average *deviation* of the scores from the mean. In Table 3-9 we have constructed a listing of such deviation scores based on the scores that were originally presented in Table 3-3. At the far left we have listed the original (or raw) scores from that table. To compute deviation scores, we have simply subtracted the mean from each of the raw scores. Hence in the second column (labeled $x_i - M$) we obtain a list of these deviation scores. Our first thought might be to just get the average (mean) deviation score and use that as our measure of variability, the idea being that a big mean deviation score would reflect comparatively high variability whereas a small mean deviation would reflect less variability. However, there is a problem with this procedure—namely, that some of the deviation scores are negative whereas others are

TABLE 3-9.   Deviation Scores Obtained from the Data of Table 3-3

x_i	$x_i - M$	$\lvert x_i - M \rvert$	$(x_i - M)^2$
$x_1 = 8$	2	2	4
$x_2 = 5$	-1	1	1
$x_3 = 7$	1	1	1
$x_4 = 9$	3	3	9
$x_5 = 8$	2	2	4
$x_6 = 1$	-5	5	25
$x_7 = 3$	-3	3	9
$x_8 = 4$	-2	2	4
$x_9 = 7$	1	1	1
$x_{10} = 7$	1	1	1
$x_{11} = 7$	1	1	1
$\displaystyle\sum_{i=1}^{11} x_i = 66$	$\displaystyle\sum_{i=1}^{11} (x_i - M) = 0$	$\displaystyle\sum_{i=1}^{11} \lvert x_i - M \rvert = 22$	$\displaystyle\sum_{i=1}^{11} (x_i - M)^2 = 60$

M = average score	average deviation score	average absolute deviation score	average squared deviation score
$= \dfrac{\displaystyle\sum_{i=1}^{n} x_i}{n}$	$= \dfrac{\displaystyle\sum_{i=1}^{11} (x_i - M)}{n}$	$= \dfrac{\displaystyle\sum_{i=1}^{11} \lvert x_i - M \rvert}{n}$	$= \dfrac{\displaystyle\sum_{i=1}^{11} (x_i - M)^2}{n}$
$= \dfrac{66}{11}$	$= \dfrac{0}{11}$	$= \dfrac{22}{11}$	$= \dfrac{60}{11}$
$= 6.0$	$= 0$	$= 2.0$	$= 5.455$
			$=$ Variance $= S^2$

positive. In our example we discover that when we add the deviation scores, their sum ends up being zero. Therefore, the mean deviation score (the sum of the deviation scores divided by *n*) also has to equal zero. We can prove that *any* set of deviation scores obtained by subtracting the mean from each score must add to zero, which makes the mean deviation score useless as a measure of variability.

Digression 3.1.

If we have a set of deviation scores around the mean, the sum of these scores must be zero.

Assume we have *n* scores. Then the sum of the deviation scores is represented by

$$\text{sum} = \sum_{i=1}^{n} (x_i - M)$$

$$= \sum_{i=1}^{n} x_i - \sum_{i=1}^{n} M$$

Since

$$\sum_{i=1}^{n} M$$

simply means "multiply M by n,"

$$\text{sum} = \sum_{i=1}^{n} x_i - nM$$

Now since

$$M = \frac{\sum_{i=1}^{n} x_i}{n}$$

$$\text{sum} - \sum_{i=1}^{n} x_i \quad \frac{n \sum_{i=1}^{n} x_i}{n}$$

Cancelling the ns in the numerator and denominator of the second term,

$$\text{sum} = \sum_{i=1}^{n} x_i - \sum_{i=1}^{n} x_i = 0$$

The second possibility is to take the mean of the *absolute value* of the deviation scores. In the third column of Table 3-9 we have listed the absolute values of the deviation scores (which involves changing any minus sign to a plus sign), and we have calculated the mean absolute value. Now we are in somewhat better shape in terms of coming up with a measure of dispersion. In this particular example, the sum of the absolute deviation scores is 22, and the average absolute deviation score is therefore 2. As we can easily see the average absolute deviation score provides us with a reasonable measure of variability. If the scores are tightly clustered around the mean, the mean absolute deviation will be small, whereas if the scores are widely spread, the mean absolute deviation will be large.

The variance. Although the average absolute deviation score provides a reasonable measure of variability, there is a better measure. The fourth column of Table 3-9 shows a listing of the *squared* deviation scores. Notice that any squared number is automatically positive, since any negative number times itself will yield a positive result, as will any positive number times itself. Therefore, squaring the deviation scores is another way of getting around the problem of some deviation scores being negative and others being positive. We have calculated the sum of the squared deviation scores, which is 60, and the average squared deviation score, which is $60/11 = 5.455$. This number—the average squared deviation score—has a name; it is called the *variance*. Notationally, the variance is generally referred to by the symbol S^2. (Do not confuse this S with the S that represented a sample space in previous chapters. They are completely different.)

The variance is the most widely used measure of variability of a set of scores. The reason for the preference is similar to the reason for preferring the mean to the median—the variance has useful mathematical properties in the sorts of inferential statistics about which we will be talking in subsequent chapters. This reason is probably unconvincing to you at present, but for the time being, please just try to accept it.

The standard deviation.   Another often-used closely related measure of variability is called the *standard deviation*. The standard deviation is nothing more than the square root of the variance. **Therefore, the formula used to obtain the standard deviation (which is commonly written as $S = \sqrt{S^2}$) is**

$$S = \sqrt{\frac{\sum_{i=1}^{n} (x_i - M)^2}{n}}$$

It may seem sort of strange to have two such closely related measures for variability; however closely related they are, though, there are various situations in which one is somewhat more useful than the other. We shall see what these situations are in subsequent chapters.

> 3.4   The most commonly used measures of dispersion are the *variance*, which is the mean squared deviation score, and the *standard deviation*, which is the square root of the variance.

Computational formulas.   As we have seen, the variance is equal to the average squared deviation score[1]

---

[1] The formula for variance is often given as the sum of the squared deviation scores divided by $n - 1$ (not n). The distinction between these cases will be described in detail in Chapter 9.

$$S^2 = \frac{\sum_{i=1}^{n} (x_i - M)^2}{n}$$

This formula is referred to as a deviation score formula for the obvious reason that it involves using deviation scores. However, it is often the case that when one is actually calculating the variance of a set of scores, it is tedious to go through the scores, subtract the mean from each score to get a deviation score, square each deviation score, and so on. In Appendix C, we start with the deviation-score formula and derive another formula that doesn't involve all these steps. This formula only involves use of the original xs—the raw scores—and appropriately enough it is referred to as a raw-score formula. Note that the set of steps in Appendix C consists simply of algebraic manipulation of the terms in the deviation-score formula to produce the raw-score formula. **As shown, the variance may be computed by the formula**

$$S^2 = \frac{\sum_{i=1}^{n} (x_i - M)^2}{n} = \frac{\sum_{i=1}^{n} x_i^2}{n} - M^2$$

The range. One more measure of dispersion is sometimes used, just as a "quick and dirty" estimate of the variability of the scores. This is called the *range* and is nothing more than the largest score in the set of scores minus the smallest score plus one. So considering the scores in Tables 3-3 and 3-9, the range would be equal to 9 (the highest score) minus 1 (the smallest score), plus 1, or 9.

Descriptive Statistics Applied to Empirical Frequency Distributions

We spent two earlier sections of this book discussing how one goes about calculating the mean and the variance from a set of raw scores. Suppose, however, a set of scores is presented to us as a frequency distribution. In this case we can use the information provided by the frequency distribution to calculate the same mean and variance of the original set of scores somewhat more efficiently. To see how this is done, let's consider an example. We will use a familiar example from the last chapter—the hypothetical data listing the number of heads in four flips of a coin obtained by you and your friends. The raw data and the corresponding frequency distribution were shown in Tables 2-1 and 2-3; this information is reproduced in Table 3-10. Notice that when listing the raw data, we have used the subscript notation we introduced earlier. Thus, the first score (3) is denoted by x_1, the second score (1) is denoted by x_2, and so on. When depicting the frequency distribution, we introduce some new notation. Specifically we have represented each *value* that the score can take on (that is,

TABLE 3-10.   Hypothetical Data: Numbers of Heads in Four Coin-Flip Sequences; Calculation of the Mean

		Frequency distribution	
Raw data		Value: v_i	Frequency: $n(v_i)$
$x_1 = 3$	$x_{11} = 1$	$v_1 = 0$	$n(v_1) = 1$
$x_2 = 1$	$x_{12} = 2$	$v_2 = 1$	$n(v_2) = 6$
$x_3 = 1$	$x_{13} = 2$	$v_3 = 2$	$n(v_3) = 6$
$x_4 = 3$	$x_{14} = 1$	$v_4 = 3$	$n(v_4) = 5$
$x_5 = 1$	$x_{15} = 2$	$v_5 = 4$	$n(v_5) = 2$
$x_6 = 2$	$x_{15} = 1$		
$x_7 = 0$	$x_{17} = 2$		
$x_8 = 2$	$x_{18} = 3$		
$x_9 = 4$	$x_{19} = 3$		
$x_{10} = 4$	$x_{20} = 3$		

$$M = \frac{\sum_{i-1}^{20} x_i}{n}$$

$$= \frac{3 + 1 + \cdots + 3}{20}$$

$$= \frac{41}{20}$$

$$= 2.05$$

$$M = \frac{\sum_{i-1}^{20} x_i}{n} = \frac{\sum_{i-1}^{5} v_i n(v_i)}{n}$$

$$= \frac{(0)(1) + (1)(6) + (2)(6) + (3)(5) + (4)(2)}{20}$$

$$= \frac{0 + 6 + 12 + 15 + 8}{20}$$

$$= \frac{41}{20}$$

$$= 2.05$$

each member of V) by v with a subscript. Hence, the first value (0) is denoted by v_1, the second value (1) is denoted by v_2, and so on, up to v_5. (The highest subscript number is 5, as there are five possible values for this score.) Continuing with this notation, we have represented the *frequency* of each value, v_i, by $n(v_i)$. So the fact that the value v_1 occurs in the data one time is denoted by $n(v_1) = n(0) = 1$. Likewise, since v_2 occurs six times, $n(v_2) = n(1) = 6$, and so on.

The mean.   At the bottom of Table 3-10, we have calculated the mean of the raw data in two different ways. The first involves using the raw data itself, whereas the second way uses the information encapsulated in the frequency distribution. On the left-hand side we can see the mean calculated using the normal procedure. We simply divide the sum of all 20 numbers by the total number of scores we have, or by 20. This procedure yields a mean of 2.05. On the right-hand side we have calculated the mean using a slightly different formula. We start with our familiar raw-score formula, the sum of the xs divided by n. However, we can see that by using our frequency distribution notation we can compute the sum of the xs somewhat more efficiently. Of the 20 numbers that we will be summing up, one is 0, six are 1s, six are 2s, and so on—this is the information provided for us by the frequency distribution. So the sum of the 20 numbers can simply be obtained by summing the products of each value and the frequency with which that value occurred; that is, the sum of the 20 numbers can be written as 0 times 1 plus 1 times 6 plus 2 times 6 and so on. Notice that now since we are summing over[2] all possible *values* (of which there are 5) rather than all our scores (of which there are 20), our subscript now only goes from 1 to 5 rather than from 1 to 20, as it did previously. We see, of course, that the sum and the mean we obtain are the same no matter which procedure we use.

The median.   Recall our discussion of percentiles (pp. 79). Consider now the score that corresponds to the fiftieth percentile. Another way of characterizing this score is that it is the middlemost score and is therefore the median. Referring to Figure 3-3, we can see that a score of 74 corresponds to the fiftieth percentile for the exam score data of Table 3-1. Therefore, 74 would be the median for this set of data.

Calculation of variance from a frequency distribution.   When calculating the mean of a frequency distribution, we use the information from the frequency distribution to make our lives somewhat easier in terms of calculating *sums*. Instead of tediously taking the sum of, say, 20 numbers, we only had to take the sum of 5 products of values times frequencies of those values. Using this technique, we can similarly simplify our lives when we calculate the variance of a set of numbers by using information contained in a frequency distribution.

Recall that the raw-score formula for variance is

$$S^2 = \frac{\sum_{i=1}^{n} x_i^2}{n} - \frac{\left(\sum_{i=1}^{n} x_i\right)^2}{n^2}$$

---

[2]We have pointed out that we represent what we are summing over by the limits on the summation sign. So for example

$$\sum_{i=1}^{10}$$

would mean "sum from 1 to 10." Another way of depicting what we want to sum over is just to write the name of the relevant set underneath the summation sign. Thus,

$$\sum_{scores}$$

would mean "sum over all scores" and

$$\sum_{values}$$

would mean "sum over all values."

This notation is somewhat more informal, but it often proves to be convenient, so we will use it from time to time.

Here we have two sums to compute. We have to compute the sum of the *squares* of the raw scores ($\Sigma_{i=1}^{n} x_i^2$), as well as the sum of the raw scores themselves ($\Sigma_{i=1}^{n} x_i$). Let's go back to the hypothetical data in Table 3-10 and see how we use the information in a frequency distribution to simplify the calculation of these sums. These data have again been reproduced in Table 3-11. Since we are concerned here with calculating the variance, we have included a listing of the squared scores (the x_i^2s) along with the scores themselves. Likewise, when listing these scores as a frequency distribution, we have listed not only the values (v_i) but also the squares of the values (the v_i^2s). At the bottom left of Table 3-11 we have calculated the variance using our raw-score formula. As we can see, this variance is 1.1475. On the right-hand side we have calculated the variance using the frequency distribution information. As was the case when we were calcu-

TABLE 3-11.   Hypothetical Data: Numbers of Heads in Four Coin-Flip Sequences; Calculation of the Variance

Raw data				Frequency distribution		
x_i	x_i^2	x_i	x_i^2	Value: v_i	v_i^2	Frequency: $n(v_i)$
$x_1 = 3$	$x_1^2 = 9$	$x_{11} = 1$	$x_{11}^2 = 1$	0	0	$n(v_1) = 1$
$x_2 = 1$	$x_2^2 = 1$	$x_{12} = 2$	$x_{12}^2 = 4$	1	1	$n(v_2) = 6$
$x_3 = 1$	$x_3^2 = 1$	$x_{13} = 2$	$x_{13}^2 = 4$	2	4	$n(v_3) = 6$
$x_4 = 3$	$x_4^2 = 9$	$x_{14} = 1$	$x_{14}^2 = 1$	3	9	$n(v_4) = 5$
$x_5 = 1$	$x_5^2 = 1$	$x_{15} = 2$	$x_{15}^2 = 4$	4	16	$n(v_5) = 2$
$x_6 = 2$	$x_6^2 = 2$	$x_{16} = 1$	$x_{16}^2 = 4$			
$x_7 = 0$	$x_7^2 = 1$	$x_{17} = 2$	$x_{17}^2 = 4$			
$x_8 = 2$	$x_8^2 = 4$	$x_{18} = 3$	$x_{18}^2 = 9$			
$x_9 = 4$	$x_9^2 = 16$	$x_{19} = 3$	$x_{19}^2 = 9$			
$x_{10} = 4$	$x_{10}^2 = 16$	$x_{20} = 3$	$x_{20}^2 = 9$			

$$S^2 = \frac{\sum_{i=1}^{20} x_i^2}{n} - \left(\frac{\sum_{i=1}^{20} x_i}{n}\right)^2$$

$$= \frac{(9 + 1 + \cdots + 9)}{20}$$

$$- \left[\frac{(3 + 1 + \cdots + 3)}{20}\right]^2$$

$$= 107/20 - (41/20)^2$$

$$= 5.35 - 4.2025$$

$$S^2 = 1.1475$$

$$S = \sqrt{1.1475} = 1.07$$

$$S^2 = \frac{\sum_{i=1}^{20} x_i^2}{n} - \left(\frac{\sum_{i=1}^{20} x_i}{n}\right)^2$$

$$= \frac{\sum_{i=1}^{5} v_i^2 n(v_i)}{n} - \left[\frac{\sum_{i=1}^{5} v_i n(v_i)}{n}\right]^2$$

$$= \frac{(0)1 + (1)6 + (4)6 + (9)5 + (16)2}{20}$$

$$- \left[\frac{(0)1 + (1)6 + (2)6 + (3)5 + (4)2}{20}\right]^2$$

$$= 107/20 - (41/20)^2$$

$$S^2 = 1.1475$$

$$S = \sqrt{1.1475} = 1.07$$

lating the mean, we start with the same formula that we applied to the raw scores. This formula involves computing the sum of the x^2s as well as the sum of the xs. We now notice that of the 20 x_i^2s, there are two 16s (that is, two 4^2s), five 9s (that is, five 3^2s), and so on. **Hence we can easily calculate the sum of the squared scores by simply taking the sum of each value times the frequency with which that squared value occurs or**

$$\sum_{i=1}^{20} x_i^2 = \sum_{i=1}^{5} v_i^2 n(v_i)$$

Likewise, to compute the sum of the xs, we simply take the sum of the values times the frequency with which each value occurred. This gives us the formula for the variance in terms of the frequency distribution that is derived on the right-hand side of Table 3-11.

3.5 When data are presented in the form of a frequency distribution, we can simplify calculations by making use of the facts that

$$\sum_{scores} x_i = \sum_{values} v_i n(v_i)$$

and

$$\sum_{scores} x_i = \sum_{values} v_i^2 n(v_i)$$

Descriptive Statistics Applied to Empirical Probability Distributions

As we saw in the last chapter, an empirical probability distribution and an empirical frequency distribution are practically the same thing. Specifically an empirical probability distribution is derived by simply dividing each frequency in an empirical frequency distribution by n, where n is the total number of scores. The crucial thing to notice is that a frequency distribution and a probability distribution are merely two different ways of characterizing the *exact same raw data*. Consequently the variance and all descriptive statistics of the data must be identical whether they are calculated from raw data or from a corresponding distribution derived from those raw data. Table 3-12 provides an illustration of this assertion. In Table 3-12 we have represented our coin-flip data both as an empirical frequency distribution and as an empirical probability distribution. Under the frequency distribution we have reproduced the formulas for calculating the mean and the variance. On the right-hand side of Table 3-12 we have shown the same values represented as a probability distribution. Directly under

TABLE 3-12.   Hypothetical Data: Numbers of Heads in Four Coin-Flip Sequences; Empirical Frequency and Probability Distributions; Calculation of Mean and Variance

Frequency distribution		Probability distribution	
Value: v_i	Frequency: $n(v_i)$	Value: v_i	Probability: $p(v_i) = n(v_i)/n$
$v_1 = 0$	$n(v_1) = 1$	$v_1 = 0$	$p(v_1) = 0.05$
$v_2 = 1$	$n(v_2) = 6$	$v_2 = 1$	$p(v_2) = 0.30$
$v_3 = 2$	$n(v_3) = 6$	$v_3 = 2$	$p(v_3) = 0.30$
$v_4 = 3$	$n(v_4) = 5$	$v_4 = 3$	$p(v_4) = 0.25$
$v_5 = 4$	$n(v_5) = 2$	$v_5 = 4$	$p(v_5) = 0.10$

Calculation of the mean

$$M = \frac{\sum_{i=1}^{5} v_i n(v_i)}{n}$$

$$= 41/20 = 2.05$$

$$M = \sum_{i=1}^{5} v_i \left(\frac{n(v_i)}{n}\right)$$

$$= \sum_{i=1}^{5} v_i \left(p(v_i) \right)$$

$$= 0(0.05) + 1(0.3) + 2(0.3) + 3(0.25) + 4(0.10)$$
$$= 2.05$$

Calculation of the variance

$$S^2 = \frac{\sum_{i=1}^{5} v_i^2 n(v_i)}{n} - \left[\frac{\sum_{i=1}^{5} v_i n(v_i)}{n} \right]^2$$

$$= 107/20 - (41/20)^2$$

$$= 1.1475$$

$$S^2 = \sum_{i=1}^{5} v_i^2 \left(\frac{n(v_i)}{n}\right) - \left[\sum_{i=1}^{5} v_i \left(\frac{n(v_i)}{n}\right) \right]^2$$

$$= \sum_{i=1}^{5} v_i^2 p(v_i) - \left[\sum_{i=1}^{5} v_i \left(p(v_i) \right) \right]^2$$

$$= 0(0.05) + 1[0.3) + 4(0.3) + 9(0.25)$$
$$+ 16(0.10) - [0(0.05) + 1(0.3) + 4(0.3)$$
$$+ 9(0.25) + 16(0.10)]^2$$
$$= 5.35 - 4.2025 = 1.1475$$

each type of distribution we have reproduced the formula for calculating the mean. Notice—and this is the crucial thing—that under the probability distribution one component of this formula, which is circled, is nothing more than the formula for obtaining a probability from a frequency. Therefore, we can just substitute $p(v_i)$ for $n(v_i)/n$ to get the second equation shown underneath the probability distribution. Hence, we can calculate the mean either in terms of frequencies (which involves dividing frequencies by n), or we calculate it in terms of probability; in either case we are doing the same thing and the answer must be exactly the same—2.05 in this case. Precisely the same argument can be made for calculation of the variance. In the formula for the variance we have two instances in which the term $n(v_i)/n$ appears. In each case, we can substitute

$p(v_i)$, thereby obtaining the formula in terms of probabilities instead of frequencies. Once again the answer has to come out exactly the same no matter which way it is calculated.

To reiterate: an empirical frequency distribution and an empirical probability distribution are simply two different ways of encapsulating the essential information in a set of raw data. When we calculate descriptive statistics (mean and variance) using either the raw data or the resulting frequency distribution of the resulting probability distribution, the results we get must be exactly the same.

> 3.6 When data are presented in the form of a probability distribution, we can further simplify calculations by making use of the fact that for any value v_i
>
> $$p(v_i) = \frac{n(v_i)}{n}$$

THEORETICAL PROBABILITY DISTRIBUTIONS

When we are concerned with theoretical probability distributions, we begin to think about descriptive statistics somewhat differently. Recall that a theoretical probability distribution is not based on data; rather, it is based on a priori considerations and it involves what we think a probability distribution *ought to* look like if we went out and collected some data. Examples of theoretical probability distributions were presented in the last chapter for two specific random variables: first the number of heads in four flips of a coin, and second the number of throws of a die before the first 6 appears.

Expected Value: The Mean Renamed

When concerned with theoretical probability distributions, we will compute a number that is akin to the mean of the distribution and that is calculated by exactly the same formula as we derived above for an empirical probability distribution. However, this number is referred to by a different name, the *expected value*. This choice of terms makes sense. A mean is something that is associated with data; it is descriptive of a set of numbers (or of a distribution that comes from a set of numbers). The corresponding measure for a theoretical probability distribution does not come from data—rather, it is what we would *expect* the mean of the data to be if we actually went out and collected data.

Table 3-13 gives some of examples of expected values of theoretical distributions we have already encountered—the theoretical distribution of the

TABLE 3-13.   Expected Values of Theoretical Distributions

(a) Number of heads in four flips of a coin

v_i	$p(v_i)$	
0	0.0625	Expected value $= E(x) = \sum\limits_{i=1}^{n} v_i p(v_i)$
1	0.2500	
2	0.3750	$= 0\,(0.0625) + 1\,(0.25) + 2\,(0.375) + 3\,(0.25)$
3	0.2500	$\quad + 4\,(0.0625)$
4	0.0625	$= 0 + 0.25 + 0.75 + 0.75 + 0.25 = 2$

(b) Number of throws of a die prior to obtaining a 2

v_i	$p(v_i)$	v_i	$p(v_i)$	
0	0.1667	11	0.0224	
1	0.1389	12	0.0187	
2	0.1157	13	0.0156	
3	0.0965	14	0.0130	Expected value
4	0.0804	15	0.0108	
5	0.0670	16	0.0090	$= E(x) = \sum\limits_{i=1}^{\infty} v_i p(v_i)$
6	0.0558	17	0.0075	$= 0\,(0.1667)$
7	0.0465	18	0.0063	$\quad + 1\,(0.1389)$
8	0.0388	19	0.0052	$\quad + 2\,(0.1157) + \ldots$
9	0.0323	20	0.0043	$= 6$
10	0.0269	.		
		.		
		.		

number of heads in four flips of a coin, and the theoretical distribution of the number of throws of a die prior to getting a 6. Note that the first distribution is finite whereas the second distribution is countably infinite (i.e., they are both discrete distributions). Let's consider how to calculate the expected values of these two distributions.

The top portion of the table concerns the distributions of the number of heads in four flips of a coin. Here we have listed the probability distribution using our new notation—the five possible values under the heading v_i, and the probabilities of the values under the heading $p(v_i)$.

The notation for expected value is $E(x)$. To briefly elaborate on the meaning of this notation, we ask you to recall that the expected value consists of the number we *expect* to be the mean of the distribution if we actually had people going out and flipping a coin four times. That is to say, suppose that we had some number of people (it doesn't matter how many) each go out and flip a coin four times. We would designate the number of heads obtained by the ith person by x_i (as usual). $E(x)$ refers to what we expect the mean of all those xs to be.

The formula for expected value of a theoretical distribution is precisely the same as the formula for the mean of an empirical distribution. It is simply

$$E(x) = \sum_{values} v_i\, p(v_i) \tag{3.1}$$

where the sum is taken over all the values listed in **V**. In this example

$$E(x) = \sum_{i=1}^{5} v_i\, p(v_i)$$

The bottom part of Table 3-13 shows how we can analogously calculate the expected value of a countably infinite probability distribution. Here we are concerned with the number of throws of a die prior to obtaining a 6. As we have seen in the previous chapter, this value can take on any value from zero to infinity. For illustrative purposes we have listed the first 20 values along with their probabilities. We can again calculate the expected value using Equation 3.1. Here, however, we must take our sum from 1 to infinity, since v can be any of an infinite number of potential values. Once again it is of course impossible to take the sum of an infinite number of products. Therefore, we have to arbitrarily cut it off somewhere, and the expected value that we get using this method will only be approximate. However, by taking the sum of enough things, we can get an arbitrarily close approximation to the real expected value. In this case the expected value turns out to be six. So on the average we would expect a person to throw a die six times before rolling the first 6.

Digression 3.2.

Mean of a Geometric Distribution

Our number-of-throws-before-the-first-6 distribution can be represented as follows: the probability that the random variable will assign any value i is obtained by the formula:

$$p(i) = \left(\frac{5}{6}\right)^{i-1} \frac{1}{6}$$

This is a specific instance of what is called a *geometric distribution*. The general form of a distribution is

$$p(i) = x^{i-1}(1 - x)$$

where $0 < x < 1$

Thus, the mean of a geometric distribution is

$$E(i) = \sum_{i=1}^{\infty} i\, p(i) = \sum_{i=1}^{\infty} i\, x^{i-1}(1 - x) \tag{D.1}$$

It is, of course, impossible to sum up an infinite number of terms. However, by mathematical techniques beyond the scope of this book, it can be shown that Equation D.1 reduces to

$$E(i) = \frac{1}{1 - x}$$

3.7 The expected value of a theoretical probability distribution is analogous to and is calculated in the same way as the expected value of an empirical probability distribution:

$$E(x_i) = \sum_{\text{values}} v_i p(v_i)$$

Lotteries.   The notion of an expected value can be applied to a lottery. To see how this is done, consider the lottery described in Table 3-14. This is a very simple lottery in which it is assumed that 100 people enter, that the winner is chosen randomly from among the 100 entrants, and that the prize is $200.

A lottery such as this takes the form of a random variable that assigns to each person a value—namely, the amount of money that person wins. Notice that the amount of money assigned can be either $200 (which will be assigned to the winner) or $0 (which will be assigned to each of the losers). In Table 3-14 we have listed these two values along with the probability that each of the two values will be assigned to any given random person. Since only one person out of the 100 will win the lottery, the probability of being assigned $200 if 1/100. Likewise, since the other 99 people will not win the lottery, the probability of being assigned $0 is 99/100. Now we can calculate the expected value using

---

TABLE 3-14.   **Expected Value To Be Won in a Lottery** (it is assumed that 100 people enter, the winner is randomly drawn, and the prize is $200)

v_i	$p(v_i)$
0	99/100
200	1/100

$E(x) = (0)(99/100) + (200)(1/100) = 2$

Equation 3.1. We have done so at the bottom of Table 3-14; it turns out to be $2.

Using this scheme it's easy to calculate one's expected value in any game of chance. Furthermore, by calculating the expected value and comparing it with how much one pays to partake in the game of chance, one can easily determine whether one expects to gain or lose money by participating. For example suppose that each person in the lottery depicted in Table 3-14 has paid $5 for a lottery ticket. In this case one pays $5; one can expect to win $2; therefore, one's expected *loss* is $3. Conversely, in the unlikely event that one had paid only $1.50 to enter the lottery, the expected *gain* would be the amount one expects to win ($2) minus the amount that one paid to enter ($1.50), or 50¢.

In virtually every game of chance an entrant expects to lose rather than gain money, because the people who run the game arrange to get some sort of a cut of their own. Being able to calculate how much you expect to lose, however, is very useful if you are trying to decide among a number of games of chance. For example if you are in a casino in Las Vegas and calculate your expected loss, you would discover that the expected loss is relatively greater with, say, slot machines than it is at the blackjack table.

Variance

The relationship between the variance of an empirical versus a theoretical probability distribution is much the same as the relationship between the mean of an empirical versus a theoretical probability distribution. In both cases the former number is something that is *obtained from data* whereas the latter is something that *we expect to get if we were to collect data.* And in both cases the computational formula is identical whether an empirical or a theoretical distribution is involved.

In particular the mean of a theoretical probability distribution is, as we have seen, the expected score, designated as $E(x)$. Likewise the variance of a theoretical probability distribution is the expected squared deviation between any score and the expected value—that is,

$$\text{Variance} = E[x - E(x)]^2$$

Digression 3.3

The Computational Formula

The computational formula for the variance of a probability distribution is as follows:

$$\text{Variance} = E[x - E(x)]^2$$

$$= E[x^2 - 2xE(x) + E(x)^2]$$

Like a summation sign, expectation is *distributive.* Hence,

$$\text{Variance} = E(x^2) - E[2xE(x)] + E[E(x)]^2$$

The expectation of a constant like $[E(x)]^2$ is simply that constant. Thus,

$$\text{Variance} = E(x^2) - 2E(x)E(x) + [E(x)]^2$$
$$= E(x^2) - 2[E(x)]^2 + [E(x)]^2$$
$$= E(x^2) - [E(x)]^2$$

Now note that

$$E(x^2) = \sum_{\text{values}} v_i^2 p(v_i)$$

and

$$E(x) = \sum_{\text{values}} v_i p(v_i)$$

Thus,

$$\text{Variance} = \sum_{\text{values}} v_i^2\, p(v_i) - \left[\sum_{\text{values}} v_i p(v_i)\right]^2$$

The variance of a theoretical probability distribution is

$$\boldsymbol{E[x - E(x)]^2} = \sum_{\text{values}} \boldsymbol{v_i^2 p(v_i)} - \left[\sum_{\text{values}} \boldsymbol{v_i p(v_i)}\right]^2$$

We reemphasize that this is exactly the same as the computational formula used when an empirical probability distribution is under consideration.

TABLE 3-15.   Variance of a Theoretical Probability Distribution: Number of Heads in Four Flips of a Coin

v_i	$(v_i)^2$	$p(v_i)$
0	0	0.0625
1	1	0.2500
2	4	0.3750
3	9	0.2500
4	16	0.0625

$$E[(x - E(x)]^2 = \sum_{\text{values}} v_i^2\, p(v_i) - [E(x)]^2$$
$$= [0\,(0.625) + 1\,(0.25) + 4\,(0.3750) + 9\,(0.25)$$
$$\quad + 16\,(0.0625)] - 2^2$$
$$= [0.25 + 1.5 + 2.25 + 1.0] - 4$$
$$= 1$$

To provide a brief example, Table 3-15 shows how the variance of our theoretical number-of-heads-in-four-flips-of-a-coin distribution is computed.

3.8 The variance of a theoretical probability distribution is the expected squared deviation and is calculated by

$$E[x - E(x)]^2 = \sum_{values} v_i^2 p(v_i) - [E(x)]^2$$

Digression 3.4

Theoretical Probability Distributions. Continuous

Recall that we had to think about continuous probability distributions somewhat differently than we thought about discrete probability distributions. In particular rather than summing things (like probabilities of discrete values), we had to *integrate* or compute areas under curves. Recall too that instead of using summation signs, we used integral signs, which essentially directed us to take the sum of an infinite number of infinitely skinny rectangles.

The formulas for computing the mean and variance of a continuous probability distribution conform to the general principles sketched above: summation signs are replaced by integral signs. Since we will not, for the purposes of this book, ever actually have to calculate the mean or variance of a continuous distribution, we will not go into the process in great detail. Rather, for completeness and aesthetics, we will simply provide the formulas. We have done that in the text in Table D-1.

TABLE D-1.   Formulas for the Mean and Variance of a Probability Distribution; Relationship Between Discrete and Continuous Distributions

	Discrete distributions	Continuous distributions
Mean	$E(x) = \sum_{values} v_i p(v_i)$	$E(x) = \int_{values} vp(v)dv$
Variance	$E[x - E(x)]^2 = \sum_{values} v_i^2 p(v_i) - \left[\sum_{values} v_i p(v_i)\right]^2$	$E(x - E(x))^2 = \int_{values} v^2 p(v)\, dv - \left[\int_{values} vp(v)dv\right]^2$

New Terms for Mean and Variance of Theoretical Distributions

One final note regarding theoretical distributions: it is somewhat awkward to always have to write out $E(x)$ for the mean and $E[x - E(x)]^2$ for the variance. There are commonly used symbols instead; these are the Greek letters μ (mu) for the mean of σ^2 (sigma squared) for the variance. We shall have more to say about these terms in a later chapter; for the time being, Table 3-16 provides a summary of notation.

TABLE 3-16.   Summary of Notation for Mean, Variance, and Standard Deviation

	Mean	*Variance*	*Standard Deviation*
1. A set of scores (data)	M	S^2	S
2. An empirical frequency or probability distribution (data)	M	S^2	S
3. A theoretical probability distribution	$E(x) = \mu$	$E[x - E(x)]^2 = \sigma^2$	σ

STANDARD SCORES

The final topic with which we shall be concerned in this chapter involves combining a particular score (x_i) from a set of scores, the mean (M) of these scores, and the standard deviation (S) of the scores to produce what is referred to as a *standard score*.

Standard Scores: The Mathematical Formula

To see what a standard score is, let's go back to the example we used at the beginning of the chapter. In particular suppose you have just received an exam back, and you got an 80. We suggested that there were two things that you would like to know about the total set of scores in order to determine how "good" your score of 80 really was. First you would want to know what the mean exam score was so that you would be able to tell whether you did generally better or generally worse than most people in the class. Second, we argued that you would want to know some measure of dispersion such as the standard deviation. Having a measure of dispersion would give you some idea of how "close" your score was to the mean. In particular suppose the class mean was 75. Then if the standard deviation of the scores were very small, your score of 80 would be "far above the mean" and thus relatively good. Conversely, if the standard deviation were large, your score of 80 would be "close to the mean" and not all that good.

The standard score is a mechanism for taking these intuitive notions and expressing them concisely and quantitatively. **Specifically, a standard score or *z*-score[3] is expressed by the following formula.**

$$z_i = \frac{x_i - M}{S}$$

Observe that this formula encapsulates the notions we discussed above. The numerator consists of the expression $x_i - M$. Therefore, the numerator (and the standard score itself) will be negative if the x_i is below the mean and positive if x_i is above the mean. Likewise, the absolute value of the standard score will be large to the extent that x_i deviates from the mean and small to the extent that x_i does not deviate from the mean.

We see, then, that the numerator is a mathematical characterization of the first notion discussed above—the comparison of a given score with the mean. The denominator, on the other hand, is the mathematical representation of the second notion. By putting the standard deviation—a measure of dispersion—into the denominator, we scale the standard score such that it is big if the scores are closely packed around the mean (corresponding to a small standard deviation) and smaller if the scores are widely spread out (corresponding to a large deviation).

In brief when we change a set of raw scores to a corresponding set of standard scores, we have simply *rescaled* the random variable in which we are interested. By rescaling we mean that typically, when we are interested in some random variable, the values that the random variable assigns are scaled in some kind of convenient units. For example if we are interested in height, the scale might be in terms of inches or feet. Likewise, if we are interested in weight, the scale might be in terms of pounds or kilograms. In the example discussed above, we are interested in exam scores, so the scale would be in terms of points. However, by taking standard scores, we rescale into units that are common (standard) across *all* measures—units corresponding to the standard deviation of the distribution of scores at hand. Thus, a standard score conveys more information than a raw score.

Suppose for example that I am telling you about Glug, my visiting Martian friend. In the course of the discussion I mention that Glug is 50 in. tall. This raw score will not tell you anything about how tall or short Glug is relative to the rest of the Martian population. If, however, I told you that Glug's height has a standard score of -1, such information would be much more meaningful to you; specifically it would tell you that Glug is a rather short Martian. I could give you additional information about Glug that would also help to tell you about what she is like relative to the rest of the Martian population. For example I might tell

---

[3]We are actually using the term "*z*-score" somewhat incorrectly here. Technically a *z*-score is a standard score of a very specific kind of distribution (a normal distribution). However, the term "*z*-score" is loosely used to refer to a standard score of *any* distribution, so we shall use it that way here.

you that her standard score for IQ is +2 (which would mean that she is a fairly intelligent Martian), that her standard score for weight is 0 (which mean that she is of average weight), that her standard score for age is +1, and so on.

In a slightly more practical vein consider the following problem. My younger sister Clarissa is trying to decide whether to become an airline pilot or a short-order cook. She takes an airline pilot aptitude test on which she receives a score of 62 and a cook aptitude test on which she receives a score of 90. What should she do?

It would be unwise for Clarissa to base her decision on a comparison of her raw scores alone. Instead she should convert to standard scores, a comparison of which will provide her with much more meaningful information. Suppose she discovers that the mean airline pilot aptitude test score is 50 with a standard deviation of 6, and that the mean cook aptitude test score is 85 with a standard deviation of 5. Thus, in terms of her potential to be an airline pilot:

$$z = \frac{62 - 50}{6} = 2.0$$

Whereas in terms of her potential to be a cook,

$$z = \frac{90 - 85}{5} = 1.0$$

The fact that she has a standard score of 2.0 on the pilot test versus a standard score of 1.0 on the cook test indicates that Clarissa is probably better suited to be a pilot than a short-order cook.

3.9 A standard score, obtained by the formula

$$z = \frac{(x - M)}{S}$$

gives the magnitude of the score in units of the standard deviation of the distribution from which x comes.

SUMMARY

In this chapter we have been primarily concerned with techniques by which we can extract the crucial and interesting information from a set of data. The points of interest that have emerged from this chapter are the following.

1. Scores can be grouped into an empirical frequency or probability distribution, a concept that was introduced in the previous chapter. If

the data are shown as a cumulative probability distribution, the percentile corresponding to any given score can be immediately calculated.
2. There are three principal measures of central tendency: the mean, the median, and the mode. Of these the mean is the most widely used measure.
3. The principal measure of dispersion is the variance, which is the average squared deviation score. The variance is small to the extent that scores are tightly clustered around the mean and large to the extent that the scores are widely spread out around the mean.
4. A set of scores is sometimes represented as an empirical frequency distribution, or an empirical probability distribution. When the data are represented this way, the mean and the variance may be calculated somewhat more efficiently by summing over all values that scores can take on (that is, over all members of V) rather than summing over all scores. (Summing over values is more efficient, because there are typically many fewer values than scores.)
5. The mean of a theoretical probability distribution is referred to as the *expected value*—it is the number that we would expect the mean to be if data were actually collected. The formula for calculating the expected value of a theoretical probability distribution is exactly the same as the formula for calculating the mean of an empirical probability distribution.
6. The variance of a theoretical probability distribution is the expected squared deviation of a score from the expected value. The variance of a theoretical probability distribution is calculated using the same formula that is used to calculate the variance of an empirical probability distribution.
7. A standard score combines the mean and standard deviation of a distribution to allow calculation of a score that is standardized across all measures. A standard score is negative if it is below the mean and positive if it is above the mean. Furthermore, the absolute value of a standard score is large to the extent that it deviates from the mean and small to the extent that it does not deviate from the mean.

PROBLEMS

1. Billy and Bobby Jones set up a lemonade stand. They have five sizes of lemonade that sell for 5¢, 10¢, 15¢, 25¢, and 30¢. The numbers of each size of lemonade they sell on the first day are:

5¢	10¢	15¢	25¢	30¢
10	7	8	5	5

Considering the number sold for each size as one observation, calculate:

   1. The mean number of lemonades sold.
   2. The median number of lemonades sold.

3. The mode number of lemonades sold.
4. The variance and standard deviation of the number of lemonades sold.
5. The range of the number of lemonades sold.
6. Suppose the 30¢ size is ignored. Now calculate the median number of lemonades sold.

2. A drugstore is keeping tabs on the number of Cokes it sells during nine hours of a business day. The data turn out as follows:

Hour	1	2	3	4	5	6	7	8	9
# Cokes	3	2	3	6	5	7	8	9	11

1. Compute the mean number of Cokes sold per hour.
2. Compute the median number of Cokes sold per hour.
3. Compute the mode number of Cokes sold per hour.
4. Compute the variance and standard deviation of the number of Cokes sold per hour. USE THE DEVIATION-SCORE FORMULA.
5. Recompute the variance USING THE RAW-SCORE FORMULA.
6. Suppose the first hour of the day is ignored, leaving only the last eight hours. Recompute the median and mode.

3. The number of Pepsi Colas (in millions) sold in the United States each year for ten past years is shown in the following table:

1963:   3.0
1964:   7.5
1965:   5.0
1966:   2.5
1967:   7.5
1968:   2.0
1969:   1.0
1970:   2.0
1971:   7.5
1972:   1.0

1. What is the *median* number of Pepsi Colas sold over the past ten years?
2. What is the *mode* number of Pepsi Colas sold over the past ten years?

4. For problems 1 and 2 of Chapter 2, compute the mean, median, mode, variance, standard deviation, and range of the scores.
5. For problems 3, 4, 5, 6, 7, 8, and 9 of Chapter 2, compute the expected value and variance of these distributions.
6. For problem 14 of Chapter 2, compute the expected value and variance of this probability distribution.
7. For problem 15 of Chapter 2, compute the expected value and variance of the probability distributions for *a*, *b*, *x*, and *y*.

8. Joe Shablotnik is playing a chess match with his friend, Bobby. Bobby gives Joe his choice: either they can play three games or they can play nine games. In either case the person winning the majority of the games wins the match. (Assume that in either case they will play all the games—that is, if they decided to play nine games, they'd play all nine even if one of them won the first five in a row.) You may ignore the possibility of a draw.

   Joe knows that his probability of winning any given chess game against Bobby is one in three.

   1. How many games does Joe expect to win if they play three games?
   2. How many games does Joe expect to win if they play nine games?
   3. What is the variance of the number of games Joe expects to win if they play three games (to three decimal places)?
   4. What is the variance of the number of games Joe expects to win if they play nine games (to three decimal places)?
   5. If they play three games, what is the probability (to three decimal places) that Joe will win the match?
   6. If they play nine games, what is the probability (to three decimal places) that Joe will win the match?

9. The city of Puyallup has a lottery. In this lottery 1000 people enter, and the following prizes are awarded to randomly selected entrants:

   > One vacation trip to Pocatello worth $600
   > Five Sears 10-speed bicycles worth $100 apiece
   > Twenty Timex wristwatches worth $10 apiece
   > Two hundred solid aluminum medallions worth $.50 apiece

   Consider a random variable which assigns to each lottery entrant some value in dollars.

   1. What are the members of V where V is the set of values that this random variable could assign?
   2. What is the expected value and variance of the probability distribution of this random variable?

10. Joe Shablotnik is playing the "Bellingham Game," which works as follows. Two dice are thrown and the sum, s, is noted. Joe now wins x dollars where

   $x = 4s$ if s is odd and 7 or greater
   $\quad = 3s$ if s is even and 8 or greater
   $\quad = s$ if s is less than 7

   1. What is V, the set of values (in dollars) that Joe can earn?
   2. Compute the probability of each member of V.
   3. Suppose that Joe plays this game ten times. What is the total amount of money that he expects to win?
   (Hint: Consider the expected value per trial.)
   4. What is the variance of the amount of money won per trial?

11. The "St. Petersburg" game works as follows. A coin is flipped until a head comes up. The amount of money paid by the flipper to the player is then equal to $\$2^N$ where N is the number of flips prior to obtaining a head.

   1. What is the expected amount of money that a player will win in this game?
   2. Suppose that the total amount of money that a flipper is in a position to pay is $\$1,073,741,842$ ($\$2^{30}$). Now what is the expected amount of money that a player will win in this game?

Binomial Distribution 4

In Chapters 2 and 3 we discussed probability distributions—what they are, how they behave, and how to deal with them. In this chapter we shall direct our focus to one particular distribution known as the *binomial distribution*. We have chosen to single out the binomial distribution for two reasons. First it is a distribution that occurs rather often in real-life situations, and second it is very similar to another theoretical distribution (the normal distribution) that will be of paramount importance when we take up subsequent topics.

WHAT A BINOMIAL DISTRIBUTION IS AND HOW TO MAKE ONE

In many types of probabilistic situations we find ourselves interested in two mutually exclusive and exhaustive outcomes—that is, an event and its complement. In this type of situation it is common to call one of the two outcomes a *success* and the other a *failure*. (There is no implication of good or bad inherent in the terms "success" and "failure"; rather, they are simply arbitrary nomenclature.) By convention we represent the probability of a success and the probability of a failure as follows:

probability of a success = p
probability of a failure = q

Notice that because a success and a failure partition the sample space (that is, they constitute two mutually exclusive and exhaustive events), their probabilities must add to 1.0. Therefore,

$$p = 1 - q$$

and likewise,

$$q = 1 - p$$

Table 4-1 provides some examples of such situations.

TABLE 4-1.   Examples of Binomial Situations

Situation	Success	p	Failure	q
Flip a coin	head	0.5	tail	0.5
Roll a die	6	0.17	not-6	0.83
Have a child	girl	0.5	boy	0.5
Draw a card	heart	0.25	not a heart	0.75

Number of Successes as a Random Variable

Suppose now that we have N trials, and each trial is independent of every other trial. Referring to the examples in Table 4-1, we might flip a coin N times, roll a die N times, have N children, and so on. Let's define a random variable, which is the *number of successes we obtain out of the* N *trials*. A binomial distribution is then simply defined as the probability distribution of this random variable.

Let's formulate this distribution as we learned in Chapter 2. The first step in the distribution-formulation process is to determine the members of V (recall that V is the set of all possible values the random variable can assign). In this case

$$V = \{0, 1, 2, \ldots, N\}$$

since we may have any integral number of successes from none (0 successes) to the total number of trials (N successes). It is evident that V contains $N + 1$ members.

Formulating the distribution of the members of V is equivalent to computing

$$p(0) = ?$$
$$p(1) = ?$$
$$\vdots$$
$$p(N) = ?$$

> 4.1. A binomial distribution is the distribution of the number of successes in N trials. A success has probability p of occurring and probability $q = 1 - p$ of not occurring on each trial.

Rolling a die: A specific example. To be more concrete about how we actually compute these probabilities, we will frame our explanation within the context of a specific example. In particular let's imagine that we plan to roll a die four times. Since we have four trials, $N = 4$. Let us now further imagine that a success is the event that the die comes up a 6. Therefore, a failure constitutes any outcome other than a 6 (that is, a not-6). It is not difficult to see that on each trial

$$p = p(6) = 1/6$$
$$q = p(\text{not-6}) = 5/6$$

Now let's compute the probability of 0—that is, the probability that the four throws will produce no 6s. No 6s requires throwing a not-6 four times in a row, and the sequence that will produce zero successes is therefore not-6, not-6, not-6, not-6. Notice that each of the four throws of the die is independent of all other throws—that is, on any given throw, the die is completely unaffected by what the outcome was on any other throw. Recall from our discussions of probability in Chapter 2 that when we have four (or any number of) independent events, the joint probability of these events is equal to the product of their unconditional probabilities. Therefore, the probability that we get zero 6s out of four throws of a die is

$$p(0) = p(\text{not-6})p(\text{not-6})p(\text{not-6})p(\text{not-6})$$

or

$$p(0) = (5/6)(5/6)(5/6)(5/6) = 625/1296 = 0.482$$

In general we can see that the probability of obtaining zero successes in N trials is:

$$p(0) = q^{N}$$

Now let's compute the probability of obtaining exactly one success out of our N rolls. This is slightly more complicated than computing the probability of a zero. Continuing with our example, let's consider some sequence that produces exactly one 6. One such sequence might be:

Sequence 1: not-6, not-6, not-6, 6.

Again since these four events are all independent, it is quite straightforward to determine that the probability of this sequence is

$$(5/6)(5/6)(5/6)(1/6) = 0.0965$$

However, this is not the *only* sequence that could contain exactly one 6. There are in fact three other sequences that would also produce one 6:

Sequence 2: not-6, not-6, 6, not-6
Sequence 3: not-6, 6, not-6, not-6
Sequence 4: 6, not-6, not-6, not-6

Each of these three other possible sequences also has a probability $(5/6)^3(1/6) = 0.0965$ of occurring.

We have now identified four distinct outcomes, each of which will produce exactly one success. Therefore, the probability of one success is the probability of the union of the four outcomes. Note that the four outcomes are mutually exclusive—if one of the outcomes occurs, none of the others can occur. Recall that the probability of the union of mutually exclusive events is the sum of the individual probabilities, so

$$p(1 \text{ success}) = 0.0965 + 0.0965 + 0.0965 + 0.0965 = (0.0965)4 = 0.386$$

Why are there exactly four sequences that involve one success out of four chances? Why not three possible sequences, or five or six? The reason there are four such sequences stems directly from the notion of unordered combinations that is discussed in Appendix D. (If the term "unordered combinations" is unfamiliar to you, you should now read Appendix D.) In this particular instance getting one success out of four possible trials is equivalent to *selecting* exactly one of the four trials to be a success. To calculate the number of ways of getting one success in four trials, we can simply use our formula for unordered combinations:

$$\binom{N}{r} = \frac{N!}{(r!)(N-r)!}$$

or in this case we are choosing $r = 1$ success from $N = 4$ throws. Therefore

$$\binom{N}{r} = \binom{4}{1} = \frac{4!}{(1!)(4-1)!} = \frac{4 \cdot 3 \cdot 2 \cdot 1}{1 \cdot 3 \cdot 2 \cdot 1} = 4$$

To compute the probability that we will get exactly one success, we simply multiply the *number* of sequences involving one success times the *probability* of each sequence. Thus

$$p(1) = (4)(5/6)^3(1/6) = 0.386$$

just as we calculated earlier.

This technique, which may be used to compute the probability of *any* number of successes, underlies the construction of the binomial distribution. To finish this example, let's quickly compute the probabilities of obtaining two, three, and four successes. There are $\binom{4}{2}$ or 6 sequences that will yield exactly two successes, and each of these sequences has probability $(5/6)^2(1/6)^2$ of occurring. So

$$p(2) = \binom{4}{2}(5/6)^2(1/6)^2 = (6)(0.694)(0.028) = 0.116$$

Similar considerations for the probabilities of three and four successes produce

$$p(3) = \binom{4}{3}(5/6)^1(1/6)^3 = (4)(0.833)(0.005) = 0.015$$

and

$$p(4) = \binom{4}{4}(1/6)^4 = (1)(0.001) = 0.001$$

As a check on our calculations, we add up all our probabilities and find

$$
\begin{array}{ll}
p(0) & = 0.482 \\
+p(1) & = 0.386 \\
+p(2) & = 0.116 \\
+p(3) & = 0.015 \\
+p(4) & = 0.001 \\
\hline
 & = 1.000
\end{array}
$$

The probabilities of obtaining either zero, one, two, three, or four heads are thus seen to add to 1.0—just as they should since they are probabilities of mutually exclusive and exhaustive events.

The formula for the binomial distribution.   Let's recapitulate the way in which a binomial distribution is constructed.

1. We first characterize any binomial distribution by two numbers (also called *parameters,* which will be discussed at length in a later chapter). The two parameters that characterize a binomial distribution are N, the number of trials, and p, the probability of a success. (Naturally once we have specified p, then we also know q, the probability of a failure, which is simply $1 - p$.)

2. The random variable with which we are concerned once we have specified p is the number of successes in the N trials. V, the set of possible values that the random variable can assign, becomes

$$V = \{0, 1, 2, \ldots, N\}$$

3. To compute the probability of any member of V (that is, of some number, i, of successes), we simply multiply the number of sequences involving i successes times the probability of each such sequence. We have seen that the number of sequences involving some number, i, of successes is $\binom{N}{i}$. Any sequence involving i successes must also involve $N - i$ failures, so the probability of such a sequence is $p^i q^{N-i}$. **The formula for calculating the probability of i successes is therefore**

$$p(i) = \binom{N}{i} p^i q^{N-i}$$

Digression 4.1

Binomial Coefficients: Pascal's Triangle

As noted, we may think of the above formula for computing $p(i)$ as being broken into two parts: the *probability* of a sequence containing i

successes, and the *number* of sequences involving exactly i successes out of N trials. The latter component, $\binom{N}{r}$, is sometimes referred to as the *binomial coefficient*.

The French philosopher Blaise Pascal took an interest in binomial coefficients (among many other things). Pascal constructed a table that systematically displayed these coefficients for various values of N and i. This table is reproduced as Table D-1 for values of N from 1 through 11.

·At first glance, this table may simply appear to contain an unwieldly array of numbers. However, a closer examination reveals an orderliness that would probably constitute nirvana for a number addict. We list here a few of these orderly characteristics.

1. Each number is the sum of the two numbers above it.
2. For any diagonal column, the *differences* of the successive members of that column are identical to the members of the just preceding diagonal column. The second diagonal column, for example, consists of the numbers 1, 2, 3, 4, 5, ... The *differences* of successive members $(2 - 1, 3 - 2, 4 - 3, \ldots)$ are 1, 1, 1, 1, ... which are simply the members of the first diagonal column. Likewise, the third diagonal column consists of the

TABLE D-1.   Pascal's Triangle: Values of $\binom{N}{i}$; Rows Represent Values of N; for Each Row, i Ranges from 0 (far left) to N (far right)

$N = 1$						1	1					
$N = 2$					1	2	1					
$N = 3$				1	3	3	1					
$N = 4$			1	4	6	4	1					
$N = 5$		1	5	10	10	5	1					
$N = 6$	1	6	15	20	15	6	1					
$N = 7$	1	7	21	35	35	21	7	1				
$N = 8$	1	8	28	56	70	56	28	8	1			
$N = 9$	1	9	36	84	126	126	84	36	9	1		
$N = 10$	1	10	45	120	210	252	210	120	45	10	1	
$N = 11$	1	11	55	165	330	362	362	330	165	55	11	1

numbers 1, 3, 6, 10, 15, ... The differences of successive
members of this column are 2, 3, 4, 5, . . . which are simply the
members of the second diagonal column, and so on.

3. A consequence of this latter characteristic is that each successive
diagonal column contains numbers which, when plotted, would
form successively higher-order polynomial functions. For
instance, 1, 2, 3, 4, 5, . . . would be linear; 1, 3, 6, 10, 15, . . .
would be quadratic: 1, 4, 10, 20, 35, . . . would be cubic, and so
on.

We leave it to the fascinated reader to stare at this table and uncover
other of its delightful patterns.

ANOTHER EXAMPLE

Just to consolidate the method used in constructing a binomial distribution, let's
consider one more example.

Opinion Polls

Suppose Senator Soot, running for reelection, claims that 70% of the
electorate will vote for him. To evaluate his claim, Acme Sampling Inc. plans to
poll a random sample of the electorate and ask each member of the sample
whether he or she plans to vote for Soot. Being a small company, Acme can only
afford to sample a total of $N - 20$ individuals. If Soot is correct that 70% of the
electorate will vote for him, then each member of the sample will claim to be
voting for Soot with a probability of $p = 0.70$. Assuming that each member's
response is independent of all other members' responses, the number of people
out of the sample who will claim to be for Soot is binomially distributed with
$N = 20$, $p = 0.70$, and $q = 0.30$.

How many people in the 20-person sample will be for Soot? Using the
techniques from the last section, we can compute the distibution, which is done
in Table 4-2. You must understand that this is the probability distribution of the
number of people in the sample who are for Soot *assuming* Soot is correct about
70% of the electorate being for him.

Looking over the information in Table 4-2, we note some interesting things.
For example the probability that exactly 70% of the sample (in this case 14
people) will be for Soot is 0.192. The probability that 15 or more people will be
for Soot is

$$p(15 \text{ or more}) = p(15) + p(16) + \cdots + p(20)$$
$$= 0.179 + 0.130 + \cdots + 0.001 = 0.416$$

TABLE 4-2.   A Binomial Distribution with Parameters $N = 20$, $p = 0.70$, $q = 1.0 - 0.70 = 0.30$

i	$p(i)$	i	$p(i)$
0	$\binom{20}{0}(0.7^0)(0.3^{20}) = 0.000$	11	$\binom{20}{11}(0.7^{11})(0.3^9) = 0.065$
1	$\binom{20}{1}(0.7^1)(0.3^{19}) = 0.000$	12	$\binom{20}{12}(0.7^{12})(0.3^8) = 0.114$
2	$\binom{20}{2}(0.7^2)(0.3^{18}) = 0.000$	13	$\binom{20}{13}(0.7^{13})(0.3^7) = 0.164$
3	$\binom{20}{3}(0.7^3)(0.3^{17}) = 0.000$	14	$\binom{20}{14}(0.7^{14})(0.3^6) = 0.192$
4	$\binom{20}{4}(0.7^4)(0.3^{16}) = 0.000$	15	$\binom{20}{15}(0.7^{15})(0.3^5) = 0.179$
5	$\binom{20}{5}(0.7^5)(0.3^{15}) = 0.000$	16	$\binom{20}{16}(0.7^{16})(0.3^4) = 0.130$
6	$\binom{20}{6}(0.7^6)(0.3^{14}) = 0.000$	17	$\binom{20}{17}(0.7^{17})(0.3^3) = 0.072$
7	$\binom{20}{7}(0.7^7)(0.3^{13}) = 0.001$	18	$\binom{20}{18}(0.7^{18})(0.3^2) = 0.028$
8	$\binom{20}{8}(0.7^8)(0.3^{12}) = 0.004$	19	$\binom{20}{19}(0.7^{19})(0.3^1) = 0.007$
9	$\binom{20}{9}(0.7^9)(0.3^{11}) = 0.012$	20	$\binom{20}{20}(0.7^{20})(0.3^0) = 0.001$
10	$\binom{20}{10}(0.7^{10})(0.3^{10}) = 0.031$		

How about the probability that 14 or fewer people will be for Soot. Since "15 or more" and "14 or fewer" are complementary events

$$p(14 \text{ or fewer}) = 1 - p(15 \text{ or more}) = 1 - 0.416 = 0.584.$$

CHARACTERISTICS OF THE BINOMIAL DISTRIBUTION

Having discussed the underlying basis for a binomial distribution and the means by which the distribution is constructed, we shall attempt to flesh out the binomial by describing some of its principal characteristics. We will first discuss what the binomial looks like, and we will then describe computation of its mean and variance.

Shape of the Binomial

Recall that a binomial distribution is completely described by the two parameters N and p. The values of N and p as well as the *relationship* between N and p determine what the binomial distribution will look like.

Effect of p on shape.   Let us consider first what happens when p and q are equal, that is, when $p = q = 0.5$. Figure 4-1 shows the resulting binomial distributions for $N = 3$, $N = 6$, and $N = 10$. Notice that these distributions are all *symmetrical*. That is, a center axis can be defined such that the left-hand side of the distribution is a mirror image of the right-hand side. Such symmetry seems reasonable: when $p = q = 0.5$, a success is simply the "mirror" of a failure; that is, we would expect the probability of i successes to be the same as the probability of i failures. (For example, in 10 flips of a coin we would expect 7

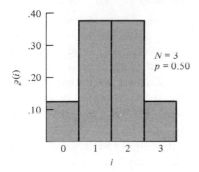

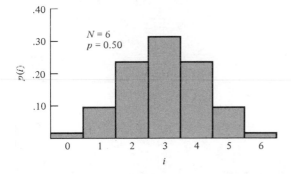

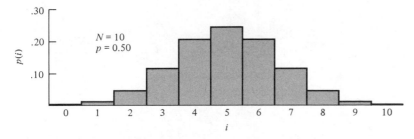

FIGURE 4-1.   Three binomial distributions: effect of changing N. As long as $p = q = 0.5$, the distribution is symmetrical, independent of N.

heads to be just as likely as 7 tails.) As is exemplified in Figure 4-1 (and, again, as should be intuitively reasonable), this symmetry does not depend on N.

Figure 4.2, on the other hand, shows what happens when p deviates from 0.5. The greater such deviation, the more *asymmetrical* is the resulting distribution. In particular when p is small, the distribution is *skewed* (stretched out) to the right. Conversely, when p is large, the distribution is skewed to the left. Again such a situation makes sense. When p deviates from 0.5, we would expect some number, r, of successes to have a different probability from the same number of failures. For instance, consider throwing a die six times. We would expect (as anyone who has needed a six to "come on" in a backgammon game is painfully aware) the probability of getting five sixes to be quite a bit smaller than the probability of getting five not-sixes.

Effect of N on shape.   We have seen that any deviation of p from 0.5 tends to produce a nonsymmetrical binomial distribution. What happens as we vary N? Figure 4-3 illustrates what happens. For a given value of p (assuming p is not 0.5) the distribution becomes *more symmetrical* as n increases. Again this makes sense.[1] Consider what happens when N is very small (for example, 4). In the case when p is small (for example 0.10), the probability of some small number of successes (say, zero) will be relatively large—but the probability of a large number of successes (say, four) will be essentially zero. However, the situation is quite different when N is large (for example 40). In this situation the probability of a large number of successes (say, 40) is still close to zero, but the

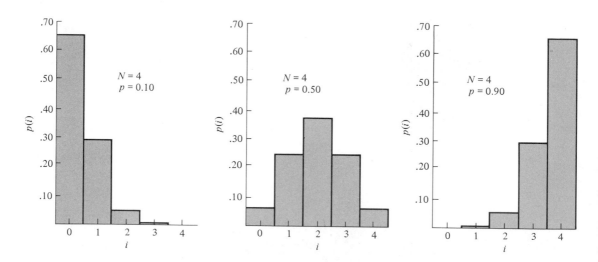

FIGURE 4-2.   Three binomial distributions: effect of changing p. To the extent that p and q deviate from 0.5, the distribution becomes asymmetrical.

---

[1] Fortunately even in statistics most things make sense.

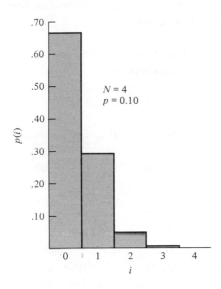

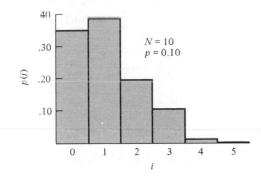

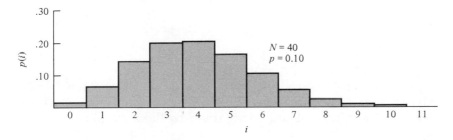

FIGURE 4-3. Three binomial distributions: effect of changing N when p deviates from 0.5 (in this case, when $p = 0.10$). The distribution becomes more symmetrical as N increases.

probability of a small number of successes (say, 0) is also quite diminished, because even though the probability of a failure, q, is quite large, the probability of obtaining *all* failures, q^N, becomes small as N increases.

Determination of shape: A summary.   The effects of the parameters p and N on the shape of the distribution may be summarized as follows.

1. When $p = q = 0.5$, the distribution is *always* symmetrical (see Figure 4-1).
2. To the extent that p (and q) deviate from 0.5, the distribution is asymmetrical (Figure 4-2).
3. As N increases, the distribution generally becomes more symmetrical (see Figure 4.3).

Expected Value (Mean) and Variance of a Binomial Distribution

Bear in mind that the mean or expected value (μ) of a distribution is a measure of the average or central tendency of that distribution. Likewise, the variance (σ^2) of the distribution is a measure of the variability or spread-outness of the scores in that distribution. In this section we will formulate equations for the mean and expected value and the variance of the binomial.

Expected value.   From our discussions of expected value in Chapter 3, we know that for a probability distribution

$$\mu = E(x) = \sum_{\substack{\text{members} \\ \text{of } V}} v_i \, p(v_i)$$

In the case of a binomial, recall that the v_i's range from 0 to N, and

$$p(i) = \binom{N}{i} p^i q^{N-i}$$

Thus,

$$\mu = \sum_{i=1}^{N} i \binom{N}{i} p^i q^{N-i} \tag{4.1}$$

Equation 4.1 appears to be a somewhat unwieldy expression to work with. Happily, however, it turns out the Equation 4.1 can be considerably simplified. **The mean of the binomial can be obtained by the formula:**

$$\mu = \sum_{i=1}^{N} i \binom{N}{i} p^i q^{N-i} = Np$$

The proof of this not particularly obvious assertion is provided in Digression 4.2. Meanwhile an intuitive view is provided as follows. Suppose we have some number N (say, 24) of trials and suppose we have some probability p (say, 1/6) of a success. (Thus, we might consider our situation to be one of throwing a die 24 times, seeking 6s.) How many successes (6s) do we *expect* to get? Intuitively, we would expect that 1/6 of our trials would turn up 6s. Thus, in general, we would expect a proportion p of our N trials, or Np trials to be successes. This is equivalent to saying that the expected value of number of successes is Np.

Digression 4.2

Derivation of the Expected Value of the Binomial Distribution

Let's assume that we have a binomial distribution with parameters N, p, and $q = 1 - p$. The expected value of the distribution may be characterized as follows.

$$\mu = \sum_{i=1}^{N} ip(i) \tag{D.1}$$

Since we know that

$$p(i) = p^i q^{N-i} \frac{N!}{i!(N-i)!} \tag{D.2}$$

we may substitute this expression for $p(i)$ into Equation D.1 to obtain

$$\mu = \sum_{i=1}^{N} \frac{ip^i q^{N-i} N!}{i!(N-i)!} \tag{D.3}$$

Now we shall take an N and a p from the right-hand side of the summation sign in Equation D.3 and put them on the left-hand side to obtain

$$\mu = Np \sum_{i=1}^{N} \frac{ip^{i-1} q^{N-i}(N-1)!}{i!(N-i)!} \tag{D.4}$$

Cancelling an i in the numerator and denominator in Equation D.4, we get

$$\mu = Np \left[\sum_{i=1}^{N} \frac{p^{i-1} q^{N-i}(N-1)!}{(i-1)!(N-1)!} \right] \tag{D.5}$$

Now we know that what we want to end up with is

$$\mu = Np$$

Therefore, we see that what we want to show is that the expression within the brakets in Equation D.5 is equal to 1. To do this, we shall define some new terms. Let

$$M = N - 1$$

and

$$j = i - 1$$

Substituting M and j into the within-brackets part of Equation D.5, we get

$$\sum_{j=1}^{M} \frac{p^{j}q^{M-j}M!}{j!(M-j)!} \tag{D.6}$$

Note that Expression 6 is simply the sum of all probabilities in a binomial distribution with parameters M and p. This sum must equal 1.0, therefore

$$\mu = Np$$

Variance.   Recall that our expression for the variance of a probability distribution is

$$\sigma^2 = \sum_{\substack{\text{members} \\ \text{of } V}} v_i^2 p(v_i) - \mu^2$$

or in the case of a binomial

$$\sigma^2 = \sum_{i=0}^{N} i^2 \binom{N}{i} p^i q^{N-i} - (Np)^2$$

Again we are happy to report that this formula reduces to a much simpler one. **Specifically the variance of a binomial distribution is obtained by**

$$\sigma^2 = \sum_{i=1}^{N} i^2 \binom{N}{i} p^{N-i} - (Np)^2 = Npq$$

The complete proof of this is presented in Digression 4.3.

Digression 4.3

Derivation of the Variance of the Binomial Distribution

Derivation of the variance (σ^2) of a binomial with parameters N, p, and $q = 1 - p$ is very similar to derivation of the expected value (Digression 4.2).

    We start with the general formula for variance as applied to the binomial distribution:

$$\sigma^2 = \sum_{i=1}^{N} i^2 p(i) - \mu^2$$

Substituting for $p(i)$ and $E(i)$,

$$\sigma^2 = \sum_{i=1}^{N} \frac{i^2 p^i q^{N-i} N!}{i!(N-i)!} - (Np)^2 \tag{D.1}$$

Bringing an Np to the left of the summation sign and canceling an i in numerator and denominator,

$$\sigma^2 = Np \sum_{i=1}^{N} \frac{i p^{i-1} q^{N-i}(N-1)!}{(i-1)!(N-i)!} - (Np)^2 \tag{D.2}$$

Letting $M = N - 1$ and $j = i - 1$ and substituting into Equation D.2,

$$\sigma^2 = Np \sum_{j=0}^{M} \frac{(j+1) p^j q^{M-i} M!}{j!(M-j)!} - (Np)^2 \tag{D.3}$$

Expanding Equation D.3,

$$\sigma^2 = Np \sum_{j=0}^{M} i \frac{p^j q^{M-j} M!}{j!(M-j)!} + Np \sum_{j=0}^{M} \frac{p^j q^{M-j} M!}{j!(M-j)!} - (Np)^2 \tag{D.4}$$

Now note that

$$\sum_{j=0}^{M} j \frac{p^j q^{M-j} M!}{j!(M-j)!}$$

is the expression for the expected value of a binomial distribution with parameters M and p; that is, it is equal to $Mp = (N - 1)p$. Likewise,

$$\sum_{j=0}^{M} \frac{p^j q^{M-j} M!}{j!(M-j)!}$$

is (as noted in Digression 4.2) simply the sum of all probabilities for a binomial distribution with parameters M and p, and therefore must equal 1.

Putting these pieces of information together, Equation D.4 becomes

$$\sigma^2 = Np(N-1)p + Np - (Np)^2$$

And the rest is just algebra:

$$\sigma^2 = (Np)^2 - Np^2 + Np - (Np)^2$$
$$= Np - Np^2$$
$$= Np(1 - p)$$
$$= Npq$$

4.2 The mean, variance, and standard deviation of a binomial distribution are determined by the formulas

$$\text{Mean} = \mu = Np$$
$$\text{Variance} = \sigma^2 = Npq$$
$$\text{Standard deviation} = \sigma = \sqrt{Npq}$$

FREQUENCIES AND PROPORTIONS

We now arrive at another of those issues that for some reason always seems confusing even though in reality it is extremely simple and straightforward. This notion involves the fact that we can speak equivalently about either *frequencies* (we can speak, for example, of a frequency of 12 successes out of $N = 20$ trials) or about *proportions* (we can speak of the proportion of successes as being 0.6). Try to bear in mind that this is a simple issue, and the ideas will not be new to you. Only the notation may be new.

Proportions

Suppose that we flip a coin 200 times and come up with 110 heads. As suggested above, we can express this outcome in either of two ways. We can, as we have just done, express it as a frequency and say that we got 110 heads. Alternatively we can express the 110 heads as a *proportion* of the total number of trials. That is, we can say that the proportion of heads we obtained is $110/200 = 0.55$.

Frequencies and proportions convey equivalent information. More generally any frequency i of N trials can also be expressed as a proportion, i/N. Note that proportions, like probabilities, can range from 0 ($i = 0$) to 1.0 ($i = N$).

Binomial Distributions Using Proportions

Since, as we have seen, proportions and frequencies are essentially equivalent, we can talk about probability distributions of either frequencies or of proportions. So for instance if $N = 200$, we can talk about the probability of one success, the probability of two successes, and so on, up to the probability of 200 successes. Likewise, we can talk about the probability that the proportion of successes will be $0/N = 0$ or $1/N = 0.005$ or $2/N = 0.01$, and so on up to $N/N = 1.0$. Figure 4-4 is a graphical representation of the relationship between probability distributions of frequencies versus proportions. As can be seen, the transition from one to the other is quite straightforward: the abscissa values are simply relabeled.

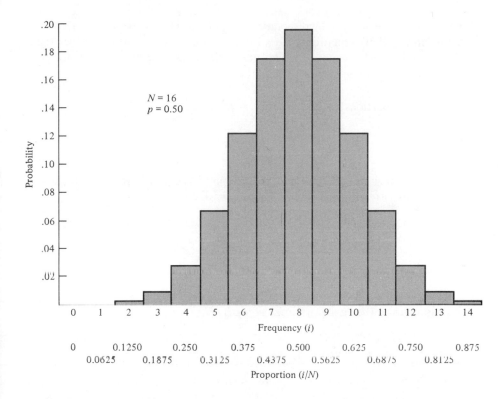

Probability

.20

.18

.16

.14

.12

.10

.08

.06

.04

.02

$N = 16$
$p = 0.50$

0   1   2   3   4   5   6   7   8   9   10   11   12   13   14

Frequency (i)

0   0.1250   0.250   0.375   0.500   0.625   0.750   0.875

0.0625   0.1875   0.3125   0.4375   0.5625   0.6875   0.8125

Proportion (i/N)

FIGURE 4-4.   A binomial distribution may be represented equally well either in terms of frequencies or in terms of proportions. Any given frequency i has a corresponding proportion of i/N.

Mean of a probability distribution of proportions.   Suppose we have a binomial distribution with parameters N and p. Recall that, in terms of frequencies,

$$\mu = Np$$

However, any proportion is equal to its corresponding frequency, i, divided by N. **Since the mean Np is a frequency, the corresponding mean in terms of proportions is $Np/N = p$.** Again this should make intuitive sense to us. If we flip a coin N times, we expect the *proportion* of heads to be p, or in this case, 0.5. Likewise, if we roll a die N times, we expect the *proportion* of sixes to be p or $1/6$.

Variance of a probability distribution using proportions.   Again dealing with the variance is slightly less intuitive than dealing with means, and we will confine our discussion to the mathematics of the situation. Recall that for a binomial distribution,

$$\sigma^2 \text{ (for frequencies)} = \sum_{i=0}^{N} i^2 p(i) - (Np)^2$$

Since, as we have seen, a corresponding proportion is i/N, we must substitute i/N for i in the above equation if we want a variance for proportions rather than for frequencies. (Also we must substitute p, the mean for proportions, for Np, the mean for frequencies.) We see then that

$$\sigma^2 \text{ (for proportions)} = \sum_{i=0}^{N} \left(\frac{i}{N}\right)^2 p(i) - p^2$$

$$= \frac{1}{N^2}\left[\sum_{i=1}^{N} i^2 p(i) - (Np)^2\right]$$

Note that the variance when proportions are used is equal to the variance when frequencies are used divided by N^2. Therefore, since

$$\sigma^2 \text{ (for frequencies)} = Npq$$

then

$$\sigma^2 \text{ (for proportions)} = \frac{Npq}{N^2} = \frac{pq}{N}$$

Note also that σ (for proportions) $= \sqrt{pq/N}$

4.3 A binomial distribution can be expressed in terms of proportions rather than in terms of frequencies. If so, then

Mean $= \mu = p$
Variance $= \sigma^2 = pq/N$
Standard deviation $= \sigma = \sqrt{pq/N}$

Frequencies Versus Proportions: An Example

Although we have assured you that the frequency/proportion business is quite simple, we think it is conceivable that you are still confused. If so, perhaps a concrete example will help at this point (see Table 4-3). We imagine ourselves flipping a coin 400 times and considering a success the occurrence of a head. Note that, when talking about *frequencies,* we expect the distribution of number of heads to have a mean or expected value of 200. However, the frequency of heads we obtain will not *always* be exactly 200. Sometimes it will be a little more (perhaps 204 heads), sometimes a little less (perhaps 192 heads). In particular the *standard deviation* of the distribution is

$$\sqrt{Npq} = \sqrt{(400)(0.5)(0.5)} = \sqrt{100} = 10 \text{ heads}$$

If we choose to speak in terms of *proportions,* on the other hand, we expect

---

TABLE 4-3.   Frequencies and Proportions: A Comparison

Binomial situation: Flip a coin 400 times. A success is a head. Thus, $N = 400$ and $p = 0.5$.

Frequencies	Proportions
Mean: $\mu = Np$	Mean: $\mu = p$
$= (400)(0.5) = 200$	$= 0.5$
(Thus we expect to get 200 heads out of our 400 flips.)	(Thus we expect to get a proportion of 0.5 heads.)
Variance: $\sigma^2 = Npq$	Variance: $\sigma^2 = pq/N$
$= (400)(0.5)(0.5) = 100$	$= (0.5)(0.5)/400 = 0.000625$
Standard deviation: $\sigma = \sqrt{Npq}$	Standard deviation: $\sigma = \sqrt{pq/N}$
$= \sqrt{100}$	$= \sqrt{0.000625}$
$= 10$	$= 0.025$

---

the proportion of heads to be 0.5. However, the proportion of heads we obtain will not always be exactly 0.5. Sometimes the obtained proportion will be a little less than 0.5 (perhaps 0.48); other times it will be a little more (perhaps 0.51). In particular the standard deviation of this distribution will be

$$\sqrt{pq/N} = \sqrt{(0.5)(0.5)/400} = \sqrt{0.000625} = 0.025$$

Notice that the mean proportion of 0.5 corresponds to the mean frequency of $(0.5)(400) = 200$. Likewise, the proportion standard deviation of 0.025 corresponds to the frequency standard deviation of $(0.025)(400) = 10$. So the mean and standard deviations coincide when we speak of frequencies or of proportions.

SUMMARY

In this chapter we have undertaken a close examination of one particular type of distribution: the binomial. The following points are of interest.

1. A binomial situation is a probabilistic situation in which two mutually exclusive and exhaustive outcomes are of interest. These two outcomes are (arbitrarily) termed a *success* and a *failure.*

2. A binomial distribution is the probability distribution of the number of successes obtained in N trials. A binomial distribution is completely specified by N, the number of trials, and p, the probability of a success.

3. The mean of a binomial distribution is Np, the variance is Npq, and the standard deviation is $\sqrt{Npq}$.

4. A binomial distribution may be expressed in terms of proportions as well as frequencies. When speaking of proportions, the mean of the distribution is p, the variance is pq/N, and the standard deviation is $\sqrt{pq/N}$.

PROBLEMS

1. Harvey the harassed househusband has to go to the laundromat, supermarket, high school, bowling alley, and fortune teller all in one day.

    1. In how many possible orders could Harvey do his five chores?
    2. Suppose Harvey could only do three of his chores today, but the fortune teller *must* be done today. How many ways are there to choose which three chores will be done today?

2. Social security numbers on the island of Pingo-Pongo consist of 3 letters followed by 4 digits followed by either an asterisk (*) or a dollar sign ($). One such number would be GRL4552*.

    1. What is the total number of different social security numbers possible?
    2. What is the number of social security numbers involving the letters ABC (in any order) followed by the digits 1234 (in any order) followed by an asterisk?
    3. What is the number of social security numbers involving the letters ABC (in that order) followed by the digits 1234 (in that order)?

3. This question concerns five-letter English-language "words." For purposes of this question, assume that each letter of the "word" can be drawn from the 26-letter alphabet except that Q is always followed by U. (This means that Q can't appear in the last position.)

    1. How many words are there that contain the letters A, B, C, D, and E in any order?
    2. How many words are there that do not contain a Q?
    3. How many words are there that do contain at least a Q?
    4. How many words are there in all?

4. Consider the set of all eight-letter strings (QBAACNBL would be such a string).

    1. How many such strings are there with vowels in the first and fifth positions and consonants in the other positions?
    2. How many such strings are there such that each string contains exactly three vowels?

5. This question refers to telephone numbers. A telephone number may be thought of as a three-digit exchange plus a four-digit "ending." *There can be no zeros in the exchange.*

    1. How many possible telephone numbers are there?
    2. How many numbers are there that satisfy the following characteristics:

 a. All digits in the exchange are the same.

 b. There are exactly two 9s and one 7 in the ending.

 3. How many numbers are there such that all digits in the exchange are different from one another?

6. A student body has the following distribution of majors:

Basket Weaving: 20
Martian Studies: 20
Psychology: 40
History of Billiards: 10
Massage: 50
California Languages: 50
Faucets: 10

   A student committee is to be formed of 20 representatives, with representation being proportional to the number of majors (2 BW, 2 MS, 4 Psych, 1 HB, 5 M, 5 CL, and 1 F).

 1. How many such committees are there?
 2. How many committees are there that contain Lee, who majors in massage, and his girlfriend, Farrah, a faucet major?
 3. What is the probability that neither Lee nor Farrah will be on the committee?
 4. What is the probability that a given person will be chosen if that person majors in psychology? In California languages?

7. Consider the word PANCREAS. Suppose four of the seven letters in this word are randomly drawn.

 1. What is the probability that the four letters will all be different?
 2. What is the probability that two of the four letters will be the same?
 3. What is the probability that two of the four letters will be the same and the other two will be different?

8. Ten people meet and shake hands with one another.

 1. How many handshakings will be required such that each person has shaken hands with every other person?
 2. Suppose that the ten persons consist of five women and five men. Women and men do not shake hands with one another, although women shake hands with women and men shake hands with men. Now how many handshakings will take place?

9. A psychology experiment contains four conditions: A, B, C, and D.

 1. Each subject is expected to participate in all four conditions. Naturally, a subject has to go through the conditions in a particular order (for example BCDA). How many subjects will be necessary such that all possible orders are gone through exactly once?

2. Suppose that due to lack of time, each subject can go through only two of the four conditions. How many subjects will be required in order that each possible pair of conditions is given to exactly one subject?

3. Under the conditions of question 2 how many subjects will be required such that each possible pair of conditions in each possible order is given to exactly one subject?

10. The residents of Carlsag, Mars, have telephones of the following sort: each number is made up of a 2-digit exchange plus four letters, such as 27-FROG. (Martians, it turns out, have the same ten digits and 26 letters that we do.)

1. How many possible telephone numbers are there?
2. How many numbers are there that contain the digits 3 and 4 and the letters A, B, C, and D in any order?
3. How many numbers are there such that all digits and letters are different from one another?
4. Imagine that Martians are order-blind; that is, to them 13 and 31 would look just the same; likewise, FROG, GROF, ROGF, GRFO, and so on would all look the same. Now how many numbers are there that are different, from the Martians' point of view? (Assume all letters and numbers must be different from one another.)

11. I have six pairs of underpants, eight pairs of socks, two pairs of pants, ten shirts, and three hats. Define an "outfit" of a combination of one pair of underpants, one pair of socks, one pair of pants, one shirt, and one hat.

1. How many different outfits can I create?
2. Suppose four of my ten shirts are identical to one another. Now how many different outfits can I create?
3. Suppose I don't care whether or not my socks match. Now how many different outfits can I create?
4. Suppose my early-rising brother has gotten up before me, made up an outfit from my clothes, and left. Now how many different outfits can I create?
5. Suppose I am in a hurry and don't have time to put on any underpants. Now how many different outfits can I create?

12. In a shooting contest shots are fired at a 64-square checkerboard. Each square has an equal probability of being hit, and each shot must hit the checkerboard somewhere.

1. Suppose ten shots are fired. What is the probability that they will all hit different squares?
2. Suppose 100 shots are fired. What is the probability that three or more squares will remain unhit?

13. Suppose a standard bridge hand of 13 cards is randomly dealt from a standard deck. What is the probability that the hand will contain

   1. 13 spades?
   2. 0 spades?
   3. 5 spades?
   4. The 6, 7, and 2 of clubs; the king and jack of diamonds, the 8, 10, and queen of hearts; and the 7, 8, 10, jack, and queen of spades? How does your answer to this part relate to your answer to the first part?

14. Compute the expected value, variance, and standard deviation of the following binomial distributions. Then plot the distributions.

   $N = 4, p = 0.10$
   $N = 4, p = 0.50$
   $N - 4, p = 0.90$
   $N = 15, p = 0.10$

15. A fair die is thrown (one with equal probability of coming up 1, 2, 3, 4, 5, or 6).

   1. What is the probability of obtaining a 6?
   2. What is the probability of *not* obtaining a 6?
   3. Suppose this die is thrown 120 times. Characterize the distribution of the number of 6s you should get. What is the *mean* and *variance* of this distribution?

16. Suppose that for the next 100 days, the probability of rain in Seattle on a given day is 4/5. Furthermore the probability of rain on any given day is independent of the probability of rain on any other day.

   1. What is the expected number of rainy days during the next 100 days?
   2. What is the variance of the probability distribution of rainy days during the next 100 days?
   3. What is the probability that *all* of the next 100 days will be rainy (to 3 decimal places)?
   4. What is the probability that between 83 and 91 (inclusive) of the next 100 days will be rainy (to 3 decimal places)?

17. Joe Shablotnik has enrolled at Central Puyallup State College. To get there from home each morning, he must go through seven traffic lights. Each light is green with a 0.3 probability and each light is independent of each other light. Compute the following probabilities.

   1. All lights are green.
   2. Between three and five (inclusive) lights are green.
   3. Fewer than four lights are green.
   4. At least one light is red or orange.
   5. At least two lights are red or orange.

18. Frank Jones claims that 60% of the people in his city will vote for him for mayor. Suppose that out of a random sample of ten people, three claim they will vote for Jones and the other seven claim they will vote for Tom Smith.

    If Jones is correct in saying that 60% of the people will vote for him:

    1. How probable is it that exactly three out of a sample of ten people will vote for him?
    2. How probable is it that three or fewer out of a sample of ten people will vote for him?

19. The Muy Munchy Mixed Nut Company puts out cans of mixed nuts that contain 40% walnuts, 30% peanuts, and 30% pecans. Suppose I dip into a Muy Munchy can and scoop out five random nuts. What are the following probabilities?

    1. At least one nut is a walnut.
    2. There is one pecan.
    3. Three of the nuts are peanuts.
    4. Four of the nuts are walnuts.

20. The Tacoma Tigers baseball team has a probability of 0.10 of winning each time they play. Furthermore each game is independent of every other game.

    Suppose the Tigers play ten games:

    1. What is the probability that they will win no games?
    2. What is the probability that they will win at least one game? At least two games?
    3. What is the probability that they will win fewer than three games?
    4. What is the probability that they will win exactly one game?

21. The Charlie Tuna Company claims that any given tuna in the ocean has only a 10% chance of ingesting mercury. But a random sample of ten tunas reveals that four of them have mercury poisoning. If the Charlie Tuna Company is correct in its claim, how probable is it that as many as four tuna would show mercury poisoning?

22. Little Wanda Welter has a probability of 0.70 of spelling any given word correctly on her spelling test.

    1. If the test contains ten words, what is the probability that Wanda will get exactly seven words correct?
    2. If the test contains 100 words, what is the probability that Wanda will get exactly 70 words correct?
    3. If the test contains 100 words, what is the probability that Wanda will get between 65 and 75 words correct?
    4. What is the relationship among your answers to the previous questions? Can you think of any reason for this relationship?

23. Joe Shablotnik has gone to Reno for vacation. Joe wanders into a casino to play blackjack. The following things are true:

   —Joe will bet $1.00 on each hand (that is, if he wins the hand, he gains $1.00; if he loses the hand, he loses $1.00).
   —Joe's probability of winning any given hand is 1/4.
   —Each hand Joe plays is independent of every other hand.

Now,

   1. If Joe plays four hands, how many hands does he expect to win?
   2. If Joe plays eight hands, how many hands does he expect to win?
   3. Define a "profit" as Joe ending up with *more* money than he started with after playing N hands. What is the probability of a profit if $N = 4$?
   4. Defining a profit as in the question above, what is the probability of a profit if $N - 8$?

5 Introduction to Hypothesis Testing: The Sign Test

In the previous chapters we have mostly been discussing abstract matters: probability theory, distributions, random variables, and so on. In this chapter our knowledge of all of this material will be brought to bear on real-world problems. Specifically we will see how this material is used in *hypothesis testing,* a procedure consisting of a set of steps that allows us to use the data from some experiment as a basis for making inferences about the nature of some aspect of reality.

DECISION MAKING

Before plunging into hypothesis testing per se, we'll digress slightly to discuss the topic of decision making. The process of making a decision is one of the core steps in the hypothesis-testing procedure.

Correct and Incorrect Decisions

We are engaged in decision making all the time. Before we leave for school or work, for example, we must decide whether or not to take an umbrella. In preparing for an exam, we must make decisions about which of the materials to study. Notice that when we make a decision about what to do, we are also (implicitly) making decisions about some aspect of the world. For example if we decide to carry an umbrella, we are also implicitly choosing to believe it will rain that day. Likewise, if we decide to focus our studying on lecture notes, we are implicitly choosing to believe that the examination will emphasize information covered in the lectures.

The object of making a decision is, of course, to make a *correct* decision, where "correct" is determined in terms of the actual state of the world. A decision to carry an umbrella will be correct if it rains that day and incorrect if it does not rain. Conversely, a decision *not* to carry an umbrella will be correct if it does not rain, but incorrect if it does.

Crime and Punishment:
A Specific Example

To discuss decision making somewhat more concretely, let's suppose that a defendant, accused of committing a crime, is being tried before a judge. In this sort of situation various pieces of evidence (data) are presented to the judge; this evidence bears on the issue of whether or not the defendent actually commited the crime. The judge's task is to weigh the evidence and to make a decision. The judge's decision will result either in the conviction of the person (a choice that there is enough evidence for the person's guilt) or in the acquittal of the person (a choice that there is not enough evidence for the person's guilt).

The four possible outcomes of this decision are depicted in Table 5-1. First there are two possible states of the world (that is, two possible realities): the defendant is either guilty or innocent. For each of these two possible states of reality there are two possible outcomes of the judge's decision: the judge can either convict or acquit the defendant. We can easily see that two of these outcomes (convicting a guilty person; acquitting an innocent person) are correct. The other two outcomes (convicting an innocent person; acquitting a guilty person) are incorrect. For ease of discussion let's label the former incorrect decision a *Type I error* and the latter incorrect decision a *Type II error*. We'll discuss these two types of error at greater length a little later in the chapter.

Evidence versus criterion.   Implicit in this decision-making situation is the notion of a *criterion* on the part of the judge. This criterion concerns the amount of evidence the judge will require before pronouncing the defendant to be guilty. The nature of the judge's criterion then determines the extent to which he or she makes Type I errors as opposed to Type II errors. Consider for example a liberal judge, that is, with a very high criterion. Such a judge would require a vast amount of evidence that the defendant is guilty before convicting. This judge will rarely convict an innocent person (that is, will rarely make a Type I error),

TABLE 5-1.   Four Possible Outcomes of a Judge's Decision
These four outcomes are obtained by combining two states of reality with two possible decisions. Two of the outcomes represent correct decisions and the other two outcomes represent incorrect decisions.

		Reality	
		Defendent Innocent	*Defendent Guilty*
Judge's Decision	*Convict Defendent*	Type I error	Correct decision
	Acquit Defendent	Correct decision	Type II error

but may well often acquit guilty people (that is, he or she will often make Type II errors). The other side of the coin is an extremely conservative judge, with a very low criterion; such a judge may well send a defendant to jail based on the merest shred of evidence. This judge will rarely acquit a guilty person (that is, will rarely make a Type II error), but could often make the Type I error of convicting an innocent person.

5.1 Any decision-making situation can be characterized as follows:

1. There are four possible outcomes to the situation: two states of reality combined with two possible decisions. Two of the outcomes are correct, and the others are erroneous.
2. There are two factors going into the decision: (1) the *evidence* for making one decision versus another, and (2) a *criterion* for determining how much evidence is needed to make one decision as opposed to the other. Notice that the evidence is external to the decision maker, whereas the criterion is determined by the decision maker and is independent of the evidence.
3. There is a trade-off between the two types of error such that changing the criterion will result in a smaller probability of one type of error but a larger probability of the other type of error.

THE ROLE OF DECISION MAKING IN HYPOTHESIS TESTING

The typical social science experiment is designed to answer a question about some aspect of reality. In an experiment we collect *data* and these data then constitute the evidence for one state of reality or another. We then make a decision based on this evidence, about which state of reality is true.

To be more specific, consider a memory experiment involving free recall procedure. In a free recall procedure a list of words—say, 20 words—is read to a subject one word at a time. The subject then recalls as many of the words as possible in any order. The dependent variable, which we measure in such a procedure, is the number of words out of 20 that the subject correctly recalls. Suppose now that we were interested in the question of whether or not free recall performance is affected by *incentive*. To investigate this question, each subject can be given two lists of words. For one list the subject is given one cent for each word correctly recalled and for the other list the subject is given 10¢ for each word correctly recalled. Incentive is therefore our independent variable, the variable that we control.

Samples and Populations: An Introduction

At this point we must anticipate an issue that will be discussed in detail two chapters hence. This issue involves the fact that we typically want any conclusions from an experiment to apply to some *population* of individuals. In our memory experiment for example, we might like to be able to conclude that incentive does or does not have an effect on memory performance for the entire population of the United States.

We do not, for obvious reasons, perform our experiment on the entire population to which we wish our conclusions to apply. Instead we randomly select a *sample* of individuals from the population, perform the experiment on the sample, and use data from the sample to infer what's happening with respect to the population as a whole.

> 5.2 We do experiments on samples, and we use the data from the samples to make inferences about larger populations.

Hypotheses About the Population

The question in which we are interested may now be phrased: does the independent variable (incentive in this example) have any effect on the dependent variable (number of words recalled in this example) in the population? To formalize this question, we will set up two mutually exclusive and exhaustive hypotheses about the nature of the population. The first, which is called the *null hypothesis* (often abbreviated "H_0") is the following:

H_0: the independent variable (incentive in this case) has no effect on the dependent variable (number of words recalled in this case).

The second hypothesis, called the *alternative hypothesis* (often abbreviated "H_1"), is:

H_1: the independent variables *does* have some effect on the dependent variable.

Now our job is to decide, based on the evidence (data), which of these two states of reality is correct. As in any decision-making situation there are four possible outcomes obtained by combining the two possible states of what is happening in the population (null hypothesis true versus alternative hypothesis true), with the two possible decisions of rejecting the null hypothesis (that is, deciding that the alternative hypothesis is true) or not rejecting the null hypothesis.[1] These outcomes are shown in Table 5-2. As with the judge situation

---

[1]The terminology "reject the null hypothesis" and "not reject the null hypothesis" may seem strange. Why not just "accept the alternative hypothesis" or "accept the null hypothesis"? In fact, "rejecting

TABLE 5-2.   Four Possible Outcomes of a Psychologist's Decision in a Hypothesis-Testing Situation
These four outcomes are defined by combining two possible states of reality with two possible decisions. Two of the outcomes represent correct decisions whereas the other two represent incorrect decisions.

		Reality	
		Null Hypothesis True	*Alternative Hypothesis True*
Psychologist's Decision	*Reject Null Hypothesis*	Type I error (probability = α)	Correct decision
	Do Not Reject Null Hypothesis	Correct decision	Type II error (probability = β)

the two types of errors have been labeled. The error of rejecting the null hypothesis when the null hypothesis is in fact correct is called a Type I error. The error of not rejecting the null hypothesis when the alternative hypothesis is actually true is called a Type II error.

We may think of the cells of Table 5-2 as representing four *conditional probabilities*. The probability of incorrectly rejecting H_0 given that H_0 is true is referred to[2] as α. It therefore follows that the probability of correctly not rejecting H_0 given that H_0 is true is $1 - \alpha$. In like fashion the probability of incorrectly failing to reject H_0 given that H_1 is true is referred to as β and the probability of correctly rejecting H_0 given that H_1 is true is therefore $1 - \beta$.

> 5.3 The null hypothesis (H_0) states that in the population, the independent variable has no effect on the dependent variable. The alternative hypothesis (H_1) states that in the population, the independent variable does have an effect on the dependent variable. The probability of erroneously rejecting H_0 when H_0 is true is α. The probability of erroneously failing to reject H_0 when H_0 is false is β.

the null hypothesis" is equivalent to "accepting the alternative hypothesis"; the two terms can be used interchangeably. It is only by convention that we tend to use the former. "Not rejecting the null hypothesis" is, however, *not* equivalent to "accepting the null hypothesis." In Chapter 8, we'll provide a detailed discussion of why the former term is more logically correct.

[2]Type I errors and Type II errors are sometimes loosely referred to respectively as α-errors and β-errors.

The Data

Keeping in mind what we want to accomplish with our experiment, let's look at the hypothetical data from ten subjects shown in Table 5-3. This table shows the number of words that each of the subjects remembered in each of the two conditions.

In general if H_1 is true, we should expect people to do better in the 10¢ condition than in the 1¢ condition. But—and here is a critical point—H_1 being true need *not* imply that *every* subject in our sample will do better in the 10¢ condition relative to the 1¢ condition. For example some subjects might be in a bad mood or tired when they go through the 10¢ condition, and they will not do as well. Or it may be that some subjects are unaffected by incentive.

We can, however, make some general kinds of statements about what will tend to make us believe that one hypothesis as opposed to the other is correct. Thus, to the extent that people remember more words in the 10¢ condition than in the 1¢ condition, we tend to believe that the alternative hypothesis is true. Conversely, to the extent that people remember about equal numbers of words in the two conditions, we tend to believe that the null hypothesis is true. To be more concrete about this, let's assign a "+" to each subject who remembers more words in the 10¢ than in the 1¢ condition and a "−" to each subject who remembers more words in the 1¢ than in the 10¢ condition (if a subject did identically in both conditions, we would just throw out that subject's data and make believe that the subject had never been in the experiment).

Having assigned our pluses and minuses, we can now count up the number

TABLE 5-3.   Data for Ten Subjects in a Hypothetical Free-Recall Experiment
Each subject is assumed to have had one list of 20 words in the 1¢/word condition and another list of 20 words in the 10¢/word condition. The data represent the number of words recalled by each subject in each of the two conditions. A "+" is assigned to each subject who remembered more words in the 10¢ condition and a "−" is assigned to each subject who remembered more words in the 1¢ condition.

Subject	Number of Words Recalled in 1¢ Condition	Number of Words Recalled in 10¢ Condition	
1	7	8	+
2	5	7	+
3	6	5	−
4	5	9	+
5	6	7	+
6	5	9	+
7	3	5	+
8	4	5	+
9	8	11	+
10	2	4	+
			$T = 9$

of pluses we have. (Naturally the maximum number of possible pluses would be ten and the minumum number would be zero.) Let's refer to the total number of pluses as *T*. Notice that in T we now have a single *summary score* that summarizes the data of the experiment in the sense that T reflects the probability that the alternative hypothesis as opposed to the null hypothesis is true. That is, to the extent that *T* is large enough, we will want to end up making the decision that the alternative hypothesis is true.

The idea of a summary score that reflects the probability of H_1 as opposed to H_0 being true is very important. In subsequent chapters the idea of a summary score will arise again and again.

But how large a summary score is "large enough"? Suppose that *T* were 10—that is, suppose that all our subjects remembered more words in the 10¢ condition than in the 1¢ condition. That would probably constitute sufficient evidence to convince us that the alternative hypothesis is true—that incentive really does have an effect on memory performance. On the other hand if *T* were only 6, this probably wouldn't constitute enough evidence. But suppose *T* were 7? Or 8? or 9? Where do we draw the line—that is, what do we use as a *criterion* number of pluses? Notice that if our criterion were high (for instance if we were to require 10 pluses before rejecting the null hypothesis), we would have a relatively low probability of making a Type I error, but a relatively high probability of making a Type II error. Conversely, if we set our criterion very low (if we were to require only 6 or more pluses before rejecting the null hypothesis), we would have a higher probability of making a Type I error, but a lower probability of making a Type II error.

Type I Versus Type II Errors

As we can see, there is a *trade-off* between the two types of error. As we shift our criterion around, we increase the probability of a Type I error but decrease the probability of a Type II error, or vice versa. How do we determine what is the optimal combination of Type I and Type II errors? Two considerations enter into this determination: error badness and error measurability.

Error badness.   It goes without saying that *any* error is by nature bad and to be avoided if possible. Sometimes, however, errors just cannot be avoided. Given that we are sometimes going to make errors, we must decide which errors we most want to avoid, as opposed to which errors we could tolerate—that is, we must decide which is the least of a number of evils.

In many decision-making situations one type of error is, for some reason or another, considered to be worse than the other type. This is exemplified fairly clearly in the judicial setting, where the error of convicting an innocent person is generally thought to be worse than the error of acquitting a guilty person. In the realm of science we have a somewhat analogous situation. Making a Type I error (erroneously rejecting the null hypothesis) is considered very bad, but making a Type II error (erroneously failing to reject the null hypothesis) is thought to be not quite so bad. The reasons why Type I errors are bad whereas Type II errors are less bad have relatively little to do with statistics per se; rather, this situation results primarily from the manner in which science progresses. In the context of our current example, suppose that in reality the null hypothesis were true—that incentive had *no* effect on memory performance—but for some reason we erroneously rejected the null hypothesis and concluded there *was* an effect. In this case the erroneous conclusion would be published, other experimenters in the discipline would read it, get excited, race into their laboratories to investigate this "effect" more thoroughly, and so on. But of course this effort would all be for nothing, since it would be based on false premises—the psychologists would be chasing theoretical will-o'-the-wisps, and all their effort would be wasted.

On the other side of the coin suppose that in fact the alternative hypothesis *is* true—incentive does affect memory performance—but the results of our experiment did not provide us with enough evidence to exceed our criterion and draw this conclusion. This sort of result is often simply not reported because (1) the lack of effect is often scientifically uninteresting, and (2) as we shall see in later chapters, if no effect is found, the statistical conclusions that can be made are generally quite weak.

Error measurability. The other consideration that goes into criterion construction is that the probability of a Type I error (α) can always be measured whereas, as we shall see in later chapters, the probability of a Type II error (β) can be measured only rarely. Therefore, it makes sense to have a generally agreed-upon value of α. We construct our criterion so as to produce this value of α and then let β fall where it will.

Construction of a Criterion

In any event social scientists have decided that a Type I error is generally to be avoided. In terms of how *much* it is to be avoided, it has been arbitrarily decided that the maximum acceptable value of α in any given experiment is 0.05. Therefore, when we do an experiment, the goal is to set some kind of criterion for rejecting the null hypothesis such that if the null hypothesis is really true, the probability of erroneously rejecting it will be 0.05 or less.[3]

Returning to our free-recall experiment (Table 5-3), remember that our problem was to choose some criterion number of pluses such that if the number of pluses we actually obtained in the experiment equaled or exceeded that criterion we could decide to reject the null hypothesis and conclude that the alternative hypothesis is true. Knowing now that our goal in choosing this criterion is to keep the probability of a Type I error less than 0.05, we proceed as follows. First we say to ourselves, "Suppose that the null hypothesis is really

---

[3]There has been an unfortunate trend among psychologists to endow the value of 0.05 with an unwarranted mystique. We reemphasize here that 0.05 as a criterion value for α is *arbitrary*. There is nothing magical about it.

true. In such a situation, how many pluses do we expect to get?" If the null
hypothesis is true—if in the population incentive really has no effect on memory
performance—then a given subject randomly selected from the population
would have a 0.5 probability of remembering more words in the 10¢ condition
and likewise and would have a 0.5 probability of remembering more words than
the 1¢ condition. With ten subjects all independent of one another, each one
having a probability of 0.5 of getting a plus, we have a binomial situation exactly
analogous to flipping a coin ten times. That is, the distribution of pluses over the
ten people should be exactly the same as the distribution of the number of heads
over the ten flips of the coin. Using what we have learned in the last chapter
about binomial distributions, we can now calculate the probability of getting
exactly ten pluses over the ten people, the probability of getting exactly nine
pluses, the probability of getting exactly eight pluses, and so on. These probabilities are shown in Table 5-4. In Table 5-4 we have also shown this probability
distribution as a *cumulative distribution*—that is, we have listed the probability
of getting exactly ten heads, the probability of getting nine *or more* heads, the
probability of getting eight or more heads, and so on.

Examination of Table 5-4 places us, at last, in a position to establish our

---

**TABLE 5-4.   How *T*, the Total Number of Pluses, Is
Distributed over Ten Subjects if the Null Hypothesis Is True**

This is simply a binomial distribution with

$$p = 0.5$$
$$q = 0.5$$
$$N = 10$$

Therefore, the probability that *T* is equal to any
number r $(0 \leq r \leq 10)$ is obtained by

$$p(T = r) = \binom{10}{r}(0.5^r)(0.5^{10-r})$$

r	$p(T = r)$	$p(T \geq r)$
10	0.001	0.001
9	0.010	0.011
8	0.044	0.055
7	0.117	0.172
6	0.205	0.377
5	0.246	0.623
4	0.205	0.828
3	0.117	0.945
2	0.044	0.989
1	0.010	0.999
0	0.001	1.000

criterion. Suppose that we were very extreme and set our criterion at 10—that is, we required that all subjects got a plus before we rejected the null hypothesis. If the null hypothesis really is true, we can see from Table 5-4 that the probability of getting 10 pluses by chance[4] would be only 0.001. Thus, if the null hypothesis were true and our criterion were 10 pluses, we would have a probability equal to 0.001 of reaching our criterion by chance and making the Type I error of erroneously rejecting the null hypothesis. It seems that this criterion would be unnecessarily high, since we have decided that we will permit the probability of a Type I error to be as high as 0.05. With a criterion of 10, α, the probability of a Type I error is small, but β, the probability of a Type II error—of failing to reject the null hypothesis when the null hypothesis is actually false—would be unnecessarily high. Suppose we move down a notch and set the criterion at 9 or more—that is, we require that nine or more subjects get pluses before we decide to reject the null hypothesis. In this case if the null hypothesis were true, the probability of equaling or exceeding the criterion by chance would be the probability of getting ten pluses by chance (0.001), plus the probability of getting nine pluses by chance (0.010) for a total of 0.011. This is still somewhat lower than 0.05 so we try creeping down yet another notch to a criterion of 8 or more. In this case the probability of exceeding the criterion by chance would be the probability of ten pluses (0.001) plus the probability of nine pluses (0.010) plus the probablity of eight pluses (0.044) for a total of 0.055. Since 0.055 exceeds our permissible probability of 0.05, we will retreat and wind up setting our criterion at 9 or more. That is, we will require that nine or more people remember more words in the 10¢ than in the 1¢ condition before we reject the null hypothesis and conclude that incentive has an effect on memory performance.[5]

Our Decision

Having established our criterion, we can now reach our conclusions very simply. We note that the number of pluses obtained in our experiment is nine, which reaches our criterion score of 9 or more. We therefore make two conclusions:

Conclusion 1. The null hypothesis is false: incentive does have some effect on memory performance in the population as a whole.
Conclusion 2. Given that incentive actually has no effect, the probability that the number of pluses would be this high is less than 0.05.

---

[4]You may have noticed that we have made an implicit assumption that if the null hypothesis is false, it will be false in the sense that people will do better in the 10¢ than in the 1¢ condition. We will have more to say about this sort of assumption in the next chapter.
[5]Conventionally if the probability of a Type I error cannot be set to exactly 0.05, we choose the conservative alternative and set it less than 0.05. This convention, like the 0.05 level itself, is, of course, arbitrary.

> 5.4 We set a criterion for how large we require our summary score (in this case T) to be such that if H_0 is true, the probability that the summary score will exceed the criterion is less than 0.05.

If the Criterion Is Not Exceeded

In the example we just described, we ended up rejecting H_0, that is, concluding that incentive does have an effect on memory performance. Suppose, however, the data had been such that we could *not* reject H_0. What would we have concluded then? Would we, in particular, have concluded that the null hypothesis is true? This turns out to be a subtle and tricky question, which we will discuss in detail later in this book. For now the general answer is that we would *not* conclude that H_0 is true. Instead we would wind up in limbo, concluding simply that we "cannot reject H_0."

HYPOTHESIS-TESTING STEPS

We have just worked our way through a prototypical set of steps used for hypothesis testing. These steps are summarized in Table 5-5. Table 5-5 also shows how these hypothesis-testing steps are applied to the sign test in the example we have just described.

TABLE 5-5.   Hypothesis-Testing Steps as They Are Applied in General and as They Are Applied to a Sign Test in the Free-Recall Example

Hypothesis-Testing Steps in General	*Hypothesis-Testing Steps as Applied to a Sign Test in the Free-Recall Example*

1. General null and alternative hypothesis:
    H_0: The independent variable has no effect on the dependent variable.
    H_1: The independent variable has some effect on the dependent variable.
	1. Null and alternative hypotheses are:
	    H_0: Incentive has no effect on memory performance.
	    H_1: Incentive does have some effect on memory performance.

2. Determine some summary score from the data that reflects the extent that the alternative hypothesis as opposed to the null hypothesis is true.	2. The summary score is T, the number of pluses obtained over the ten subjects. To the extent that T is large (close to 10), we believe that H_1 is true. To the extent that T is smaller (close to 5), we believe that H_0 is true.

3. Determine how this summary score is distributed if the null hypothesis is true.	3. T would be binomially distributed with
	    $p = 0.5$
	    $q = 0.5$
	    $N = 10$

4. Determine a criterion for the score such that if H_0 is true, the criterion will be reached at most only 5% of the time.	4. The criterion is 9 or more pluses. If H_0 is true, the probability that T will be 9 or more (9 or 10) is 0.011.

TABLE 5-5. continued

Hypothesis-Testing Steps in General	Hypothesis-Testing Steps as Applied to a Sign Test in the Free-Recall Example
5. If the obtained score is as extreme or more extreme than the criterion score, reject H_0 and make the decision that H_1 is correct. If the obtained score is not as extreme as the criterion score, do not reject H_0.	5. The obtained number of pluses (9) equals the criterion (9 or more); therefore, our decision is to reject H_0 and conclude that incentive does affect memory performance.

It is crucial to note that this basic procedure—of setting up null and alternative hypotheses, obtaining some summary score from the data that reflects the extent to which the alternative as opposed to the null hypothesis is true, determining how the summary score would be distributed if the null hypothesis were true, and determining a criterion such that if the null hypothesis were true, that criterion would be exceeded only 5% of the time—is a procedure common to *all* the different types of statistical tests we will be considering in future chapters. The only thing that changes from test to test is the type of distribution with which we are concerned. In this chapter the summary score of interest (T) was distributed binomially. In future chapters we will be concerned with scores that form z-distributions, t-distributions, F-distributions, and χ^2 distributions.

SUMMARY

In this chapter we have described some of the basic mechanisms by which probability theory is applied to data to produce conclusions. The following major conclusions emerged:

1. Hypothesis testing is a decision-making process.
2. In a prototypical decision-making situation there are four possible outcomes: two possible states of reality combined with two possible decisions. Two of these outcomes are correct responses and the other two outcomes are errors.
3. A decision is based both on the available *evidence* and on the decision maker's *criterion*.
4. In a statistical hypothesis-testing procedure, two possible states of reality are those in which an independent variable does or does not have an effect on some dependent variable with respect to the population as a whole. A Type I error consists of concluding the existence of such an effect when none in fact exists. A Type II error consists of not concluding the existence of such an effect when the effect does exist.
5. By convention a criterion is constructed so that the probability of A Type I error is less than 0.05.

PROBLEMS

1. List ten instances of decision-making situations. For each situation, state the outcome analogous to a Type I error and the outcome analogous to a Type II error.

2. An experiment is done to test the hypothesis that a person's reaction time is slower under the influence of alcohol than under the influence of marijuana. A subject sits at a table with a light in front of him. When the light goes on, the subject pushes a button as quickly as possible. The time he takes to do this is measured. Subjects perform this task twice: once under the influence of alcohol and once under the influence of marijuana.

   Of 36 subjects, 26 were slower in the alcohol condition and ten were slower in the marijuana condition. Perform a sign test on these data using the following steps:

   1. What are the null and alternative hypotheses?
   2. Define a "success" as "person is slower with alcohol as opposed to marijuana." What should the distribution of number of successes be if the null hypothesis is true?
   3. Should you reject the null hypothesis? Why or why not?

3. The Newark Muggers football team plays the Palo Alto Eggshells eight times in one season. The scores are as follows:

Game	Muggers	Eggshells
1	28	12
2	20	20
3	13	15
4	17	5
5	10	3
6	95	2
7	13	0
8	2	2

   Can the Muggers' coach use these data to support her claim that the Muggers are a better team than the Eggshells (use a 0.10 α-level)?

4. Dr. Chop is interested in whether people eat more food on rainy or on sunny days. So he gets individuals to participate in his study by coming to the laboratory on one rainy and one sunny day. A bowl of peanuts is put in front of them while they are waiting for the experiment to begin. Dr. Chop counts the number of peanuts each person eats on both the rainy and the sunny days. Here are the data:

	Rainy Day	Sunny Day
S_1	12	9
S_2	24	11
S_3	21	17
S_4	19	18
S_5	4	0
S_6	7	6

Evaluate the hypothesis that more food is eaten on rainy days (use a 0.01 α-level).

5. The Department of Motor Vehicles is trying to see whether alcohol increases the time it takes to react to some stimulus. To test this issue, the reaction times (RT) of ten subjects to a single stimulus are compared both when the subjects are sober and after they have each had two martinis. The data are as follows:

	Reaction Times (in thousandths of a second)	
Subject	Sober	Martinis
1	180	200
2	159	158
3	201	193
4	180	195
5	204	230
6	185	521
7	224	232
8	219	220
9	220	219
10	143	156

Can you reject (at the 0.05 α-level) the hypothesis that alcohol does not increase RT?

6. An experiment is done to test the hypothesis that rats prefer warm milk to a sugar solution. Each of 14 rats is given its choice of either of the two solutions. Twelve rats prefer the milk, one prefers the sugar, and one falls asleep without expressing a preference for either.

   Can these data be used to conclude that rats prefer warm milk to the sugar solution?

7. Fizzy-Cola has outsold Matzoh-Cola in 8 of the past 11 years. Can Fizzy use this datum to support its claim that people tend to prefer Fizzy to Matzoh?

8. A developmental psychologist finds 12 sets of identical twins brought up in different environments. In each case one twin was brought up in a high socioeconomic status (high-SES) environment, whereas the other was brought up in a low-SES environment. All 24 children are given an IQ test, and the following data are obtained.

Twin Pair	Low SES Member	High SES Member
1	100	98
2	95	115
3	80	101
4	98	125
5	120	120
6	98	102
7	80	92
8	103	110
9	104	105
10	68	75
11	111	112
12	110	90

1. Use a sign test to evaluate the hypothesis that the high-SES twin tends to score higher on the IQ test than does the low-SES twin.
2. Evaluate the hypothesis that the high-SES twins have IQs above the population mean of 100.
3. Evaluate the hypothesis that the low-SES twins have IQs below the population mean of 100.

9. Stoors Beer claims it's the best, but Hood Beer, manufactured in Portland, claims to be just as good as Stoors. To check this out, eight University of Kelso students agree to take a blind test. That is, blindfolded, they each drink a sip of each type of beer and indicate which tastes better.

How many of the eight students would have to claim Stoors to be better in order to reject the null hypothesis that there is no difference between the two?

1. Use the 0.10 α-level.
2. Use the 0.05 α-level.

10. Suppose you were doing an experiment in which each of four subjects participated in each of two conditions. Your alternative hypothesis is that Condition A will produce higher scores than Condition B.

Could ypu perform a sign test on the resulting data, using an α-level of 0.05? Why or why not?

11. An occultist has two coins, a quarter and a dime. She claims that the quarter exerts a force field over the dime such that when the two coins are flipped, the dime will tend to turn up the same way (heads or tails) as the quarter. In 15 trials in which the quarter and dime were simultaneously flipped, the following pattern of results emerged:

| | | Quarter | |
		Heads	Tails
Dime	Heads	5	1
	Tails	2	7

Use a sign test to evaluate the occultist's hypothesis.

Normal Distribution 6

For the last two chapters we have been discussing one particular type of theoretical probability distribution, the binomial. In Chapter 4 we discussed what the binomial is, what it looks like, how it works, and so on; in the previous chapter we described one of the ways in which it is used. In this chapter we will delve into another type of distribution—the *normal* (often referred to as a Gaussian distribution after Carl Friedrich Gauss, who was one of its developers). We shall first describe what the normal distribution is and how it is used. Then we shall continue our discussion of hypothesis testing, using the normal distribution as an example.

PRINCIPAL CHARACTERISTICS OF THE NORMAL DISTRIBUTION

Unlike the binomial, which deals with discrete random variables (such as number of heads in ten flips of a coin), the normal distribution deals with continuous random variables (such as height or time). The general shape of the normal distribution is illustrated in Figure 6-1. This bell-like shape may look familiar to you, because it frequently appears in nature. For example, volcanos such as the state of Washington's fabled Mt. Rainier are typically shaped like a normal distribution, as depicted on the cover of this book.)

Mathematical Form of the Normal Distribution

As indicated in Chapter 2, any continuous probability distribution can be characterized as a mathematical function, relating $p(x)$ (probability density) to x, the random variable under consideration. The mathematical form of a normal distribution, such as the normal distribution shown in Figure 6-1, is the following:

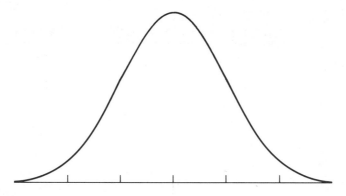

FIGURE 6-1.   The form of a normal distribution.

$$p(x) = \frac{e^{-(x-\mu)^2/2\sigma^2}}{\sqrt{2\pi\sigma^2}}$$

(6.1)

That is, if you took various values of x, and churned them through Equation 6.1 to compute corresponding values of $p(x)$, the resulting function would look like the curve shown in Figure 6-1. Equation 6.1 looks somewhat formidable—not one that you would ever want to meet in a dark alley. As it happens, we will never actually have to use Equation 6.1 since the essential information that it contains has been transferred into easy-to-use tables. However, some of the characteristics of Equation 6.1 are interesting in terms of understanding what the normal distribution is all about. Let's go through some of these characteristics.

Parameters.   Notice that most of the terms in Equation 6.1—specifically the 1, the 2, π, and e—are all *constants*. The x, of course, is whatever random variable we are concerned with (that is, the variable on the abscissa). What about μ and σ^2? These are called *parameters*. Recall that in Chapter 4 we noted that N and p are parameters of the binomial distribution—once specified, they completely determine the distribution. Likewise, the parameters μ and σ^2 in Equation 6.1, once specified, completely determine the form of the normal distribution. In particular μ (which can be any real number) turns out, once specified, to be the *mean* of the distribution; and σ^2 (which can be any positive number) turns out, once specified, to be the variance. This is just as you might expect, since the terms μ and σ^2 have been used in the past to represent mean and variance.

Because its equation has parameters, the normal distribution is not a single distribution but rather a *family* of distributions. Each specific pair of values corresponding to μ and σ^2 produces one specific member of the family. Figure 6-2 shows what the normal distribution looks like for various values of μ and σ^2. As can be seen, all of the distributions depicted in Figure 6-2 have the same

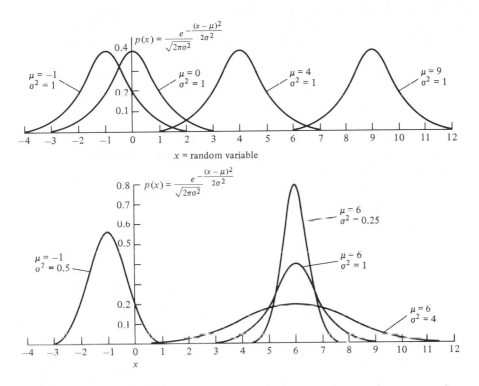

FIGURE 6-2.   Various members of the normal distribution family. Each member corresponds to a specific value of μ and σ^2. Changing μ does not change the shape of the distribution, but changes the position of the distribution on the x-axis. Changing σ^2 changes the shape of the distribution from tall and skinny (for a small σ^2) to short and squat (for a large σ^2).

general bell-shaped form. However, they differ in ways that correspond systematically to the values chosen for μ and σ^2. In particular changing μ has the effect of "sliding" the distribution back and forth along the x axis without changing its shape. Likewise, the effect of changing σ^2 is to flatten the distribution (increase the variance) as σ^2 becomes larger and to narrow the distribution (decrease the variance) as σ^2 becomes smaller. (Needless to say, the distributions in Figure 6-2 are all probability distributions in the sense that the area under each of the curves is equal to 1.0.)

> 6.1 Any values of μ and σ^2, when inserted into Equation 6.1, turn Equation 6.1 into a normal probability density function relating $p(x)$ to x. The value of μ will be the mean of the distribution and the value of σ^2 will be the variance.

Symmetry of the normal distribution.   As can be seen in Figure 6-2, a normal distribution is symmetrical around μ, the mean. "Symmetrical" means of course that a vertical line can be drawn through μ such that each half of the distribution is the mirror image of the other half.

In terms of Equation 6.1 this symmetry around μ is directly attributable to the term $(x - \mu)^2$. That is, any x that is some particular amount less than μ must produce the same value of $p(x)$ as the corresponding x that is the same amount greater than μ. For example suppose μ were 5. In that case if x were 3 (2 below the mean) then $(x - \mu)^2$ would be $(-2)^2$ or 4. On the other hand if x were 7 (2 above the mean) then $(x - \mu)^2$ would be 2^2, which is also 4. Therefore, $p(x)$ would be the same for these two values of x.

The Role of the Normal Distribution in the World

A large number of random variables in the world are normally distributed, at least to a first approximation. For example, the heights of females in the United States are normally distributed, as are the heights of males. The weights of males and females are normally distributed; exam scores are often normally distributed; IQ scores tend to be normally distributed; the time it takes a person to run the 100-yard dash tends to be normally distributed; and so on. The fact that so many things are normally distributed makes the normal distribution very important. For our purposes the normal distribution is important because it forms the basis for many statistical tests, which we will discuss later.

In this section we'll discuss why the normal distribution is so pervasive. To do this, we shall first describe a general mathematical situation (or mathematical model) that produces a normal distribution. Then we shall argue that many naturally occurring situations are formally equivalent, or at least very similar, to this general mathematical model.

A mathematical scenario: the central limit theorem.   Suppose we have a giant caldron filled with an essentially infinite number of slips of paper, each slip with one number on it. The numbers on the slips of paper can be any numbers. For purposes of illustration let's assume that each is an integer between 0 and 4. The distribution of the numbers can also be anything. Once again for purposes of illustration let's assume that there are equal numbers of 0s, 1s, 2s, 3s, and 4s in the giant cauldron.

We can now define a random variable in the following way. We dip into the giant cauldron and draw out random samples, each sample of N slips of paper. Suppose for our example that $N = 100$—that is, there are 100 slips (and therefore 100 numbers) in each sample. We now compute the *sum* of each sample of N numbers, and we let this sum be our random variable. We now ask what the probability distribution of this random variable is. That is, what is the probability that the sum of our N numbers will equal 0, that it will equal 1, and so on?

As usual when formulating a probability distribution, we first specify V, the set of all values that this random variable can assign. The way we have set things up, it is not particularly difficult to define V, which is

$$V = \{0, 1, 2, \ldots, 400\}$$

That is, the members of V range from 0 (which is what the random variable would assign when all 100 of the numbers in the sample happen to turn up 0) to 400 (which is what the random variable would assign when all 100 of the numbers in the samples happen to turn up 4). We can guess some of the characteristics of this distribution. First we would suppose that the probability of coming up with a sum of 0 is extraordinarily small, because for us to get a 0, *every single one* of the 100 numbers in our sample would have to be a 0. The same holds true for the probability that the sum will turn out to be 400. In order for us to come up with 400, every single one of the numbers in our sample would have to be a 4, and this also would seem to be an extraordinarily unlikely event. It therefore seems likely that a sum will fall somewhere in between 0 and 400. More specifically, we would intuit that the average number in our sample should be about 2—so the sum of a 100-number sample should be around 200.

With these loose intuitions as a background we turn to the central limit theorem, which says: "As N, the size of our sample, increases, the distribution of sample sums approaches a normal distribution." (For most practical purposes if the sample size is greater than about 20, then the distribution of sample sums can be construed as normal.)

> 6.2 If the sample size is large enough, then the sum of any sample of random numbers is normally distributed.

Pine forests and examinations: The central limit theorem in nature. We will now try to demonstrate why many random variables in nature can be viewed as equivalent to the giant-cauldron scenario just described. Imagine as an example a forest of pine trees, and suppose that the random variable in which we're interested is the *height* of the pine trees in the forest. Suppose that the average height (μ) of the trees is 50 ft. Now let's consider one particular tree, whose height is 54 ft. What made the tree 54 ft tall, or 4 ft above the mean? Why didn't it grow up to be 50 ft high, or 48 ft, or 60 ft? Let's enumerate some aspects of the tree's history which might have contributed to its height.

1. Suppose that the tree was planted at a particularly good time of the year, which added an extra foot to the height of the tree.
2. Suppose that its location causes this tree to get a little extra fertilizer, which added an extra 2.3 ft to the height.
3. Suppose that the tree was accidentally bruised by the planter, which subtracted 1.4 ft.

4. Suppose that the tree got some extra sunshine, which added an additional 2.0 ft.
5. Suppose that a group of antagonistic insects decided to nest in the tree, which subtracted 0.6 ft.

And so on. We could continue at length enumerating all the various influences either adding or subtracting height from the tree. The point is that the ultimate height attained by the tree (relative to μ, the mean) would be the *sum of all these influences.* For the particular tree we've been describing, the sum of all these little influences was 4; thus, this tree had added to it 4 ft over and above the mean of 50 ft. We see then that the height of any particular tree in the forest may be viewed as determined by the sum of a collection of random influences. In other words we can view each individual tree as having "dipped into a giant cauldron and pulled out a sample of random numbers," and the sum of these random numbers determined the height of the tree. Therefore, according to the central limit theorem, all the trees in the forest should have heights that, over trees, are normally distributed.

The same argument can be made for many other random variables in the world. Consider for example a set of examination scores. One can view any particular exam score (say, Irving's exam score) as the sum of a number of random influences. For example, the fact that Irving had the flu the night before the exam might subtract 5 points from his score. On the other hand Irving did make a few lucky guesses, which added 2 points. Unfortunately Irving had missed a crucial lecture, and his failure to get the notes subtracted 5 points. Happily, however, the professor was in a good mood at the time she graded Irving's exam, and that added a few points. And so on. Again we can see that Irving's exam score is made up of the sum of a number of random influences, as is each of the scores in the class. As a consequence of the central limit theorem, the exam scores in the class should be normally distributed.

> 6.3 Many random variables are such that any given score can be viewed as the sum of a collection of random influences (random numbers). Therefore, they will be normally distributed.

The normal approximation to the binomial distribution.   Let's consider one more very important example. Suppose we take a fair coin; on the head side we write a 1 and on the tail side we write a 0. Now we flip this coin 100 times, and we compute the sum of the 1s and 0s (the equivalent of the number of 1s we got). This sum might be 53, 42, 50, or whatever. We consider the number of 1s a random variable, and we ask the question: What is the distribution of this random variable? We can determine what the distribution is in either of two ways. First we note that it is simply the distribution of number of heads out of 100 flips of a coin, and we already know (from Chapter 4) that this random variable is binomially distributed (with parameters $N = 100$ and $p = 0.5$). But this situation may also be viewed as equivalent to the central limit theorem scenario; that is, we can consider ourselves to be reaching into a giant cauldron filled with equal numbers of 1s and 0s, picking out a random sample of 100 of these 1s and 0s, and computing their sum. According to the central limit theorem, this sum will be normally distributed. We therefore arrive at an interesting conclusion: with a large enough N, a binomial distribution and a normal distribution are similar. We say that the binomial distribution is *approximated* by the normal distribution. This is a very useful discovery because, as we shall see, the normal distribution is much easier to work with than the binomial distribution, particularly when N is large. We will have more to say about this in a later section of this chapter.

THE Z DISTRIBUTION

Recall that the formula for a normal distribution is described by the equation

$$p(x) = \frac{e^{-(x-\mu)^2/2\sigma^2}}{\sqrt{2\pi\sigma^2}} \tag{6.2}$$

As we have noted, this formula is for arbitrary values of μ and σ^2. Now consider a specific instance of a normally distributed random variable, for example a set of exam scores with a mean of 75 and a variance of 25. Plugging $\mu = 75$ and $\sigma^2 = 25$ into Equation 6.2, the formula for describing this particular normal distribution of exam scores would be

$$p(x) = \frac{e^{-(x-75)^2/50}}{\sqrt{50\pi}} \tag{6.3}$$

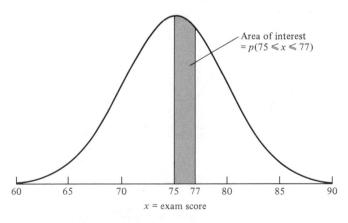

FIGURE 6-3.   The probability $p(x_1 \le x \le x_2)$ is equal to the area under the curve in the interval between x_1 and x_2. In this example $x_1 = 75$ and $x_2 = 77$.

This function is plotted in Figure 6-3. Suppose now that we want to find the probability that an exam score falls between $x_1 = 75$ and $x_2 = 77$. Recall that to do this, we must compute the *area under the curve* in the interval from $x_1 = 75$ to $x_2 = 77$, as depicted in Figure 6-3. One way of computing this area would be to *integrate* Equation 6.3 above, with 75 and 77 as the limits of integration. That is, the probability we are seeking would be computed by performing the following integration:

$$p(75 \leq x \leq 77) = \int_{75}^{77} \frac{e^{-(x-75)^2/50}}{\sqrt{50\pi}}\, dx$$

z-Scores: Normal Distribution Standard Scores

Fortunately we can make life much easier for ourselves in terms of calculating this probability. Recall our discussion of standard scores in Chapter 3: Any particular raw score (x_i) can be transformed into a corresponding standard score (z_i) by subtracting the mean from the raw score and then dividing this difference by the standard deviation. Since to any raw score there corresponds a z-score, we can transform an entire distribution of raw scores into a corresponding distribution of z-scores. Figure 6-4 shows our example distribution of exam scores both in terms of raw scores (x_i) and in terms of z-scores (z_i). Recall further from our discussion of standard scores that each standard

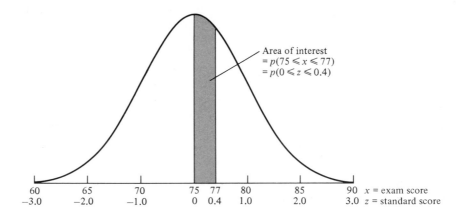

FIGURE 6-4.   An interval can be represented either in terms of raw scores or in terms of z-scores. In this example $x_1 = 75$ and $x_2 = 77$. The corresponding z-scores are $z_1 = 0$ and $z_2 = 0.4$.

deviation corresponds to a standard score of 1. Since in our example distribution the mean is 75 and the standard deviation is 5, a score of 80 corresponds to a standard score of $+1$, a score of 70 corresponds to a standard score of -1, and so on. Now we make a useful claim. Considering this new distribution of *z*-scores, the mean is 0 and the standard deviation is 1. (And as a corollary if the standard deviation is 1, then the variance is also 1.) These claims are almost self-evident, stemming from our way of constructing a *z*-distribution.

Digression 6.1

The Mean and Variance of a Distribution of Standard Scores

The goal of this digression is to prove mathematically that the mean of a distribution of standard scores is 0 and that the variance is 1.0. Let's first establish some notation:

μ_x = mean of raw scores

σ_x^2 = variance of raw scores

μ_z = mean of z-scores

σ_z^2 = variance of z-scores

Mean of Standard Scores

The mean of a distribution of standard scores is obtained by the equation

$$\mu_z = \int_{-\infty}^{\infty} zp(z)dz \tag{D.1}$$

Note that for any given z,

$$z = \frac{x - \mu_x}{\sigma_x}$$

And substituting for z in Equation D.1,

$$\mu_z = \int_{-\infty}^{\infty} \left[\frac{x - \mu_x}{\sigma_x}\right] p(x)dx$$

$$= \frac{1}{\sigma_x} \left[\int_{-\infty}^{\infty} xp(x)dx - \int_{-\infty}^{\infty} \mu_x p(x)dx\right] \tag{D.2}$$

Note that, considering the first term within the brackets of Equation D.2,

$$\int_{-\infty}^{\infty} xp(x)dx = \mu_x$$

and considering the second term within the brackets of Equation D.2,

$$\int_{-\infty}^{\infty} \mu_x p(x)dx = \mu_x \int_{-\infty}^{\infty} p(x)dx = \mu_x(1) = \mu_x$$

Therefore, Equation D.2 boils down to

$$\mu_z = \frac{1}{\sigma_x}[\mu_x - \mu_x] = 0$$

Variance of Standard Scores

We begin with the equation for the variance of a distribution of standard scores.

$$\sigma_z^2 = \int_{-\infty}^{\infty} z^2 p(z)dz - \mu_z^2 \tag{D.3}$$

Substituting

$$z = \frac{x - \mu_x}{\sigma_x}$$

into Equation D.3, and noting that $\mu_z = 0$,

$$\sigma_z^2 = \int_{-\infty}^{\infty} \frac{(x - \mu_x)^2}{\sigma_x^2} p(x)dx$$

$$= \frac{1}{\sigma_x^2} \int_{-\infty}^{\infty} (x - \mu_x)^2 p(x)dx$$

Note that

$$\int_{-\infty}^{\infty} (x - \mu_x)^2 p(x)dx$$

is simply one formula for σ_x^2. Therefore,

$$\sigma_z^2 = (1/\sigma_x^2)\sigma_x^2 = 1.0$$

Note of course that the z-distribution is still a normal distribution—in fact it is exactly the same distribution as the original raw-score distribution, but with the abscissa relabeled. The formula for this probability distribution of z-scores (as opposed to raw scores) is relatively simple. Since $\mu = 0$ and $\sigma^2 = 1$, Equation 6.2 reduces to

$$p(z) = \frac{e^{-z^2/2}}{\sqrt{2\pi}} \tag{6.4}$$

In summary, then **a z distribution is one specific member of the family of normal distributions.** It is that member corresponding to the parameter values of $\mu = 0$ and $\sigma^2 = 1$.

Let's now reformulate our problem of computing the probability that an exam score falls between 75 and 77 into a problem involving z-scores rather than raw scores. Let's first compute the z-scores corresponding to 75 and 77. **To do this, we simply use our equation for z-scores to discover that**

$$z_1 = \frac{x_1 - \mu}{\sigma} = \frac{75 - 75}{5} = 0$$

and

$$z_2 = \frac{x_2 - \mu}{\sigma} = \frac{77 - 75}{5} = \frac{2}{5} = 0.4$$

Therefore, to calculate the probability, we now have to integrate Equation 6.4 or

$$p(0 \le z \le 0.4) = \int_0^{0.4} \frac{e^{-z^2/2}}{\sqrt{2\pi}}\, dz$$

Although this looks easier than integrating Equation 6.3, it still doesn't seem as if it would be particularly easy to do. Fortunately, as noted earlier, we never really do have to perform this integration because somebody already did it for us a long time ago and put the results into convenient, easy-to-use tables for us.

The z-Distribution Tables

The tables used to perform the sorts of integrations we have just discussed appear in many places, including Appendix F (pp 601). Look them up, because we'll be referring to them in the forthcoming section.

The table is set up with pairs of columns. For any particular pair the left-hand column refers to z-scores, is labeled z, and goes from 0 to 5.5. (The upper limit of 5.5 is arbitrary. A z-score could actually go up to infinity, but the tables have to stop someplace.) The right-hand corresponding column is something that we have labeled $F(z)$. **Formally, $F(z)$ is defined as follows.**

$$F(z) = \int_{-\infty}^{z} \frac{e^{-z^2/2}}{\sqrt{2\pi}}\, dz$$

That is, **$F(z)$ corresponds to the area under the z-distribution curve from negative infinity to z.** Another way of stating this is that $F(z)$ is the probability that the random variable will assign a value that is less than a particular z-score. Figure 6-5 is a graphical representation of this notion.

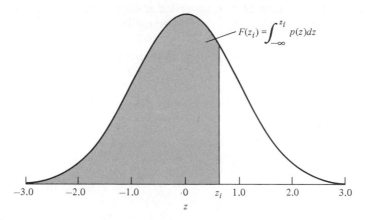

FIGURE 6-5.   A representation of $F(z)$, which is simply the area under the z distribution from $-\infty$ to z.

Using the z-Distribution Tables to Compute Probabilities

Let's go back to our example of computing exam-score probabilities. Recall that we wished to find the probability of an exam score falling between 75 and 77, which is equivalent to finding the area of the z distribution between zs of 0 and 0.4. For notational convenience we shall always refer to the larger z (0.4 in this case) as z_2 and the smaller z (0 in this case) as z_1. **Now the process of finding the desired area is quite simple and consists of three steps.**

1. **Look up $F(z_2)$ in the z-tables. In this case $F(z_2) = F(0.4) = 0.655$.**
2. **Look up $F(z_1)$ in the z-tables. In this case $F(z_1) = F(0) = 0.500$.**
3. **Now notice that the *difference* between the two areas we have just looked up corresponds to the area of the interval in which we are interested.**

So the final step consists simply of subtracting $F(z_1)$ from $F(z_2)$. In our example

$$p(0 \leq z \leq 0.4) = F(z_2) - F(z_1)$$
$$= F(0.4) - F(0)$$
$$= 0.655 - 0.500$$
$$= 0.155$$

Thus, the probability that the score will be between $x_1 = 75$ and $x_2 = 77$ (or equivalently between $z_1 = 0$ and $z_2 = 0.4$) is 0.155.

6.4 To find $p(x_1 \le x \le x_2)$

first convert x_1 and x_2 to z-scores by $z_i = (x_i - \mu)/\sigma$

Then

$$p(x_1 \le x \le x_2) = p(z_1 \le z \le z_2) = F(z_2) - F(z_1)$$

Let's consider another example. Assume that the heights of U.S. males are normally distributed with a mean, μ, equal to 70 in. and a variance, σ^2, equal to 4. (Therefore, the standard deviation, σ, is equal to 2.) Now suppose that we pluck a random male out of the population. What is the probability that his height would fall between $x_1 = 69$ and $x_2 = 72$ in.; that is, what is the probability

$$p(69 \le x \le 72)$$

Figure 6-6 depicts the situation graphically. To compute the probability of interest (that is, the area in the interval of interest), we first must convert our raw scores (xs) to corresponding z-scores so as to be able to make use of the z-tables. This process produces

$$z_1 = \frac{x_1 - \mu}{\sigma} = \frac{69 - 70}{2} = \frac{-1}{2} = -0.5$$

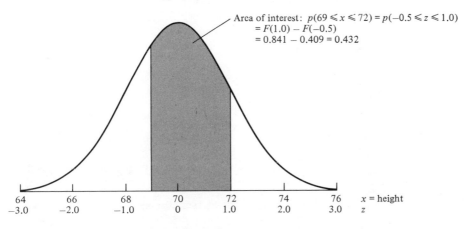

Area of interest: $p(69 \le x \le 72) = p(-0.5 \le z \le 1.0)$
$= F(1.0) - F(-0.5)$
$= 0.841 - 0.409 = 0.432$

FIGURE 6-6. The area under the curve in the interval from $x_1 = 69$ to $x_2 = 72$ is equivalent to the area from $z_1 = -0.5$ to $z_2 = 1.0$.

and

$$z_2 = \frac{x_2 - \mu}{\sigma} = \frac{72 - 70}{2} = \frac{2}{2} = 1.0$$

Therefore, the probability that we are seeking can be expressed in terms of z-scores as

$$p(-0.5 \leq z \leq 1.0)$$

To compute $F(z_2)$, we simply consult our tables to discover that $F(1.0)$ equals 0.841. Now how about $F(z_1) = F(-0.5)$? We seem at this point to be in some difficulty, since the tables begin with z equal to 0—that is, the tables do not seem to list values for negative zs. What are we to do?

The answer is that we make use of the symmetrical nature of the normal distribution and in particular the symmetry between $F(z)$ when $z = 0.5$ and $F(z)$ when $z = -0.5$. Figure 6-7 depicts this symmetrical relationship and how we use it. We note that the area to the left of $z = -0.5$ (which is what we want) is the same as the area to the right of $z = +0.5$. Now we note that the area to the right of $+0.5$ is simply equal to 1 (the area under the entire curve) minus the area to the left of $+0.5$. We can look up the area to the left of $+0.5$ in our z-tables. This is $F(0.5)$, which is equal to 0.691. Subtracting 0.691 from 1.0 gives us 0.309, which is the area to the right of $z = 0.5$ which, by symmetry, is the area to the left of $z = -0.5$. **In summary whenever we have some negative z-score, we can make use of the general equation**

$$F(-z) = 1 - F(+z)$$

So in this particular example

$$F(z_1) \approx F(-0.5) = 1 - F(0.5) = 1 - 0.691 = 0.309$$

And to complete the example,

$$p(69 \leq x \leq 72) = p(-0.5 \leq z \leq 1)$$
$$= F(z_2) - F(z_1)$$
$$= 0.841 - 0.309$$
$$= 0.532$$

6.5 $F(-z) = 1 - F(z)$

Let's briefly compute two more probabilities using our height example. First what is the probability that a random male will be *taller than* 71.5 in., and

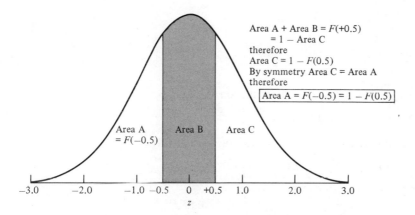

FIGURE 6-7.   The technique for using the z-tables to find $F(z)$ when z is negative. This technique makes use of the symmetry of the distribution. In particular the area we wish to calculate, $F(-z)$ = Area A, is equal to Area C, which in turn is equal to $1 - F(+z)$.

second what is the probability that a random male will be *shorter than* 69.5 in.? Figure 6-8 depicts these probabilities.

Let's concern ourselves with the first question: How probable is it that a random male is taller than 71.5 in.? Clearly x_1 here—our lower limit—is 71.5. But what is x_2? In this case, x_2 must be $+\infty$ since we are essentially asking: What is the probability that a random male is between 71.5 and $+\infty$ in. tall? To compute the z-scores that correspond to x_1 and x_2,

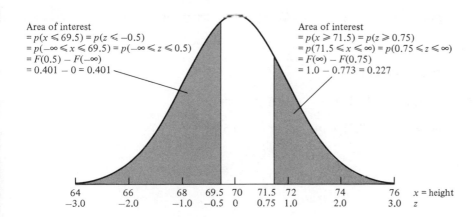

FIGURE 6-8.   Calculation of "less than" and "greater than" probabilities. To calculate a "less than" probability, we let $z_1 = -\infty$, in which case $F(z_1) = 0$. To calculate a "greater than" probability, we let $z_2 = \infty$, in which case $F(z_2) = 1.0$.

$$z_2 = \frac{x_2 - \mu}{\sigma} = \frac{+\infty - \mu}{2} = +\infty$$

$$z_1 = \frac{x_1 - \mu}{\sigma} = \frac{71.5 - 70}{2} = \frac{1.5}{2} = 0.75$$

Now to carry through our area computation steps,

$$F(z_2) = F(+\infty) = 1.0$$

(since the area under the curve to the left of $+\infty$ is 1.00).

$$F(z_1) = F(+0.75) = 0.773$$

$$p(71.5 \leq x \leq \infty) = p(0.75 \leq z \leq \infty)$$

$$= F(z_2) - F(z_1)$$

$$= F(\infty) - F(0.75)$$

$$= 1.00 - 0.773 = 0.227$$

The second example is quite similar. To compute the probability that height is less than 69.5, we let $x_1 = -\infty$ and $x_2 = 69.5$ in. Thus,

$$z_1 = \frac{-\infty - 70}{2} = -\infty$$

and

$$z_2 = \frac{69.5 - 70}{2} = \frac{-0.5}{2} = -0.25$$

So, to compute the area under the curve

$$F(z_2) = F(-0.25) = 1 - F(0.25) = 1 - 0.599 = 0.401$$

and

$$F(z_1) = F(-\infty) = 0$$

(since the area to the left of $-\infty$ is 0). Therefore,

$$p(-\infty \leq x \leq 69.5) = p(-\infty \leq z \leq -0.5)$$

$$= F(z_2) - F(z_1)$$

$$= 0.401 - 0$$

$$= 0.401$$

Looking at Figure 6-8, you can probably see that some of the steps we just went through may be short-circuited. We include all the steps in order to demonstrate how all examples may be fitted into the same mold.

Calculations Involving the
Normal Approximation to the Binomial

As discussed in an earlier section, the binomial distribution can usually be approximated by the normal distribution. Using this normal approximation can make our lives somewhat easier when we are faced with calculating probabilities, because calculating exact binomial probabilities is quite tedious when N is any larger than about 5. (It's not even particularly easy when N is less than 5.) On the other hand as we have just seen, it is quite easy to calculate probabilities based on the normal distribution since we have convenient tables. Let's go through an example of how we would use the normal approximation to do a binomial-type problem.

Cocktail-party treats.   Suppose that the Muy Munchy Mixed Nut Company produces a brand of mixed nuts consisting of 1/4 peanuts and 3/4 walnuts. Muy Munchy produces bags with 200 nuts in each using the following technique. First huge numbers of nuts are thoroughly mixed in a vat at the Muy Munchy plant. (As noted 3/4 of all these nuts are walnuts and the other 1/4 are peanuts.) To produce a 200-nut bag, a mechanical hand dips into the vat, grabs a 200-nut random sample, and dumps these nuts into a bag. These bags are then sold. Now suppose that you buy a bag of Muy Munchy mixed nuts. What is the probability that this bag will contain between 47 and 56 peanuts?

We can easily see that this is a binomial situation with $N = 200$ nuts and $p = 0.25$. (A success is defined here as getting a peanut.) We know how to calculate probabilities using the binomial distribution. Letting x be the number of peanuts in a bag, the probability of getting between 47 and 56 peanuts would simply be

$$p(47 \le x \le 56) = p(x = 47) + p(x = 48) + \cdots + p(x = 56)$$

$$= \binom{200}{47}(0.25)^{47}(0.75)^{153} + \binom{200}{48}(0.25)^{48}(0.75)^{152}$$

$$+ \cdots + \binom{200}{56}(0.25)^{56}(0.75)^{144}$$

These probabilities could be computed but it would be a long, agonizing, thankless task.

Using the normal distribution.   Let's see how we would use the normal approximation to the binomial to make our lives a little simpler. We first calculate the mean and standard deviation of the binomial distribution with which we are working.

$$\mu = Np = (200)(0.25) = 50$$

$$\sigma = \sqrt{Npq} = \sqrt{(200)(0.25)(0.75)} = \sqrt{37.5} = 6.12$$

We now make believe that we are dealing with a normal distribution whose mean is 50 and whose standard deviation is 6.12, and we wish to calculate the probability of getting a score between $x_1 = 47$ and $x_2 = 56$. We now can solve this problem quite easily. We first calculate the z-scores corresponding to $x_1 = 47$ and $x_2 = 56$, or

$$z_1 = \frac{x_1 - \mu}{\sigma} = \frac{47 - 50}{6.12} = -0.49$$

$$z_2 = \frac{x_2 - \mu}{\sigma} = \frac{56 - 50}{6.12} = 0.98$$

We then use the normal probability tables to find the corresponding $F(z)$s and go through our subtraction procedure to arrive at the area of interest. That is

$$F(z_2) = F(0.98) = 0.836$$

$$F(z_1) = F(-0.49)$$

$$= 1 - F(0.49)$$

$$= 1 - 0.688$$

$$= 0.312$$

$$p(47 \leq x \leq 56) = p(-0.49 \leq z \leq 0.98)$$

$$= F(z_2) - F(z_1)$$

$$= 0.836 - 0.312$$

$$= 0.524$$

The correction for continuity. The above example gives the general idea of what we do when we use the normal approximation to the binomial, but this example is not quite correct, because it has overlooked one annoying detail called the correction for continuity. The correction for continuity is needed to adjust for the fact that the normal distribution is continuous, whereas the binomial distribution is discrete. We shall illustrate the necessity for and the mechanics of using this correction with a simple example.

Suppose we flip a coin four times and compute the distribution of the number of heads. We have dealt with this example many times before; it is a binomial situation with $N = 4$ and $p = 0.5$. Figure 6-9 shows the resulting distribution along with the normal approximation to it.

Suppose we now want to calculate the probability that the number of heads we get is in the interval between $x_1 = 1$ and $x_2 = 2$ (inclusive). That is, we want

$$p(1 \leq x \leq 2)$$

Let's first use the binomial to calculate exactly what this probability is. It is of course simply the possibility of getting either 1 or 2 heads in our 4 flips, which in

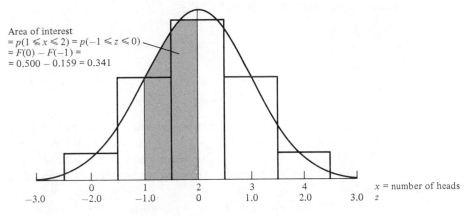

Area of interest
$= p(1 \leqslant x \leqslant 2) = p(-1 \leqslant z \leqslant 0)$
$= F(0) - F(-1) =$
$= 0.500 - 0.159 = 0.341$

x = number of heads

FIGURE 6-9.   Binomial distribution with $N = 4$ and $p = 0.5$, along with normal approximation to this binomial. The area under the normal approximation between $x_1 = 1$ and $x_2 = 2$ is also shown.

turn is the sum of the probabilities of getting 1 head and getting 2 heads. To calculate these probabilities,

$$p(1 \text{ head}) = \binom{4}{1} (0.5)^1 (0.5)^3 = 0.250$$

$$+$$

$$\frac{p(2 \text{ heads}) = \binom{4}{2} (0.5)^2 (0.5)^2 = 0.375}{p(1 \leq x \leq 2) \qquad = \qquad 0.625}$$

Now let us calculate this probability using the normal approximation. First we see that the mean of the distribution is

$$\mu = Np = (4)(0.5) = 2$$

and the standard deviation is

$$\sigma = \sqrt{Npq} = \sqrt{(4)(0.5)(0.5)} = \sqrt{1} = 1$$

Now from our two raw scores, $x_1 = 1$ and $x_2 = 2$, we calculate the corresponding z-scores:

$$z_1 = \frac{x_1 - \mu}{\sigma} = \frac{1 - 2}{1} = -1$$

$$z_2 = \frac{x_2 - \mu}{\sigma} = \frac{2 - 2}{1} = 0$$

Now calculate the corresponding $F(z)$s

$$F(z_2) = F(0) = 0.500$$

$$F(z_1) = F(-1) = 1 - F(1) = 1 - 0.841 = 0.159$$

and by subtraction we find that the probability is

$$p(1 \leq x \leq 2) = p(-1 \leq z \leq 0)$$
$$= F(z_2) - F(z_1)$$
$$= 0.500 - 0.159$$
$$= 0.341$$

This is not a very good approximation. The actual value is 0.625, but what we calculated using the normal approximation is 0.341—not even close! What has gone wrong? An examination of the correspondence between the binomial and the normal distribution (Figure 6-9) reveals the problem which is that the limits we have selected for our xs—$x_1 = 1$ and $x_2 = 2$—are not exactly what we want. Rather, an approximation of the two bars corresponding to $x = 1$ and $x = 2$ involves slightly different limits, since the xs themselves (the 1 and the 2) are in the middle of the bars. We can see that, rather than taking the area under the normal approximation between $x_1 = 1$ and $x_2 = 2$, we actually want the area between $x_1 = 1 - 0.5 = 0.5$ and $x_2 = 2 + 0.5 = 2.5$, as shown in Figure 6-10. Let's now recompute our probability using these new limits. We get new z-scores:

$$z_1 = \frac{x_1 - 0.5 - \mu}{\sigma} = \frac{1 - 0.5 - 2}{1} = -1.5$$

$$z_2 = \frac{x_2 + 0.5 - \mu}{\sigma} = \frac{2 + 0.5 - 2}{1} = 0.5$$

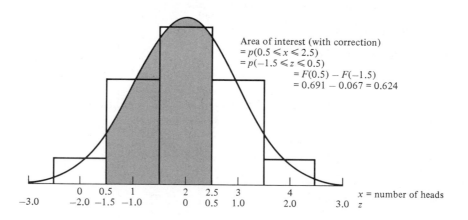

Area of interest (with correction)
$= p(0.5 \leqslant x \leqslant 2.5)$
$= p(-1.5 \leqslant z \leqslant 0.5)$
$= F(0.5) - F(-1.5)$
$= 0.691 - 0.067 = 0.624$

0	0.5	1		2	2.5	3		4		x = number of heads

-3.0   0   0.5   1   2   2.5   3   4 x = number of heads
     -2.0 -1.5 -1.0   0   0.5   1.0   3.0 z

FIGURE 6-10. Use of the correction for continuity. We want to approximate the area of the two binomial-distribution bars corresponding to $x_1 = 1$ and $x_2 = 2$. To establish the appropriate limits, we must subtract 0.5 from x_1 and add 0.5 to x_2. The proper approximation to the interval of interest then results.

We get two new $F(z)$s:

$$F(z_2) = F(0.5) = 0.691$$

$$F(z_1) = F(-1.5) = 1 - F(1.5) = 1 - 0.933 = 0.067$$

And by subtraction we find the probability we are looking for is

$$p(1 \leq x \leq 2) = p(-1.5 \leq z \leq 0.5)$$

$$= F(z_2) - F(z_1)$$

$$= 0.691 - 0.067$$

$$= 0.624$$

Since the actual probability, we recall, is 0.625, this is a much better approximation.

To reiterate, when we are using the normal approximation to the binomial, and we wish to compute $p(x_1 \leq x \leq x_2)$ we compute our z-scores, using the equations:

$$z_1 = \frac{x_1 - \mu - 0.5}{\sigma}$$

and

$$z_2 = \frac{x_2 - \mu + .05}{\sigma}$$

6.6 Suppose we have a binomial distribution with parameters N and p, and suppose further that we want to find the probability that a score is between x_1 and x_2, inclusive. (Note that x_1 and x_2 must, of course, be integers if we are dealing with a binominal.) We find our z-scores in the following way:

$$z_1 = \frac{x_1 - \mu - 0.5}{\sigma}$$

and

$$z_2 = \frac{x_2 - \mu + 0.5}{\sigma}$$

That is, we *subtract* 0.5 from the smaller raw score (x_1) and we *add* 0.5 to the larger raw score (x_2).

MORE ABOUT HYPOTHESIS TESTING:
THE z-TEST

In Chapter 5 we briefly outlined the rationale behind hypothesis-testing procedures, using the sign test as an example. We are now going to discuss several other issues involved in hypothesis testing. We will use an example involving a binomial situation, and the normal approximation to the binomial will provide us with a powerful computational tool.

An Illustrative Experiment:
Gourmet Rats

Imagine an experiment in which we inquire as to the types of food preferred by rats. To address this issue, we construct a T-maze. A T-maze is simply a box shaped like a T in which the rat can wander around (see Figure 6-11). As can be seen, we begin by putting the rat in the bottom part of the T. The rat can trot up the box and proceed to turn either left or right, thereby arriving at one of the two

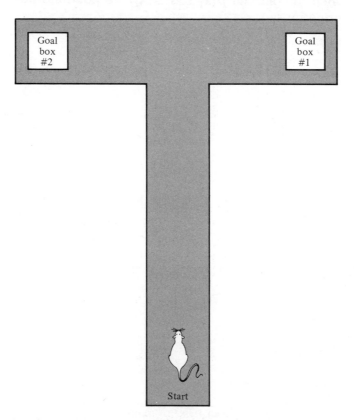

FIGURE 6-11.   A T-maze. The rat starts in the bottom and then can either turn right to goal box 1 or left to goal box 2.

goal boxes. Because one goal box contains one type of food and the other goal box contains another type of food, we will ultimately be interested in which direction the rat turns. Suppose, for example, we wish to determine whether rats prefer cabbage to rat pellets. To answer this question, we put a rat pellet in one goal box and a cabbage in the other goal box. We then put 100 rats in the T-maze, one at a time, and for each rat we record whether the rat turns toward the pellet or toward the cabbage.

Three Types of Alternative Hypotheses

We now want to talk about the three different types of hypothesis-testing procedures that we might use to draw conclusions from our data. These three procedures all involve the same type of test, referred to as a z-test. However, they differ with respect to the nature of the alternative hypothesis that is entertained.

A specific (quantitative) alternative hypothesis.   Suppose we have a specific quantitative theory that we are attempting to test with this experiment. In particular the theory predicts that, with a probability of 0.6, any rat should prefer the cabbage to the pellet.

In this experiment the dependent variable (that we record) is "turning behavior" (either toward the cabbage or toward the pellet). The independent variable (that we vary) is the type of food (either pellet or cabbage). We are now in a position to formulate a null hypothesis and an alternative hypothesis that we can pit against one another. Recall that the null hypothesis always states that the independent variable has no effect on the dependent variable or in terms of our particular experiment

H_0: The type of food has no effect on turning behavior. Specifically the rat should turn toward the cabbage with probability $p = 0.5$.

The rat should, according to H_0, turn toward the cabbage with a probability $p = 0.5$. On the other hand the alternative hypothesis always says that the independent variable does have an effect on the dependent variable. In the situation we have concocted, the alternative hypothesis is very specific about the nature of the effect that the independent variable should have on the dependent variable, namely

H_1: Type of food does have an effect on turning behavior. Specifically the rat should turn toward the cabbage with a probability $p = 0.6$.

We will now decide, based on how many of the 100 rats actually turn toward the cabbage, whether or not to reject the null hypothesis in favor of the alternative hypothesis. As we discussed in Chapter 5, there are four possible outcomes of our decision process; these are depicted in Table 6-1. We have noted

TABLE 6-1. Four Possible Outcomes of the Hypothesis-Testing Procedure

| | | Reality | |
		$H_0 True$	$H_0 False\ (H_1 True)$
Decision	Reject H_0	Type I error	Correct decision
	Fail To Reject H_0	Correct decision	Type II error

that two of these outcomes—rejecting the null hypothesis when the alternative hypothesis is true and failing to reject the null hypothesis when the null hypothesis is true—are correct decisions. The other two decisions are errors. If the null hypothesis is really true and we erroneously reject it, we have made a Type I error. Conversely, if the alternative hypothesis is really true and we fail to reject the null hypothesis, we have made a Type II error. Recall that we wish to arrange the decision-making process such that the probability of a Type I error will be less than 0.05.

Let's first assume that in reality the null hypothesis is true and determine what the resulting distribution of number of turns for cabbage will be. If the null hypothesis is true, then we have a binomial situation with N equal to 100 (trials) and p equal to 0.5 (where a "success" on any given trial is defined to be a cabbage turn). We can easily calculate the mean and standard deviation of this distribution.

$$\mu = Np = 100(0.5) = 50$$
$$\sigma = \sqrt{Npq} = \sqrt{(100)(0.5)(0.5)} = \sqrt{25} = 5$$

This distribution is depicted in Figure 6-12.

Under what circumstances will we reject the null hypothesis? It would seem reasonable to reject the null hypothesis if *enough* rats were to turn toward the cabbage. For example if 90 of the 100 rats were to turn toward the cabbage, then we would feel reasonably certain that the null hypothesis was false. On the other hand if only 52 of the rats were to turn toward the cabbage, then we would probably not be inclined to reject the null hypothesis. But how many rats is enough? Where do we set our criterion? How many rats turning toward the cabbage will it take to convince us that the null hypothesis is false?

This is where our α-level of 0.05 enters the picture. We want to construct a criterion number of cabbage turns such that if the null hypothesis is really true a probability of erroneously rejecting it will be less than 0.05. This corresponds to setting a criterion that chops off the upper 5% of the null hypothesis distribution shown in Figure 6-12. Whatever that criterion turns out to be will be the minimum number of cabbage turns that we will require before rejecting the null hypothesis.

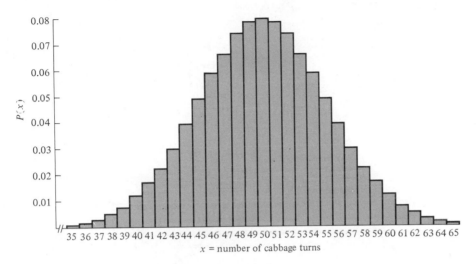

FIGURE 6-12.   Distribution of number of cabbage turns when H_0 is true. This is a binomial distribution with $N = 100$ and $p = 0.5$.

To calculate what this criterion should be, we make use of what we know about the normal approximation to the binomial distribution. Figure 6-13 depicts the normal approximation. Note that we have also provided a z-score labeling of the abscissa in Figure 6-13.

We can see that the z-score which will chop off the upper 5% of the distribution falls somewhere between 1 and 2. To calculate it precisely, we reason that if there is 0.05 of the distribution *above* the criterion, then 0.95 of the distribution must fall *below* the criterion. Therefore, if we can find a probability in our z-tables that is equal to 0.95, the corresponding z-score must be the one that we are looking for. To do this, we simply search along a right-hand column rather than a left-hand column in our z-tables and find that the z-score corresponding to a probability $[F(z)]$ of 0.95 is 1.64.

We almost have our criterion. All we need to do now is to translate our criterion z-score back into a raw score. To do this, we set up our equation for getting z-scores from raw scores and rearrange terms so that with a particular z-score, we can find the corresponding raw score.

$$z_2 = 1.64 = \frac{x_2 - \mu + 0.5}{\sigma} = \frac{x_2 - 50 + 0.5}{5} = \frac{x_2 - 49.5}{5}$$

Thus

$$(1.64)5 + 49.5 = x_2$$

or

$$x_2 = 57.7$$

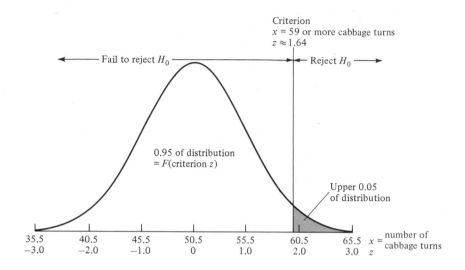

FIGURE 6-13.   Establishment of a criterion that cuts off the upper 5% of the z distribution. The criterion z-score is $z = 1.64$, which corresponds to a criterion raw score of 57.7.

Solving[1] for x, we find that the x corresponding to a z of 1.64 equals 57.7. Rounding off upward (to be conservative[2]), we will call it 58. This means that if 58 or fewer turns toward the cabbage are made, we will fail to reject the null hypothesis. Thus, we require 59 or more turns toward the cabbage before we will reject the null hypothesis. This criterion is determined by our desire to keep the probability of a Type I error at 0.05 or less.

We are now in a position to calculate the probability of a Type II error. That is, we can now calculate the probability of failing to reject the null hypothesis given that the alternative hypothesis is actually true. To do so, we imagine that the alternative hypothesis is actually true and we construct the distribution of cabbage turns. This is still a binomial distribution, with $N = 100$ trials; now, however, $p = 0.6$, as dictated by the alternative hypothesis. Therefore, the mean and standard deviation of the distribution are:

---

[1]Note that the correction for continuity dictates that $+0.5$ be *added* in the equation. This is because the area of 0.95 constitutes the area between an x_1 of $-\infty$ and the x_2 that we are trying to calculate. So the equation is relating the bigger x (x_2) to the bigger z (z_2).

[2]Since "number of turns" is a discrete variable, we are constrained to either 57 or 58 turns as our criterion. A criterion of 57 turns would correspond to an α-level of slightly greater than 0.05, whereas a criterion of 58 turns would correspond to an α-level of slightly less than 0.05. As noted in Chapter 5, when we must depart from an α-level of exactly 0.05, we conventionally depart in the direction of making the α-level smaller. The smaller the α-level, the more conservative the test is said to be.

$$\mu = Np = (100)(0.6) = 60$$

$$\sigma = Npq = (100)(0.6)(0.4) = 4.89$$

Figure 6-14 shows this distribution. We have included the criterion of 58 that we determined previously. Note that if 58 or fewer rats make cabbage turns, we will fail to reject the null hypothesis; and assuming, as we are doing, that the alternative hypothesis is true, this outcome will constitute β, the probability of a Type II error. Therefore, referring to the distribution shown in Figure 6-14, the area under the curve to the *left* of the criterion represents β. It is not difficult to figure out what this area is. To do so, we once again make use of the normal approximation to the binomial and proceed as follows. First, we let $x_1 = -\infty$ and $x_2 = 58$. We then determine the corresponding z-scores:

$$z_1 = \frac{x_1 - \mu - 0.5}{\sigma} = \frac{\infty \quad 60 \quad 0.5}{5} = -\infty$$

$$z_2 = \frac{x_2 - \mu + 0.5}{\sigma} = \frac{58 - 60 + 0.5}{5} = \frac{-.03}{5} = -0.5$$

and the $F(z)$s:

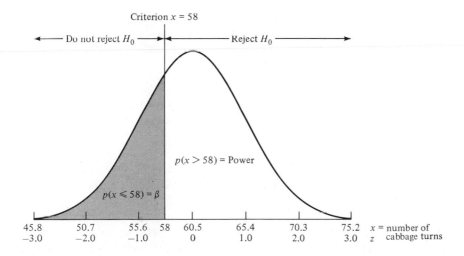

Criterion $x = 58$

$\longleftarrow$ Do not reject H_0 $\longrightarrow$ $\longleftarrow$ Reject H_0 $\longrightarrow$

$p(x > 58) =$ Power

$p(x \leqslant 58) = \beta$

| 45.8 | 50.7 | 55.6 | 58 | 60.5 | 65.4 | 70.3 | 75.2 | x = number of |
| -3.0 | -2.0 | -1.0 | | 0 | 1.0 | 2.0 | 3.0 | z | cabbage turns |

FIGURE 6-14. To consider the probability of a β error, we must determine the distribution of scores that occurs when H_1 is true. Any score that is lower than our preestablished criterion will lead to a β error. The probability of a β error, therefore, is the area under the curve to the left of the criterion.

$$F(z_2) = F(-.03) = 1 - F(.03) = 1 - .512 = 0.488$$

$$F(z_1) = F(-\infty) = 0$$

By our subtraction procedure we determine that

$$p(\text{Type II error}) = F(z_2) - F(z_1) = 0.488 - 0 = 0.488$$

We see therefore that the probability of making a Type II error is 0.488.

At this point we will introduce the term *power*. The power of a statistical test is simply $1 - \beta$, or the probability of *correctly* rejecting H_0 when H_1 is really true. So β and power are directly related. Power is, in general, a good thing, and the bigger the better. Beta, on the other hand, is a bad thing, and the smaller the better. In this case power is $1 - 0.488$ or 0.512.

A directional alternative hypothesis. The situation we have just described is somewhat artificial, in that we assumed an *exact* alternative hypothesis—that the rat will prefer cabbage to pellets with the *specific* probability of 0.6. For the most part, however, social science being what it is, alternative hypotheses are typically not exact; theories are typically not sufficiently specific and quantitative to predict such precise probabilities. Thus, alternative hypotheses usually are much vaguer. In this section we will discuss a typical vague alternative hypothesis. In the next section we will discuss an even vaguer alternative hypothesis.

Let's reformulate our null and alternative hypotheses. The null hypothesis will not change (it rarely does—it is almost always that the independent variable has no effect on the dependent variable). So once again in this experiment,

H_0: The independent variable (food) has no effect on the dependent variable (turning behavior). That is, the rats should turn toward the cabbage with a probability of $p = 0.5$.

In terms of the alternative hypothesis, we will now assume that our theory simply says that the rat should *prefer* the cabbage to the rat pellet, although it does not commit itself to the exact degree of preference it did in the last example. Therefore,

H_1: The independent variable (food) does have an effect on the dependent variable (turning behavior). In particular the rat should turn toward the cabbage more than toward the rat pellet, or $p > 0.5$.

Once again, our decision-making task is to choose under what circumstances we will reject the null hypothesis and under which circumstances we will not reject the null hypothesis. Again this task corresponds to setting some *criterion* number of cabbage turns such that if the observed number of rats turning toward the cabbage exceeds this criterion, we will reject the null hypothesis. To establish our criterion, we go through precisely the same steps we

Digression 6.2

Statement of the Null Hypothesis

Referring to the probability of a cabbage turn as p, we have stated our null and alternative hypotheses as

H_0: $p = 0.5$
H_1: $p > 0.5$

Note that these two hypotheses are not exhaustive in the sense that they exclude the possibility that $p < 0.5$. The very notion of a directional alternative hypothesis renders this possibility unlikely. That is, we set up a directional hypothesis only if we think any possible relationship between type of food and turning behavior is one in which cabbage will be preferred over pellets.

However, some statisticians take the view that hypotheses should be constructed so as to exhaust all possible states of reality and therefore that the null hypothesis should be set up as

H_0: $p \leq 0.5$

What does this way of stating the null hypothesis do to the probability of a Type I error? Note that if we set up the criterion such that the probability of a Type I error is 0.05 when $p = 0.5$, then when $p < 0.5$ the probability of a Type I error must be less than 0.05. Therefore, stating the null hypothesis this way, the *maximum* probability of a Type I error is 0.05.

did in the last example. That is, we first construct the distribution of cabbage turns assuming that the null hypothesis is true. Since the null hypothesis has not changed, this distribution does not change either: it is a binomial distribution with a mean (μ) equal to 50 and a standard deviation (σ) equal to 5. Our goal of maintaining an α-level of 0.05 is similarly unchanged; so the criterion number of cabbage turns that we will require for an α-probability of less than 0.05 is still 59 or more. The decision-making procedure is therefore precisely the same as it was in the last example. In each of the two examples, the test we do is called a *one-tailed test* in reference to the single tail of the distribution that represents α, the probability of a Type I error.

What has changed from the first example to the second? That is, what are the consequences of having a nonspecific alternative hypothesis? To answer these questions, let's consider the issue of Type II errors. To determine the probability of a Type II error, recall that we must first construct the distribution of cabbage turns that occurs when the alternative hypothesis is true. What is this distribution with the directional (but inexact) alternative hypothesis with which we are now dealing? The answer is that we can't tell. We know that if $p > 0.5$, then μ, the mean of the distribution, must be greater than 50. But we don't know exactly what μ is, so we can't draw out what the distribution would look like if the alternative hypothesis were true. Therefore, we don't know the probability of a Type II error. This is an unfortunate although common characteristic of the hypothesis-testing procedure. We can always set our criterion so that the probability of a Type I error is known and controlled; but in the very common situation that we do not have a precise alternative hypothesis, we do not know what the probability of a Type II error is. We will have still more to say on this issue in Chapter 8.

6.7 With an exact alternative hypothesis, the probability of a Type II error can be calculated. In the absence of an exact alternative hypothesis the probability of a Type II error cannot be calculated. But in all cases, the probability of a Type I error remains at whatever we choose it to be, usually 0.05.

Nondirectional alternative hypotheses.   The third type of situation we shall discuss is one in which we are extremely vague about our alternative hypothesis. In this situation the alternative hypothesis claims only that there will be a *difference* in preference between cabbage and pellets. But it doesn't specify which will be preferred—the cabbage or the pellets.

The null hypothesis does not change. It is still

H_0: The independent variable (type of food) has no effect on the dependent variable (turning behavior); that is, the probability of turning toward the cabbage is $p = 0.5$.

The alternative hypothesis is now stated as

H_1: The independent variable (type of food) does have an effect on the dependent variable (turning behavior); that is, the probability of turning toward the cabbage is $p \neq 0.5$.

In this situation, the criterion-setting business takes on a new twist. This is because there are now two sets of circumstances under which we will want to reject the null hypothesis—if we have (a) a large number of turns toward the cabbage, *or* (b) a large number of turns toward the pellet. Either one of these two experimental outcomes would provide evidence against the null hypothesis and in favor of the alternative hypothesis.

Therefore, we now need two criteria—a high criterion and a low criterion.

Our goal in establishing these criteria is the same as it always has been: to keep the probability of a Type I error below 0.05. How are we going to establish these two criteria? One possibility would be to extrapolate from our previous situations. That is, we might argue that if one criterion is set at 59 or more cabbage turns, then the other criterion ought to be "symmetrical" and therefore should be set at 59 or more pellet turns (that is, 41 or fewer cabbage turns), as depicted in Figure 6-15. However, such a procedure would be based on faulty reasoning. Figure 6-15 illustrates that if the null hypothesis is actually true, then the probability of erroneously rejecting it and making a Type I error would be 0.05 (in the upper tail) plus another 0.05 (in the lower tail) for a total of 0.10, which is unacceptable. What we are actually looking for is depicted in Figure 6-16. We want to set our criteria in such a way as to only have a total of 0.025 in each of the two tails, thereby producing a *total* α of 0.05. So let's find the z corresponding to an upper tail of 0.025. Once again we argue that if there is 0.025 above the upper criterion, then there must be 0.975 below the criterion. We therefore look up a z in our z-tables such that $F(z)$ is equal to 0.975. This z turns out to be 1.96. In terms of establishing a *raw score* criterion,

$$z = \frac{x - \mu + 0.5}{5} = \frac{x - 50 + 0.5}{5} = \frac{x - 49.5}{5} = 1.96$$

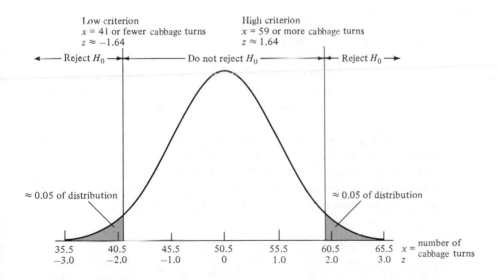

FIGURE 6-15.   A faulty way of setting up criteria for a situation involving a nondirectional alternative hypothesis. Here the total area in the two tails—the total probability of making an α error—is 0.05 + 0.05 = 0.10.

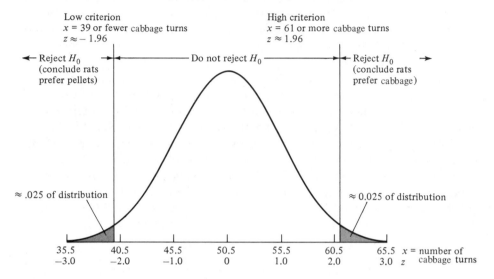

FIGURE 6-16. The correct way of setting up criteria to produce a 0.05 α-level. The area contained in each tail is a maximum of 0.025, for a total α error probability that is less than 0.05.

or

$$(1.96)(5) + 49.5 = x = 59.3$$

which we will round up to 60 to be conservative. Since 60 or fewer constitute the nonrejectory region, we see that we will require 61 or more turns toward the cabbage. By symmetry we argue that our lower criterion must be 61 or more turns toward the pellet (or fewer than 40 turns toward the cabbage). To reiterate, we will now reject the null hypothesis either if the rat makes 61 or more turns toward the cabbage or if the rat makes 61 or more turns toward the pellet. In either of these cases we will conclude that rats do not have equal preference for the two types of food. This type of test is called a *two-tailed test* in reference to the two tails of the distribution that together constitute α.

6.8 With a directional alternative hypothesis there is just one criterion that chops off 0.05 of the null hypothesis distribution. With a nondirectional alternative hypothesis there are two criteria, each chopping off 0.025 of the null hypothesis distribution.

The One-Tailed/Two-Tailed "Paradox"

Suppose that the experiment we have been discussing is performed and 59 rats turn toward the cabbage. Should we reject the null hypothesis or not? As we have seen in the last two sections, the answer to this question depends on the nature of our original alternative hypothesis. If we had a directional alternative hypothesis, then we should reject the null hypothesis; if we had a nondirectional alternative hypothesis, then we should fail to reject the null hypothesis.

This seems paradoxical. Data are data! Why should exactly the same data lead to one conclusion in one case and another conclusion in another case? To answer this question, we shall describe a hypothetical situation in which Sinbad, a psychologist, goes about the hypothesis-testing procedure improperly and gets into trouble. Let's use the same example we have been using in the previous sections and let us suppose Sinbad goes into the experiment with no preconceptions about whether the rat should prefer cabbage or rat pellets. Therefore, he should use a nondirectional alternative hypothesis and ultimately perform a two-tailed test on the data.

Suppose now that in reality the null hypothesis is true—in fact, the rat has absolutely no preference about the cabbage versus rat pellets. The probability of turning toward the cabbage would therefore be 0.5, and the number of cabbage turns should be distributed as depicted in Figures 6-12, 6-13, 6-15 and 6-16. Let's see the various ways in which Sinbad could end up making an erroneous conclusion.

First let's imagine that 59 out of the 100 rats turn toward the cabbage merely by chance. Sinbad will look at the data, say, "Aha! Rats prefer cabbage to rat pellets," and proceed on the spot to make up a theory predicting that rats prefer cabbage to rat pellets. Along with the theory Sinbad will make up the directional alternative hypothesis that the probability of turning toward the cabbage is greater than 0.5. Sinbad will therefore set a one-tailed criterion at 59 or more cabbage turns, reject the null hypothesis, publish these findings in a journal article, and conclude that his Type I error probability is less than 0.05.

However, this is not the only way in which Sinbad could go astray. Still imagining that the null hypothesis is really true, suppose that 59 rats turned toward the pellet by chance. What will Sinbad do now? He will look at the data and say, "Aha! Rats prefer rat pellets to cabbage." He will then make up a theory predicting that rats prefer rat pellets to cabbage and along with the theory he will make up an alternative hypothesis that the probability of turning toward the pellet is greater than 0.5. He will then set his criterion at 59 or more turns toward the rat pellet, will erroneously reject the null hypothesis, and will publish this result in a journal article, claiming that the probability of a Type I error is less than 0.05.

In either of these two cases, Sinbad would conclude that the probability he has made a Type I error is 0.05 or less. But he will be fooling himself. By going about things as he has done, he will have *two* ways of making an error, assuming the null hypothesis is correct. He will make a Type I error if 59 or more rats turn toward the cabbage *or* if 59 or more rats turn toward the rat pellet. Each of these two events occurs with probability of approximately 0.05, and they are mutually exclusive events, so the total probability of making an error is approximately 0.10. The improper procedure, of course, consisted of doing a one-tailed test where a two-tailed test should have been done. With a two-tailed test, Sinbad would have constructed stricter criteria (61 or more turns in either of the two directions) and only then would α really have been about 0.05.

A Shortcut

In the three hypothesis-testing examples we've discussed, we have actually included one superfluous step, that of determining what the criterion value (or criteria values) ought to be *in terms of raw scores*. More typically criteria are simply stated in terms of z-scores. As we have seen, when the α-level is 0.05, the criterion z is 1.64 in a one-tailed test and ± 1.96 in a two-tailed test. Our procedure then consists of getting an obtained raw score (in this case an obtained number of cabbage turns) from our data, translating that obtained raw score into an obtained z-score, and then comparing the obtained z-score to the criterion z-score. If the obtained z-score is more extreme than the criterion z-score, then the null hypothesis is rejected. For example suppose that we were in a directional alternative hypothesis (one-tailed) situation, and we observed that 62 rats turned toward the cabbage. In this case, we would compute an obtained z-score in the usual way:

$$z = \frac{x - \mu + 0.5}{5} = \frac{62 - 50 + 0.5}{5} = \frac{12.5}{5} = 2.5$$

We would then determine that our obtained z-score of 2.5 is more extreme than our criterion z-score of 1.64, and we would end up rejecting the null hypothesis.

> 6.9 Typically we calculate an obtained z and see whether that obtained z is more extreme than the criterion z of 1.64 (for a one-tailed test) or 1.96 (for a two-tailed test). If the obtained z is more extreme than the criterion z, then the null hypothesis is rejected.

SUMMARY

In this chapter we have discussed the most pervasive distribution that occurs in social science statistics: the normal distribution. We discussed both the nature of the normal distribution and its application in a prototypical hypothesis-testing situation. In particular,

1. The normal distribution is a continuous distribution, shaped like a bell. The normal distribution actually consists of the *family* of distributions, each member of the family corresponding to one value of μ (the mean of the distribution) and of σ^2 (the variance of the distribution).

2. Many random variables in the world are normally distributed. This is a consequence of the fact that for many random variables, any specific value consists of the sum of a series of random influences. According to the central limit theorem any random variable that produces values made up of the sum of random numbers is normally distributed. One specific and important consequence of this is that the binomial distribution can be approximated by the normal distribution if the N is large enough.

3. A very important member of the normal distribution family is the z distribution. The z distribution is a normal distribution with a mean of 0 and a variance and standard deviation of 1. Any normal distribution can be transformed into a z distribution by simply transforming the raw scores into z scores.

4. Because any area under the z distribution curve can be determined quickly and easily using tables of the z distribution, it is very easy to perform calculations involving any normal distribution.

5. The normal distribution is very pervasive in the world of hypothesis testing. A hypothesis-testing situation can involve any one of the different types of alternative hypotheses: an *exact* alternative hypothesis, a *directional* alternative hypothesis, and a *nondirectional* alternative hypothesis. Only an exact alternative hypothesis allows us to compute the probability of a Type II error.

PROBLEMS

1. Graduate Record Exam (GRE) scores are normally distributed with a mean (μ) of 500 and a standard deviation (σ) of 100.

   1. Write the equation for $p(x)$ where $p(x)$ is the probability density function for GRE scores.

   Suppose a random student takes the GRE. Compute the probabilities that x, the student's exam score, will fall in the following ranges:

   2. $x > 500$
   3. $x < 400$
   4. $x > 620$
   5. $490 < x < 530$
   6. $305 < x < 480$
   7. $530 < x < 600$
   8. $x > 460$
   9. $x < 590$

2. The time lengths of films produced by a French film company are found to be normally distributed with a mean length of 92 minutes and a standard deviation of 23 minutes. Sketch the normal curve depicting the distribution of film lengths. Find what percentage of films last

   1. Less than 65 minutes
   2. More than 77 minutes
   3. Less than 112 minutes
   4. More than 120 minutes
   5. Between 69 and 115 minutes
   6. Between 92 and 100 minutes
   7. Between 100 and 108 minutes
   8. Speculate as to *why* the films in question 4 might be normally distributed. Be specific in your speculations.

3. It is known that the amount of tartar sauce served in tartar sauce cups at Photi's Fish House is normally distributed with a mean of 50 cc and a *variance* of 25. What is the probability that a random cup of Photi's Fish House tartar sauce contains:

   1. Between 45 and 55 cc of tartar sauce
   2. Between 39 and 48 cc of tartar sauce
   3. Between 52 and 61 cc of tartar sauce
   4. Less than 43 cc of tartar sauce

4. Joe Shablotnik is growing a Merkin plant. His *Handbook of Merkin Plants* tells him that the heights of adult Merkin plants are normally distributed with a mean of 65 in. and a standard deviation of 3 in. What is the probability that when Joe's Merkin plant grows up it will be

   1. Between 64 and 67 in. tall
   2. Between 61 and 63 in. tall
   3. Between 66 and 67 in. tall
   4. Less than 64 in. tall
   5. Greater than 60 in. tall
   6. Less than 40 in. tall
   7. Less than 100 in. tall

5. The Know-Not Rope Company manufactures climbing ropes. The breaking strength of these ropes is normally distributed with a mean of 300 lbs and a standard deviation of 76 lbs.

   What is the probability that a Know-Not Rope will break if used for climbing by:

   1. A 150 lb climber
   2. A 200 lb climber
   3. A 200 lb climber with his 80 lb backpack
   4. A 150 lb climber with her 160 lb boyfriend hanging onto her
   5. A 50 lb macaque monkey

6. The lifetime of Aardvark lightbulbs is normally distributed with a mean of 10 hours and a variance of 6.

   Suppose I buy an Aardvark lightbulb. What is the probability that it will last

   1. Between 9 and 12 hours
   2. Between 7 and 8 hours
   3. More than 11 hours
   4. Fewer than 9.5 hours
   5. Exactly 10 hours

7. Consider the population of all marijuana cigarettes ("joints"). Assume that the amount of THC (the active ingredient) in a joint is normally distributed with a *mean* of 5 gm and a *variance* of 4 gm.

   1. What is the probability that a random joint will contain between 4.5 and 7.5 gms of THC?
   2. Suppose it takes at least 2 gms of THC to make a person high. What is the probability that a person will become high if he or she smokes a random joint?

8. Assume that the amount of beer served in beer mugs at Ratso's Tavern is normally distributed with a mean, μ, of 25 oz and a standard deviation of 1.6 oz. What is the probability that a beer purchased at Ratso's will consist of

   1. More than 24.2 oz?
   2. Between 24 and 25.5 oz?
   3. Exactly 25 oz?

9. Suppose that each day in the month of October the probability of rain in Seattle is 7/10. Suppose further that the weather on one day is independent of the weather on any other day.

   What is the probability that there will be *at least one* sunny day during the month of October? Answer this question in two ways: using the binomial, and the normal approximation to the binomial, distribution.

10. (Use the correction for continuity in this problem.)

    Joe Shablotnik has taken up squash. Every day for 900 days, Joe drives to the squash courts and parks in front of a parking meter. Joe never puts any money in the meter and every day the probability is 0.1 that a policeman will check the meter and give Joe a ticket.

    Over the 900 days,

    1. How many tickets does Joe expect to get?
    2. What is the *standard deviation* of the distribution of the number of tickets Joe expects to get?
    3. What is the probability that Joe will get 87 or more tickets?

    4. What is the probability that Joe will get between 95 and 100 tickets (inclusive)?

    5. What is the probability that Joe will get *exactly* 90 tickets?

11. Joe Shablotnik is shooting fish in a Penny Arcade at the fair. Each time he shoots, he wins

—a rag doll with probability 0.30
—a medallion with probability 0.20
—nothing with probability 0.50

Suppose Joe has 20 shots.

    1. What is the probability that Joe wins between 5 and 7 (that is, 5, 6, or 7) rag dolls? Compute using the binomial distribution.

    2. Compute the probability that Joe wins between 5 and 7 rag dolls using the normal approximation to the binomial *without* the correction for continuity.

    3. Recompute the probability using the correction for continuity.

    4. Suppose Joe has 50 shots. What is the probability that he wins exactly 15 rag dolls?

12. A very nasty professor decides to flunk roughly a quarter of his class on their final examination. If the exam scores are *normally distributed* with a *mean* of 75 and a *variance* of 25, what score does a student need to pass?

13. It is known that the amount of liquor drunk by college students on weekends is normally distributed with

$$\mu = 10 \text{ oz (per student)}$$
$$\sigma = 2 \text{ oz}$$

It is hypothesized that on the weekend of final exams students will drink *more* than they ordinarily would. A random student is selected, and the amount she drinks on the weekend during finals turns out to be 13.5 oz. We now wish to test the hypothesis that this is more than the average of 10.0 oz.

    1. What is the null hypothesis?

    2. What is the alternative hypothesis?

    3. Assume the null hypothesis is correct. Characterize the distribution that our observed sample of 13.5 comes from. What are the *mean, variance,* and *standard* deviation of this distribution?

    4. What is the *z*-score corresponding to our observed score of 13.5? What probability does this *z*-score correspond to?

    5. Should we reject the null hypothesis? Why or why not?

14. Suppose you are tutoring a student and decide to give him a test to find out if he's learned anything. You give him a 100-question test; each question is multiple choice, with four possible answers.

His score on the test turns out to be 30 out of 100. You want to test the alternate hypothesis that he has learned something against the null hypothesis that he has learned nothing.

1. What are four possible outcomes of your decision? Which outcome corresponds to a Type I error and which corresponds to a Type II error?
2. What is the null hypothesis?
3. What is the alternative hypothesis?
4. Assume the null hypothesis is true. How should the number of correct answers in this 100-question test be distributed? What are the mean and standard deviation of the distribution?
5. Can you reject the null hypothesis at the 0.005 alpha level?

15. Suppose you are in Las Vegas playing roulette. You decide you will play 100 times, and each time you will bet on the numbers 1–9. (Thus, if the wheel stops at 1–9, you win; if it stops at 10–36, you lose. Assume there are *only* the numbers 1–36 on the wheel). After 100 plays you have won 20 times and lost 80 times. You now wish to test the alternative hypothesis that the wheel is biased against you against the null hypothesis that the wheel is fair.

1. What are four possible outcomes of your decision? Which outcome corresponds to a Type I error, and which corresponds to a Type II error?
2. What is the null hypothesis?
3. What is the alternative hypothesis?
4. Assume the null hypothesis is true. How should the number of your wins in 100 spins of the wheel be distributed? What is the mean and standard deviation of this distribution?
5. Should you reject the null hypothesis at the .001 α-level?

16. Suppose you are doing an experiment in ESP. You sit in front of a subject with an opaque screen separating you, and you begin to throw a die. Each time you throw a die, the subject has to guess how the die came up. After 1296 throws, you stop and compute how many correct answers the subject got.
   You now wish to answer the question: Does the subject have ESP?

1. What are four possible outcomes of your decision? Which outcome corresponds to a Type I error, and which corresponds to a Type II error?
2. What is the null hypothesis?
3. What is the alternative hypothesis?
4. Assume the null hypothesis is true. Characterize the distribution of correct answers in the sample of 1296. (What are the mean and the standard deviation?)
5. You want to make sure that your probability of incorrectly rejecting

the null hypothesis is 0.05. How do you set your criterion (or criteria)?

17. It is known that 53% of the pupils in a very large school system are male. Use a two-tailed test at the 0.05 significance level to test the hypothesis that a *random sample* of 450 pupils was obtained if the number of males it contains is (a) 180, (b) 168, (c) 220, (d) 210, (e) 225, (f) 300.
    (*Hint:* Remember that for a binomial distribution mean = Np and standard deviation = $\sqrt{Npq}$.)

18. A student is taking a test with 1000 questions; each question has four alternatives. Unfortunately the student knows absolutely nothing about the material, so she guesses randomly on each question. What is the probability that her proportion of correct answers will be between 0.26 and 0.35?

19. A thousand babies are born in the Sheikdom of Shirk. Suppose that the probability is 0.5 that any given baby is a boy.

    1. What is the probability that exactly 500 babies will be boys?
    2. What is the probability that between 495 and 510 babies will be boys? (Use the correction for continuity.)
    3. Suppose that 470 of the babies are boys. Test the null hypothesis that p, the probability of a boy, is 0.5 against the alternative hypothesis that $p \neq 0.5$.

20. The Muy Munchy Mixed Nut Company has three types of nuts—cashews, peanuts, and pecans—in its cans of mixed nuts. There are equal quantities of each type of nut, and they are thoroughly mixed.
    Joe Shablotnik draws 200 nuts from a Muy Munchy can. Compute the following probabilities.

    1. Joe will get between 68 and 75 (inclusive) pecans.
    2. The sum of Joe's peanuts and cashews will be 133.
    3. Suppose Joe draws three nuts. What is the probability that he will get one cashew, one peanut, and one pecan?

21. Suppose that you are buying banjo strings for your five-string banjo. You know that the manufacturer says each banjo string has a mean breaking tension of 12 lbs with a standard deviation of 1 lb. You also know that your strings will be under the following tensions:

    String 1: 11.26 lbs
    String 2: 11.15 lbs
    String 3: 10.72 lbs
    String 4: 10.72 lbs
    String 5: 10.72 lbs

    1. What is the probability that each of the five strings will break?

2. Assuming that strings break independently, what is the probability that *at least* one string will break?
3. What is the probability that *all* the strings will break?

22. Gail Goody comes in for her statistics final exam knowing 80% of the material. The exam consists of n questions, and thus on any given question Gail has a probability of 0.8 of being correct. Naturally the ideal outcome of the exam is for Gail to achieve a score of around 80% so that her performance will perfectly reflect her knowledge of the course material.

Suppose n = *10*

1. What is the probability that Gail will get *exactly* 80% on the exam?
2. What is the probability that Gail will get between 70% and 90% on the exam?

Suppose n — *100*

1. What is the probability that Gail will get exactly 80% on the exam?
2. What is the probability that Gail will get between 70% and 90% on the exam?
   (*Hint:* If you use the normal approximation, use the correction for continuity.)

7 Sampling Distributions and Hypothesis Testing with Means

Lately we have been discussing two specific types of theoretical distributions: binomial distributions and normal distributions. We have dealt with the mathematics of these distributions, their principal characteristics, and most importantly how they are applied in hypothesis-testing situations. In this chapter we are going to focus on a type of distribution called a *sampling distribution,* which is derived from some original probability distribution. We'll first describe what a sampling distribution is and how certain sampling distributions behave. We will then go on to discuss the role of sampling distributions in the hypothesis-testing process.

The concept of a sampling distribution is closely related to the issue of populations versus random samples drawn from populations (alluded to in Chapter 5). We will therefore begin with a discussion of populations and samples.

POPULATIONS AND SAMPLES

A social scientist is typically concerned with making conclusions that are applicable to some general population—for example to the population of all people in the United States or perhaps to the population of all people in the world. As an illustration imagine that we have developed a computer-assisted instruction (CAI) program designed to improve a student's mathematical ability. Having developed this program, we naturally want to market it—but before doing so it is necessary to test the program to see whether it does in fact raise math ability. One possible way to test the program would be to administer it to all people in the population to which it will ultimately be applied (let's say to all college students in the United States) and then test the math ability of these students by giving them the math Graduate Record Exam (GRE).[1] Since we

---

[1] For purposes of this discussion we can assume that math GRE scores are normally distributed in the population of U.S. college students with a mean $\mu = 500$ and a variance $\sigma^2 = 10,000$.

know that the population mean GRE score had been 500, we could see if this mean had been raised. However, such a procedure is clearly preposterous—it would be impossibly expensive and time consuming. A much more efficient strategy would be to administer the program to a random sample of students drawn from the U.S. college population. We could then observe whether the average GRE score of the students *in that sample* was higher than 500. The results obtained from the sample could then be used as a basis for *inferring* the effects of the program on the population as a whole. In short we do experiments on samples and then use the sample results to make inferences about the population from which the sample is drawn.

Three Types of Probability Distributions

Let's consider some variable under consideration—in this case math GRE score. There are three types of probability distributions with which we will typically be concerned when we do experiments involving this (or any) variable: a *population* (or parent) distribution, a *sample* distribution, and a *sampling* distribution. We have dealt with population and sample distributions in one form or another in past chapters, so their description will be familiar to us, but the concept of a sampling distribution will be new.

The population distribution.   As the term suggests, the population distribution is simply the probability distribution of the random variable in the population. We have purposely chosen to deal with math GRE as an example because we happen to know the population distribution of this random variable. As noted, it is a normal distribution with

$$\text{mean} = \mu = 500$$
$$\text{variance} = \sigma^2 = 10,000$$
$$\text{standard deviation} = \sigma = 100$$

As we discussed in the last chapter, the mean, the variance, and the standard deviation are all *parameters* of the distribution. Parameters describe the population distribution of some random variable and, as the above notation suggests, they are designated by *Greek letters*.

The sample distribution.   Suppose that we take a random sample (say, a sample of $n = 9$) people from the population and measure the math GRE score of each person in the sample. We then calculate the mean, the variance, and the standard deviation of the math GRE scores from the sample. Suppose that for our example

$$\text{sample mean} = M = 513$$
$$\text{sample variance} = S^2 = 7569$$
$$\text{sample standard deviation} = S = \sqrt{7569} = 87$$

The mean, the variance, and the standard deviation of the sample are examples of *sample statistics*. Sample statistics are like population parameters, except they describe samples instead of populations. Notice that sample statistics are designated by *Roman letters* rather than by Greek letters.

Let's pause at this point to pull together some loose ends that have been lying around from previous chapters. You probably noticed in the past that when we discussed means, variances, and standard deviations, we were somewhat ambiguous about the terms we used to designate them. Sometimes we used Roman letters such as M and S (as was the case in Chapter 3 on descriptive statistics) and other times we used Greek letters such as μ and σ (as has been the case in the last few chapters). As we have just seen, however, the proper use of Greek versus Roman letters is actually quite clear. Greek letters represent parameters used to describe population distributions, whereas Roman letters represent descriptive statistics used to designate distributions of random samples drawn from the population.

The sampling distribution.   Let's suppose we are drawing random samples from the population of math GRE scores and for each sample we consider some particular sample statistic—for example, S, the sample standard deviation. We now ask: How is this sample statistic distributed *over samples*? That is, we imagine ourselves drawing huge numbers of samples, each sample of some size (for example, of size $n = 9$) and measuring the standard deviation of each sample. Taking the sample standard deviation as the random variable under consideration, we consider what the probability distribution of this random variable would be. In general we would guess that the standard deviation of each sample would be somewhere around 100—the standard deviation of the population. However, we also intuit that, because of random fluctuation, not every sample will have a standard deviation of exactly 100. Some sample standard deviations will be greater than 100; other sample standard deviations will be less. In short sample standard deviations will form some distribution. As it happens, the distribution of sample standard deviations that we have just described looks like the distribution depicted in Figure 7-1. By looking at this figure, we can make some statements about the sample standard deviation—for example, we can see that the probability is somewhere around 0.2 that any given sample standard deviation will be 107 or greater. Or we can say that the probability is about 0.4 that any given sample standard deviation will be 87 or lower. And so on. In other words knowing this distribution, we can make fairly precise statements regarding our expectations about what sample standard deviations will be and how deviant any given sample standard deviation is.

The distribution of some sample statistic *over samples* is called a *sampling distribution*. Many different types of sampling distributions stem from the original population distribution, one corresponding to a given sample statistic along with a given sample size. In the example we just depicted, we have considered the sampling distribution of sample standard deviations for samples of size $n = 9$. We could also talk about the sampling distribution of sample

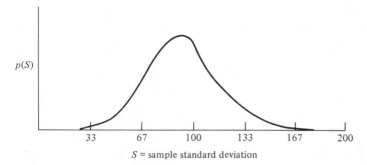

FIGURE 7-1.   An example of a sampling distribution. This is the sampling distribution of sample standard deviations for random samples of size $n=9$. The population (parent) distribution is assumed to be normal with a mean $\mu = 500$ and a standard deviation $\sigma = 100$.

variances—that is, we could inquire as to the way in which sample variances would be distributed over samples. Or we could talk about the sample distribution of sample means, or the sampling distribution of sample medians, or whatever strikes our fancy.

Relationship of These Distributions to Data

We would now like to clarify a point that often leads to confusion. To do this, let's consider the distribution of areas (say, in square centimeters) of all oak leaves in the United States. Without doing anything, we know that this distribution exists and that it has some mean μ and variance σ^2. We may not know what μ and σ^2 are, but we know they are both *something*. The point we want to make is that *a population distribution along with the parameters of this distribution do not depend on the collection for data for their existence.*

The situation is quite different when we consider a sample distribution. A sample *is* data, and a sample distribution along with sample statistics such as M and S^2 exist only in the presence of data. Thus we can see that a population distribution corresponds roughly to what we characterized as a theoretical distribution in Chapter 2 and a sample distribution corresponds to what we characterized as an empirical distribution.

Nowhere is the crucial point. *Any sampling distribution,* like its parent population distribution, *does not depend on data for its existence.* For example, a sampling distribution for sample standard deviations of random samples of $n = 25$ oak leaf areas exists. Again we may not know the mean and standard deviation of that distribution, but they exist too. The fact that sampling distributions exist in the absence of data is less intuitively obvious than the fact that population distributions exist in the absence of data. But it is a very important fact to realize and to understand.

7.1 Three types of distributions are (a) population distributions, (b) sample distributions, and (c) sampling distributions. A sampling distribution is the distribution of a particular sample statistic for a particular sample size. (Thus, any population distribution gives rise to an infinite number of sampling distributions.) Population distributions and sampling distributions do not depend on data for their existence and have associated with them parameters (such as μ, σ) designated by Greek letters. Sample distributions require data to exist and have associated with them sample statistics (such as M, S) designated by Roman letters.

THE SAMPLING DISTRIBUTION OF SAMPLE MEANS

We have just noted that, given a population distribution and a particular sample size, we can speak of the resulting sampling distribution of any given sample statistic. As we shall see, one specific sampling distribution—the sampling distribution of sample means—is particularly useful in a number of common hypothesis-testing situations, and we will therefore consider this distribution in some detail. To do so, we once again consider all college students in the United States and the corresponding population distribution of math GRE scores. As noted above, we happen to know the parameters of this distribution, which are

$$\mu = 500$$

$$\sigma^2 = 10{,}000$$

$$\sigma = 100$$

Suppose we concern ourselves with samples of some size n, and we consider the sampling distribution of sample means. What will be the mean and the variance of this distribution?

The Mean of a Distribution of Sample Means

Let's refer to the mean of the sampling distribution of sample means as μ_M. Digression 7.1 proves that μ_M **is equal to** μ, **the original population mean.** The fact that μ_M equals μ should be intuitively reasonable. To see why this is so, imagine yourself dipping into the U.S. population, drawing out samples of students, and calculating the mean math GRE score of each sample. The first sample that you draw might have a sample mean of $M_1 = 510$. (That is, the people in this first sample happen to have a slightly higher math ability than the average.) The second sample you get might have a mean $M_2 = 495$. The third sample might have a mean $M_3 = 504$, and so on. But in general, you would expect the mean of the samples to be 500. There is no reason to think that the mean of the samples should differ systematically from the mean of the original population. Since μ_M is always equal to μ, we shall use μ and μ_M pretty much interchangeably.

Digression 7.1

Mean (Expected Value) of a Sampling Distribution of Sample Means Based on Samples of Size n

The expression for the mean of a sampling distribution of sample means is

$$\mu_M = E(M)$$

we know that M is the sum of the scores in the sample divided by the sample size, so

$$\mu_M = E\left[\frac{\sum_{i=1}^{n} x_i}{n}\right]$$

The expectation of a sum is equivalent to the sum of expectations. So we can shift E to the right side of the summation sign, or

$$\mu_M = \left(\frac{1}{n}\right) \sum_{i=1}^{n} E(x)$$

Since $E(x)$ equals μ, the population mean,

$$\mu_M = \left(\frac{1}{n}\right) \sum_{i=1}^{n} \mu$$

And since μ is a constant,

$$\mu_M = \left(\frac{1}{n}\right)(n\mu) = \mu$$

The Variance of a Distribution of Sample Means

Unlike μ, σ_M^2, the variance of the sampling distribution of sample means is dependent on n, the sample size. **Specifically,**

$$\sigma_M^2 = \frac{\sigma^2}{n}$$

and correspondingly

$$\sigma_M = \sqrt{\sigma_M^2} = \sqrt{\sigma^2/n} = \sigma/\sqrt{n}$$

That is, the variance of the sampling distribution of sample means is *directly* related to σ^2, the original population variance, but is *inversely* related to n, the sample size. The larger the size of the sample, the smaller is the variance of the resulting sampling distribution of sample means. Why should this be? Digression 7.2 provides a mathematical justification for this assertion. We will attempt to provide an intuitive view here.

Digression 7.2

Variance of a Sampling Distribution of Sample Means Based on Samples of Size *n*

We know that the general expression for the variance is the expected value of the squared difference between scores in the distribution and the mean of the distribution. In the present instance we are dealing with a distribution of means. We have already seen that the mean of this distribution is μ, the population mean. Therefore,

$$\sigma_M^2 = E(M - \mu)^2$$

which, as we have seen in the past, may be written as

$$\sigma_M^2 = E(M^2) - \mu^2 \tag{D.1}$$

Let us concentrate on the term $E(M^2)$ in Equation D.1.

$$E(M^2) = E\left(\sum_{i=1}^{n} x_i/n\right)^2$$

$$= (1/n^2)E\left(\sum_{i=1}^{n} x_i\right)^2$$

$$= (1/n^2)E(x_1 + x_2 + \cdots + x_n)^2$$

Consider now the expression $(x_1 + x_2 + \cdots + x_n)^2$: Table D-1 shows what this expression looks like when it is expanded. Specifically, we see that the expression is the sum of n^2 terms. Of these n^2 terms n of them (the diagonal in the matrix of Table D-1) are

$$x_1^2 + x_2^2 + \cdots + x_n^2$$

The other $n(n - 1)$ (off-diagonal) terms consist of two each of the cross-products $x_i x_j$ (where $i \neq j$, $i \leq n$, and $j \leq n$). Thus, $E(M^2)$ may be written as

$$E(M^2) = \frac{1}{n^2}\left[E\left(\sum_{i=1}^{n} x_i^2\right) + 2\,E\left(\sum_{i<j} x_i x_j\right)\right]$$

$$= \frac{1}{n^2}\left[\sum_{i=1}^{n} E(x_i^2) + 2\sum_{i<j} E(x_i x_j)\right]$$

(D.2)

Now we have shown in previous chapters that

$$\sigma^2 = E(x^2) - \mu^2$$

or

$$E(x^2) = \sigma^2 + \mu^2 \tag{D.3}$$

Additionally if x_i and x_j are independent (which they are presumed to be) then

$$E(x_i x_j) = E(x_i)E(x_j) = (\mu)(\mu) = \mu^2 \tag{D.4}$$

Substituting the expressions for $E(x^2)$ and $E(x_i x_j)$ from Equations D.3 and D.4 into Equation D.2,

---

TABLE D-1.   Representation of $\left(\sum\limits_{i=1}^{n} x_i\right)^2$

$$\left(\sum_{i=1}^{n} x_i\right)^2$$

$= (x_1 + x_2 + x_3 + \cdots + x_n)^2 =$

$x_1(x_1 + x_2 + x_3 + \cdots x_n)$
$+$
$x_2(x_1 + x_2 + x_3 + \cdots x_n)$
$+$
$x_3(x_1 + x_2 + x_3 + \cdots x_n)$
$\cdot\quad\quad\cdot$
$\cdot\quad\quad\cdot$
$\cdot\quad\quad\cdot$
$+$
$x_n(x_1 + x_2 + x_3 + \cdots x_n)$

$$\begin{array}{l}
x_1^2 + x_1 x_2 + x_1 x_3 + \cdots + x_1 x_n \\
+ \\
x_2 x_1 + x_2^2 + x_2 x_3 + \cdots + x_2 x_n \\
+ \\
x_3 x_1 + x_3 x_2 + x_3^2 + \cdots + x_3 x_n \\
\cdot \quad \cdot \quad \cdot \quad\quad\quad \cdot \\
\cdot \quad \cdot \quad \cdot \quad \cdot \quad\quad \cdot \\
\cdot \quad \cdot \quad \cdot \quad \cdot \quad\quad \cdot \\
+ \\
x_n x_1 + x_n x_2 + x_n x_3 + \cdots + x_n^2
\end{array}$$

Note that this matrix contains n x_i^2 terms (along the diagonal) and $n(n-1)\,x_i x_j$ terms.

$$E(M^2) = (1/n^2)\left[\sum_{i=1}^{n}(\sigma^2 + \mu^2) + 2\sum_{i<j}\mu^2\right]$$

or

$$E(M^2) = (1/n^2)[n(\sigma^2 + \mu^2) + n(n-1)\mu^2]$$
$$= (1/n^2)[n\sigma^2 + n\mu^2 + n^2\mu^2 - n\mu^2]$$
$$= \sigma^2/n + \mu^2$$

Finally, substituting the expression for $E(M^2)$ from Equation D.5 into Equation D.1,

$$\sigma_M^2 = \sigma^2/n + \mu^2 - \mu^2 \tag{D.5}$$

or

$$\sigma_M^2 = \sigma^2/n$$

Small samples.   Imagine first that we are drawing very small samples—for instance, samples of size $n = 1$. Notice that when we draw any given sample of size $n = 1$, the mean, M, of the sample is simply equal to the single score (x) we have drawn. Now we imagine ourselves drawing a large number of these little samples and plotting the mean of each sample, which of course is equivalent to plotting the distribution of the scores. The first "sample" might have a mean $M_1 = x_1 = 700$. That is, we might happen to pick a person with fairly high math ability as our first sample. Likewise our second sample might consist of a score $M_2 = x_2 = 456$; our third sample might consist of $M_3 = x_3 = 498$, and so on. Taking a very large number of samples and plotting the distribution of means (that is, of scores), we see that we are essentially plotting the original population distribution. Therefore, the variance of this sampling distribution must be equal to the variance of the original population distribution, or 10,000. The intuitive path that we have just taken to arrive at the variance of this particular sampling distribution coincides very nicely with the result we would get if we were to compute the variance mathematically. Since n, the sample size, is 1, we see that

$$\sigma_M^2 = \frac{\sigma^2}{n} = \frac{10,000}{1} = 10,000 = \sigma^2$$

Bigger samples.   Now suppose we were to choose larger samples, say, samples of size $n = 25$. If the mean of our first sample turned out to be 700, would we be surprised? The answer ought to be yes; it would be quite shocking. Having picked 25 random people, we might expect some of the scores to be higher than 500, but we would likewise expect other scores to be lower than 500. We would therefore expect the *mean* of all 25 scores in the sample to be somewhere fairly close to the population mean of 500. If a sample of 25 people turned out to have a mean of 700, this would indicate we must have miraculously picked 25 people *all* of whom had very high math ability. In our previous example it wouldn't have surprised us to find a sample mean (that is, one score) of 700. However, it would surprise us to find a truly random sample as large as 25 people with a mean math GRE score of 700.

Close clustering is small variance.   A way of summarizing the above intuitions is to say that we expect any given large sample mean to be rather close to the population mean. With a sample of 25 people a sample mean of 496, 503, or 489, or even 525 would not be particularly surprising. Recall that to the extent that scores in a distribution are expected to be close to (clustered around) the mean of the distribution, that distribution has a small variance. Therefore, to say that we expect the mean of a sample of 25 people to be close to the population mean is to say that *the variance of the sampling distribution of sample means is relatively small.* And in fact when we compute the variance and standard

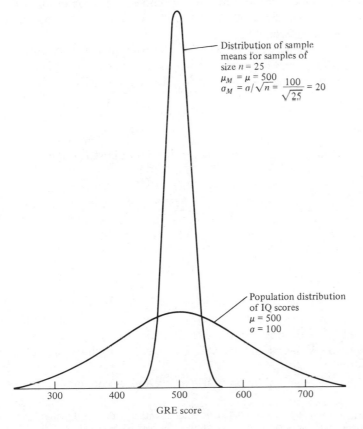

Distribution of sample means for samples of size $n = 25$

$\mu_M = \mu = 500$
$\sigma_M = \sigma/\sqrt{n} = \dfrac{100}{\sqrt{25}} = 20$

Population distribution of IQ scores
$\mu = 500$
$\sigma = 100$

300   400   500   600   700

GRE score

FIGURE 7-2.   Relationship between a population distribution and a sampling distribution of sample means based on size $n = 25$. The variance of the sampling distribution is smaller by a factor of 25; thus, the standard deviation is smaller by a factor of 5. The means of the two distributions are the same.

deviation of the sampling distribution of sampling means with samples of size 25, we would discover that

$$\sigma_M^2 = \sigma^2/n = 10,000/25 = 400$$

and

$$\sigma_M = \sigma/\sqrt{n} = 100/\sqrt{25} = 100/5 = 20$$

So we would see that when the sample size is 25, the resulting sampling distribution of sample means is still a normal distribution with a mean $\mu = 500$ but now with a standard deviation of 20. Figure 7-2 shows what this distribution looks like and how it relates to the original population distribution.

Huge samples.   Finally let's consider samples of 4 million people per sample. Note that such a sample would encompass the entire U.S. student population. Therefore, the mean math GRE score of a 4-million-person sample would have to be 500, the population mean. Since any time we choose a sample of this size, the sample mean would be the same (500), the variance of the sampling distribution of sample means would be *zero*.[2]

We see therefore that σ_M^2, the variance of the sampling distributions of sample means, ranges from σ^2, the population variance (when the sample size is 1), to zero (when the sample size is equivalent to the population size).

Digression 7.3

Relationship of a Sampling Distribution to a Binomial Distribution

Consider the distribution of GRE scores we have been discussing. In the population there are scores such as 495, 520, 500, 603, and so on. The random variable under consideration assigns one of these numbers to each person. We have seen that the population mean of these scores is 500 and the population variance is 10,000.

The Binomial Population

Now let us consider a binomial distribution. Suppose for example that the random variable now under consideration assigns a 1 to each person if the person is left-handed and a 0 if the person is right-handed. Suppose further that there are N people in the entire population and that k of them are left-handed.

Table D-2 provides a derivation of the mean and variance of such a

---

[2]When we compute σ_M^2, it is actually $10,000/4,000,000 = 0.0025$ rather than zero. The discrepancy stems from the assumption that the population size is infinite. The population sizes that we deal with are generally not infinite, but they are large enough so that for all practical purposes they can be considered to be infinite.

TABLE D-2. Representation of the Mean and Variance of a Population of Binomial Scores (1s and 0s)

Score (x)	x^2
$x_1 = 1$	$x_1^2 = 1$
$x_2 = 0$	$x_2^2 = 0$
$x_3 = 0$	$x_3^2 = 0$
$x_4 = 1$	$x_4^2 = 1$
$x_5 = 0$	$x_5^2 = 0$
.	.
.	.
.	.
$x_N = 1$	$x_N^2 = 1$
$\Sigma x = k$	$\Sigma x^2 = k$
population	population

Mean: $= k/N = p$ (by notational convention)
Variance: $\sigma^2 = \Sigma x_i^2/N - \mu^2$
$\sigma^2 = k/N - \mu^2$
$= p - p^2 = p(1 - p) = pq$

binomial population. As we have constructed the distribution, the mean is k/N. By definition this mean (in this case the proportion of left-handers in the population) is p. It is also shown that the variance of the population distribution is

$$\sigma^2 = pq$$

Samples from a Binomial

What we have been concerned with in the past, however, are *samples* of size n taken from a binomial population. Let's consider the mean of such a sample. From what we have just discussed about the sampling distribution of sample means, we expect that the mean of the sample should be

$$E(M) = \mu = p$$

and the variance of the sample mean should be

$$\sigma_M^2 = \sigma^2/n = pq/n$$

which is another way of showing what we asserted in Chapter 6 about the mean and variance of a binomial distribution: when we choose a sample of size n from a binomial population, we expect that the proportion of successes we get to be distributed with a mean of p and a variance of pq/n.

Calculating Probabilities Involving Means

The sampling distributions with which we are concerned in this chapter arise from normal parent (population) distributions. The sampling distributions of sample means will be normal distributions as well, and we will be able to use sampling distributions of sample means to calculate various probabilities involving sample means using exactly the same techniques we described in the last chapter. Let's just go through a few brief examples of how such calculations are performed. We will stick with our convenient math GRE score example, and we will suppose that we intend to dip into the U.S. population, pull out a random sample of $n = 100$ students, and measure the math GRE score of all the students in the sample. What would we expect M, the mean score of the sample, to be?

To begin with, let's calculate the mean, μ_M, the variance, σ^2_M, and the standard deviation, σ_M, of the sampling distribution of sample means that would result. In particular, we know that

$$\mu_M = \mu = 500$$

$$\sigma^2_M = \sigma^2/n = 10{,}000/100 = 100$$

and

$$\sigma_m = \sigma/\sqrt{n} = 100/\sqrt{100} = 100/10 = 10$$

So we see that the sample mean we will obtain will come from a normal distribution whose mean, μ, is equal to 500 and whose standard deviation, σ_M, is equal to 10. Suppose that we now ask questions like: What is the probability that the mean of the sample will be greater than 513? What is the probability that the mean of the sample will be between 490 and 505? And so on. Using what we know about the normal distribution, we can answer these questions without difficulty. Figure 7-3 shows what the sampling distribution of sample means would look like along with the areas under the distribution corresponding to the probabilities about which we have inquired. We have also included the z-scale, since calculation of probabilities from a normal distribution always involves an initial transformation to z-scores.

Let's first compute the probability that our sample mean will be greater than or equal to 513—that is, the probability $p(513 \leq M \leq \infty)$. To do this, we first establish the xs (or in this case Ms) that constitute our limits:

$$M_1 = 513$$

and

$$M_2 = +\infty$$

and from these Ms we compute corresponding z-scores:

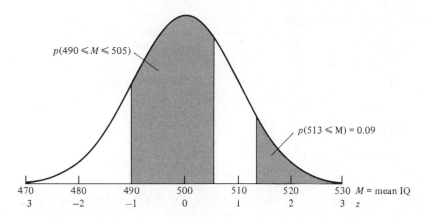

FIGURE 7-3.   Computation of various probabilities based on a sampling distribution of sample means uses the same procedures as computation of probabilities from any other normal distribution.

$$z_1 = \frac{M_1 - \mu}{\sigma_M} = \frac{513 - 500}{10} = 1.30$$

$$z_2 = \frac{M_2 - \mu}{\sigma_M} = \frac{+\infty - 500}{1.5} = +\infty$$

We get our $F(z)$s from the z distribution tables in the usual way:

$$F(z_1) = F(1.30) = 0.90$$

$$F(z_2) = F(+\infty) = 1.00$$

and finally, the probability that we wish to compute is

$$p(102 \le M \le +\infty) = p(1.33 \le z \le +\infty)$$

$$= F(+\infty) - F(1.33)$$

$$= 1.00 - 0.90$$

$$= 0.10$$

Therefore, with a probability of 0.10, the mean GRE score of 100 randomly chosen students will be 513 or higher. To compute the probability that the sample mean will fall between 490 and 505, or $p(490 \le M \le 505)$, we proceed in an analogous fashion. We first establish our limits,

$$M_1 = 490$$

$$M_2 = 505$$

and our corresponding z-scores,

$$z_1 = \frac{M_1 - \mu}{\sigma_M} = \frac{490 - 500}{10} = \frac{-10}{10} = -1.00$$

$$z_2 = \frac{M_2 - \mu}{\sigma_M} = \frac{505 - 500}{10} = \frac{5}{10} = 0.50$$

To obtain $F(z)$s,

$$F(z_2) = F(0.50) = 0.69$$

$$F(z_1) = F(-1.00) = 1 - F(1.00) = 1 - 0.84 = 0.16$$

And finally to compute the probability (area under the curve) in question:

$$p(490 \leq M \leq 505) = p(-1.00 \leq z \leq 0.50)$$

$$= F(0.50) - F(-1.00)$$

$$= 0.69 - 0.16$$

$$= 0.53$$

Thus, the probability is 0.53 that the mean math GRE score of 100 randomly chosen people will fall between 490 and 505.

7.2 The mean, M, of a sample of size n will come out of a sampling distribution of sample means for samples of size n whose parameters are:

$$\mu_M = \mu$$

$$\sigma_M^2 = \sigma^2/n$$

$$\sigma_M = \sigma/\sqrt{n}$$

HYPOTHESIS TESTING WITH MEANS

Having spent the last section discussing various characteristics of the distribution of sample means, let's now go on to discuss how such a distribution is used in hypothesis testing. We shall continue using the math GRE score as our dependent variable.

CAI Programs: Testing a Mean Against a Constant

Suppose we're research psychologists for the Acme Computer Software Company. One day the company president informs us that Acme has developed a new Computer-Assisted Instruction program called SMARTIUM which is supposed to raise math GRE scores. As yet, however, SMARTIUM is untested; thus, our job is to check whether or not SMARTIUM really does raise math GRE scores. So we perform the following experiment: We pick a random sample of 25 students from the country and administer SMARTIUM to each. We then administer the math GRE and measure each student's score. Let's refer to the mean score of the sample as M. Suppose M turns out to be 540.

What are we to conclude about SMARTIUM as a result of this experiment? Initially we note that the sample mean of 540 is higher than the population mean of 500. We are therefore somewhat inclined to believe that SMARTIUM is effective in terms of raising math ability of students in the population from which the sample was drawn. However, there are other possibilities which might have led to the sample mean of 540—for example, the 25 students in our sample might just by chance be better than average in terms of math ability. Our problem is the same that we have faced in previous hypothesis-testing situations: we want to be able to ascertain whether the result of our experiment is a "real" result (a result due to the program and reliably generalizable to the entire population) or whether the result we obtained is merely due to chance factors. This brings us to our hypothesis-testing steps.

Null and alternative hypotheses.   Our null hypothesis is (as usual) that the independent variable (in this case administration of SMARTIUM) has no effect on the dependent variable (in this case math GRE score). That is to say,

> H_0: Administering SMARTIUM has no effect on math ability. The math GRE scores from the sample of people who have been administered SMARTIUM come from a population with a mean $\mu = 500$.

The alternative hypothesis is (as always) that the independent variable does have an effect on the dependent variable, or in this case,

> H_1: Administering SMARTIUM raises math ability. The math GRE scores from the sample of people who have been administered SMARTIUM come from a population with a mean $\mu > 500$.

Choosing an appropriate summary score.   As usual we want to obtain from our experiment some summary score that will reflect the degree to which we believe the alternative as opposed to the null hypothesis is true. A reasonable score to use in this experiment would simply be the sample mean, M. To the extent that M is higher than the population mean of 500, we tend to think that the program improves math ability, and we tend, therefore, to want to reject the null hypothesis. On the other hand to the extent that M is close to 500, we tend to think that the program did not raise math ability, and we would therefore tend to *not* want to reject the null hypothesis.

Naturally, however, we are somewhat worried about erroneously rejecting the null hypothesis. For suppose that SMARTIUM really has no effect on math ability (that the null hypothesis were really true), but we erroneously concluded that SMARTIUM did have an effect. In this case SMARTIUM would go into production, but it would not work. Acme would have wasted large sums of money. As if this weren't enough, the Federal Trade Commission would probably swoop down to charge fraud and false advertising; the president of the company would go to jail; we would be fired, and so on. In short many unpleasant things would result. So we want to take care to ensure that the probability of a Type I error is low, and in this case we will consider the traditional value of 0.05 to be sufficiently low.

Distribution of the summary score.   The next step in the hypothesis-testing procedure is to determine how the sample mean is distributed if the null hypothesis is true. This should not be difficult for us, since we have just seen in the last section how sample means are distributed. In this particular case if the null hypothesis is true,

$$\mu_M = \mu = 500$$

$$\sigma_M^2 = \sigma^2 n = \sigma^2/25 = 10{,}000/25 = 400$$

and

$$\sigma_M = \sqrt{\sigma_M^2} = \sqrt{400} = 20$$

A criterion.   Our next step is to establish some criterion sample mean which we will use as a cutoff for rejecting or not rejecting H_0. As noted, we wish to establish this criterion such that if the null hypothesis is true, then the obtained sample mean will exceed the criterion (and cause us to erroneously reject H_0) with a probability of only 0.05. Note that we have a directional alternative hypothesis here—that is, if SMARTIUM does anything at all, we expect it to *improve* math ability. This means that we should set up a one-tailed criterion. Figure 7-4 shows what the distribution of sample means would look like if the null hypothesis were true. Along with this distribution we have included the appropriate criterion that chops off the upper 5% off the distribution. As always the criterion z-score is 1.64.

An obtained z-score.   To determine how deviant our obtained mean is under the assumption that H_0 is true, we compute our obtained z-score in the usual way:

$$\text{Obtained } z = \frac{M - \mu}{\sigma_M} = \frac{540 - 500}{20} = 2.0$$

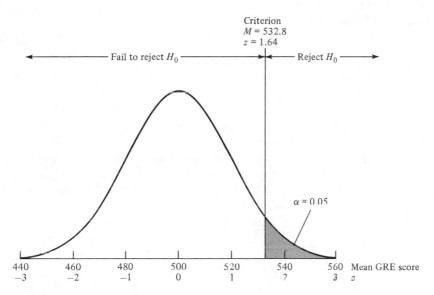

Criterion
$M = 532.8$
$z = 1.64$

◄─────── Fail to reject H_0 ───────►│◄─── Reject H_0 ───►

$\alpha = 0.05$

| 440 | 460 | 480 | 500 | 520 | 540 | 560 | Mean GRE score |
| −3 | −2 | −1 | 0 | 1 | ? | 3 | z |

FIGURE 7-4. Distribution of sample means if H_0 is true. The criterion z of 1.64 chops off the upper 5% of the distribution. This criterion z corresponds to a criterion mean of 532.80.

A comparison. It's easy to see that our obtained z-score of 2.0 is more extreme than our criterion z-score of 1.64. We therefore reject the null hypothesis and conclude that SMARTIUM really does raise math ability.

To summarize, we have gone through precisely the same hypothesis-testing steps that we formulated at the end of the last chapter. The only difference between that situation and the present one is that the summary score we are presently using is a sample mean (of a sample of 25 people). Whereas previously we were using a single score (the number of rats turning toward the cabbage). In both situations, however, the central issue has been the determination of how the summary score is distributed assuming the null hypothesis to be true. In the last chapter the score was distributed binomially. In the present situation the score, a sample mean, comes from a sampling distribution of sample means which, in turn, is a normal distribution. In either case knowledge of how the summary score is distributed if H_0 is true allows us to determine how deviant our obtained summary score is.

7.3 In any hypothesis-testing procedure we choose a summary score that reflects whether H_1 as opposed to H_0 is true, and we determine the distribution of that summary score assuming H_0 to be true.

Competing CAI Programs: Testing the Difference Between Two Means

Suppose we're still research psychologists at Acme, but we face a somewhat different problem. It's still true that Acme has developed a program called SMARTIUM which is intended to raise math ability; this time, however, the plot has a new twist. Company spies have reported that Acme's arch-rival, Consolidated Software, has come out with a program to compete with SMARTIUM. This competing program is referred to as "PROGRAM X." Now Acme's president calls us in and says, "PROGRAM X is being developed to compete with SMARTIUM. It's important for us to know which, if either, program is better—that is which, if either, improves math ability more. Your mission is to investigate this question."

We now manage to get a copy of PROGRAM X as well as a copy of SMARTIUM. We find 64 random students from the U.S. student population and randomly divide them into two groups of 32 students apiece. We administer SMARTIUM to the first group and PROGRAM X to the second group. We then measure the math GRE score of all 64 people in order to see which group has the higher mean score. Suppose that the mean for the SMARTIUM group is

$$M_S = 600$$

and the mean for the PROGRAM X group is

$$M_X = 533$$

A comparison of these two means appears to indicate that SMARTIUM has won in the sense that SMARTIUM has apparently elevated the mean score by an average of 67 points more than has PROGRAM X. Once again, however, we want to be careful before concluding that SMARTIUM is really better than PROGRAM X. The difference between the two means could still arise from chance factors. For example, it may be that the random sample of SMARTIUM students just happened to start off with better math ability than did the PROGRAM X students. Our task, once again, is to determine whether we can *reliably* conclude that SMARTIUM and PROGRAM X differ in terms of the degree to which they improve math ability. Let's see how our hypothesis-testing steps will apply in this particular situation.

Formulation of null and alternative hypotheses.   The null and alternative hypotheses state respectively that the independent variable (in this case, which program) will not or will have an effect on the dependent variable (in this case, math GRE score). Specifically the null hypothesis is:

H_0: Which program is administered has no effect on how much math ability is improved. The math GRE scores of students who are administered SMARTIUM come from the same distribution as do the math GRE scores of students who are administered PROGRAM X. That is, $\mu_S = \mu_X$ where μ_S is the population mean GRE score of all students administered SMARTIUM and μ_X is the population mean GRE score of all students who have been administered PROGRAM X.

The alternative hypothesis is:

H_1: Which program is administered does have an effect on how much math ability is improved—that is, $\mu_S \neq \mu_X$.

The choice of a summary score.   Now we need some kind of summary score that reflects the extent to which we believe the alternative as opposed to the null hypothesis to be true. In this situation a reasonably useful summary score would be the difference between the mean of the SMARTIUM group and the mean of the PROGRAM X group, $(M_S - M_X) = 600 - 533 = 67$. To the extent that this score is large (either positively or negatively), we would tend to believe the alternative hypothesis is true. To the extent that the score is near zero, we would tend to believe the null hypothesis to be true.

How is the score distributed if the null hypothesis is true?   Now we have a summary score. This score happens to be a difference score, but it is still a score; our next task is to determine how it is distributed if the null hypothesis is true.

To do this, we must briefly explore the topic of *distributions of difference scores.* Suppose in particular that we have two distributions, A and B, as depicted in Figure 7-5. Distribution A has mean, μ_A, variance, σ_A^2, and standard deviation, σ_A. Likewise Distribution B has mean, μ_B, variance, σ_B^2, and standard deviation, σ_B. Now let's imagine that we form a new distribution in the following way. We randomly dip into Distribution B and pick a score which we call x_B. We then dip into Distribution A and pick a score which we call x_A. We then take the difference between these two scores, $(x_B - x_A)$. We repeat this process numerous times, getting a large number of $(x_B - x_A)$s, and we ask the question: What is the distribution of these difference scores? This distribution is depicted at the bottom of Figure 7-5, and we should be able to guess a few things about it. First let's consider the *mean,* μ_{B-A}, of this distribution. We can assume that, in general, the average score we get from Distribution B is its mean, or μ_B. Likewise, the average score we get from Distribution A should be its mean, or μ_A. Therefore, on the average we should get a difference score that is equal to the mean of Distribution B minus the mean of Distribution A, or $\mu_{B-A} = \mu_B - \mu_A$. The mean, μ_{B-A}, of the distribution of difference scores is thus

$$\mu_{B-A} = E(x_B - x_A) = E(x_B) - E(x_A) = \mu_B - \mu_A$$

What about the variance of this distribution of difference scores? It turns

Original distributions

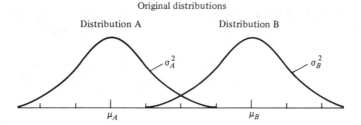

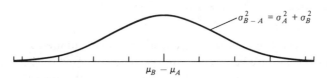

Distribution of difference scores ($x_B - x_A$)

FIGURE 7-5.   Illustration of a distribution of difference scores. The original distributions, A and B, are assumed to have means μ_A and μ_B and variances of σ_A^2 and σ_B^2. The resulting distribution of difference scores $(X_B - X_A)$s has a mean of $\mu_B - \mu_A$ and a variance of $\sigma_A^2 + \sigma_B^2$.

out that this variance σ_{B-A}^2 of a distribution of difference scores, is equal to

$$\sigma_{B-A}^2 = \sigma_B^2 + \sigma_A^2$$

The mathematical proof that the variance of the distribution of difference scores is equal to the sum of the variances of the two original distributions[3] is given in Digression 7.4. The best intuitive grasp we can give on this fact is to ask you to recall that any score we pluck from Distribution A has some variance associated with it. Likewise, a score obtained from Distribution B has some variance associated with it. When we combine these two scores by subtracting them (or adding them, for that matter), then the resulting score has variability that arises from both of the original scores. Therefore, the variability of this combined (difference) score should be greater than the variability of either of the two original scores.

---

[3]This assumes that the two scores constituting the difference scores are *uncorrelated*. Correlation is a topic we defer to a subsequent chapter. However, those familiar with it may be interested to know that the complete formula is

$$\sigma_{B-A}^2 = \sigma_B^2 + \sigma_A^2 - 2r\sigma_B\sigma_A$$

where r is the population correlation of pairs of scores X_A and X_B. When the two scores are uncorrelated, $r = 0$.

Digression 7.4

Variance of a Distribution of Difference Scores

We start with one of our normal expressions for variance:

$$\sigma^2_{B-A} = E[(x_B - x_A) - \mu_{B-A}]^2$$

$$= E(x_B - x_A)^2 - \mu^2_{B-A}$$

Letting $\mu_{B-A} = \mu_B - \mu_A$ (as we have proved already)

$$\sigma^2_{B-A} = E[(x_B - x_A)^2 - (\mu_B - \mu_A)^2]$$

Expanding both terms,

$$\sigma^2_{B-A} = E[(x_B^2 - 2x_A x_B + x_A^2) - (\mu_B^2 - 2\mu_A\mu_B + \mu_A^2)]$$

$$= E(x_B^2) - 2E(x_A x_B) + E(x_A^2) - \mu_B^2 + 2\mu_A\mu_B - \mu_A^2 \qquad (D.1)$$

If x_A and x_B are independent, which they are presumed to be, then

$$E(x_A x_B) = E(x_A)E(x_B) = \mu_A\mu_B \qquad (D.2)$$

Substituting the expression $E(x_A x_B)$ from Equation D.2 into Equation D.1,

$$\sigma^2_{B-A} = E(x_B^2) - 2\mu_A\mu_B + E(x_A^2) - \mu_B^2 + 2\mu_A\mu_B - \mu_A^2$$

$$= [E(x_A)^2 - \mu_A^2] + [E(x_B^2) - \mu_B^2] \qquad (D.3)$$

Notice that the two terms in the brackets of Equation D.3 are

$$E(x_A^2) - \mu_A^2 = \sigma_A^2$$

$$E(x_B^2) - \mu_B^2 = \sigma_B^2$$

Therefore

$$\sigma^2_{B-A} = \sigma_A^2 + \sigma_B^2$$

In any case knowing these facts about distributions of difference scores puts us in the position of being able to calculate what the distribution of our present summary score—$(M_S - M_X)$—should look like, assuming that the null hypothesis is true. Consider first the mean of this distribution, $\mu_{M_S-M_X}$. If the null hypothesis is true, then M_S and M_X should come from the same distribution. When we are taking differences of two scores drawn from the same distribution, the mean of the resulting distribution of difference scores should be zero. **That is, when two distributions are the same, the mean difference score is**

$$\mu_{M_S-M_X} = 0$$

(This should be intuitively obvious.) Now the variance of the distribution of difference scores must be

$$\sigma^2_{M_S - M_X} = \sigma^2_{M_S} + \sigma^2_{M_X}$$

We know the distribution from which M_S comes is simply the sampling distribution of sample means based on sample size of $n = 32$. Therefore,

$$\sigma^2_{M_S} = \sigma^2/n = 10{,}000/32 = 312.5$$

Exactly the same argument can be made for the distribution from which M_X comes. M_X is also a mean coming from M, the sampling distribution of sample means based on samples of size $n = 32$. Therefore,

$$\sigma^2_{M_X} = \sigma^2/n = 10{,}000/32 = 312.5$$

And finally,

$$\sigma^2_{M_S - M_X} = \sigma^2_{M_X} + \sigma^2_{M_X} = 312.5 + 312.5 = 625$$

$$\sigma^2_{M_S - M_X} = \sqrt{\sigma^2_{M_S - M_X}} = \sqrt{625} = 25$$

So we see that if the null hypothesis is true—if program type has no effect in terms of how much math ability is improved—then the difference between mean GRE scores of two 32-person samples should be distributed with a mean, $\mu_{M_S - M_X}$, of 0 and a standard deviation, $\sigma_{M_S - M_X}$, of 25, as shown in Figure 7-6.

7.4 When testing two means against each other, the appropriate summary score is $(M_1 - M_2)$. This summary score comes from a distribution of summary scores with parameters.

$$\mu_{M_1 - M_2} = \mu_{M_1 - \mu M_2}$$

and

$$\sigma^2_{M_1 - M_2} = \sigma^2_{M_1} + \sigma^2_{M_2}$$

and

$$\sigma_{M_1 - M_2} = \sqrt{\sigma^2_{M_1} + \sigma^2_{m_2}} = \sqrt{\frac{\sigma^2}{n_1} + \frac{\sigma^2}{n_2}}$$

Two notes: First $\mu_{M_1} - \mu_{M_2}$ is typically O assuming H_0 is true. Second when

$$n_1 = n_2 = n,$$

$$\sigma_{M_1 - M_2} = \sqrt{\frac{\sigma^2}{n} + \frac{\sigma^2}{n}} = \sqrt{\frac{2\sigma^2}{n}} = \sigma\sqrt{\frac{2}{n}}$$

How extreme is our score? We now determine how deviant our obtained difference score is, assuming that the null hypothesis is actually true. Again to

Distribution M

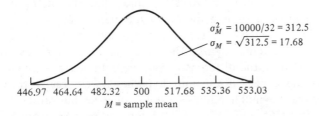

Sampling distribution of sample means: assuming H_0 is true, this is the distribution of both M_s and of M_x

$\sigma_M^2 = 10000/32 = 312.5$

$\sigma_M = \sqrt{312.5} = 17.68$

446.97 464.64 482.32 500 517.68 535.36 553.03

M = sample mean

Distribution $M_s - M_x$

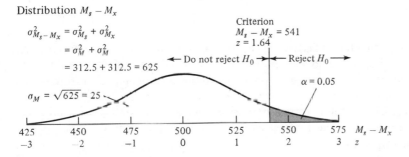

$\sigma_{M_s - M_x}^2 = \sigma_{M_s}^2 + \sigma_{M_x}^2$

$= \sigma_M^2 + \sigma_M^2$

$= 312.5 + 312.5 = 625$

$\sigma_M = \sqrt{625} = 25$

Criterion
$M_s - M_x = 541$
$z = 1.64$

← Do not reject H_0 → ← Reject H_0 →

$\alpha = 0.05$

| 425 | 450 | 475 | 500 | 525 | 550 | 575 $M_s - M_x$ |
| -3 | -2 | -1 | 0 | 1 | 2 | 3 z |

FIGURE 7-6. If H_0 is true and $\mu_S = \mu_X$ then a difference score $(M_S - M_X)$ should be distributed with a mean of 0 and a variance of $2\,\sigma_M^2$.

do so we compute an obtained z-score by taking our obtained score $(M_X - M_S)$, subtracting the mean of the distribution of that score $(\mu_{M_s - M_x} = 0)$, and dividing by the standard deviation of the distribution of that score $(\sigma_{M_s - M_x} = 25)$. The z-score is therefore

$$z = \frac{(M_S - M_X) - \mu_{M_s - M_x}}{\sigma_{M_s - M_x}} = \frac{67 - 0}{25} = \frac{67}{25} = 2.68$$

Making the comparison. We see that our obtained z of 2.68 is greater (more extreme) than our standard, two-tailed criterion z of ± 1.96, and we are in a position to reject the null hypothesis. We conclude that SMARTIUM improves math ability more than does PROGRAM X.

In summary we have once again undertaken precisely the same hypothesis-testing steps that we described in the last two chapters. The only thing different in this particular example concerns our choice of a summary score and the mechanism by which we construct the distribution of that summary score, assuming that the null hypothesis is true. In the example we just described, the score we obtained was a difference score (actually the difference between two sample means). This selection of a score necessitated going into the new topic of how difference scores are distributed. Once we discovered how difference scores were distributed, it was no problem to determine what the distribution of difference scores would be, assuming the null hypothesis to be true. We then formulated a criterion z (which is ± 1.96 for a two-tailed test), computed an obtained z from our data, and compared the two. The obtained z was greater than the criterion z; that is, the difference score we obtained was much more deviant than we would expect if the null hypothesis were really true. We therefore concluded that the null hypothesis is not really true, but rather that the alternative hypothesis is true, and SMARTIUM wins out over Program X.

SUMMARY

We began this chapter by discussing the issue of samples versus populations, and we saw that whereas we do an experiment on a sample of subjects, we typically wish to make conclusions that are applicable to the entire population from which that experimental sample was drawn. We then discussed an important type of distribution—the sampling distribution of sample means—and we saw how this distribution is used in the hypothesis-testing procedures. The following points are of some importance.

1. There are three major types of distributions with which we must be concerned. The first is the population distribution, about which we wish to make inferences or conclusions. Population distributions are characterized by population parameters such as μ and σ^2. The second type of distribution is a sample distribution which is the distribution of scores in some random sample. This distribution is characterized by sample statistics such as M and S^2. The third type of distribution is a sampling distribution of some sample statistic (such as the sample mean). This is the theoretical distribution that would result if we were to obtain large numbers of samples, get a particular statistic (such as the mean) from each sample, and plot the resulting probability distribution of this statistic. The sampling distribution of sample means is a distribution whose parameters can be derived directly from n, the sample size, and the parameters of the population distribution. Specifically the sampling distribution of sample means is normal with parameters $\mu_M = \mu$ and $\sigma_M^2 = \sigma^2/n$.

2. The first type of hypothesis-testing situation we discussed using the sampling distribution of sample means involved testing a sample mean against some constant. In this situation the summary score we obtained

from our data was a sample mean, and the distribution of this score was simply the sampling distribution of sample means.

3. The second major type of hypothesis-testing situation we described involved testing the difference between two sample means. In this situation we had two conditions pitted against one another, and we obtained a mean from each condition. The summary score with which we were concerned was the difference between these two sample means $(M_1 - M_2)$. If the null hypothesis were true, this score comes from a distribution of difference scores whose parameters are

$$\mu_{M_1 - M_2} = 0$$

$$\sigma^2_{M_1 - M_2} = \sigma^2_{M_1} + \sigma^2_{M_2} = \sigma^2/n_1 + \sigma^2/n_2$$

$$\sigma_{M_1 - M_2} = \sqrt{\frac{\sigma^2}{n_1} + \frac{\sigma^2}{n_2}}$$

PROBLEMS

1. Acme lightbulbs have a mean lifetime of $\mu = 150$ hours and a variance of $\sigma^2 = 75$. Suppose random samples of Acme bulbs are tested. How probable is it that the mean lifetime of a sample will be:

   1. Greater than 155 hours if the sample size is 1
   2. Greater than 155 hours if the sample size is 5
   3. Greater than 155 hours if the sample size is 10
   4. Less than 150 hours if the sample size is 1 billion
   5. Less than 149.9 hours if the sample size is 1 billion
   6. Between 149 and 151 hours if the sample size is 100

2. The distribution of scores on a driving skill test has a mean of 80 and a standard deviation of 9.

   1. Find the mean and standard deviation for a distribution of means of random samples of size 100.
   2. Find the percentage of random samples of size 100 that would be expected to have a mean above 87.
   3. Find the percentage of random samples of size 30 that would be expected to have a mean below 82.
   4. Find an interval centered at the mean that would be expected to include the means of 50% of random samples of size 50.

3. Weights of Mississippi sheriffs are known to be distributed with a mean, $\mu = 250$ lbs.

   An experiment is done to test whether Texas sheriffs weigh less in general than Mississippi sheriffs. Thus, two Texas sheriffs are randomly selected and have the following weights:

Sheriff	Weight
1	$x_1 = 240$
2	$x_2 = 250$

Assume weights of Texas sheriffs are distributed with a standard deviation $\sigma = 20$ lbs. Test the hypothesis that Texas sheriffs come from a population whose mean is less than that of Mississippi sheriffs.

4. A string manufacturer claims his product has a mean breaking strength of 60 lbs. with a standard deviation of 3.5 lbs. Test the hypothesis that $\mu = 60$ lbs. if a random sample of 49 pieces of this string is found under test to have a mean breaking strength of 55 lbs. Use a test at the 0.01 level of significance.

5. It is known that armed robbers receive sentences normally distributed with a mean μ of 7 years and a variance of 5.

   A criminologist is interested in whether rapists tend to receive sentences different from those of armed robbers. A sample of ten rapists turns out to have a mean sentence M of 5 years.

   Test the hypothesis that rapists tend to receive sentences different from those of armed robbers.

6. It is known from long experience that Boeing 747 airliners fly from San Francisco to New York in times that are normally distributed with $\mu = 4.7$ hours and $\sigma = 0.5$ hour.

   Jupiter Airlines tries a new brand of wax on its 747s. The wax is supposed to increase air speed. After 100 flights with the new wax, the mean San Francisco/New York flying time is 4.45 hours.

   Test the hypothesis that the new wax really does increase airspeed.

7. Two hundred students are randomly assigned to two statistics classes taught by Dr. A and Dr. B (100 students per class). At the end of the course all 200 students take a standard statistics exam known to have a population mean of 500 and a standard deviation of 100. The 100 students from Dr. A's class get an average score of 520, while the 100 students from Dr. B's class get an average score of 490.

   1. Test the hypothesis that Dr. A's students are *different* from the average.
   2. Test the hypothesis that Dr. B's students are *different* from the average.
   3. Test the hypothesis that Dr. A's students are *different* from Dr. B's students.

8. Scores on the Miller Analogies Test (MAT) are normally distributed with a *population variance* $\sigma^2 = 22.5$.

   An experiment is done to determine whether University of Washington students and Harvard students differ in terms of MAT performance. A sample of 5 Harvard students and 5 U.W. students are given the MAT and the following scores are obtained:

UW	Harvard
85	81
82	83
84	77
88	79
86	80

Do Harvard and U.W. students differ significantly?

9. A test of spatial abilities is known to have a standard deviation $\sigma = 6$. Groups of $n_1 = 10$ left-handed children and $n_2 = 15$ right-handed children take the test. The left-handed children get a mean $M_1 = 78$ on the test, whereas the right-handed children get a mean $M_2 = 69$. Can it be concluded that left- and right-handed children differ in terms of spatial ability?

10. The time it takes aardvarks to run down a runway is known to be distributed with $\sigma^2 = 4$

It is of interest to see whether aardvarks run faster to peanut butter or to jelly. A group of $n_1 = 3$ aardvarks is put in a peanut-butter condition and a second group of $n_2 = 5$ aardvarks is put in a jelly condition. The mean times for the aardvarks are as follows:

Peanut butter ($n_1 = 3$)	Jelly ($n_2 = 5$)
$x_{11} = 3$	$x_{12} = 7$
$x_{21} = 2$	$x_{22} = 11$
$x_{31} = 4$	$x_{32} = 9$
	$x_{42} = 5$
	$x_{52} = 8$

Test whether the two groups differ.

11. Joe Shablotnik is employed as a dental researcher and is interested in whether Ultra-Brite toothpaste produces more sex appeal that does Crest. Joe thus gets one group of 5 people to brush with Ultra-Brite and a second group of 5 people to brush with Crest. Sex appeal is then measured with an appealometer. Unfortunately the appealometer malfunctions while measuring 2 of the 5 Crest subjects and while measuring 3 of the 5 Ultra-Brite subjects. So Joe is left with the following data:

Crest	Ultra-Brite
5	2
3	2
7	

(*Note:* The greater the sex appeal, the higher the appealometer score.)

It is well known that appealometer scores are distributed with a population variance $\sigma^2 = 1.5$.

Test the hypothesis that Ultra-Brite leads to more sex appeal than does Crest. List all hypothesis-testing steps. Use the 0.01 α-level.

12. Joe Shablotnik is working to try to make his Merkin plants grow taller. He suspects that one type of plant food, Miracugrow, leads to taller Merkin

plants than does another kind, Amazofood. To test this hypothesis, he gives Miracugrow to one group of 10 Merkin plants and Amazofood to another group of 10 Merkin plants. Unfortunately 9 of the 10 Amazofood plants die.

At adulthood, the one remaining Amazofood plant is 48′ tall, and the 10 Miracugrow plants have a mean height of 52″.

It is known that heights of Merkin plants are distributed with a variance of $\sigma^2 = 4$.

Carry out hypothesis-testing steps to decide on the original issue (whether Miracugrow leads to taller Merkin plants than does Amazofood). Use all hypothesis-testing steps and *use the 0.01 α-level.*

13. It is known that the amount of beer served in Dante's Tavern is normally distributed with a variance of 4. Of interest is whether the mean amount served in men's mugs differs from the mean amount served in women's mugs. Suppose we have samples of:

    $n_1 = 1$ male whose amount of beer is M_1
    $n_2 = 4$ females whose mean amount of beer is M_2

    1. How big a difference between M_1 and M_2 would be required such that we could reject (at the 0.05 level) the null hypothesis that males and females do not differ in terms of amount of beer served. Assume a one-tailed test with an α-level of 0.05.
    2. Recompute the above, but assume a two-tailed test.

14. Assume 80% of Seattle voters favor legalization of marijuana and 50% of Bellingham voters favor legalization. A random sample of 25 voters is drawn from Seattle and another random sample of 25 voters is drawn from Bellingham. What is the probability that the percentage of the Bellingham sample that favor marijuana will *exceed* the percentage of the Seattle sample that favor marijuana?

    (*Hint:* Let the proportion of the Seattle sample that favor marijuana be P_S. Likewise, let the proportion of the Bellingham sample be P_B. Now the question may be rephrased: What is the probability that $P_B - P_S$ exceeds zero?)

15. A musical aptitude test was given to all third-grade classes in Fun City. The mean score on this test was 75 and the standard deviation of children's scores was 15 points. Suppose that the classes are composed of 36 children each, so that in effect there are a vast number of samples of 36 scores each.

    1. Draw and label the graph of the distribution of the means of these samples (assuming that each sample is a random collection of 36 scores).
    2. In what percentage of the classes would we expect the class mean to be as high as 80?
    3. In what percentage would the mean be as low as 70?

16. Consider a population consisting of only the scores 3, 6, 8, 11, 15.

1. Find the mean and standard deviation of these scores.
2. Find all the possible samples of size 2 (that is, 3 and 6, 3 and 8, 3 and 3, and so on) which can be drawn from this population *with replacement* and list these (there are 25 such samples).
3. Find the means of all the samples of size 2.
4. Find the mean of the sample means.
5. Find the standard deviation of the sample means. Compare the mean of all the samples of size 2 and the standard deviation with the results in question 1 of this problem.

8 Power

In the previous three chapters we have been primarily concerned with hypothesis testing. As we have seen, the principal goal of the hypothesis-testing procedure is to formulate a criterion score such that we will or will not reject the null hypothesis depending on whether the score obtained from the data is more or less extreme than the criterion score. It has been evident that the establishment of such a criterion is based solely on considerations having to do with the probability of a Type I error. In general the criterion is set such that the probability of a Type I error is less than 0.05.

We have only briefly considered the other type of error to which the hypothesis-testing procedure can potentially lead. Recall that a Type II error occurs when the alternative hypothesis is actually true but our obtained score fails to exceed the criterion, and we therefore fail to reject the null hypothesis. The probability of making this sort of error has been referred to as β and we have declared that *power* or $1-\beta$ is the probability of correctly rejecting the null hypothesis when the alternative hypothesis is actually true.

Figure 8-1 depicts the relationships among α, β, and power.[1] In Chapter 6 we indicated that one of the unfortunate aspects of most hypothesis-testing situations is an inability to calculate a single, specific value for power. The reason for this shortcoming rests on the fact that calculation of power requires formulating the distribution of our summary score assuming the alternative hypothesis to be true. But the lack of an exact alternative hypothesis characteristic of most hypothesis-testing situations renders us unable to do this.

The situation, however, is not quite as grim as we have painted it so far, and in this chapter we'll be considering power somewhat more explicitly. We will start by discussing what are called *power curves*. Power curves, as we will see, provide us with a way of looking at the power of an entire experimental situation despite the lack of an exact alternative hypothesis. After discussing power curves, we will go on to explore a number of specific issues involving power.

---

[1]For the most part we shall be talking about one-tailed situations in this chapter, because it is within the context of one-tailed situations that power is most easily explainable. Later in the chapter we shall specifically consider a two-tailed situation.

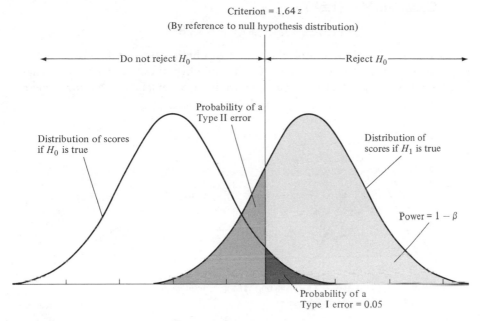

Criterion = 1.64 z
(By reference to null hypothesis distribution)

←——————Do not reject H_0——————→ ←————————Reject H_0————————→

Probability of a
Type II error

Distribution of scores
if H_0 is true

Distribution of
scores if H_1 is true

Power = $1 - \beta$

Probability of a
Type I error = 0.05

FIGURE 8-1.   Relationships among α, β, and power. The left-hand distribution assumes that the null hypothesis is true. This distribution dictates establishment of the criterion which is placed so as to chop off the upper 5% of the distribution. The right-hand distribution assumes that the alternative hypothesis is true. The area under this distribution to the left of the criterion is β, and the area to the right of the criterion is power.

CONSTRUCTION OF POWER CURVES: SMARTIUM REVISITED

Let's go back to SMARTIUM, the program that Acme Software Company has developed to improve math ability. Let's assume, as we did in one of our examples, that Acme plans to administer SMARTIUM to a sample of $n = 25$ students, measure their math GRE scores, and then see if the mean GRE score of the sample is above 500. As we saw in the last chapter, the summary score that we use in this situation is a sample mean. To set our criterion score, we consider what the distribution of means would be if the null hypothesis were true. This is a sampling distribution of sample means, which in turn is a normal distribution with a mean of

$$\mu_M = \mu = 500$$

and a standard deviation of

$$\sigma_M = \sigma/\sqrt{n} = 100/\sqrt{25} = 100/5 = 20$$

We know that in terms of z-scores, our criterion is $z = +1.64$. **Therefore, we can determine our raw-score criterion to be**

$$\text{Criterion } M = (1.64)\, \sigma_M + \mu$$

$$= (1.64)(20) + 500 = 532.8$$

Consequently, if the sample mean that we obtain from our data turns out to be greater than 532.8, we will reject the null hypothesis and conclude that SMARTIUM does in fact improve math ability. Conversely, if the sample mean

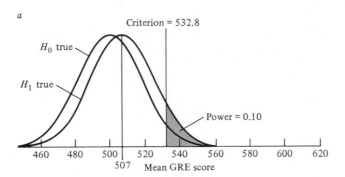

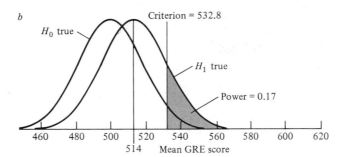

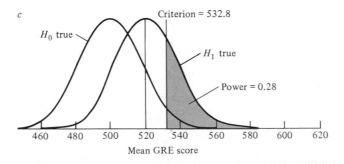

FIGURE 8-2.   Six different alternative hypotheses. The six hypotheses represent successively greater differences between the means specified by the alternative versus the null hypothesis. They therefore represent successively greater power.

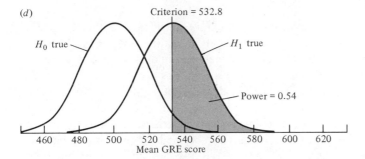

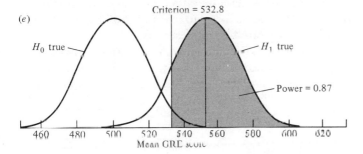

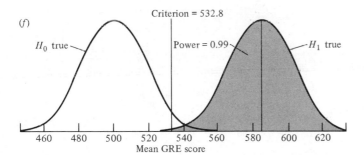

FIGURE 8.2. Continued.

we obtain turns out to be *less* than 532.8, we will fail to reject the null hypothesis.

Now let us assume that the alternative hypothesis is really true—that SMARTIUM really does improve math ability. In order to examine the power of our test, we will specify several exact alternative hypotheses, arbitrarily chosen. We will then compute power assuming each of these null hypotheses in turn to be true. A plot of power as a function of the means corresponding to the various alternative hypotheses will constitute a power curve.

Calculation of Power for
Specific Alternative Hypotheses

Figure 8-2 illustrates the means by which we consider a number of specific alternative hypotheses. First we assume that in reality SMARTIUM raises math GRE scores by 7 points. Therefore, as depicted in diagram *a* of Figure 8-2, the distribution that our sample mean comes from has a mean $\mu_M = 507$ and a standard deviation,[2] $\sigma_M = 20$. In the first diagram of Figure 8-2 we have also included the criterion score of 532.8. Since this criterion was determined with reference to the null hypothesis distribution, it is invariant. No matter what the actual distribution of scores is, if we get an obtained mean greater than this criterion, we will reject the null hypothesis and if we get an obtained mean less than the criterion, we will fail to reject the null hypothesis. Assuming, as we are doing now, that the null hypothesis is actually false, obtaining a sample mean that is less than the criterion will cause us to fail to reject the null hypothesis and will therefore constitute a Type II error. As we can see, the probability of failing to reject the null hypothesis is equal to the area under the alternative hypothesis distribution shown in the first diagram in Figure 8-2 to the left of the criterion. Calculation of this area is quite simple. First we compute the appropriate z-scores

$$z_1 = -\infty$$

$$z_2 = \frac{c - \mu}{\sigma_M} = \frac{532.8 - 507}{20} = 1.29$$

and the $F(z)$s

$$F(z_1) = 0$$

$$F(z_2) = F(1.29) = 0.90$$

By subtraction, we find that

$$p(-\infty \leq M \leq 104.92) = p(0 \leq z \leq 1.31)$$

$$= F(z_2) - F(z_1)$$

$$= 0.90 - 0$$

$$= 0.90$$

Therefore, the probability of making a Type II error in this situation is 0.90, and power is $1 - \beta$ or $1 - 0.90 = 0.10$.

---

[2]We have smuggled in an assumption here which is that, although SMARTIUM may increase the mean GRE score of people who take it, *it does not affect the variance*. Therefore, σ_M is equal to 20 for the alternative hypothesis distribution as well as for the null hypothesis distribution. This assumption is called the homogeneity of variance assumption, and we shall return to it in later chapters.

Now let's assume the other specific alternative hypotheses illustrated in Figure 8-2. We will calculate power assuming that SMARTIUM actually raises math ability by 14 points, by 21 points, by 35 points, by 55 points, and by 84 points. The calculations of β and the power for each of these particular alternative hypotheses are shown in Table 8-1.

Power Curves: Power as a Function of the Alternative Hypothesis

Now we are ready to construct a power curve, which is simply a graph of power as a function of the means corresponding to specific alternative hypotheses. Figure 8-3 shows this power curve. As can be seen, power starts from a minimum equal to the probability of a Type I error (in this case 0.05) when the alternative hypothesis mean is very close to the null hypothesis mean (in this case 500). The power curve then rises to a maximum of 1.0 when the difference between the null hypothesis mean and the alternative hypothesis mean becomes large. Notice that a power curve provides us with a feeling for the power of the entire experimental situation. Figure 8-4 elaborates on this notion, showing two hypothetical power curves—one representing a high-power situation and the other representing a low-power situation. Both power curves start at the same place (the α-level) and end at the same place (1.0). The difference between the

TABLE 8-1.   Construction of a Power Curve: Calculations of β and Power Assuming Several Different Alternative Hypotheses

				Calculation of β				
H_1	x_1	x_2	z_1	z_2	$F(z_1)$	$F(z_2)$	β	Power = $1-\beta$
$\mu = 507$	$-\infty$	532.8	$-\infty$	$\dfrac{532.8 - 507}{20} = 1.29$	0	0.90	0.90	0.10
$\mu = 514$	$-\infty$	532.8	$-\infty$	$\dfrac{532.8 - 514}{20} = 0.94$	0	0.83	0.83	0.17
$\mu = 521$	$-\infty$	532.8	$-\infty$	$\dfrac{532.8 - 521}{20} = 0.59$	0	0.72	0.72	0.28
$\mu = 535$	$-\infty$	532.8	$-\infty$	$\dfrac{532.8 - 535}{20} = -0.11$	0	0.46	0.46	0.54
$\mu = 555$	$-\infty$	532.8	$-\infty$	$\dfrac{532.8 - 555}{20} = -1.11$	0	0.13	0.13	0.87
$\mu = 584$	$-\infty$	532.8	$-\infty$	$\dfrac{532.8 - 584}{20} = -2.56$	0	0.01	0.01	0.99

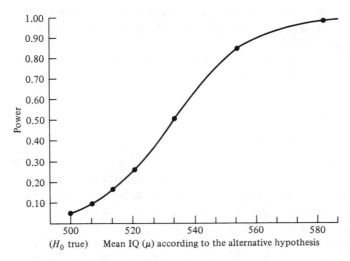

FIGURE 8-3. A power curve showing power as a function of the mean specified by the alternative hypothesis.

two situations is reflected by how fast the curves rise. A curve representing a high-power situation rises very quickly, indicating that even for a relatively small difference between the alternative hypothesis mean and the null hypothesis mean, the power is quite high. In a low-power situation on the other hand the curve rises more slowly, and it therefore requires a relatively large difference between the null hypothesis mean and the alternative hypothesis mean before power becomes high.

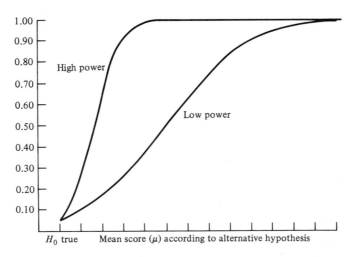

FIGURE 8-4. Two power curves representing a high-power and a low-power situation.

8.1 A power curve is a graph of power as a function of the mean (μ) of the alternative-hypothesis distribution. A high-power power curve rises faster than a low-power power curve.

We'll explore the sorts of factors that determine the degree to which a given experimental situation is a high-power or a low-power situation in the next section.

THINGS THAT AFFECT POWER

Three major factors affect power in any given experimental situation. First anything that decreases σ_M, the standard deviation of the distribution of sample means, will increase power. Second, the smaller is α (the probability of a Type I error), the smaller the power. And finally different types of tests have different powers. We shall discuss these three factors in turn.

Size of σ_M

Figure 8-5 provides a general illustration of why decreasing σ_M increases power. The two panels in Figure 8-5 depict two situations, in both of which the alternative hypothesis is assumed to be true. Additionally both situations involve the same alternative-hypothesis mean (as well as the same null-hypothesis mean). The difference between the two situations is that the one in the bottom panel portrays a smaller σ_M. With a smaller σ_M two things are true. First, the criterion mean needed to reject the null hypothesis becomes smaller (since the variance of the null-hypothesis distribution is smaller). Second, given that the alternative hypothesis is true, a sample mean has less chance of straying down below any criterion; that is, since the variance of the alternative hypothesis distribution is small, any sample mean has a high probability of being relatively close to the alternative distribution mean.

What affects the size of σ_M? Since $\sigma_M = \sigma/\sqrt{n}$, σ_M can be made smaller either by increasing n or by decreasing σ.

Number of subjects.   Staying with our SMARTIUM example, let's consider one specific alternative hypothesis: that SMARTIUM raises math GRE scores by 14 points. Now let's compute power for two different situations: one in which the sample size (n) is equal to 25, and the other in which the sample size is equal to 100. We have already computed power for the first instance; it was 0.17 (see diagram b of Figure 8-2 and Table 8-1). An illustration of the power that obtains when $n = 25$ is provided in the top panel of Figure 8-6.

What happens when n is increased to 100? As we have seen, an increase in n

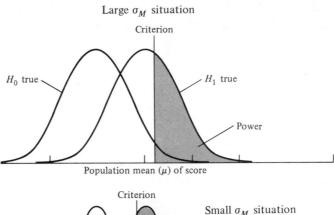

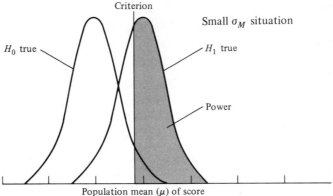

FIGURE 8-5.   Effect on power of decreasing σ_M. The upper and lower panels depict two situations that are identical except that σ_M is smaller in the lower panel. A lower σ_M leads to greater power.

leads to a decrease in σ_M, specifically,

$$\sigma_M = \sigma/\sqrt{n} = 100/\sqrt{100} = 100/10 = 10$$

Thus, when n is increased from 25 to 100, σ_M is decreased from 20 to 10.

What will our criterion score be when the sample size is 100? The criterion *z-score* is still 1.64. However, the criterion raw score (sample mean) will now be equal to

$$\text{Criterion mean} = (1.64)\ \sigma_M + 500$$

$$= (1.64)\ (10) + 500 = 516.4$$

The first benefit of decreasing σ_M is now apparent: it will require a less extreme sample mean for us to reject H_0 than it did previously. Now let us assume that the alternative hypothesis is true and compute the power. Computation of power is depicted in the lower panel of Figure 8-6. If the alternative hypothesis is true,

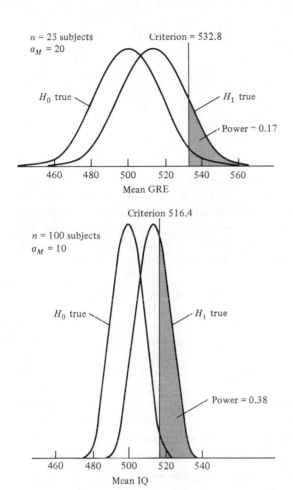

FIGURE 8-6.   Increasing sample size (lower panel) leads to a smaller σ_M which in turn leads to greater power.

the distribution of sample means still has a mean of 514 (since we still are assuming that SMARTIUM raises GRE scores by 14 points) and α, the probability of a Type I error (the area under the null-hypothesis distribution to the right of the criterion) is still 0.05. However, power is considerably greater with a sample size of 100. We compute power as follows: First β is equal to the area under the alternative hypothesis curve to the left of the criterion. To compute β, our z-scores are

$$z_1 = -\infty$$

$$z_2 = \frac{C - M}{\sigma_M} = \frac{516.4 - 514}{10} = 0.24$$

Our $F(z)$s are

$$F(z_1) = 0$$

$$F(z_2) = F(0.24) = 0.59$$

To compute β,

$$\beta = F(z_2) - F(z_1) = 0.59 - 0 = 0.59$$

And finally power, on $1 - \beta$, is $1 - 0.59 = 0.61$.

The increase in power that results from increasing n can be shown more generally using power curves. Figure 8-7 shows the two power curves, one corresponding to a sample size of 25, and the other to a sample size of 100. As is readily apparent, the power curve corresponding to a sample size of 100 increases much more rapidly than the power curve corresponding to a sample size of 25; that is, the power of the situation is greater with larger sample size.

Manipulating σ^2.   We have just seen that one way in which σ_M can be decreased is by increasing n, the number of subjects. The second way in which σ_M can be decreased is by decreasing σ^2, the population variance. "How can the population variance be decreased?" you are probably asking yourself. The population variance, as we have been describing it, seems to be a "given." It does not seem to be something we can tamper with, but rather something that we simply must contend with.

In some cases in fact we *cannot* do anything about the population variance. For example the population variance of math GRE scores permanently exists in the population. However, in many experimental situations we will be dealing

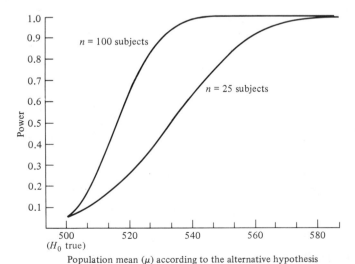

FIGURE 8-7.   Power curves for situations involving either $n = 100$ subjects (high power) or $n = 25$ subjects (low power).

with dependent variables over which we do have some control. As an illustration let's suppose that the Department of Transportaton is concerned with the question of whether ingestion of alcohol increases the amount of time it takes a person to react to a red light. The following experiment could be implemented to answer this question. A group of subjects is randomly divided into two groups. The first group is given alcohol in the form of a gin and tonic, whereas the second group is given only tonic. All subjects are then given a *reaction time* test. In such a test the subject sits down at a table that contains a red light and a button. At various random times the red light flashes on, and the subject's task is to push the button as quickly as possible following the onset of the light. The dependent variable in this procedure is the amount of time it takes the subject to press the button following the light's onset. Table 8-2 sketches some hypothetical results that might arise from this experiment.

The population variance in this situation is reflected by the variation across subjects in reaction-time scores *within* each of the two groups. That is, within the alcohol group, the average reaction time for the first subject (x_1 = 202) differs from the average reaction time of the second subject (x_2 = 184) and so on. The same sort of variation occurs within the tonic-only group. We now concern ourselves with the issue of *what causes* these scores within a group to differ from one another. Table 8-3 lists some of the influences that may cause such a variation. These influences consist of things such as the subject's mood, the subject's inherent ability, as well as factors such as the temperature and humidity of the room, the time of day, the ages and sex of the subjects, and so on. As indicated in Table 8-3, some of these factors are not under experimental control. That is, we are unable to do anything about inherent differences in reaction time among subjects; nor are we able to do much about subjects' moods, subjects' attitudes about the experiment, and things of that nature.

We are, however, able to control other factors. If variation in the room temperature causes variation in performance, we can hold temperature constant.

TABLE 8-2.   Results of a
Hypothetical Reaction-Time
Experiment

Reaction Times (milliseconds)

Alcohol Condition	Water Condition
x_1 = 202	x_1 = 168
x_2 = 184	x_2 = 184
x_3 = 212	x_3 = 150
x_4 = 196	x_4 = 188
.	.
.	.
.	.
x_n = 200	x_n = 175

TABLE 8-3.   Possible Causes of Variation in Reaction-Time Scores Within a Group

---

A. Causes that cannot be controlled:
  1. Subject's mood during experiment
     (bad mood = longer reaction time)
  2. Inherent ability of subject
     (better inherent ability = shorter reaction time)
  3. Attitude of subject toward the experiment
     (annoyed = longer reaction time)

B. Causes that can be controlled
  1. Temperature of experimental room
     (extreme temperatures = longer reaction time)
  2. Humidity of experimental room
     (higher humidity = longer reaction time)
  3. Sex of subjects
     (females = shorter reaction time)
  4. Age of subjects
     (older subjects = longer reaction time)
  5. Time of day that experiment is run
     (later in the day = longer reaction time)

If subjects tend to be faster at one time of day than another, we can run everyone at the same time of day. And so on. There are a number of precautions we can take involving *control of extraneous variables,* all of which will decrease σ^2. And of course the smaller is σ^2, the smaller is σ_M, and, therefore, the greater is power.

> 8.2 Anything that decreases σ_M increases power. In general since $\sigma_M = \sigma \sqrt{n}$, we can decrease σ_M either by increasing n or by decreasing σ.

Alpha-Level

In the past we have typically assumed the α-level to be set at 0.05, in accord with convention. Sometimes, however, experimenters wish to be even more conservative about erroneously rejecting the null hypothesis. Therefore, the criterion in a hypothesis-testing situation is sometimes established such that the α-level is less than 0.05—say, 0.01 or 0.005.

Figure 8-8 illustrates the effect of changing the criterion, which is quite simple: if the probability of a Type I error is decreased, then β, the probability of a Type II error is thereby increased. Or equivalently, as the probability of a Type I error decreases, power decreases. This fact stems directly from our original

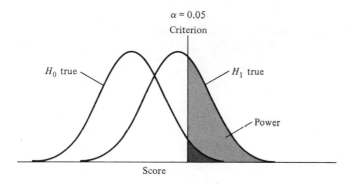

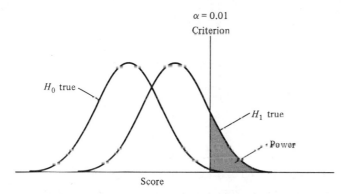

FIGURE 8-8. Effect on power of changing α. As α decreases (bottom panel), β increases, and power decreases.

discussion of decision making, and can be summed up by the old adage, "There's no such thing as a free lunch." All else being equal, the probability of one type of error can be decreased only at the expense of increasing the probability of another type of error.

> 8.3 As α decreases, β increases. Therefore, as α decreases, power = $1 - \beta$ also decreases.

Type of Test

We occasionally have a *choice* of what type of test to use in a hypothesis-testing situation. Consider the following example. The Federal Trade Commission (FTC) is investigating an advertising claim made by the Dazzle Toothpaste Company that Dazzle makes teeth brighter than does its rival White-O Tooth-

TABLE 8-4. Hypothetical Data from the FTC's Experiment Testing Dazzle Toothpaste Against White-O Toothpaste

	Photometer Scores			
Subject	Dazzle Score x_D	White-O Score x_W	Difference $x = x_D - x_W$	Sign
1	9	5	4	+
2	7	8	−1	−
3	6	3	3	+
4	7	3	4	+
5	5	6	−1	−
6	8	6	2	+
7	7	5	2	+
8	7	2	5	+
9	8	10	−2	−
10	4	3	1	+
11	6	3	3	+
12	5	6	−1	−
13	8	4	4	+
	$M_D = 6.69$	$M_W = 4.92$	$M = 1.77$	

Sign test: 9 plusses
         4 minuses

paste. To investigate this claim, the FTC draws a random sample of 13 people from the population. The investigators then administer both Dazzle and White-O (at different times) to each of the 13 subjects. For each subject brightness of teeth is measured immediately following the application of each type of toothpaste, using a photometer. Table 8-4 shows hypothetical data that might be obtained in this situation.

As indicated in Table 8-4, the mean Dazzle score ($M_D = 6.69$) is higher than the mean White-O score ($M_W = 4.92$). Knowing about chance fluctuation and being good statisticians, however, the FTC experimenters plan to go through the hypothesis-testing steps to determine whether the difference between Dazzle and White-O is a reliable one.

Results of a sign test. The FTC initially does a sign test on the data, as depicted in Table 8-4. In conjunction with the sign test the appropriate hypothesis-testing steps are as follows:

1. First, the null and alternative hypotheses are set up.
    H_0: The independent variable (type of toothpaste) has no effect on the dependent variable (photometer scores).
    H_1: Type of toothpaste does have an effect on photometer scores. In particular photometer scores are higher for Dazzle than for White-O.

2. Since a sign test is being performed, the appropriate summary score is the number of subjects who have higher Dazzle than White-O scores. Nine of the 13 subjects fit this criterion.
3. If the null hypothesis is true, then any subject has a 0.5 probability of obtaining a higher Dazzle than a White-O score. Therefore, with $n = 13$ subjects, this summary score is binomially distributed with parameters $n = 13$ and $p = 0.5$.
4. We can determine, using the appropriate binomial probabilities, that the criterion score is 10. That is, if 10 or more subjects have higher Dazzle than White-O scores, the FTC will be in a position to reject the null hypothesis and conclude that Dazzle really does lead to whiter teeth than does White-O.
5. However, since the obtained number of scores (9) does not exceed the criterion number of scores (10), the FTC fails to reject the null hypothesis. That is, the FTC must conclude that the results of its experiment do not provide enough evidence to conclude that Dazzle increases teeth whiteness more than White-O.

A z-test instead. Suppose the FTC decided to use a different test. In particular each subject is now viewed as having a *difference score* equal to his or her Dazzle minus White-O brightometer score. The mean of the 13 difference scores is $M = 1.77$. From old records, the FTC happens to know that these difference scores have a population variance of $\sigma^2 = 4.00$.

The FTC is therefore in a position to carry out the hypothesis-testing procedure using a z-test rather than a sign test. The appropriate steps are as follows:

1. The null and alternative hypotheses are the same as they were above (the only change is that a different type of test will be used to decide between them).
2. The appropriate score is now the mean difference score, M, which is 1.77. To the extent that M is large, we tend to believe that the alternative hypothesis is true; and to the extent that M is close to 0, we tend to believe that the null hypothesis is true.
3. If the null hypothesis is true, M is distributed normally with parameters

$$\mu = 0$$

$$\sigma_M^2 = \frac{\sigma^2}{n} = \frac{\sigma^2}{13} = \frac{4.00}{13} \quad 0.308$$

$$\sigma_M = \sqrt{\sigma_M^2} = \sqrt{0.308} = 0.555$$

4. Since a one-tailed z-test is being performed, the criterion z is 1.64.

5. The obtained z is

$$z = \frac{M - \mu}{\sigma_M} = \frac{1.77 - 0}{0.555} = 3.19$$

6. The obtained z exceeds the criterion z; using a z-test, the FTC will end up rejecting the null hypothesis and concluding that Dazzle leads to whiter teeth than does White-O.

Different powers.   We have seen that, starting with exactly the same data, we have arrived at two different conclusions by doing two different tests. The results of the sign test did not permit rejection of the null hypothesis at the 0.05 α-level, whereas the z-test did. What is the difference between these two types of tests? The answer is that the power of the two tests is different—the power of a z-test is greater than the power of a corresponding sign test. The reason for this is that the sign test throws information away. That is, the sign test ignores the information corresponding to the *magnitudes* of the Dazzle/White-O difference scores and concerns itself only with their *signs*. The z-test on the other hand does not throw away this information, but makes use of it; and it is the inclusion of this additional information that provides the z-test with its additional power.

In subsequent chapters we will discover other situations in which more than one type of test may appropriately be performed and where the power of one test is greater than the power of another test. Note that if the less-powerful test is done and produces a nonsignificant effect, then it is to the experimenter's advantage to go on and do the more powerful test. This is because, if the null hypothesis is actually false, its falseness is more likely to be detected by the more-powerful as opposed to the less-powerful test.

> 8.4 Given exactly the same data, two different statistical tests (such as a sign test and a z-test) may have different powers.

POWER IN A TWO-TAILED SITUATION

For the most part the logic underlying power in a two-tailed situation is no different from the one-tailed logic we've been describing. But there are several exceptions to this rule, and we shall explore them here.

A Weird and Unusual Error: Correctly Rejecting H_0 for the Wrong Reasons

Consider the situation depicted in Figure 8-9, which shows a two-tailed situation in which H_1 is correct in the sense that the actual distribution is slightly to the right of the H_0 distribution. As can be seen, one of three outcomes could

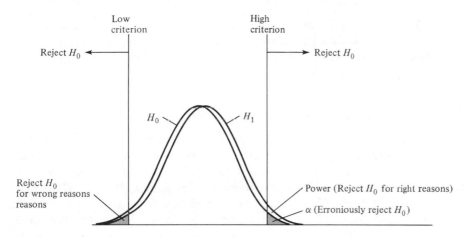

FIGURE 8 9.   Power and β in a two-tailed situation.

reasonably occur. First the area to the right of the right-hand criterion represents the outcome of correctly rejecting H_0 and correctly concluding that $\mu > \mu_0$. This area can unambiguously be labeled power; it is the probability of correctly accepting H_1 given that H_1 is true. Next the area between the two criteria is unambiguously β; it is the probability of incorrectly failing to reject H_0 given that H_1 is true. So far, so good.

But what about the area to the left of the left-hand criterion? This area is the probability of correctly rejecting H_0 but erroneously winding up with the conclusion that $\mu < \mu_0$. It could in some sense be labeled power, as it involves the outcome of rejecting H_0 when H_0 is false. But we usually think of power as a good thing, and this outcome is very bad.

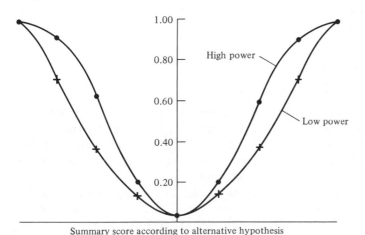

FIGURE 8-10.   Two power curves in a two-tailed situation.

This outcome will rarely constitute a problem in real life because its occurrence requires a highly unlikely conjunction of circumstances—that μ be very close to μ_0 and then that M be drawn from an area of the distribution that is itself very small (smaller than 0.025 with an α-level of 0.05).

Power Curves in a Two-Tailed Situation

Figure 8-10 shows a high-power and a low-power power curve that results from a two-tailed situation. In both cases power is defined to be the probability of rejecting H_0 *and* making the correct decision about the direction of μ relative to μ_0. As can be seen, the curves are symmetrical around μ_0. Power declines to a minimum of 0.025 when μ is very close to μ_0 and rises to an asymptote of 1.0 in both directions. Essentially, going in either direction, a two-tailed power curve corresponding to a 0.05 α-level is identical to the power curve that would result from a one-tailed power curve in which the α-level was 0.025.

> 8.5 Power in a two-tailed situation has all the same properties as power in a one-tailed situation. The main difference is that the power curves are symmetrical around μ_0.

ACCEPTING THE NULL HYPOTHESIS

Recall that we gave the two outcomes of a hypothesis-testing procedure as being either "rejecting the null hypothesis" or "failing to reject the null hypothesis." We do not talk about "accepting the null hypothesis." In this section we'll try to describe *why* we generally do not accept the null hypothesis. However, we shall also discuss situations where, for all practical purposes, we *do* accept the null hypothesis.

If You Can't Find Something, That Doesn't Mean It's Not There

In March of 1977 ski season finally arrived in the Pacific Northwest. Shortly after the slopes opened, one of the authors of this book (GL) decided to take his first skiing trip of the year. In preparation for this adventure GL rummaged around his house, gathering together ski gear that had been lying dormant since the previous season. This search turned up all crucial items except for GL's ski goggles, which were nowhere to be found. GL concluded that his goggles had probably been lost during a previous ski trip, or perhaps over the summer and announced in some exasperation that he was planning to go out and spend $20 for a new pair. The other author (EL), horrified at such an idea, demanded that GL undertake a more thorough (powerful) search. Reluctantly GL did so and finally found the ski goggles in a remote corner of his closet.

The point of this anecdote is to demonstrate something about searches, which is that the logical implications of finding something are quite different from the logical implications of not finding something. If a search results in finding the object sought, then you *know* that object is there; it exists. But not being able to find the object does not demonstrate that the object is not there. It *may* be that the object is not there. However, not finding the object may also result from the fact that the object is there but the search has not been sufficiently powerful to uncover it.

Hypothesis-testing as a search.   Like GL's quest for his goggles, the process of hypothesis-testing may be viewed as a search—not for a physical object, but rather for the effect of some variable (the independent variable) on another (the dependent variable). The search for an effect, however, is subject to the same logic as the search for a physical object. If you end up *finding* the effect (rejecting the null hypothesis), then you "know" that the effect is there—just as after GL finally found his ski goggles, he knew that they had been in the house all along. But failing to find an effect does not logically mean that the effect is not there, just as the early failure to find the ski goggles did not necessarily mean that the goggles had been lost.

We are therefore on fairly safe ground when we reject the null hypothesis and assume that the effect exists—that the alternative hypothesis is really true. But we are typically on shaky ground if we accept the null hypothesis and assume that some sought-after effect does not exist. Usually when we fail to reject the null hypothesis, we do not make very strong conclusions. Rather, we waffle and say something like, "The results of this experiment do not provide us with enough evidence to conclude that the null hypothesis is false."

Failing to Reject the Null Hypothesis When There Is a Great Deal of Power

What if GL's second, more thorough search for his wayward goggles had failed? Suppose he had totally ransacked the house—that is, conducted an extremely powerful search for the goggles—and still failed to find them. The goggles might still have been hidden in some remote nook or cranny, but for all practical purposes it would make sense to conclude that the goggles were simply not there. From a pragmatic standpoint, it would make sense to suspend the search and purchase some new ski goggles.

Likewise suppose that an experiment is done with a tremendous amount of power. If the results of some statistical test still do not permit rejection of the null hypothesis, it might make sense to accept the null hypothesis "for all intents and purposes."

Supermemory. Suppose that ChemCo Drug Company has invented a drug called "Remembrin" which is designed to prevent forgetfulness. To test Remembrin, two groups of subjects are given a list of unrelated words (say, 20 words) to study. Subsequently, all subjects are tested for their memory of these words. The test (called a *free recall test*) consists simply of asking the subjects to recall and write down as many of the 20 words as they can. The first group of subjects is tested immediately after studying the words, and the other group is tested a week later. Therefore, the independent variable is the *retention interval*—the length of time intervening between studying the words and attempting to remember them. The dependent variable is simply the number of words recalled out of the 20.

Typically we observe forgetting in this type of situation. That is, the independent variable (retention interval) has an effect on the dependent variable (number of words recalled) in that subjects tend to remember fewer words after a week than they remember immediately. However, if Remembrin really works, then this typical finding should be absent; there should be no difference between the two conditions. Number of words recalled should not depend on retention interval.

Effect of power on our conclusions. Suppose in fact no effect of retention interval is found. That is, a test between the means of the two groups fails to allow rejection of the null hypothesis. What are we to conclude? Well, if the power of the situation is fairly low—if for example only five subjects were run in each of the two conditions—then we would not be able to conclude very much. Although it *may* be that no forgetting is taking place, it may also be that forgetting actually *is* taking place, but we are simply failing to detect it.

On the other hand suppose that the power of the situation were very high. Suppose for example tht we had 10,000 subjects in each of the two conditions. In such a situation we would probably end up assuming that Remembrin works, that for all intents and purposes there is no forgetting. This would be an instance where acceptance of the null hypothesis would be a reasonable conclusion.

> 8.6 We only can accept the null hypothesis "for all intents and purposes" when there is a great deal of experimental power.

CHOOSING NUMBER OF SUBJECTS

Let's return to our familiar SMARTIUM example, and let's suppose that Acme is not only worried about *whether* SMARTIUM improves math ability—but for practical reasons is also concerned with *how much* SMARTIUM improves math ability. Acme decides that if SMARTIUM only raises GRE scores by less

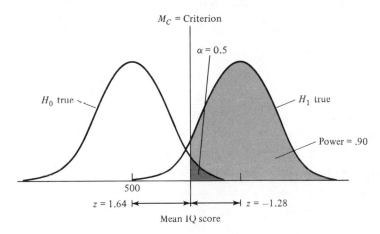

FIGURE 8-11.   A situation in which $\alpha = 0.05$ and $\beta = 0.10$ for an alternative hypothesis corresponding to $\mu - 114$. To make this situation come true, n must equal 480 subjects. This leads to a σ_M of 4.795 and a criterion of 507.86.

than 14 points, then it's not worthwhile to market it. On the other hand if SMARTIUM raises GRE scores by 14 points or more, then it is worthwhile to market it.

Acme therefore wants to arrange its hypothesis-testing situation such that if SMARTIUM raises GRE scores by 14 points or more, then the power of the resulting test—that is, the probability of correctly rejecting the null hypothesis—will be at least 0.90.

We manipulate power by choosing the number of subjects we will run in our experiment. Therefore, "arranging the situation" boils down to choosing the appropriate number of subjects. Figure 8-11 depicts the situation that we want and depicts two distributions. The left-hand distribution is the one corresponding to the null hypothesis and dictates where the criterion is set. We have drawn the criterion, labeled M_C, so that as usual it chops off the upper 5% of the null-hypothesis distribution. The right-hand distribution in Figure 8-10 corresponds to the alternative hypothesis being true.[3] We have drawn this distribution such that the probability of a Type II error—the area of the alternative hypothesis distribution to the left of the criterion—is 0.10. This, of course, means that power is 0.90, which is what we want. We also have depicted the variance of both of these distributions to be σ_M^2.

Now we want to calculate what n, the sample size, should be such that this whole situation obtains. To do this, let's first write down expressions for various

---

[3]We have set things up such that with an alternative hypothesis of $\mu = 514$, power will be 0.90. Note that any alternative hypothesis of $\mu > 514$ will lead to a value of power that is *greater* than 0.90.

z-scores. First referring to the null-hypothesis distribution, we know that

$$\text{Criterion } z = 1.64 = \frac{M_C - 500}{\sigma_M}$$

This, in turn, means that

$$1.64\sigma_M = M_C - 500 \tag{8.1}$$

Now in terms of the alternative hypothesis distribution, we determine from our z distribution tables that the z-score corresponding to an $F(z)$ of 0.10 is -1.28. So we can set up another z-score as follows:

$$-1.28 = \frac{M_C - 514}{\sigma_M}$$

This, in turn, means that

$$-1.28\sigma_M = M_C - 514 \tag{8.2}$$

We now have two simultaneous equations involving the two unknowns, σ_M and M_C. Using fairly simple algebraic techniques, we can solve these equations for both unknowns. To do so, we first subtract Equation 8.1 from Equation 8.2 to obtain the following

$$1.64\,\sigma_M = M_c - 500$$
$$\underline{-1.28\,\sigma_M = M_c - 514}$$
$$2.92\,\sigma_M = 14$$

We can therefore solve Equation 8.3 for σ_M to get

$$\sigma_M = \frac{14}{2.92} = 4.795 \tag{8.3}$$

Knowing σ_M, we are in a position to calculate what n should be. To do this, we make use of the fact that the population standard deviation, σ, is equal to 100, and

$$\sigma_M = 4.795 = \frac{\sigma}{\sqrt{n}} = \frac{100}{\sqrt{n}}$$

Therefore,

$$4.795\sqrt{n} = 100$$

or

$$\sqrt{n} = \frac{100}{4.795} = 20.86 \tag{8.4}$$

Squaring both sides of Equation 8.3,

$$n = 20.86^2 = 435$$

or n must equal about 435 subjects. Additionally we can, if we like, solve for M_C by plugging our value for σ_M either into Equation 8.1 or Equation 8.2. Plugging σ_M into Equation 8.1 we come up with

$$(1.64)(4.795) = M_C - 500$$

or

$$M_C = 500 + 7.86 = 507.86$$

Thus, given certain specifications—a desired α-level, a desired degree of power for a specific alternative hypothesis, and a known population variance—we can calculate the appropriate number of subjects for the specifications to be met.

> 8.7 To determine the number of subjects necessary for a certain amount of power given a certain H_1, we set up two equations and solve for the two unknowns of n and criterion.

SUMMARY

This chapter has concentrated on *power*, which is the probability of correctly rejecting the null hypothesis when the null hypothesis is actually false. We have seen that power is equal to $1 - \beta$, so power and β may be discussed interchangeably. We have concentrated on several aspects of power.

1. A power curve is a graph relating power to values representing specific alternative hypotheses. Power curves represent the power of an entire hypothesis-testing situation.
2. Power is affected by three major factors: first the size of σ_M, which in turn is affected both by the size of n, the number of subjects, and by the size of σ^2, the population variance. Second decreasing α-level decreases power. And finally different powers are associated with different tests.
3. Power in a two-tailed situation follows the same rules as power in a one-tailed situation.
4. Typically we do not accept the null hypothesis, because the null hypothesis specifies the nonexistence of something (an effect of the independent variable on the dependent variable) and failing to find an effect does not logically imply that the effect does not exist. However, in some instances where a great deal of power exists, we may accept the null hypothesis for all practical purposes.
5. When the population variance is known, we are able to compute how many subjects will be required for some particular degree of power.

PROBLEMS

1. Use the data from Chapter 7, Problem 4, to solve the following:

   1. Suppose the mean breaking strength is actually 59 lb. What is the power of the test?
   2. Suppose the mean breaking strength is actually 50 lb. and a 0.15 significance level is used. What is the power of the test?

2. Use the information from Chapter 7, Problem 3, to solve the following:

   1. Suppose that the mean weight of Texas sheriffs is actually 245 lb. What is the power of the test?
   2. Suppose the mean weight of the Texas sheriffs is actually 230 lb. What is the power of the test?

3. Use the data from Chapter 7, Problem 6, to solve the following:

   1. Suppose the wax actually decreases flying time to 4.4 hours. What is the power of the test?
   2. Suppose the wax actually decreases flying time to 4.65 hours. What is the power of the test?
   3. Suppose the wax actually decreases flying time to 4.0 hours. What is the power of the test?
   4. Repeat parts 1–3 but assume that the test is based only on 50 flights.
   5. Repeat parts 1–3 but assume the test is based only on 10 flights.
   6. Use parts 1–3, then 4, then 5 to generate power curves.

4. Use the information from Chapter 7, Problem 11, to solve the following:

   1. Suppose that, in terms of population means, Ultra Brite is actually 3 points higher than Crest. What would the power of the test be (again, assume a 0.01 α-level)?
   2. Suppose that Ultra Brite has a population mean that is eight points higher than that of Crest. What would the power of the test be?

5. A meteorologist hypothesizes that Portland receives more inches of rainfall per year than does Seattle. It is well known that inches of rainfall in both cities is distributed normally over years with a standard deviation $\sigma = 2$.

   Unfortunately many of the meteorological records have been destroyed in a flood, and data can be found only for a sample of six years for Seattle and a sample of ten years for Portland. (You may assume that these are random samples, as it was a random flood.)

   From these data the meteorologist calculates that the mean rainfall for the six years in Seattle is $M_S = 14''$ and the mean rainfall for the ten years in Portland is $M_P = 15.5''$.

   1. Test the hypothesis that Portland receives more rain than does Seattle.

2. Suppose that *in fact* Portland receives an average of 2″ more rain a year than does Seattle. What is the power of the test in question 1?

6. Use the information from Problem 12 in Chapter 7 to solve the following:

   Suppose that in reality, Miracugrow makes Merkin plants grow 3″ taller than Amazofood. What is the power of the test you performed in the last chapter?

7. Weights of Martians are distributed with a mean $\mu = 3$ ft and a variance $\sigma^2 = 4$. A sample of $n = 4$ Martians is fed Rice Krispies from birth to see if the Rice Krispie diet will increase their heights above the population mean.

   Against what alternative hypothesis will the power of the resulting test be 0.90? Assume $\alpha = 0.05$.

8. A coin collector is looking for "Butte pennies." Butte pennies resulted from a slip-up in the minting process and are only detectable in that their probability of turning up tails is 0.60 rather than the usual 0.50.

   When the coin collector comes across a penny he suspects to be a Butte penny, he sets up the following hypotheses:

   H_0: The penny is a normal penny.
   H_1: The penny is a Butte penny.

   To evaluate these hypotheses, the collector will flip the penny some number of times and observe the number of tails.

   1. Describe what is meant by an α error and by power in this experiment.
   2. Suppose that the α-level is set at 0.10. How many flips will be required such that the power of the experiment is 0.95.
   3. Repeat part 2 assuming an α-level of 0.01.

9. An anthropologist knows that the volume of a human head is normally distributed with the following parameters: $\mu = 100$ cubic in.^{3}; $\sigma^2 = 150$.

   1. The anthropologist is interested in determining whether heads of Neanderthal Man are distributed with a mean smaller than 100. How many Neanderthal heads would he have to test in order for his probability of successfully detecting a difference of 10 cubic in.^{3} or more to be 0.90 (assume on α-level of 0.05)?
   2. Recompute the number of heads needed using an α-level of 0.15.

10. It is known from census bureau records that females marry at an age that has a variance $\sigma^2 = 16$.

    Suppose a sociologist is interested in whether females from Seattle tend to marry at an *earlier* age than do females from Puyallup. Samples of ten Seattle and ten Puyallup females are selected and show the following means:

Seattle: $M_S = 22.2$ years
Puyallup: $M_P = 25.8$ years

1. Should the sociologist reject the hypothesis that Seattle and Puyallup females marry at the same age? Show all hypothesis-testing steps and *use the 0.10 α-level.*

2. Suppose that Seattle females actually marry at an age $\mu_S = 22$ years and Puyallup females actually marry at an age $\mu_p = 25$ years. What is the power of the test in question 1?

3. Suppose the sociologist wants to ensure that the power of her test is 0.98 against the alternative hypothesis that Seattle females marry two years earlier than do Puyallup females. How many subjects would be in each sample (assume an equal number of subjects in each sample)?

Parameter Estimation

The hypothesis-testing situations we have been discussing so far have been somewhat contrived in the sense that we have always assumed the values of some population variance (σ^2) to be *known* in some way or another. In the binomial situation for example the population variance is known because it can be computed using the expression

$$\sigma^2 = Npq$$

We have also made heavy use of examples involving math GRE scores as a dependent variable. This choice of an example was no accident: rather, GRE score was selected because we conveniently happen to know that the population mean and variance of these scores are $\mu = 500$ and $\sigma^2 = 10,000$.

Although there are other situations in which population parameters are known, such situations constitute the exception rather than the rule. When we are faced with the more typical situation in which the population parameters are *not* known, the best we can do is estimate them. Of particular concern will be an estimation of σ^2, the population variance, because knowing the population variance is crucial in terms of carrying out our hypothesis-testing steps. Therefore, prior to engaging in hypothesis testing, we must somehow estimate the population variance of the dependent variable under consideration.

The principal focus of this chapter is on the mechanism by which we go about estimating population parameters. As we shall see, estimation of population parameters is based on sample statistics.

POPULATION PARAMETERS FROM SAMPLE STATISTICS

Let's consider a typical experimental situation that would necessitate estimation of some population variance. Suppose we are psychologists interested in memory, and as part of our research we do a free-recall experiment. As described previously, a free-recall experiment involves reading a list of unrelated words (say, 20 words) to a subject, one word at a time. Following the presentation of these words the subject is asked to recall as many of the words as possible. The dependent variable in this sort of experiment is the number of words the subject remembers.

Suppose that a random sample of $n = 5$ people is drawn from the population. Each of these five individuals is administered a free-recall test and each gets a free-recall score. Hypothetical scores are shown in Table 9-1. Descriptive statistics have been computed from this sample of five scores: the sample mean $M = 13$, and the sample variance $S^2 = 2$.

We realize that this sample of five scores comes from some population about which we wish to make inferences or conclusions. This is the population of scores that would emerge if we were to obtain a free-recall score from each and every person to whom we would like to generalize our conclusions. We have no way of knowing, of course, what the mean and variance of this population of scores are.

> 9.1 When population parameters are unknown, we can estimate them from sample statistics.

We can make educated guesses based on our sample statistics about what the population parameters are. Before discussing how we make these educated guesses, however, let's ponder a bit about what our goals are. What does it mean to estimate a parameter? More specifically what kinds of *criteria* do we want to use for what constitutes a good estimate of some population parameter? We shall focus on two such criteria, which are that a parameter estimate should be (a) *unbiased* and (b) *consistent*. To facilitate matters, we will couch our discussion within the example of estimating μ, the mean, of the population of free-recall scores we described above.

TABLE 9-1.   Hypothetical Data Obtained by Five Subjects in a Free-Recall Experiment. Scores Represent Number of Correctly Recalled Words out of Twenty

Subject	Score (x)	$(x - M)$	$(x - M)^2$
1	13	0	0
2	11	-2	4
3	12	-1	1
4	15	2	4
5	14	1	1
	$\Sigma x = 65$		$\Sigma(x - M)^2 = 10$
	$M = \dfrac{\Sigma x}{n}$		$S^2 = \dfrac{\Sigma(x - M)^2}{n}$
	$= \dfrac{65}{5} = 13$		$= \dfrac{10}{5} = 2$

The only clue we have about μ is that M, the mean of our *sample,* turned out to be 13. Assuming that the sample of scores constitutes all of our information, this sample mean of 13 is the best estimate of the population mean that we can obtain. That is, M is both an unbiased and a consistent estimate of μ. Let us see what is meant by this.

Unbiased Estimates

Mathematically to say that the sample mean, M, is an unbiased estimate of the population mean, μ, is to say that

$$E(M) = \mu$$

That is, the *expectation* of any given sample mean is the population mean.

Let's try to explain a bit more concretely what we mean by this. Suppose that we were to take a great many samples from our population of scores. Consider, in fact, *all possible samples* of scores that we could take from the population and suppose that we were able to compute the mean of each and every sample. Now the term "unbiased" may be illustrated as follows. The sample mean, M, is an unbiased estimate of the population mean, μ, because the mean of all sample means (to which we shall refer to as M_M) will be equal to the population mean.

This notion can be stated in a slightly different way. Suppose we begin taking samples, computing the mean, M, of each sample, and computing M_M, the mean of all sample means. Now suppose we choose some arbitrarily small (but nonzero) number, ϵ. It is then the case that M is an unbiased estimate of μ because the more samples we have, the smaller we expect the difference to be between M_M and μ—that is,

$$p(|M_M - \mu| > \epsilon)$$

approaches zero as the number of samples increases.

> 9.2 A sample statistic is an unbiased estimator of the corresponding population parameter if the expectation of the sample statistic is equal to the population parameter.

Consistent Estimates

The notion of consistency bears a close relationship to the notion of unbiasedness, which we have just discussed, and rests on the following train of logic. Suppose we draw a "sample" consisting of the entire population. In this case any sample statistic (such as M) must by definition be *equal* to the corresponding population parameter (μ, in this example). If on the other hand we were to take a very small sample, we would not have much reason to expect that the sample statistic will be particularly close to the corresponding population parameter. Now for a sample statistic to be consistent, it must be true that the larger the sample size, the closer we expect the value of the sample statistic to be to the corresponding population parameter. Considering once again our sample mean, M, and our arbitrarily small number, ϵ, this notion is summarized mathematically as

$$p(\,|\,M - \mu\,|\, > \epsilon)\qquad \text{approaches zero as } n \text{ increases}$$

where n is the sample size. That is, the larger the sample size, the closer we expect M, the sample mean, to be to μ, the population mean.

> 9.3 A sample statistic is a consistent estimator of the corresponding population parameter if the expected difference between the sample statistic and the population parameter steadily decreases as n, the sample size, increases.

Sampling Distributions Revisited

The fact that the sample mean is an unbiased and consistent estimate of the population mean may be related to our discussion in Chapter 7 of the sampling distribution of sample means. In the course of that discussion, we demonstrated that

$$E(M) = \mu$$

which, as we now realize, is equivalent to saying that M is an unbiased estimate of μ. Additionally in Chapter 7 we paid great notice to the fact that the variance of the distribution of sample means, σ_M^2, decreases as n, the same size, increases. That is, as n becomes larger and larger, we expect M to be closer and closer to μ. As we now know, this is equivalent to saying that M is a consistent estimate of μ.

ESTIMATION OF THE POPULATION VARIANCE

Dealing with variances is slightly more complicated than dealing with means. If we want to estimate the population variance, σ^2, using the sample variance, S^2, our first inclination may be simply to assume that the sample variance S^2 is our best estimate of σ^2 (just as M is the best estimate of μ). To check this inclination, let's inquire about the expectation of the sample variance. Digression 9.1 shows that the expectation of S^2, the sample variance, is actually not σ^2, but rather is $\sigma^2 - \sigma_M^2$ (where σ_M^2 is a variance of the distribution of sample means based on samples of size n). It is immediately evident, therefore, that S^2 is not our best estimate of σ^2.

Digression 9.1

Expected Value of S^2

Let's initially establish a few facts we shall need to use. First σ^2, the population variance, may be expressed as

$$\sigma^2 = E(x - \mu)^2 = E(x^2) - \mu^2$$

or

$$E(x^2) = \sigma^2 + \mu^2 \tag{D.1}$$

Second σ_M^2, the variance of a distribution of sample means based on samples of size n, may be expressed as

$$\sigma_M^2 = E(M - \mu)^2 = E(M^2) - \mu^2$$

or

$$E(M^2) = \sigma_M^2 + \mu^2 \tag{D.2}$$

Now we know that

$$S^2 = \frac{\Sigma(x - M)^2}{n} - \frac{1}{n}\Sigma x^2 \quad M^2$$

and therefore

$$E(S^2) = E\left[\frac{1}{n}\Sigma x^2 - M^2\right]$$

$$= E\left[\frac{1}{n}\Sigma x^2\right] - E(M^2)$$

$$= \frac{1}{n}\Sigma E(x^2) - E(M^2) \tag{D.3}$$

Substituting Equations D.1 and D.2 into Equation D.3,

$$E(S^2) = \frac{1}{n}\Sigma(\sigma^2 + \mu^2) - (\sigma_M^2 + \mu^2)$$

Since $(\sigma^2 + \mu^2)$ is a constant,

$$E(S^2) = \frac{1}{n}[n(\sigma^2 + \mu^2)] - \sigma_M^2 - \mu^2$$

$$= \sigma^2 + \mu^2 - \sigma_M^2 - \mu^2$$

or finally

$$E(S^2) = \sigma^2 - \sigma_M^2$$

Unbiasing the Estimate of σ^2

Unbiasing our estimate of σ^2 is not difficult, however, and the logic of the unbiasing process is as follows. We have shown that

$$E(S^2) = \sigma^2 - \sigma_M^2$$

Now, since $\sigma_M^2 = \sigma^2/n$

$$E(S^2) = \sigma^2 - \sigma^2/n$$

$$= \frac{n\sigma^2 - \sigma^2}{n}$$

Factoring out a σ^2,

$$E(S^2) = \frac{\sigma^2(n-1)}{n}$$

Therefore,

$$\left[\frac{n}{n-1}\right] E(S^2) = \sigma^2$$

or

$$E\left[\frac{nS^2}{n-1}\right] = \sigma^2 \tag{9.1}$$

Now, since

$$S^2 = \frac{\Sigma(x-M)^2}{n} \tag{9.2}$$

we can substitute the expression for S^2 from Equation 9.2 into Equation 9.1 to obtain

$$E\left[\frac{\Sigma(x-M)^2}{n-1}\right] = \sigma^2$$

Let's refer to

$$\frac{\Sigma(x-M)^2}{n-1}$$

as "*est* σ^2" (referring[1] to "estimate of σ^2"). We see that

---

[1] *est* σ^2 is actually a sample statistic since it is calculated based on data from the sample. Therefore, by convention it ought to be represented by a Roman letter and in fact is often referred to as "$\hat{S}^2$." We prefer the notation "*est* σ^2" since it provides a convenient mnemonic. However, it does violate our notational convention, since it involves a Greek letter referring to a sample statistic.

$$E(est\ \sigma^2) = \sigma^2$$

or that *est* σ^2 is an unbiased estimate of σ^2, the population variance.

> 9.4 The best estimate of σ^2, the population variance, is
>
> $$est\ \sigma^2 = \frac{\Sigma(x - M)^2}{n - 1} = \frac{n}{n - 1} S^2$$

The Meaning of "Variance"

Let's take this opportunity to forestall some potential confusion about the meaning of the term *variance*. Different statistics books and different people have differing conceptions of whether this term means

$$\frac{\Sigma(x - M)^2}{n}$$

or whether it means

$$\frac{\Sigma(x - M)^2}{n - 1}$$

In other words, to appropriately obtain "variance" should one divide the sum of the squared deviation scores by n or by $n - 1$?

As we have just seen, we divide either by n or by $n - 1$ depending on what we want. The sample variance, S^2, obtained by dividing by n is only *descriptive*. So if we have a sample of scores and our only wish is to describe the characteristics of this sample, it makes sense to compute a sample variance—the average squared deviation score. Imagine, for example, that Professor Ziplock is describing the distribution of midterm exam scores as he returns the exams to his class. Since Ziplock's interest is only in characterizing *this particular* set of scores, the appropriate statistic for him to use would be S^2 (or S).

If, on the other hand, we were primarily interested in the population from which a sample was drawn, then we would probably want the best estimate of the population variance, and in such an instance the sum of the squared deviation scores should be divided by $n - 1$ rather than by n. An example of this sort of situation would be the free-recall experiment described above. Here, we are not interested in the sample of five people per se; rather, we are interested in the population from which the sample comes.

We see, then, that the term "variance" has two quite distinct meanings. In an attempt to emphasize the distinction, we have given the two meanings quite different notation. To reiterate, S^2 is a descriptive statistic—the variance of a sample. On the other hand *est* σ^2, while also computed from a sample, is the best estimate of the population variance. In the free-recall example described

$$S^2 = \frac{\Sigma(x - M)^2}{n} = \frac{10}{5} = 2.0$$

whereas

$$est\ \sigma^2 = \frac{\Sigma(x - M)^2}{n - 1} = \frac{10}{4} = 2.5$$

9.5 Variance has two meanings.
    As a descriptive statistic for a set of scores,

$$\text{Variance} = S^2 = \frac{\Sigma(x - M)^2}{n}$$

As an estimate of the population variance,

$$\text{Variance} = est\ \sigma^2 = \frac{\Sigma(x - M)^2}{n - 1}$$

Digression 9.2

Why Is S^2 an Underestimate of σ^2?

Going back to our free-recall example, suppose we somehow knew that the population mean were $\mu = 12$. Suppose further that we still don't know what the variance is but that we want to estimate it. In such a situation we could compute a sort of a "sample variance" by subtracting from each score *not* the sample mean M, but rather the population mean (since μ is assumed to be known). Now as depicted in the first column of Table D.1, the sum of the squared deviation scores would be equal to 15 instead of 10, and the "variance" that emerges is 15/5, or 3. Now we make a very important assertion, which is that

$$\frac{\Sigma(x - \mu)^2}{n}$$

or the average squared deviation of a set of scores from the *population mean* is an unbiased estimate of σ^2. The proof of this assertion is quite simple.

We know from Chapter 3 that

$$E\left[\frac{\Sigma(x - \mu)^2}{n}\right] = E(x^2) - \mu^2 \tag{D.1}$$

TABLE D.1.   Examples of "Sample Variances" Obtained When Deviation Scores Are Taken Around a Number that Is Slightly Greater or Slightly Less than the Sample Mean

	Deviation Scores Around 12		Deviation Scores Around 13		Deviation Scores Around 14	
x	$(x-12)$	$(x-12)^2$	$(x-M)$	$(x-M)^2$	$(x-14)$	$(x-14)^2$
13	1	1	0	0	−1	1
11	−1	1	−2	4	−3	9
12	0	0	−1	1	−2	4
15	3	9	2	4	1	1
14	2	4	1	1	0	0
	$\Sigma(x-12)^2 = 15$		$\Sigma(x-M)^2 = 10$		$\Sigma(x-14)^2 = 15$	

$$\text{``}S^2\text{''} = \frac{\Sigma(x-12)^2}{n} = \frac{15}{5} = 3 \qquad S^2 = \frac{\Sigma(x-M)^2}{n} = \frac{10}{5} = 2 \qquad \text{``}S^2\text{''} = \frac{\Sigma(x-14)^2}{n} = \frac{15}{5} = 3$$

and we know from Digression 9.1 that

$$E(x^2) = \sigma^2 + \mu^2 \tag{D.2}$$

Substituting Equation D.2 into Equation D.1,

$$E\left[\frac{\Sigma(x-\mu)^2}{n}\right] = \sigma^2 - \mu^2 + \mu^2 = \sigma^2$$

Continuing with our fantasies, let's now suppose that we knew the population mean to be 14 rather than 12. The third column of Table D.1 shows what our "sample variance" would now be. We see that the sum of squared deviations turns out to be 15, and our "sample variance" again turns out to be 3.

Now we make an interesting observation about the data in Table D.1. When we used the sample mean ($M = 13$) to compute S^2, we came up with a sample variance, S^2, of 2. However, when we computed a "sample variance" using deviation scores not around the sample mean but around some number that is either slightly smaller (12) or slightly larger (14) than the sample mean, we came up with something that was larger than the "real" sample variance obtained using deviation scores around the sample mean. Now we make another important assertion that stems from these illustrations. Suppose we have a set of scores, x_1, $x_2, \ldots, x_n$, and their mean M. Now we take any number (let's call it Q) and compute a "sample variance":

$$\text{``}S^2\text{''} = \frac{\Sigma(x-Q)^2}{n}$$

We assert that "S^2" will be smallest when Q is equal to M, the sample mean. The proof of this assertion is as follows.

First let

$$Y = \frac{\Sigma(x - Q)^2}{n}$$

Our strategy will be to take the derivative of Y with respect to Q. We will then set this derivative equal to zero and show that the resulting Q is equal to M. Taking the derivative,

$$\frac{dY}{dQ} = \frac{d[\Sigma(x - Q)^2/n]}{dQ}$$

$$= \frac{2\Sigma(x - Q)}{n}$$

Setting the derivative equal to zero,

$$\frac{2\Sigma(x - Q)}{n} = 0$$

$$\frac{\Sigma x}{n} - \frac{\Sigma Q}{n} = 0$$

Since Q is a constant,

$$\frac{\Sigma x}{n} - \frac{nQ}{n} = 0$$

$$\frac{\Sigma x}{n} - Q = 0$$

or finally

$$Q = \frac{\Sigma x}{n} = M$$

The reason we are interested in the assertion we just proved is that we want to bring together this rambling collection of facts to demonstrate why S^2 is an underestimate of σ^2. We have known since Chapter 3 that the sample variance, S^2, is computed by the expression

$$S^2 = \frac{\Sigma(x - M)^2}{n}$$

However, we have also seen that an unbiased estimate of the population variance is

$$est \ \sigma^2 = \frac{\Sigma(x - \mu)^2}{n}$$

Finally we have seen that whenever M is not equal to μ, the value computed by the expression

$$\frac{\Sigma(x - M)^2}{n}$$

will be smaller than the value computed by the expression

$$\frac{\Sigma(x - \mu)^2}{n}$$

Since in general M is not equal to μ, we therefore expect S^2 to be smaller than σ^2.

Sums of Squares and Degrees of Freedom

We are now going to introduce some new terms that will be used quite a bit in subsequent chapters. First the sum of a set of squared deviation scores

$$\Sigma(x - M)^2$$

is referred to as a *sum of squares* (for obvious reasons). Second the denominator by which a sum of squares is divided to get an estimate of a population variance ($n - 1$ in the examples that we have been discussing) is referred to as *degrees of freedom*. The rationale behind the term "degrees of freedom" is best conveyed by the following example. Suppose I ask you to make up a set of any five scores. This seems like a fairly easy exercise, and you rattle off the five numbers, 1, 69, 314, 32, and 100, which happen to be your favorite numbers. "There," you say, "are five numbers."

In this example you are under no constraints; you are perfectly at liberty to produce whatever five numbers you wish. Another way of putting it is that you have five degrees of freedom in this little exercise. Now, however, suppose I ask you to produce a set of five numbers whose mean is 25. This task is a bit different, because although you are perfectly at liberty to give me anything you wish for the *first four* numbers—for example 1, 69, 314, and 32—you would not have any liberty whatsoever when you arrive at the last score. The requirement that the mean of the scores be 25 means that the *sum* of the scores must be $25 \times 5 = 125$. Since the first five numbers you gave already add up to 416, the last number has no other choice but to be

$$125 - 416 = -291$$

So we see that if we have some number, n, of scores with a prespecified mean, then we only have $n - 1$ degrees of freedom left to fool around with. This is the situation we encounter when we estimate the population variance. We start off with a set of n scores, and we use all n scores to obtain the sample mean, M.

However, once we specify M, we only have $n - 1$ degrees of freedom left with which to estimate the population variance.

> 9.6 Degrees of freedom is equal to the number of observations minus the number of estimated parameters.

The Computational Formula for Variance

In Appendix D we derived a computational formula for variance, part of which was a computational formula for a sum of squares. This computational (or raw-score) formula for a sum of squares was:

$$SS = \Sigma(x - M)^2 = \Sigma x^2 - \frac{(\Sigma x)^2}{n}$$

To demonstrate the use of this formula, let's use it to compute $est\ \sigma^2$ from the data in Table 9-1. We first see that

$$SS = \Sigma(x - M)^2 = \Sigma x^2 - \frac{(\Sigma x)^2}{n}$$

Note that

$$\Sigma x^2 = 13^2 + 11^2 + 12^2 + 15^2 + 14^2 = 855$$

and

$$(\Sigma x)^2 = 65^2 = 4225$$

Thus

$$SS = \Sigma(x - M)^2 = 855 - 4225/5 = 855 - 845 = 10$$

From there, we proceed as before:

$$est\ \sigma^2 = SS/df = SS/n - 1 = 10/4 = 2.5$$

This raw-score formula for sum of squares may seem more tedious to use than the original deviation-score formula. That may be true in this simple example, but in more common, real-life examples with a large n and bigger numbers the raw-score formula is typically much easier to use. We'll use the raw-score formula quite often in this book from now on, and you should ultimately feel comfortable with it.

OBTAINING AN UNBIASED ESTIMATE OF THE VARIANCE OF A DISTRIBUTION OF SAMPLE MEANS

Let's go back to the free-recall experiment we have been using as an example during this chapter. We saw that the unbiased estimate of the population variance, σ^2, was 2.5. Suppose we are interested in the mean, M, of this sample of five scores. What is our best estimate of σ_M^2 — the variance of the distribution of sample means? Obtaining this best estimate is very simple. Just as

$$\sigma_M^2 = \frac{\sigma^2}{n}$$

it is also true that

$$est\ \sigma_M^2 = \frac{est\ \sigma^2}{n} = \frac{2.5}{0.5} = 0.5$$

Naturally, it is also true that to estimate the *standard deviation* of the distribution of sample means,

$$est\ \sigma_M = \sqrt{est\ \sigma_M^2} = \sqrt{0.5} = 0.707$$

> 9.7 Our best estimates of the variance and standard deviation of a sampling distribution of sample means are:
>
> $$est\ \sigma_M^2 = \frac{est\ \sigma^2}{n}$$
>
> and
>
> $$est\ \sigma_M = \sqrt{est\ \sigma_M^2} = \frac{est\ \sigma}{\sqrt{n}}$$

Suppose now that we wanted to go through a hypothesis-testing procedure using the sample mean. In the past we determined how extreme a sample mean is by computing a z-score. That is, we determined what the mean, variance, and standard deviation of the distribution of sample means would be if some null hypothesis were true, and then we computed a z-score using the standard equation,

$$z = \frac{M - \mu}{\sigma_M}$$

It might seem that we can do exactly the same thing in the situation involving an *estimated* population variance. That is, it seems as though we should be able to compute a z-score using the equation

$$z = \frac{M - \mu}{est\ \sigma_M}$$

but this turns out not to be true. We shall see that when σ^2 is estimated rather than known, then we can get something that is sort of like a z-score but is not really a z-score. Rather it's called a t-score, and instead of doing a z-test on our data we do a t-test instead. A t-test is one of the most commonly used statistical tests and is the topic of the next chapter.

SUMMARY

The major focus in this chapter has been on the ways in which sample statistics are used to estimate population parameters. The most important things to remember are the following:

1. The best estimate of a population mean, μ, is simply the sample mean, M.

2. The best estimate of a population variance, σ^2_M, is

$$est\ \sigma^2 = \frac{\Sigma(x - M)^2}{n - 1}$$

The estimate of the population variance—a sum of squares divided by its associated degrees of freedom—is to be distinguished from the sample variance, S^2, which is computed by

$$S^2 = \frac{\Sigma(x - M)^2}{n}$$

3. To estimate the variance of a distribution of sample means, we treat an estimated variance in exactly the same way as we have treated a known variance in previous chapters. That is, we simply divide our estimate of the population variance (as opposed to our known population variance) by n, where n is the number of subjects in the sample.

PROBLEMS

1. A statistics quiz was given which had a maximum score of 10. Students received the following scores:

Student	Score
1	3
2	9
3	6

Student	Score
4	5
5	7
6	4
7	5
8	2
9	7
10	1
11	8
12	3
13	5
14	6
15	9
16	4
17	5
18	1
19	7
20	4
21	2

1. Compute the mean and the variance *of this sample.*
2. Assume that the sample is a random sample from a very large population. Estimate the population mean and variance from the sample. How do these estimates differ from your answers in question 1? If they differ, why do they differ?

2. Suppose we are interested in the amount of beer served in mugs at O'Banion's Tavern. We obtain a sample of eight mugs of beer and discover that they contain the following amounts of beer.

Mug	Amount of beer (oz)
1	$x_1 = 15.5$
2	$x_2 = 16.2$
3	$x_3 = 14.8$
4	$x_4 = 16.0$
5	$x_5 = 16.0$
6	$x_6 = 15.9$
7	$x_7 = 16.3$
8	$x_8 = 16.5$

1. Compute the mean, variance, and standard deviation of this sample.
2. Estimate the mean, variance, and standard deviation of the *population* of amounts of beer served at O'Banion's.
3. Estimate the variance of the distribution of sample means that this sample mean comes from.

3. Suppose you have the following scores that come from some population:

$$x_1 = 2$$
$$x_2 = 4$$
$$x_3 = 4$$

   1. Compute the variance (S^2) of this sample.
   2. Estimate the variance (σ^2) of the population distribution that this sample was drawn from.
   3. Estimate the variance (σ_M^2) of the population distribution that the *mean* of this sample comes from.

4. For the data of Chapter 2, Problem 1, estimate the variance of the population that each set of scores comes from.

5.
   1. Use the data from Chapter 7, Problem 8. For each of the two groups, estimate *both* the variance of the population that the scores come from and the variance of the distribution that the sample mean comes from.
   2. Do the same using the data from Chapter 7, Problems 10 and 11.

6. Joe Shablotnik is test manager for Fonda motorcycles and is testing the amount of time it takes Fondas to do the standing quarter-mile.

   1. A prototype Fonda shows the following times in six different trials:

   Trial 1: 10 sec
   Trial 2: 11 sec
   Trial 3: 10 sec
   Trial 4: 9 sec
   Trial 5: 10 sec
   Trial 6: 12 sec

   What is your best estimate of the variance of standing quarter-mile times for this particular prototype Fonda?

   2. Joe is now given a random sample of five production Fondas which show the following standing quarter-mile trials:

   Fonda 1: 9 sec
   Fonda 2: 12 sec
   Fonda 3: 15 sec
   Fonda 4: 8 sec
   Fonda 5: 10 sec

   What is your best estimate of the variance of standing quarter-mile times for production Fondas?

t-Test

10

In the previous chapter we discussed the process by which we estimate unknown parameters of some distribution, and we saw that such estimation is based on sample statistics. We noted that the most important parameter we typically need to estimate is σ^2, the population variance. **This estimation of σ^2 which is obtained using the formula**

$$est \; \sigma^2 = \frac{\Sigma(x - M)^2}{n - 1} = \frac{\text{sum of squares } (ss)}{\text{degrees of freedom } (df)}$$

constitutes a preliminary step that must be carried out prior to hypothesis testing.

We ended the last chapter by stating that the appropriate test to be used in a situation involving an estimated rather than a known population variance is sort of like a z-test but is not exactly a z-test. Rather, the test is called a t-test and in place of the formula

$$z = \frac{x - \mu}{\sigma}$$

we use the formula

$$t = \frac{x - \mu}{est \; \sigma}$$

In either case μ refers to the mean of the population distribution of our summary score, assuming some null hypothesis is true, and σ refers to the standard deviation of this distribution. In either case we want to see how extreme our obtained z or t is. If it is extreme enough, we will decide that the null hypothesis is not correct and will decide to reject the null hypothesis in favor of the alternative hypothesis.

This chapter will discuss several aspects of t distributions and corresponding t-tests. We'll start by describing a relatively simple situation in which a sample mean is tested against a constant. This initial section will also serve as a forum in which general characteristics of t-tests and t distributions will be discussed. We'll then go on to describe several other, somewhat more complex, situations requiring use of a *t*-test.

> 10.1 A *t*-test is similar to a *z*-test except that it involves an unknown (estimated) as opposed to a known population variance.

MEMORY FOR THE MUELLER-LYER ILLUSION: TESTING A MEAN AGAINST A CONSTANT

A fairly well-known illusion in perceptual research is the Mueller-Lyer illusion, shown in Figure 10-1. Although the two horizontal lines are the same length, the top one is typically seen as longer.

Suppose we want to determine whether *memory* for these two lines is illusory as well. To investigate this, we perform an experiment in which a 6 cm line with the "outward" arrows (part *a* of the figure) is presented to a subject for 10 sec and then removed. The subject's task is then to draw a line the same length as the original line. Let's assume that this experiment is run on a random sample of $n = 5$ subjects. The data for these subjects (average length of the line drawn) are shown in Table 10-1. We know that if the illusion persists in memory, then the lines drawn by the subjects should be *longer* than 6 cm.

The First Step: Estimating the Population Variance

Since we have no idea what the population variance of our dependent variable (drawn-line length) is, our first step is to estimate it, using the technique described in the previous chapter. The estimation procedure for this set of scores is carried out in Table 10-1. As you can see, the best estimate based on four degrees of freedom is that σ^2 is equal to 5.0.

Hypothesis-Testing Steps

Having estimated σ^2, we are now in a position to go through our hypothesis-testing steps. As usual, the first step is to establish null and alternative

FIGURE 10-1.   An illustration of the Mueller-Lyer illusion.

TABLE 10-1.   Line Lengths (in cm) Reflecting Memory for a 6-cm Mueller-Lyer Illusion Line

Person	x_i = Line length (cm)	x_i^2
1	$x_1 = 8$	64
2	$x_2 = 11$	121
3	$x_3 = 9$	81
4	$x_4 = 5$	25
5	$x_5 = 7$	49
	$\Sigma x = 40$	$\Sigma x^2 = 340$

$$M = \frac{\Sigma x}{n} = \frac{40}{5} = 8$$

$$SS = \Sigma (x - M)^2$$

$$= \Sigma x^2 - (\Sigma x)^2/n$$

$$= 340 - (40)^2/5$$

$$= 20$$

$$df = n - 1 = 5 - 1 = 4$$

$$est\, \sigma^2 = \frac{SS}{df} = \frac{20}{4} = 5$$

hypotheses, and as is usually the case, the null hypothesis states that the independent variable (in this case whether the line as first presented is seen to be long, an illusion) has an effect on the dependent variable (in this case length of the line the subject draws). That is,

H_0: The length of the line that is drawn shows no bias. That is, the sample of scores shown in Table 10-1 comes from a population whose mean, μ, is equal to 6 cm.

The alternative hypothesis on the other hand maintains that the independent variable *does* have an effect on the dependent variable, or

H_1: The sample of scores comes from a population whose mean, μ, is greater than 6 cm.

At this point we need some kind of a summary score that reflects the extent to which the alternative hypothesis as opposed to the null hypothesis is true. An obvious summary score to use would be the *mean* length, M, of lines drawn by members of the random sample, which has been calculated in Table 10-1 to be 8.0 cm. The mean, of course, provides a clue as to which hypothesis is correct in the sense that, if the null hypothesis were true, then M is expected to be around 6.0. But to the degree that the alternative hypothesis is true, the sample mean should be greater than 6.0.

As noted, the actual sample mean is 8.0. The fact that 8.0 is greater than 6.0 suggests that the null hypothesis is false—that the Mueller-Lyer illusion is maintained in memory. However, the sample mean could of course be higher than 6.0 due just to chance factors. Therefore, it is crucial to determine how the summary score—the sample mean—would be distributed if the null hypothesis were true. Knowing this distribution will permit an assessment of how extreme is the obtained sample mean.

We know that if the null hypothesis is true, then M is distributed with a mean, μ, equal to 6. We have estimated that the population variance (that is, the variance of the distribution of scores) in the population is equal to 5.0. Having estimated σ^2, the population variance, we are now able to estimate σ_M^2, the variance of the distribution of our *summary score*—of our sample mean—which is obtained in the usual way:

$$est\ \sigma_M^2 = \frac{est\ \sigma^2}{n} = \frac{5.0}{5} = 1.0$$

And to estimate the standard deviation of the distribution of sample means,

$$est\ \sigma_M = \sqrt{est\ \sigma_M^2} = \sqrt{1.0} = 1.0$$

As noted above, we can now form something that looks sort of like a z-ratio but is actually a t-ratio. This t-ratio is

$$t = \frac{M - \mu}{est\ \sigma_M} = \frac{8.0 - 6.0}{1.0} = 2.0$$

Having obtained a t, we would now like to determine how extreme the t is (assuming the null hypothesis is true). If it turns out to be sufficiently extreme that it would only occur with a probability of 0.05 or less, then we will conclude that the null hypothesis is actually false. The criterion z that chopped off the upper 5% of the z distribution is 1.64. But what is the criterion t? Let's digress a bit to consider this question.

Intuitions about the criterion t.   Until about 50 years ago, people didn't really know about what the t distribution looked like and were therefore unable to determine what the criterion t should be. However, they did have some intuitions. "First of all," they noted, "if we were dealing with a z rather than with a t—that is, if the thing in the denominator were a *known* variance rather than an estimated variance—then the criterion would be 1.64. However, since the population variance is *not* known, but rather is estimated, there is probably somewhat more variability in a t than there is in a z.

"Therefore," they reasoned, "the criterion t should probably be set somewhat higher than 1.64 to be on the safe side." How *much* higher? This raises another consideration. The fewer the degrees of freedom used to estimate σ^2 (that is, the smaller the sample size), the more variable is the estimate of σ^2 and, therefore, the higher the criterion ought to be. The other side of the coin is that if the estimate of σ^2 is based on a large number of degrees of freedom—and is therefore a very good estimate—then the denominator of the t becomes much like a known variance, and the t itself becomes much like a z. Therefore, the higher the sample size, the closer the criterion t should be to the criterion z value of 1.64. In any case one can read old articles and find statements such as "$(M - \mu)/est\ \sigma_M$ was 3.8 with a sample size of $n = 5$. This pretty much seems high enough, so we'll reject the null hypothesis."

> 10.2 A criterion t is expected to be larger than a criterion z. This is because the former involves more uncertainty than the latter (an unknown rather than a known population variance).

The situation remained in this somewhat lackluster state until an English statistician named William Sealy Gosset decided to tackle the problem. Gosset determined what the form of the t distribution should look like, published it,[1] and thereby rescued the field of small-sample statistics from the embarrassing position in which it had been floundering.

Form of the t distribution.   We have just argued that even without knowing what the t distribution is like, we can reasonably infer a few things about it: that the better our estimate of the population variance, the more the t distribution should resemble the z distribution, and in the extreme, if we have a huge number of degrees of freedom, then the t distribution should be just about identical to the z distribution. If on the other hand we have only a very few scores on which to base our estimate of σ^2, then the t distribution will probably differ quite substantially from the z distribution. The way in which it should differ is that it should have greater variance—that is, the criterion t necessary to chop off any given portion of the distribution (for example the upper 5%) should be greater. This in turn means that the fewer the degrees of freedom, the shorter and squatter the t distribution should be relative to the z distribution.

An important implication of this train of logic is that the t distribution should not be just one unitary distribution, but rather should be a *family* of distributions; there should be one distribution corresponding to each number of degrees of freedom on which the estimate of σ^2 is based. Several members of the t distribution family are shown in Figure 10-2. The relative shapes of these distributions bear out what we have been saying: the fewer the degrees of freedom on which the estimate of σ^2 is based, the greater the variance of the corresponding t distribution. Likewise when the sample size gets quite large,

---

[1]Gosset was working for a brewery at the time he developed the t distribution and was not supposed to publish anything under his own name. Thus, he published under the pseudonym of "Student" instead and the t distribution is often referred to as "Student's t."

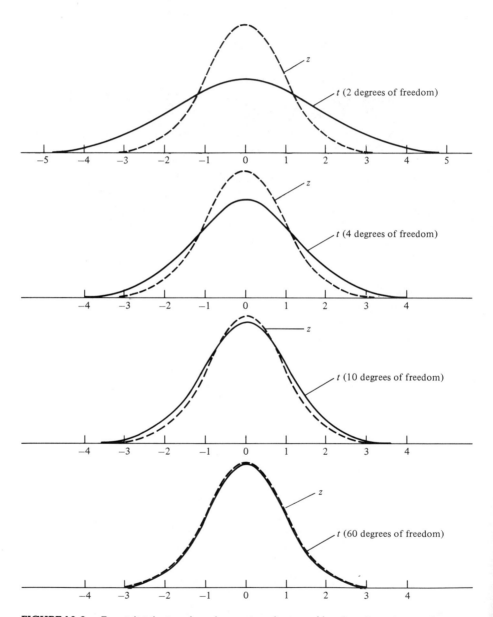

FIGURE 10-2. Four t distributions based on various degrees of freedom. In each case, the z distribution is shown for comparison. As degrees of freedom increase, the t distribution becomes more like a z distribution.

then the t distribution is virtually indistinguishable from the z distribution. Notationally a t based on some number, x, degrees of freedom is written as $t(x)$. In the example we discussed above, our t was based on 4 degrees of freedom and was 2.0. We would represent this as:

Obtained $t(4) = 2.0$

Tables of the t distribution.   Recall that our z distribution table permitted us to calculate any area under the normal distribution curve in which we happened to be interested. A z-table can be constructed and displayed with relatively little difficulty because there is only one z distribution to worry about. As we have just seen, however, such is not the case with the t distribution; there are as many t distributions as there are potential degrees of freedom—an infinity. Therefore, t distribution tables must be displayed in a somewhat abbreviated form, and in fact are usually collapsed into a single table. A t distribution table is shown on Appendix F of this book. It is abbreviated in that it shows only the criterion ts that chop off various selected portions of t distributions based on various selected degrees of freedom. The table is organized so that each row corresponds to some number of degrees of freedom. The first row corresponds to one degree of freedom, the second row corresponds to two degrees of freedom, and so on. Each column, corresponding to the chopping off of some particular percentage of the distribution, is labeled in two ways. The first way (labeled Q) corresponds to the α-level for a one-tailed test, and the other way (labeled $2Q$) corresponds to the a-level for a two-tailed test. Figure 10-3 gives examples of how these tables are used. All the distributions shown in Figure 10-3 are t distributions based on 4 degrees of freedom. Suppose, as depicted in part a of Figure 10-3, we wanted to do a one-tailed test using the 0.05 α-level. In such an instance we would want to find the criterion $t(4)$ that chops off the upper 5% of the distribution. To find this criterion, we would go to the row corresponding to 4 degrees of freedom in our t-tables and find a one-tailed (Q) α-level of 0.05. At the intersection of the appropriate row and column we would find a criterion $t(4)$ of 2.13. Now suppose, as shown in part b of Figure 10-3, we wanted to do a two-tailed t-test with an α-level of 0.05. Thus, we wish to find the criteria that chop off the upper and lower 0.025 of the distribution, so we'd look once again at the row corresponding to 4 degrees of freedom, but we would find the column corresponding to a two-tailed $(2Q)$ α-level of 0.05. There, we would find the entry 2.78. Due to the symmetry of the t-distribution we would know that ± 2.78 are the criteria we're seeking.

Notice that for any given column the one-tailed (Q) α-level is equal to half the two-tailed $(2Q)$ α-level. This makes sense. Suppose we wanted a two-tailed test using a 0.10 α-level as shown in part c of Figure 10-3. To do this, we would consult the $2Q = 0.10$ column of our table where we would find a criterion of 2.13. This indicates that criteria $t(4)$s of ± 2.13 chop off the upper 5% and the lower 5% of the distribution. But of course the criterion of 2.13 is exactly the same one-tailed criterion that chops off the upper 5% of the distribution.

a

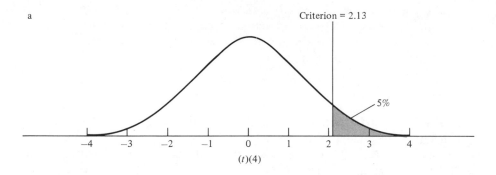

b

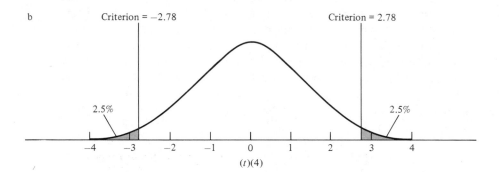

c

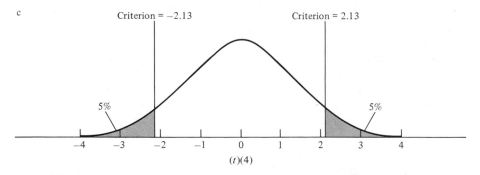

FIGURE 10-3.   Various criterion levels for a $t(4)$ distribution. Panel *a* shows a one-tailed (Q) 0.05 criterion. Panel *b* shows two-tailed $(2Q)$ 0.05 criteria. Panel *c* shows two-tailed $(2Q)$ 0.10 criteria. Note that the upper criteria for panels *a* and *c* are the same.

Let's look at one other noteworthy aspect of the t distribution table. The bottom row of the table refers to a t distribution based on infinite degrees of freedom. "Infinite degrees of freedom"is just another way of saying that the population variance is known rather than estimated, and, as indicated earlier, this particular t distribution is actually a z distribution traveling under an assumed name. Therefore, the various criteria are seen to be identical to the corresponding z distribution criteria. For example, the familiar old one-tailed test at the 0.05 α-level would involve a criterion z of 1.64. As can be seen in the t-table, this is exactly the criterion for a one-tailed, 0.05 α-level t-test for a t distribution based on infinite degrees of freedom.

10.3 To find the criterion t, do the following:

1. If you are doing a one-tailed test, refer to the top row labeled Q in the t-table. If you are doing a two-tailed test, refer to the row labeled $2Q$. In either case move across to the column corresponding to the desired α-level.
2. Find the row corresponding to the number of degrees of freedom on which the t-test is based (that is, the number of degrees of freedom on which $est\ \sigma^2$ is based).
3. The criterion will be at the intersection of the column from step 1 and the row from step 2.

What about the Mueller-Lyer illusion?   Knowing how to use the t distribution table allows us to quickly finish our example. We wish to do a one-tailed test at the 0.05 α-level, using a t distribution based on 4 degrees of freedom, so as we have seen, the criterion $t(4)$ is 2.13. Since our obtained $t(4)$ of 2.0 does not exceed the criterion $t(4)$ of 2.13, we fail to reject the null hypothesis and end up concluding that we have insufficient evidence to justify concluding that the Mueller-Lyer illusion is maintained in memory.

WITHIN-SUBJECTS DESIGNS

A common experimental situation that involves testing a sample mean against a constant is a within-subjects design. In a within-subjects design each subject participates, at one time or another, in all experimental conditions.

Within- Versus Between-Subjects Designs

A within-subjects design is to be distinguished from a *between-subjects design,* in which separate groups of subjects participate in the various conditions. So far in this book we have described examples of both within-subjects and between-subjects designs without explicitly referring to them as such. For example the SMARTIUM/PROGRAM X experiment (Chapter 7) was a between-subjects design in that one group of subjects was administered SMARTIUM and an entirely different group of subjects was administered PROGRAM X. In contrast, as part of our discussion of power (Chapter 8) we described an experiment involving the effect on teeth brightness of two different types of toothpaste (Dazzle and White-O). This experiment was a within-subjects design in that each subject was administered both different types of toothpaste (that is, was in both conditions) at different times.

In this section we will describe specifically how one goes about analyzing a within-subjects design.

An Example of Within-Subjects Designs: Marijuana and Heart Rate

Suppose we are researchers with the Department of Health and Human Services and we wish to investigate the effects of marijuana on various physiological responses. In particular we set out to determine whether ingestion of marijuana tends to change heart rate. To deal with this issue, we draw a random sample of $n = 9$ people from the population and perform the following experiment. At various times we give each of the nine subjects a brownie into which has been baked some amount of marijuana. At other times the same subjects are given a placebo consisting of a brownie spiked with oregano instead of marijuana.[2] Shortly after the subjects ingest the marijuana or the placebo, we measure their heart rate. Hypothetical heart rate scores from each subject in each of the two conditions are shown in Table 10-2.

The dependent variable: Each subject generates a difference score.   In this experiment our primary interest is on the *difference* between a given subject's heart rate in the marijuana versus the placebo condition; thus, for each subject, we accordingly obtain such a difference score. These scores (which are labeled x_i) are also shown in Table 10-2.

Statistical analyses performed on difference scores.   This set of difference scores—one for each subject—is all we have to concern ourselves with when we go about performing our statistical analyses. That is, once we have the set of difference scores, we do not worry about the original scores which gave rise to the difference scores. What we will ultimately do is calculate a mean difference score, M, and test M against a constant of 0.

Estimating the population variance.   As we have just observed, we will be concerned for statistical purposes only with the set of difference scores—the

---

[2]Placebos, used in conjunction with studies involving drugs, are substances with the appearance and other attributes of the drug under investigation but without actual relevant physiological effects. A placebo is used to control for any attitudinal effects. So for example if subjects show an increased heart rate in the marijuana relative to in the oregano condition, it is possible to conclude that this effect is due to the physiological effect of the marijuana rather than to the excitement of subjects thinking they're being given marijuana.

TABLE 10-2.   Hypothetical Results from Within-Subjects Design; the
Experimental Question Is Whether Ingestion of Marijuana Affects Heart Rate

Heart Rate in Beats Per Second

Subject	Condition 1 (marijuana)	Condition 2 (placebo)	$x_i = x_{i1} - x_{i2}$	x_i^2
1	$x_{11} = 76$	$x_{12} = 70$	$x_1 = +6$	36
2	$x_{21} = 61$	$x_{22} = 60$	$x_2 = +1$	1
3	$x_{31} = 52$	$x_{32} = 49$	$x_3 = +3$	9
4	$x_{41} = 71$	$x_{42} = 72$	$x_4 = -1$	1
5	$x_{51} = 81$	$x_{52} = 70$	$x_5 = +11$	121
6	$x_{61} = 70$	$x_{62} = 66$	$x_6 = +4$	16
7	$x_{71} = 55$	$x_{72} = 55$	$x_7 = 0$	0
8	$x_{81} = 61$	$x_{82} = 54$	$x_8 = +7$	49
9	$x_{91} = 89$	$x_{92} = 80$	$x_9 = +9$	81
			$M = 4.44$	$\Sigma x_i^2 = 314$

$$SS = \Sigma(x_i - M_d)^2 = \Sigma x_i^2 - \frac{(\Sigma x_i)^2}{n}$$

$$= 314 - \frac{(40)^2}{9} = 136.22$$

$$df = n - 1 = 9 - 1 = 8$$

$$est\ \sigma^2 = \frac{SS}{df} = \frac{136.22}{8} = 17.03$$

x_is—in Table 10.2. The first step in our statistical analysis is to estimate the
variance of the population distribution from which these x_is emerged. We do
this in a very straightforward way, computing the sum of squares (136.22) and
the degrees of freedom (*9-1*, or 8). Our estimate of the population variance is
therefore computed to be 136.22/8 = 17.03.

Hypothesis-testing steps.   Having estimated σ^2, we are now in a position to
carry out our hypothesis-testing steps. The first step is to set up the null and
alternative hypothesis, a process which at this point ought to be second nature.
In particular,

H_0: The independent variable (in this case marijuana versus placebo) has
no effect on the dependent variable (in this case heart rate). This
means that the sample of difference scores should come from a
population whose mean, μ, is equal to 0.

H_1: Marijuana does have an effect on heart rate. Thus, the scores from the
sample should come from a distribution whose mean, μ, is not equal to 0.

An appropriate summary score to examine would be the mean, M, which in this case is 4.44. To the extent that the null hypothesis is true, M should be around 0; to the extent that the alternative hypothesis is true, M would be expected to be different from 0. (Since the alternative hypothesis is nondirectional, the score could be either positive or negative if the alternative hypothesis were true.)

How is this score distributed if the null hypothesis is true? If H_0 is true, we know that M will come from a distribution whose mean, μ, is equal to 0. The variance of this sampling distribution of sample means is estimated to be

$$est\ \sigma_M^2 = \frac{est\ \sigma^2}{n} = \frac{17.02}{9} = 1.89$$

or

$$est\ \sigma_M = \sqrt{est\ \sigma_M^2} = \sqrt{1.89} = 1.37$$

To compute how extreme our obtained mean is, we calculate a t. Since the estimate of the population variance is based on 8 degrees of freedom, the obtained t will have 8 degrees of freedom as well. This $t(8)$ turns out to be

$$\text{Obtained } t(8) = \frac{M - \mu}{est\ \sigma_M} = \frac{4.44 - 0}{1.37} = 3.24$$

The criterion $t(8)$ is obtained as follows. We have 8 degrees of freedom, we are doing a two-tailed test ($2Q$), and we are using a 0.05 α-level. Looking at the t-table, we discover the criteria $t(8)$ to be ± 2.31. Since our obtained $t(8)$ of 3.24 is more extreme than the criterion $t(8)$ of 2.31, we reject the null hypothesis and conclude that ingestion of marijuana does have an effect on heart rate—specifically, marijuana appears to increase heart rate.

Difference Scores Versus Difference Between Sample Means

We should reiterate here that in a within-subjects design such as the one we have just described, we obtain for purposes of statistical analyses only a single score per subject—the difference between that subject's scores in the two conditions. Hence, we end up with only one sample of scores (the difference scores) and one sample mean. The population of scores with which we are then concerned is the population of difference scores. We therefore *directly estimate* the variance of this population, and we test our sample mean against a constant of 0 which would be μ, the mean of this difference score population if the null hypothesis were true. We would like to emphasize that the set of difference scores involved in a within-subjects design is not the same as a difference between two sample means, which is what we obtain in a between-subjects design. When dealing with a difference between sample means ($M_1 - M_2$), we did not directly compute the variance of the distribution from which ($M_1 - M_2$)

came. Rather, we computed the variance of each distribution of sampling means, $\sigma^2_{M_1}$ and $\sigma^2_{M_2}$. We then added these two variances together to get the variance of the distribution of mean differences.

> 10.4 In a within-subjects design, we compute a single score—a differ-
> ence score—for each subject. The mean of these difference scores
> is then tested against a constant, which is typically zero. The test is
> therefore based on $n - 1$ degrees of freedom where n is the
> number of subjects.

TESTING THE DIFFERENCE BETWEEN TWO SAMPLE MEANS

We have just discussed the mechanics of performing a t-test in situations involving a single sample mean that is to be tested against a constant. Our concern in this section is with a between-subjects design, in which two groups or conditions are to be compared. We described such a situation earlier in the context of a z-test. Recall, for example, the SMARTIUM/PROGRAM X experiment in which SMARTIUM was administered to one group of students and PROGRAM X was administered to a second group. The two groups were then compared to see which had a higher average math GRE score. The summary score obtained in this situation consisted of a difference score—the difference between the means of the two samples. We then established a null hypothesis stating that these two sample means came from the same distribution. Therefore, if the null hypothesis were true, then our summary score—the difference between the sample means—would come from a population whose mean, $\mu_{M_S - M_X}$ was 0 and whose variance, $\sigma^2_{M_S - M_X}$ was equal to $\sigma^2_{M_S} + \sigma^2_{M_X}$. The test we did in this situation was a z-test, which was appropriate because the relevant population variances were known rather than estimated.

When we deal with a situation that is analogous but involves an unknown population variance, we go through much the same steps except that (a) we must estimate the population variance prior to carrying out our hypothesis-testing steps, and (b) we perform a t-test rather than a z-test. Let's consider such a situation.

Hypnosis and Memory

In recent years hypnosis has become a respectable topic for scientific study. One long-standing hypothesis about hypnosis is that hypnotized people are able to remember things better. For example there have been reports of hypnotized individuals who are able to "mentally place themselves" back into their elementary school classes and apparently dredge up very accurate memories of events that had occurred long ago.

Suppose that we set out to test this "memory enhancement" hypothesis in a controlled laboratory situation using a simple free-recall test. To do this we first pluck a ten-person random sample out of the population which we wish to generalize. To this sample of subjects we read a list of 20 unrelated words. At this point we randomly divide the sample into two groups of five people each. All people in Group 1 (the experimental group) are hypnotized and given the suggestion that they will be able to remember all 20 of the words they just heard. The people in Group 2 (the control group) are not hypnotized. All ten subjects then attempt to recall the words.

Hypothetical data from this experiment are represented in Table 10-3. The mean number of words recalled from each of the two groups has been computed: the hypnosis group remembers an average of 8 of the 20 words, whereas the control group remembers an average of 5 of the 20 words. Since hypnotized people have on the average remembered more words than unhypnotized people, our inclination is to think that hypnosis has a genuine effect in terms of boosting memory performance. However, the advantage of the hypnosis group could be due just to chance factors, and we prudently plan on carrying out the hypothesis-

TABLE 10-3. Hypothetical Results of an Experiment in Which Either Hypnotized or Unhypnotized Subjects Were Tested for Memory Performance Using a Free-Recall Procedure

Group 1 (Hypnosis)		Group 2 (Control)	
x_{i1}	x_{i1}^2	x_{i2}	x_{i2}^2
$x_{11} = 9$	81	$x_{12} = 6$	36
$x_{21} = 7$	49	$x_{22} = 4$	16
$x_{31} = 10$	100	$x_{32} = 4$	16
$x_{41} = 6$	36	$x_{42} = 3$	9
$x_{51} = 8$	64	$x_{52} = 8$	64
$\Sigma x_{i1} = 40$	$\Sigma x_{i1}^2 = 330$	$\Sigma x_2 = 25$	$\Sigma x_{i2}^2 = 141$

$M_1 = \Sigma x_{i1}^2/n = 40/5 = 8$

$SS_1 = \Sigma(x_{i1} - M_1)^2$

$\quad = \Sigma x_{i1}^2 - (\Sigma x_{i1})^2/n$

$\quad = 330 - (40)^2/5$

$\quad = 10$

$df_1 = n - 1 = 5 - 1 = 4$

$est_1 \, \sigma^2 = \dfrac{SS_1}{df_1} = \dfrac{10}{4} = 2.5$

$M_2 = \Sigma x_{i2}^2/n = 25/5 = 5$

$SS_2 = \Sigma(x_{i2} - M_2)^2$

$\quad = \Sigma x_{i2}^2 - (\Sigma x_{i2})^2/n$

$\quad = 141 - (25)^2/5$

$\quad = 16$

$df_2 = n - 1 = 5 - 1 = 4$

$est_2 \, \sigma^2 = \dfrac{SS_2}{df_2} = \dfrac{16}{4} = 4.0$

testing steps before making a decision about the matter. Prior to our hypothesis testing, however, we must estimate the variance of the population distribution from which the free-recall scores came. Let's see how this is done.

The population variance: Combining estimates.   As indicated in Table 10-3, having two separate groups puts us in the position of being able to compute not one but two estimates of σ^2, the population variance—one estimate from each group. We see in Table 10-3 that the estimate from the hypnosis group, $est_1 \sigma^2$ is 2.5. The estimate of the population variance from the control group, $est_2 \sigma^2$ is 4.0.

We now have two separate, independent estimates of the population variance. The first estimate is based on 4 degrees of freedom (from the five subjects in the experimental group) and the second estimate is also based on 4 degrees of freedom (from the five subjects in the control group). However, we do not need, nor do we want, two estimates of the population variance. We would simply like to have the one best estimate. To obtain a best estimate, we do the obvious thing: we take the average (mean) of the two estimates we have. Thus,

$$est\ \sigma^2 = \frac{est_1\ \sigma^2 + est_2\ \sigma^2}{2} = \frac{2.5 + 4.0}{2} = \frac{6.5}{2} = 3.25$$

The one best estimate is based on 4 degrees of freedom from the hypnosis group plus 4 more degrees of freedom from the control group, or 8 degrees of freedom in all. Since our one best estimate of the population variance is based on 8 degrees of freedom, we will ultimately do a t-test based on 8 degrees of freedom.

10.5 When two groups of subjects are run, a separate estimate of σ^2 is obtained from each group. These two estimates, $est_1\ \sigma^2$ and $est_2\ \sigma^2$ are then averaged to obtain one best estimate of σ^2.

Digression 10.1

Within- Versus Between-Groups Variance: A Preview

You may be wondering why we've gone to the trouble of computing two separate estimates of the population variance and averaging them. You may think it seems reasonable to merely compute one best estimate directly from all ten scores. To get a feeling for this issue, consider the data in Table D.1, assumed to be from some experiment involving two groups with two scores per group. As can be seen in part a of Table D.1, the two estimates of σ^2 from the two groups, $est_1\sigma^2$ and $est_2\sigma^2$ are 2.0 and 0.5, respectively; thus, the combined (best) estimate is 1.25.

TABLE D.1.   Illustration of Why Separate Estimates of the Population Variance Are Obtained from Separate Groups

---

a Treating the two groups separately:

Condition 1 ($n = 2$)		Condition 2 ($n = 2$)	
$x_{11} = 0$	$x_{11}^2 = 0$	$x_{12} = 100$	$x_{12}^2 = 10{,}000$
$x_{21} = 2$	$x_{21}^2 = 4$	$x_{22} = 101$	$x_{22}^2 = 10{,}201$
$\Sigma x_{i1} = 2$	$\Sigma x_{i1}^2 = 4$	$\Sigma x_{i2} = 201$	$\Sigma x_{i2}^2 = 20{,}201$

$$est_1\, \sigma^2 = \frac{\Sigma x_{i1}^2 - (\Sigma x_{i1})^2/n}{1} \qquad\qquad est_2\, \Sigma^2 = \frac{\Sigma x_{i2}^2 - (\Sigma x_{i2})^2/n}{1}$$

$$= 2.0 \qquad\qquad\qquad\qquad\qquad\qquad = 0.50$$

$$est\, \sigma^2 = \frac{est_1\, \sigma^2 + est_2\, \sigma^2}{2} = \frac{2 + 0.5}{2} = 1.25 \ (\text{based on 2 } df)$$

---

b Combining all scores into one big group ($n = 4$):

x	x^2
0	0
2	4
100	10,000
101	10,201
$\Sigma x = 203$	$\Sigma x = 20{,}205$

$$SS = \Sigma x^2 - (\Sigma x)^2/n = 20205 - (203)^2/4 = 9902.75$$

$$df = n - 1 = 3$$

$$est\, \sigma^2 = \frac{SS}{df} = \frac{9902.75}{3} = 3300.92$$

However, combining all four scores into one group (part *b* of Table D.1), we "estimate" σ^2 to be over 3000. Why the discrepancy? The answer lies in the fact that there is obviously a good deal of difference *between the two groups*—a difference presumably due to whatever differential treatments we have applied to Group 2 versus Group 1. This treatment effect adds a huge amount to the total variance among our four scores.

The idea we are trying to convey here is that there are two sources of variance contributing to differences among the four scores in the experiment. The first source is the population variance, σ^2, which causes variation of scores *within* each of the two groups (that is, it underlies the fact that the two scores in Group 1 differ from each other and that the two scores in Group 2 differ from each other). The second source of variance is the treatment, which is applied differentially to the groups and therefore causes variation of scores *between* the two groups. To estimate σ^2, we must consider only variation within a single group; this is why we obtain a separate estimate from within each group and then average the two estimates.

In general our principal task in this or any experiment is to consider the total variance among all the scores in the experiment and then determine which of it arises as a consequence of the population variance (within-group variance) versus which arises as a consequence of our between-group treatments (between-group variance). This *analysis of variance* will be the topic of several subsequent chapters.

Hypothesis-testing steps.   Having an estimate of the population variance all tucked away and ready to use, we are ready to embark on the now-familiar hypothesis-testing steps. The first step is to set up the null and the alternative hypotheses. The null hypothesis is that the independent variable (in this case hypnosis versus control) has no effect on the dependent variable (in this case number of words recalled). That is,

H_0: Scores from the experimental (hypnosis) group come from the same population as do the scores from the control group. Thus, if μ_1 is the mean of the hypnosis population scores and μ_2 is the mean of the control population scores,

$$\mu_1 = \mu_2$$

The alternative hypothesis is that hypnosis does have an effect on memory performance—specifically that hypnosis will enhance memory performance. Therefore, the alternative hypothesis may be stated as

H_1: The scores from the hypnosis group come from a population whose mean, μ_1, is higher than μ_2, the mean of the population of control-group scores.

The hypothesis-testing step is to obtain a summary score. As suggested, an appropriate summary score in this situation would be the difference between the means of the hypnosis and the control groups, or

$$(M_1 - M_2) = (8 - 5) = 3$$

This score is appropriate in the sense that we would expect it to be around zero if the null hypothesis were true, whereas we would expect it to be greater than zero if the alternative hypothesis were true.

Now we inquire as to how this difference score would be distributed if the null hypothesis were true. In such an event, M_1 and M_2 should come from the same distribution, and $(M_1 - M_2)$ should come from a distribution of difference scores whose mean, $\mu_{M_1-M_2}$, is equal to zero. What about the variance of this distribution of difference scores? As we have seen, the variance of a distribution of difference scores is equal to the sum of the variances of the distributions of the original scores. In this present example we have only estimates of the relevant population variances, so the difference score comes from a distribution whose variance is also estimated, or

$$est \ \sigma^2_{M_1-M_2} = est \ \sigma^2_{M_1} + est \ \sigma^2_{M_2}$$

Estimates of the two variances on the right-hand side of this equation can easily be obtained. These estimates are

$$est \ \sigma^2_{M_1} = \frac{est \ \sigma^2}{n} = \frac{3.25}{5} = 0.65$$

and likewise,

$$est \ \sigma^2_{M_2} = \frac{est \ \sigma^2}{n} = \frac{3.25}{5} = 0.65$$

Therefore,

$$est \ \sigma^2_{M_1-M_2} = 0.65 + 0.65 = 1.30$$

To get the standard deviation of this distribution,

$$est \ \sigma_{M_1-M_2} = \sqrt{est \ \sigma^2_{M_1-M_2}} = \sqrt{1.30} = 1.14$$

An obtained t-score can now be formed which, as noted, is based on 8 degrees of freedom.

$$\text{obtained } t(8) = \frac{(M_1 - M_2) - \mu_{M_1-M_2}}{est \ \sigma_{M_1-M_2}} = \frac{3 - 0}{1.14} = 2.63$$

What is the criterion $t(8)$? Since it is based on 8 degrees of freedom, we go to row 8 in the t-table. The test is one-tailed (since the alternative hypothesis was directional, and the traditional 0.05 α-level is being used). Therefore we go to the $Q = 0.05$ column in the table. At the intersection of this column and row 8, we find a criterion $t(8)$ equal to 1.86. Since the obtained $t(8)$ of 2.63 is more extreme than the criterion $t(8)$ of 1.86, we reject the null hypothesis and conclude that hypnosis really does have a facilitative effect on memory performance.

As a parenthetical note, if you were writing up this result in a journal article, you might say something like, "The difference between the two means is statistically significant, $t(8) = 2.63$, p < 0.05." Such a statement would be a shorthand way of saying that (a) the hypnosis group is superior to the control group, and (b) given that the hypnosis and the control group really do not differ, the probability that the difference between the means would be as large as it turned out to be is less than 0.05.

To summarize, we began with two groups—a hypnosis and a control group—and inquired as to whether the two groups differed from one another in terms of memory performance. A mean score, and also an estimate of σ^2, the population variance, was obtained from each group. We then averaged (took the mean of) these two estimates to get one best estimate. This one best estimate was based on the number of degrees of freedom from the first group (4) plus the number of degrees of freedom from the second group (also 4) for a total of 8 degrees of freedom in all. At this point we went through the hypothesis-testing procedure. Since the population variance was estimated rather than known, the appropriate test to perform was a t-test rather than a z-test. The t-test was based on 8 degrees of freedom, since 8 degrees of freedom went into the estimate of σ^2.

COMPLICATING MATTERS: UNEQUAL NUMBERS OF SUBJECTS

The example we just described made things easy in that there were equal numbers of subjects in each of the two groups. (Specifically the hypnosis group and the control group each had five subjects.)

In real life, however, an experiment will not always end up with equal numbers of subjects in the various conditions. Thinking realistically rather than idealistically, there is a good deal of potential for things to go wrong in most experiments. Sometimes, for example, subjects fail to show up for the experiment. Other times the equipment fails. Occasionally the experimenter oversleeps or has a nervous breakdown. In these and many other situations the experimental design may not end up exactly the way it was planned. For purposes of the present discussion let's deal with the situation in which there are unequal numbers of subjects in the various conditions.

The general rules for doing a between-subjects t-test with unequal ns are unchanged, but a few of the details involved in some of the calculations must be modified. Let's describe an example that will serve to illustrate these modifications.

Control of Sexual Behavior

Suppose we are physiological psychologists interested in sexual behavior. In particular we believe that one of the primary brain areas responsible for sexual arousal is the posterior hypothalamus. We therefore reason that if we were to remove the posterior hypothalamus, we would discover a decrease in the amount of copulation engaged in by the unfortunate subject.

The ideal.   To test this hypothesis, we plan to have two groups of subjects with ten subjects in each group. We will operate on each subject in Group 1—the experimental group—and excise the posterior hypothalamus. We will also operate on each subject in Group 2—the control group. We will anaesthetize each subject, open the skull, and poke around, to control for the effects of surgery, but we will not remove any part of the brain.

Since we are unable to find any human beings willing to act as subjects in this experiment (particularly in the experimental group), we instead plan to perform the experiment using psychology's favorite organism, the white rat. Thus, we select 20 virile male rats. Ten of them are to be randomly assigned to the experimental group, the other ten to the control group, and the appropriate surgery carried out. After allowing the rats a week or so to recover from their operations, we intend to investigate their degree of sexual arousal by taking each rat, putting it in a cage with a collection of female rats for an hour, and seeing how many copulatory acts each rat performs.

The real. That was our plan, and it all sounded just fine. Unfortunately, however, the surgery was performed by an inexperienced graduate student, and some of the rats perished on the operating table. Specifically five of the experimental group rats died, leaving $n_1 = 5$ survivors. Similarly, three of the control group rats died, leaving $n_2 = 7$. Although we are somewhat annoyed at the all-thumbs student, we plow ahead with the experiment anyway. Table 10-4 shows the data, depicting the number of copulatory acts performed in an hour by

TABLE 10-4. **Hypothetical Results from Experiment on the Effect of Posterior Hypothalamus Removal on Sexual Activity**

Group 1 (Experimental Group) ($n_1 = 5$)		Group 2 (Control Group) ($n_2 = 7$)	
x_{i1}	x_{i1}^2	x_{i2}	x_{i2}^2
$x_{11} = 0$	0	$x_{12} = 5$	25
$x_{21} = 1$	1	$x_{22} = 7$	49
$x_{31} = 4$	16	$x_{32} = 4$	16
$x_{41} = 4$	16	$x_{42} = 3$	9
$x_{51} = 1$	1	$x_{52} = 4$	16
$\Sigma x_{i1} = 10$	$\Sigma x_{i1}^2 = 34$	$x_{62} = 6$	36
$M_1 = 10/5 = 2$		$x_{72} = 6$	36
		$\Sigma x_{i2} = 35$	$\Sigma x_{i2}^2 = 187$
		$M_2 = 35/7 = 5$	

$$SS_1 = \Sigma(x_{i1} - M_1)^2 \qquad SS_2 = \Sigma(x_{i2} - M_2)^2$$

$$= \Sigma x_{i1}^2 - (\Sigma x_{i1})^2/n_1 \qquad = \Sigma x_{i2}^2 - (\Sigma x_{i2})^2/n_2$$

$$= 34 - 10^2/5 = 14 \qquad = 187 - (35)^2/7 = 12$$

$$df_1 = n_1 - 1 = 5 - 1 = 4 \qquad df_2 = n_2 - 1 = 6$$

$$est_1 \sigma^2 = \frac{SS_1}{df_1} = \frac{14}{4} = 3.5 \qquad est_2 \sigma^2 = \frac{SS_2}{df_2} = \frac{12}{6} = 2.0$$

the five surviving experimental rats and the seven surviving control rats. Computing the mean for each group, we discover that rats in the experimental group performed a mean of $M_1 = 2$ copulatory acts. The control condition rats performed a mean of $M_2 = 5$ copulatory acts. Since the experimental group performed fewer copulatory acts on the average than did the control group, our inclination is to believe that removal of the posterior hypothalamus really is effective in terms of decreasing the amount of sexual behavior. As usual, however, we want to ensure that the obtained results are really due to the manipulation—removing the posterior hypothalamus—and not simply due to chance. In particular the hypothesis-testing steps should be carried out, but first we need to estimate the variance of the population from which the scores emerged.

Estimation of the population variance.   As in the last example, we have two groups and therefore are able to calculate two separate, independent estimates of the population variance—one estimate from each of the two groups. These estimates have been computed in Table 10-4. The estimate from the experimental group, $est_1 \, \sigma^2$, is 3.5, whereas the estimate from the control group, $est_2 \, \sigma^2$, is 2.0.

Recall from the previous example that, once we had our two estimates of the population variance, we merely averaged them to arrive at our one best estimate. Might this tactic be a reasonable one to carry out in the present situation as well? Should we compute the mean of $est_1 \, \sigma^2 = 3.5$ and $est_2 \, \sigma^2 - 2.0$ to arrive at a single estimate of 2.75 for the population variance? Such a procedure, attractive though it might seem, would overlook one important facet of our data—namely that there are more degrees of freedom going into $est_2 \, \sigma^2$ than into $est_1 \sigma^2$ owing to the greater number of surviving rats in the control group relative to the experimental group. This means that the estimate obtained from the control group is a better estimate than the one obtained from the experimental group, and we ought somehow to pay more attention to it. Although we still want to combine the estimates (since the estimate from the experimental group is, after all, worth *something*), we want to give heavier weight to $est_2 \, \sigma^2$. We actually implement this differential weighting in terms of the relative degrees of freedom. Consider: we have 4 degrees of freedom from the experimental group plus 6 degrees of freedom from the control group, for a total of 10 degrees of freedom in all. We therefore want to give relative weights of 4/10 to the experimental group and 6/10 to the control group. Our one best estimate of the population variance is then obtained by the formula

$$est \, \sigma^2 = (4/10) \, est_1 \, \sigma^2 + (6/10) \, est_2 \, \sigma^2$$

$$= (0.4) \, (3.5) + (0.6) \, (2) = 2.6$$

Thus, our best estimate of σ^2 is 2.6, which, as we would guess, is a little closer to 2 ($est_2 \, \sigma^2$, which got the somewhat greater weighting) than it is to 3.5 ($est_1 \, \sigma^2$, which got the somewhat lesser weighting).

A general formula for estimating the population variance. Let's take this opportunity to make some general remarks about how we go about obtaining our one best estimate of the population variance from two (or actually, as we shall see in the next chapter, any number of) separate groups. Consider the situation in which we have two groups of subjects, with n_1 scores in the first group and n_2 scores in the second group. In such a situation, there are $n_1 - 1$ degrees of freedom from the first group plus $n_2 - 1$ degrees of freedom from the second group for a total of $n_1 + n_2 - 2$ degrees of freedom in all. We now calculate two estimates, $est_1\ \sigma^2$ and $est_2\ \sigma^2$ of the population variance from Groups 1 and 2, respectively. To get our one best estimate, as we have seen, we weight the two estimates by their relative degrees of freedom; that is, we use the formula

$$est\ \sigma^2 = \left[\frac{n_1 - 1}{n_1 + n_2 - 2}\right] est_1\ \sigma^2 \quad + \quad \left[\frac{n_2 - 1}{n_1 + n_2 - 2}\right] est_2\ \sigma^2$$

which in turn is equal to

$$est\ \sigma^2 = \frac{(n_1 - 1)\ est_1\ \sigma^2 + (n_2 - 1)\ est_2\ \sigma^2}{n_1 + n_2 - 2} \tag{10.1}$$

Now notice that

$$est_1\ \sigma^2 = \frac{SS_1}{df_1} = \frac{SS_1}{n_1 - 1}$$

or

$$(n_1 - 1)\ est_1\ \sigma^2 = SS_1 \tag{10.2}$$

Similarly,

$$est_2\ \sigma^2 = \frac{SS_2}{df_2} = \frac{SS_2}{n_2 - 1}$$

or

$$(n_2 - 1)\ est_2\ \sigma^2 = SS_2 \tag{10.3}$$

Substituting Equations 10.2 and 10.3 into Equation 10.1,

$$est\ \sigma^2 = \frac{SS_1 + SS_2}{n_1 + n_2 - 2}$$

This best estimate of the population variance is once again seen to be a sum of the squares (in this case the sum of the sums of squares from both groups) divided by an associated number of degrees of freedom (in this case the sum of the degrees of freedom from both groups). As an illustration of this formula, we find in the case of our present example that

$$est\ \sigma^2 = \frac{SS_1 + SS_2}{n_1 + n_2 - 2} = \frac{14 + 12}{5 + 7 - 2} = \frac{26}{10} = 2.6$$

which is just exactly what it was the last time we calculated it.

> 10.6 With unequal numbers of subjects *est* σ^2 is a *weighted* average of $est_1\ \sigma^2$ and $est_2\ \sigma^2$ where the weighting is in terms of relative degrees of freedom.

Hypothesis-testing steps. Now with our population variance finally estimated and tucked away, we are ready to go through our hypothesis-testing steps. The null and alternative hypotheses may be established as follows:

H_0: The independent variable (in this case whether or not the posterior hypothalamus is removed) does not affect the dependent variable (in this case number of copulatory acts in an hour). Thus, μ_1, the mean of the population that the scores in the experimental group come from is equal to μ_2, the mean of the population from which the control group scores come.

The alternative hypothesis would be

H_1: The removal of the posterior hypothalamus will lead to fewer copulatory acts—that is, $\mu_1 < \mu_2$.

The appropriate score to use in this situation would be the difference between the means of the two groups,

$$(M_1 - M_2) = 2 - 5 = -3$$

How is this score distributed if the null hypothesis is true? As usual, if there is actually no difference between the two groups—that is, if the scores from the two groups actually come from the same population—then the two means must come from the same population as well, and the difference between the two means should come from the distribution of difference scores whose mean, $\mu_{M_1-M_2}$, is 0. What about the variance of this distribution of difference scores? Recall that

$$est\ \sigma^2_{M_1-M_2} = est\ \sigma^2_{M_1} + est\ \sigma^2_{M_2} \tag{10.4}$$

The two terms on the right-hand side of Equation 10.4 are simply variances of sampling distributions of sample means, and we know how to compute them. First,

$$est\ \sigma^2_{M_1} = \frac{est\ \sigma^2}{n_1} = \frac{2.6}{5.5} = 0.520$$

and likewise,

$$est\ \sigma^2_{M_2} = \frac{est\ \sigma^2}{n_2} = \frac{2.6}{7.7} = 0.371$$

Thus

$$est\ \sigma^2_{M_1-M_2} = 0.520 + 0.371 = 0.891$$

and

$$est\ \sigma_{M_1-M_2} = \sqrt{est\ \sigma^2_{M_1-M_2}} = \sqrt{0.891} = 0.944$$

In summary we see that if the null hypothesis is true, this difference score—$(M_1 - M_2)$, the difference between the means of the two groups—is distributed with a mean of 0 and an estimated standard deviation equal to 0.944.

Now we are in a position to calculate a t-score that will indicate how extreme our obtained $(M_1 - M_2)$ is, assuming that the null hypothesis is true. This obtained t will be based on 10 degrees of freedom, because a total of 10 degrees of freedom went into our best estimate of the population variance. To calculate this $t(10)$,

$$\text{Obtained } t(10) = \frac{(M_1 - M_2) - \mu_{M_1-M_2}}{est\ \sigma_{M_1-M_2}} = \frac{(-3) - 0}{0.944} = -3.18$$

What about our criterion $t(10)$? We note that we are doing a one-tailed test at the 0.05 α-level, based on 10 degrees of freedom. Looking at the appropriate place in our t-table, we find an entry of 1.81. (Actually, our criterion t is -1.81 because our directional alternative hypothesis predicts that $M_1 - M_2$ will be negative.) In any case we see that our obtained $t(10)$ of -3.18 is more extreme than our critierion $t(10)$ of -1.81, so we reject the null hypothesis and conclude that removing the posterior hypothalamus does indeed decrease amount of sexual activity.

Reiterating This Somewhat Complicated Business

Summarizing the procedure we just used, we can say we started with two groups; from each group we obtained a mean, M_1 and M_2, and additionally we obtained from each group an estimate of the population variance, $est_1\ \sigma^2$ and $est_2\ \sigma^2$. We calculated a one best estimate of the population variance using the weighted average of the two estimates obtained from the two groups—the weighting was in terms of degrees of freedom. Having obtained our best estimate of the population variance, we were then ready to go through the hypothesis-testing steps, using a t-test because we had an estimated rather than a known population variance.

ASSUMPTIONS

At this stage of the proceedings we feel it wise to point out that most of the statistical tests discussed in this book (specifically the z-test, the t-test, and the soon-to-be-described analysis of variance) are based on certain *assumptions*.

Consider, for example, a t-test. As noted above, the self-effacing Mr. Gossett accomplished the task of determining how the score

$$\frac{x - \mu}{est\ \sigma}$$

is distributed. However, to derive the form of the t distribution, Gossett was forced to make a number of assumptions about the nature of the original data. The major assumptions are as follows.

1. The scores constituting the raw data (that is, the xs) are *independent* of one another—that is, each x is uninfluenced by the values of all other xs. This assumption would for example be violated in a free-recall experiment, if a subject being tested cheated and looked at the answer sheet of another subject. This assumption is called, appropriately enough, the *independence* assumption.
2. The original xs are drawn from a *normal distribution*. This is the *underlying normality* assumption.
3. When there is more than one condition, scores from the different conditions are drawn from populations having variances that are all equal to one another. This assumption has been mentioned previously and is often referred to as the homogeneity of variance assumption.

> 10.7 The validity of the statistical tests depends on the validity of certain assumptions regarding the experimental situation.

Violation of Assumptions

Suppose one or more of these assumptions is false in some experiment. What then? Well, in such an event the distribution of

$$\frac{x - \mu}{est\ \sigma}$$

will not be distributed as we expect it to be. Specifically the published α-level criteria will not be correct. For example, the criterion that chops off the upper 5% of a t distribution may chop off the upper 6% of our actual distribution. What is more, we will not be able to formulate the actual distribution of our summary score, so we will be unable to determine what the appropriate criterion should be.

But Not to Worry (Usually)

The preceding was the bad news; now for the good news. It turns out that even fairly gross violations of the underlying normality and the homogeneity of variance assumptions generally have a surprisingly small effect on the summary score distribution. Violation of the independence assumption has a rather serious effect, and therefore it is important to avoid violating it. But a prudent experimenter can generally arrange his or her experiment such that the independence assumption is met.

A detailed discussion of the assumption violation effects is beyond the scope of this book. We provide here a few rules of thumb.

Only worry if your obtained score is near the criterion score.   The actual distribution will only rarely depart substantially from the assumed (or published) distribution. For example, assumptions would have to be very grossly violated for the assumed 5% criterion to actually chop off, say, 15% of the distribution. Therefore, if an obtained score *far* exceeds a criterion score, you are probably on safe ground rejecting H_0.

Check your distributions.   A quick plot of obtained raw-score distributions will provide a rough idea of whether the underlying normality and/or homogeneity of variance assumptions are being violated.

Comparison of *est* σ^2s.   To the degree that the *est* σ^2s from two groups ($est_1\,\sigma^2$ and $est_2\,\sigma^2$) differ, the homogeneity of variance assumption is being violated. However, a rule of thumb is that when the ns are approximately equal, the *est* σ^2s can differ by as much as a four to one ratio without great cause for alarm.

Large ns and equal ns are good.   In general violation of assumptions have less serious consequences to the degree that (a) there are large numbers of observations (20 or more) and (b) there are equal numbers of observations in the various conditions.

> 10.8 For the most part, less than horrendous assumption violations have quite small effects on the conclusions of the test.

The Scaling Assumption

This assumption has been deferred and given a subsection of its own because it raises problems that are frightfully thorny, difficult to explain, under debate, and generally beyond the scope of this book. But we feel you should be aware of what the problem is, so we describe it briefly.

The scaling assumption is that the dependent variable is measured on at least an *interval scale* (that is, an interval or a ratio scale). We recall from the introductory chapter, that an interval scale may be defined as follows. Suppose we have four values of our dependent variable, $x_1, x_2, x_3,$ and x_4. Suppose further that

$$x_2 - x_1 = x_4 - x_3$$

Then the *x*s are measured on an interval scale if these *differences* mean the same thing no matter what the *x*s are. For instance, *speed* may be thought of as an interval (actually a ratio) scale. That is, for example, the difference between 10 and 20 mph may be thought of as equivalent to the difference between 60 and 70 mph. But suppose in a psychological experiment a subject rates on a scale from 1–7 how much he or she likes the taste of spinach. Is the difference between ratings of 6 and 7 psychologically equivalent to the difference between ratings of 3 and 4? This is a debatable point. Thus, the scaling assumption may be taken to mean that there can be doubts as to the equivalence of scale ratings when the ratings have a subjective source.

Nonparametric Tests

If any of the assumptions are violated to the degree that a *z* test or a *t*-test is untenable, the solution is to use nonparametric tests, the validity of which does not depend on these assumptions. Nonparametric tests will be discussed further in Chapter 17.

CONFIDENCE INTERVALS

Confidence intervals are involved in the estimation of an unknown population mean. Consequently, the topic of confidence intervals could appropriately have been described in Chapter 9 along with the rest of the material on parameter estimation. However, some of our discussion of confidence intervals will require a familiarity with the *t* distribution; therefore, we have deferred the topic to this chapter instead.

To get a flavor for what confidence intervals are, let's go back to the first example that we discussed in this chapter involving the drawn lines representing a Mueller-Lyer line. Suppose we wanted to estimate μ, where μ now refers to the population mean line length. Recall from our discussions in Chapter 9 that the best estimate of an unknown population mean is simply the mean of a random sample drawn from that population. In our present example a random sample of $n = 5$ has been drawn from the population, and the mean length of lines drawn by people in the sample was $M = 8.0$ cm. Our best estimate of the population mean would therefore be 8.0.

If pressed, however, we would not want to stake our lives or our reputation on the assertion that the population mean is *exactly* 8.0 cm. Although 8.0 cm is our best guess, we would not be surprised if the *actual* population mean were a bit above or a bit below 8.0. Suppose, however, that instead of committing ourselves to an *exact value* as our estimate we hedge a little and establish some *range*—say, the range from 7 cm to 9 cm. Now we could assert with somewhat more certainty that the population mean falls somewhere within this range. Such a "hedging range" around a sample mean is referred to as a confidence interval, and as we shall see, we can be quite precise about what we mean by a confidence interval. That is, we can select the size of the confidence interval such that with any arbitrary probability, the population mean will fall somewhere within it. For example, if we established a confidence interval such that the population mean fell within the confidence interval with a probability of 0.95, we would refer to it as a 95% confidence interval. We could also, if we wished, establish a somewhat larger interval, with a 99% chance of capturing the population mean, or a somewhat smaller interval, with a 50% chance of capturing the population mean.

We shall discuss confidence intervals involving both a known and an unknown population variance. As you might suspect, the calculations involved in these two situations are very similar to one another except that the former situation will involve a z distribution, whereas the latter situation will involve a t distribution.

> 10.9 A confidence interval around a sample mean will include the population mean with some probability. The larger the confidence interval, the greater the probability that it will include the population mean.

Calculation of Confidence Intervals When the Population Variance Is Known

Suppose that the Chancellor of Aardvark University is interested in the mean GRE score obtained by all graduate students accepted at Aardvark during the past ten years. (Recall that GRE scores are normally distributed with a population variance of 10,000.) Rather than going through all the records, the Chancellor finds it more expedient to draw a random sample (say, of size $n = 50$) of graduate students and to use the sample data to make the inferences about the Aardvark graduate student population as a whole. Let's imagine that the mean GRE score of this 50-person random sample is 600.

The standard error of the mean and the sampling distribution of the sample means. Consider now this sample mean, $M = 600$. We know that M comes from a sampling distribution whose mean, μ, is the same as the mean of the original distribution of Aardvark GRE scores. Therefore, to estimate μ, it's sufficient to estimate the mean of the sampling distribution of sample means. Let's compute the standard deviation of this distribution:

$$\sigma_M = \frac{\sigma}{\sqrt{n}} = \frac{100}{\sqrt{50}} = 14.14$$

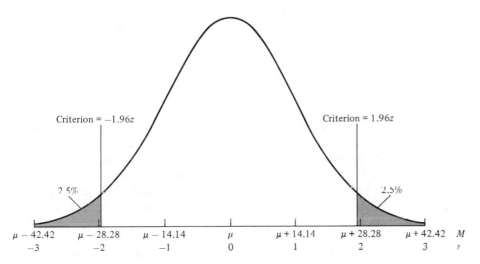

FIGURE 10-4. A sampling distribution of sample means with a standard deviation (σ_M) of 14.14 and an unknown mean of μ. Criteria of $z = \pm 1.96$ have been established around μ such that 95% of all sample means are included.

This standard deviation of the sampling distribution of sample means sometimes goes by another name—the *standard error of the mean*.[3]

Now let's characterize the sampling distribution of sample means, which is shown in Figure 10-4. We know that this distribution is a normal distribution with a mean of μ (μ being what we are trying to estimate) and a standard deviation of 14.14. The z-scale has been shown in addition to the raw-score scale. In Figure 10-4 we have established symmetrical criteria such that 95% of the area under the distribution is contained between the two criteria. We know from previous dealings with the z distribution that these criterion z-scores are ± 1.96; thus, in terms of raw scores (Ms) the criteria may be computed as follows:

$$\pm 1.96 = \frac{M - \mu}{14.14}$$

or

$$(\pm 1.96)\,(14.14) + \mu = M$$

---

[3]People have a tendency to confuse σ_M, the standard error of the mean, and σ, the standard deviation. Try to resist this tendency and constantly bear in mind that the standard error of the mean is smaller than the standard deviation by a factor of $\sqrt{n}$.

And finally,

$$M = \mu \pm 27.72$$

Thus, the criterion raw (M) scores are 27.72 above and below μ.

Establishment of the confidence interval.   The curve shown in Figure 10-4 is a hypothetical one, in the sense that μ is unknown. The way we have set up this hypothetical distribution, however, is such that 95% of all potential sample means are included within the confidence interval of $\mu \pm 27.72$. Now what we actually have to work with is some obtained sample mean—one of the means that will emerge from this distribution. Suppose we set up our confidence interval of ± 27.72 around our obtained *sample mean*, as has been done in Figure 10-5. Notice that due to the symmetry of the situation, if the confidence interval is set up around the sample mean, then 95% of all possible sample means will include the population mean within their confidence intervals. The only sample means that will *not* capture the population mean within their confidence intervals are those 5% of the sample means that fall in the two tails of Figure 10-5. Therefore, with a probability of 0.95, such a confidence interval around a sample mean will capture the population mean.

All of this is a somewhat roundabout way of describing a fairly straightforward process. To compute a confidence interval around a given sample mean, M, we first compute σ_M, the standard error of the mean. Then we find the two-tailed criterion z that will include an area corresponding to the desired confidence

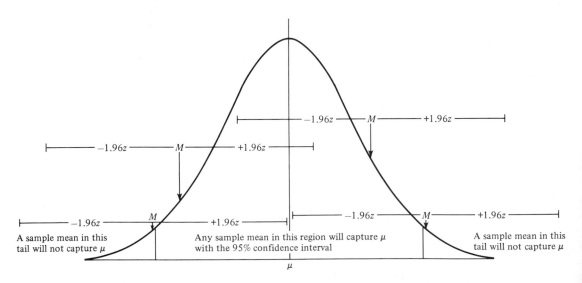

FIGURE 10-5.   Confidence intervals are actually established around the sample mean. With a 95% confidence interval 95% of all sample means will capture the population mean (μ) within the confidence interval. The only sample means that will not capture μ are those 5% that fall in the tails of the distribution.

interval. So for example a 95% confidence interval would be obtained using the criteria $z = \pm 1.96$.

The confidence interval is then simply

$$M \pm (\text{criterion } z) \, \sigma_M$$

In the particular example of interest the confidence interval would be 600 ± 27.72. Therefore, we would be 95% certain that the actual mean GRE score of Aardvark students falls within the interval ranging from $600 - 27.72 = 572.28$ to $600 + 27.72 = 627.72$.

10.10 When σ^2 is known, the $x\%$ confidence interval around a sample mean, M, is

$$\text{Confidence interval} = M \pm (\text{criterion } z)\,(\sigma/\sqrt{n})$$

where the criterion z is two-tailed corresponding to the $(100 - x)\%$ α level. Note that $\sigma/\sqrt{n}$ is of course σ_M, the standard error of the mean.

Confidence Intervals When a Population Variance Is Unknown

To find the confidence interval when a population variance is unknown, we go through almost exactly the same procedure except that (1) we initially have to estimate σ^2, the population variance, and (2) we eventually use a criterion t rather than a criterion z. The number of degrees of freedom on which the criterion t is based is of course the number of degrees of freedom that went into the estimate of the population variance. To continue with the example sketched at the beginning of this section, let's suppose we are trying to compute the 95% confidence interval around the 8.0 cm sample mean, M, of line lengths representing memory for the Mueller-Lyer illusion. As we saw in the first section of this chapter, the estimate of the population variance, est σ^2, was 5; therefore, the standard error of the mean was estimated to be

$$est \; \sigma_M = \sqrt{est \; \sigma_M^2} = \sqrt{\frac{est \; \sigma^2}{n}} = \sqrt{\frac{5}{5}} = \sqrt{1} = 1.0$$

Since we are interested in a 95% confidence interval, we find in our t-table the two-tailed ($2Q$) criterion t at an α-level of 0.05, with 4 degrees of freedom. This criterion is 2.78. **Our confidence interval is now obtained by the expression**

$$M \pm (\text{criterion } t) \; est \; \sigma_M$$

or in the present example,

Confidence interval $= 8.0 \pm (2.78)1$
$$= 8.0 \pm 2.78$$

Thus, we would be 95% certain that the population mean line length falls somewhere between $8.0 - 2.78 = 5.22$ cm and $8.0 + 2.78 = 10.78$ cm.

10.11 When σ^2 is unknown, an x% confidence interval is computed as:

Confidence interval $= M \pm [\text{criterion } t][\text{est } \sigma / \sqrt{n}]$

where the criterion t is based on the number of degrees of freedom going into the estimate of σ^2 is two-tailed and corresponds to the $(100\text{-}x)$% α-level.

Power and Confidence Intervals

There is a direct relationship between the power of a given experimental situation and the size of the confidence intervals produced in that situation: in particular the greater the power of the situation, the smaller will be the confidence intervals. Let's expound on this relationship just a bit.

Sample size.   Recall from our discussion of power (Chapter 8) that manipulation of n, the sample size, is the most direct way of affecting the power in some experiment: the greater the n, the smaller will be σ_M and the greater will be the power. Likewise, reference to our formulas for confidence intervals indicates that the greater the n, and thus the smaller the σ_M, the smaller the confidence interval. Note that the smaller the confidence interval, the better—the smaller the confidence interval, the more accurately we have pinpointed μ.

Experimental control.   Likewise, recall that the greater the control we exercise over the experimental situation, the smaller will be σ^2 and, therefore, the smaller will be σ_M. By the same reasoning we applied above, greater experimental control will therefore yield smaller confidence intervals.

Alpha-level.   Alpha-level is akin to the confidence level for which we opt. If for example we choose a 95% confidence level, then we have a 0.05 chance of excluding μ from the confidence interval—this is analogous to having an α of 0.05. If we raised our confidence level to 0.99, we would decrease the probability of excluding μ from the confidence interval, but at the expense of increasing the size of the confidence interval. Conversely, if we lowered our confidence level to 0.90, we would decrease the size of the confidence level at the cost of increasing the μ-exclusion probability.

10.12 Confidence intervals give us information about experimental power. The greater the power, the smaller the confidence interval.

Uses of Confidence Intervals

When we do an experiment we would ideally like to know the μ's—the population means of concern. But ordinarily such knowledge does not emerge from an experiment. Rather, an experiment will yield sample means, which are only estimates of the μ's. Confidence intervals may be used to illustrate the goodness of these estimates—and thus in turn may be used to evaluate how seriously the results of the experiment should be taken. As an example suppose we are doing an experiment to determine the shape of forgetting functions. To do this, we present some to-be-remembered material and then test memory performance after various periods of time (retention intervals). The function relating memory performance to retention interval is a forgetting function.

Imagine now that we have two competing theories that yield different predictions about the shape of the forgetting function. The predictions of the theories are shown in Figure 10-6. Theory A predicts a linear function whereas Theory B predicts a curvilinear function. The purpose of our experiment is to distinguish between these theories. Hypothetical data from our experiment are shown in Figure 10-7. The sample means appear to decrease more or less linearly

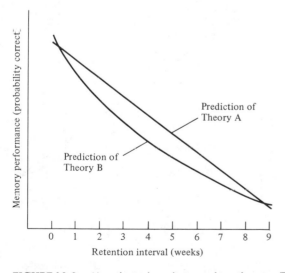

FIGURE 10-6.   Hypothetical predictions of two theories. The curves are forgetting curves.

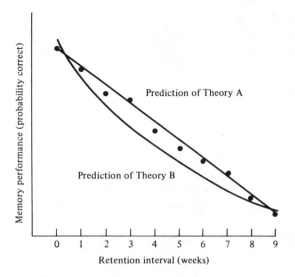

FIGURE 10-7.   Comparison of predicted curves with data. On the surface it appears as if Theory A has won out.

as a function of retention interval. Therefore Theory A is apparently supported, whereas Theory B is disconfirmed.

However, suppose the experiment is based on a small number of subjects, and therefore the confidence intervals around the means are relatively large as depicted in part *a* of Figure 10-8. In such a situation either forgetting function

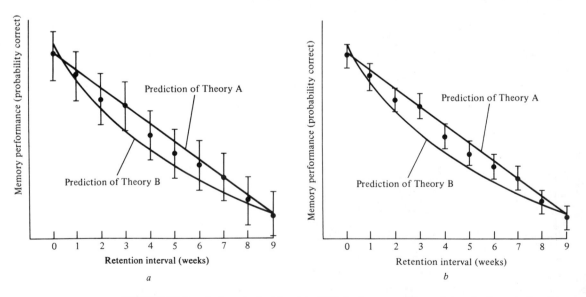

FIGURE 10-8.   A closer look at the data. With the large confidence intervals shown in Panel *a*, Theory B could not be rejected. However, with the smaller confidence intervals shown in Panel *b*, Theory B could be rejected.

can be fit to the data—either function falls within the confidence intervals. So this experiment would not really distinguish between the two theories even though it appeared to do so on the surface. But now imagine that the experiment had more power and produced smaller confidence intervals, as depicted in part *b* of Figure 10-8. Now the situation is quite different. Theory A's prediction still fits the data, but Theory B's prediction now fails to fit the data, and thus Theory B falls by the wayside.

In short we see that we must base conclusions from an experiment not only on the sample means that emerge, but also on how *good* the sample means are as estimates of the relevant population means.

SUMMARY

In this chapter we have discussed the *t*-test, which is one of the most common types of statistical tests. We initially saw that a *t*-test was much like a *z*-test, except that a *z*-test was performed in a situation involving a known population variance and a *t*-test is performed in a situation involving an unknown population variance. In a *t*-test situation a necessary initial step is to obtain an estimate of the population variance. We described several different *t*-test situations.

1. The simplest situation involves testing a single sample mean against a constant. Here a single estimate of σ^2 is obtained from the sample; this estimate is based on $n - 1$ degrees of freedom, where n is the sample size. The formula used to test a mean against a constant is

$$t(n - 1) = \frac{M - \mu}{est\ \sigma_M}$$

where M is the sample mean and μ is the constant.

2. A similar situation is a within-subjects design. Here each subject participates in both conditions of an experiment and each subject obtains a difference score. The population variance, σ^2, of this set of difference scores is estimated and the mean, M, of the set of difference scores is tested against the constant of zero.

3. In a between-subjects design there are two independent groups of subjects. From each group is obtained the mean of that group and an estimate of the population variance. The one best estimate of the population variance is then obtained by computing a weighted average of the two estimates from the two groups. The weighting is in terms of degrees of freedom; and the final estimate is based on $n_1 - 1 + n_2 - 1 = n_1 + n_2 - 2$ degrees of freedom where n_1 and n_2 are the sample sizes of the two groups. In the case where $n_1 = n_2$, the weighting is equal, and therefore the one best estimate of σ^2 is simply the mean of the two estimates from the two groups.

   Finally the hypothesis-testing steps involve testing the difference between the two sample means against zero using the formula,

$$t(n_1 + n_2 - 2) = \frac{(M_1 - M_2) - 0}{est \; \sigma_{M_1 - M_2}}$$

4. The t-test as well as the z-test and the analysis of variance are based on certain assumptions. Violation of these assumptions is generally not a cause for alarm, but when it is, one may do nonparametric tests, which are the topic of a later chapter.

5. Confidence intervals are ranges established around sample means and designed to include the corresponding population means with some predetermined probability. Confidence intervals reflect the accuracy with which the sample means estimate the population means and therefore determine how useful are the results of the experiment.

PROBLEMS

1. In the three questions below, the mean, M, of a random sample from an approximately normal population is given, and the value of the sample standard deviation, S, computed from that sample is also given. In each problem test the specified hypothesis about the population mean, μ. Use (a) a one-tailed t-test at the 0.05 significance level; (b) a two-tailed t-test at the 0.05 significance level; (c) a one-tailed t-test at the 0.01 significance level; and (d) a two-tailed t-test at the 0.01 significance level. Also use the given data to find (e) a 95% confidence interval for the mean of the population and (f) a 99% confidence interval for the mean of the population.

    1. The hypothesis is that a sample of 20 clinical interview tapes have $M = 52$ minutes and $S = 19$ minutes comes from a population having $\mu = 50$ minutes.
    2. The hypothesis is that a sample of 12 weights of males having $M = 182$ lbs. and $S = 16$ comes from a population having $\mu = 170$ lbs.
    3. The hypothesis is that a sample of 64 heights having $M = 5'9''$ and $S = 3''$ comes from a population having $\mu = 5'6''$.

2. Gail Goody reads in the newspaper that Americans see an average (mean) of 10 homicides a night on TV. Wondering whether her sorority sisters differ from this national average, she asks each of the four other members of her sorority how many homicides each of them saw on TV the previous night. Their answers are as follows:

Sister	Number of Homicides
1	9
2	8
3	10
4	9

1. What is the mean number of homicides and the 95% confidence interval?
2. Perform the relevant t-test on these data.

3. Consider Chapter 5, Problems 3, 4, 5, and 8.

   1. Reevaluate the data from these problems performing a t-test rather than a sign test.
      (*Hint:* Each of these t-tests should be a within-something t test.)
   2. Compute 95% confidence intervals around the sample mean difference score, in each case.

4. In questions 1 and 2 below the means M_1 and M_2 of random samples of sizes N_1 and N_2, respectively, and the corresponding sample standard deviations, S_1 and S_2, are given. In each problem test the null hypothesis is that the two samples came from the same normal population. Use (a) a one-tailed t-test at the 0.05 significance level; (b) a two-tailed t-test at the 0.05 significance level; (c) a one-tailed t-test at the 0.01 significance level; and (d) a two-tailed t-test at the 0.01 significance level. Also find the 95% confidence interval around each of the two means.

   1. For students' graduating averages: $M_1 = 78$, $n_1 = 16$, $s_1 = 6$; $M_2 = 84$, $n_2 = 16$, $S_2 = 6$.
   2. For female heights: $M_1 - 5'9''$, $n_1 - 81$, $S_1 - 3''$; $M_2 = 5'4''$, $n_2 = 11$, $S_2 = 4.5''$.
   3. For weights of U.S. males: $M_1 = 170$ lbs., $n_1 = 16$, $S_1 = 15$; $M_2 = 185$ lbs., $n_2 = 10$, $S_2 = 12$.

5. A developmental psychologist is interested in whether dating behavior is different for first-born versus second-born adolescent males. He gathers two samples of six first-born and six second-born teenagers and measures how many dates each has in a month. He gets the following data:

Number of Dates in a Month

First-born	Second-born
6	2
4	5
5	4
7	2
3	1
5	4

   1. Plot the two means, including 95% confidence intervals around the means.
   2. Perform a t-test on the data.

6. Consider the following data (assume that 10 Ss each participate in both conditions):

Subject	Condition 1	Condition 2
1	9	7
2	2	2
3	7	4
4	12	13
5	14	13
6	10	7
7	6	4
8	7	6
9	12	8
10	10	9

1. Get the set of difference scores (condition 1 minus condition 2 for each S).
2. Compute the mean difference score and the 95% confidence interval around the mean difference score.
3. Use a t-test to test the hypothesis that condition 1 is better than condition 2.
4. Do the same thing with a sign test.

7. An experiment is done on the effect of marijuana on memory. Two groups of six subjects per group are given a free-recall test on a 15-word list. The experimental group has been given two joints to smoke, whereas the control group has been given two cigarettes. The results (number of correct responses out of 15) are as follows:

Number of Words Remembered	
Experimental (marijuana)	Control (cigarette)
6	10
5	5
8	8
7	8
5	9
7	8

1. Plot the two sample means along with 95% confidence intervals around the means.
2. Test the hypothesis that the marijuana causes a decrease in memory performance.

8. Photi's Fish Shop claims that it sells heavier salmon than does Ivan's across the street. Samples of $n_1 = 4$ salmon and $n_2 = 3$ salmon are obtained from Photi's and Ivan's, respectively. The weights (in pounds) are as follows:

Group 1 (Photi's)	Group 2 (Ivan's)
6	3
8	5
7	4
7	

1. Plot the means for the two groups along with the 95% confidence intervals.
2. Test the hypothesis that Photi's salmon weigh more, in general, than do Ivan's.

9. Plot means and 95% confidence intervals for the data from Chapter 8, Problems 7, 8, 9, 10, 11, 12, and 13. Assume for each that σ^2, the population variance, is *unknown*.

10. An experiment is done to test whether Vitamin C is effective in preventing colds, using two groups of four subjects per group. Each subject in Group 1 takes 1000 mg of Vitamin C per day for a year whereas each subject in Group 2 takes a 1000 mg sugar tablet every day. The number of colds during the year is recorded for each subject.

   Unfortunately, one of the Group 2 subjects catches pneumonia and has to leave the experiment.

   The data are as follows:

Group 1 (Vitamin C)	Group 2 (sugar tablet)
$x_{11} = 1$	$x_{12} = 3$
$x_{21} = 0$	$x_{22} = 3$
$x_{31} = 1$	$x_{32} = 3$
$x_{41} = 2$	

1. Plot the two sample means along with 95% confidence intervals.
2. Can you conclude from these data that Vitamin C reduces the number of colds? Carry out the appropriate test.

11. A developmental psychologist believes that if a child is forced to eat a certain food, he or she will grow to like that food. An experiment is designed to test that notion. Two groups of 10 five-year-old children per group are picked. In the "liver group" each child is made by his or her parents to take two bites of liver a week for a year. In the control group liver is offered to the child, but he or she is not forced to eat it. Five years later, a liver-preference test is given to the two groups. Each child is asked to rate how much he or she likes liver on a scale ranging from 1 = very much to 5 = not at all. The families of two of the children who were in the liver group have moved away from town. Four of the families of children who were in the control group

have moved away. The data are as follows:

Rating for How Much Liver Is Liked

Liver Group	Control Group
4	3
5	3
3	2
3	1
4	2
5	5
2	
4	

1. Plot the two sample means along with 95% confidence intervals.
2. Test the hypothesis that the liver group likes liver *more* than the control group.

12. An experiment is done to test the effect of communicator "peer similarity" on attitudes. Subjects (college students) are randomly assigned to two groups: a "similar peer" (SP) and "dissimilar peer" (DP) group, with eight students per group. In both groups a 19-year-old communicator gives a lecture on why the presidential incumbent should be reelected. In the SP condition the communicator is long-haired and dressed in sandals, jeans, and a tie-dyed shirt. In the DP condition the communicator has a crew-cut and is wearing glasses.

At the end of the talk the students are asked to rate on a 1–7 scale how strongly they think the incumbent should be reelected. Unfortunately in the DP condition two subjects ask to be excused in the middle of the lecture due to nausea; while in the SP condition one subject leaves to go to the bathroom and never returns. This leaves six subjects in the DP condition and seven subjects in the SP condition.

The results are as follows:

Rating that incumbent should be reelected (1 means definitely should be reelected; 7 means definitely should not be reelected)

SP Group	DP Group
4	7
3	5
3	7
2	6
5	6
4	7
6	

1. Plot the two means along with 95% confidence intervals.
2. Test the hypothesis that a similar peer produces *different* attitudes from a dissimilar peer.

13. Marine engineer Joe Shablotnik is interested in spills that occur when oil tankers sink. In particular Joe is interested in the question of whether Liberian tankers tend to spill more oil when they sink than do American tankers. He looks at records from six sunken Liberian tankers, but he is only able to find one sunken American tanker. The data are as follows.

Amount of Oil Spilled (Millions of Gallons)

Liberian tankers	American tanker
$x_{11} = 125$	$x_{12} = 112$
$x_{21} = 120$	
$x_{31} = 131$	
$x_{41} = 132$	
$x_{51} = 119$	
$x_{61} = 126$	

1. Test the hypothesis that more oil is spilled when a Liberian tanker sinks than when an American tanker sinks. Use the 0.05 α-level.
2. Compute a 98% confidence interval around the mean in each condition.

14. Dr. Joe Shablotnik is doing an experiment to determine whether aspirin will reduce fever. Twelve feverish hospital patients are randomly divided into six experimental group and six control group subjects. The experimental group subjects are given aspirin and the control group subjects are given aspirin-like sugar pills. The change in fever is then noted.

Unfortunately five of the control group subjects and three of the experimental group subjects are mistakenly given cyanide instead of sugar pills or aspirin and have to be eliminated from the experiment. The data from the remaining subjects are as follows. (Negative sign indicates decrease in temperature.)

$n_1 = 3$	$n_2 = 1$
Experimental Group	Control Group
$x_{11} = -2°$	$x_{12} = -1°$
$x_{21} = 0°$	
$x_{31} = -4°$	

1. Can Joe conclude that aspirin reduces fever more than do sugar pills? Perform the appropriate test.

2. Do you think the homogeneity of variance assumption is being violated in this experiment?
3. Suppose it is known that the population variance of fever reduction under these conditions is $\sigma^2 = 1$. Test whether aspirin and sugar pills differ in their effect on fever reduction.
4. What is the power of the test against the alternative hypothesis that aspirin reduces fever 1° more than sugar pills?

15. An experiment is done to test the effects of marijuana on time perception. The experiment involves 18 subjects and consists of two phases:

Phase 1: The subjects are randomly divided into two groups: nine subjects to a marijuana group (Group 1) and nine subjects to a control group (Group 2). To get baseline measures, each subject in the two groups is placed alone in a room and asked to estimate when 10 minutes has elapsed. The dependent variable is thus the subject's estimate of 10 minutes. Note that no manipulation has been performed between the two groups yet.
     The data may be represented as follows:

Group 1 (Experimental)	Group 2 (Control)
x_{11}	x_{12}
x_{21}	x_{22}
.	.
.	.
.	.
x_{91}	x_{92}

Phase 2: The subjects in the experimental group are given marijuana cigarettes to smoke, while subjects in control group are given oregano, which smells and tastes enough like marijuana to be taken for it. The subjects then go through the time-estimation procedure again. The data may be represented as follows (here we will use the notation x' to distinguish the data from the baseline data)

Group 1 (Experimental)	Group 2 (Control)
x'_{11}	x'_{12}
x'_{21}	x'_{22}
.	.
.	.
.	.
x'_{91}	x'_{92}

1. How would you test the hypothesis that the time estimates would be shorter under the influence of marijuana than under the influence of oregano? Be sure to describe

—what you would use as your scores

—what means you would calculate

—what estimates of the population variance you would calculate and how you would combine them into one best estimate

—how many degrees of freedom your test would be based on.

2. In similar fashion describe how you would test the hypothesis that time estimates under the influence of oregano are shorter than the baseline estimates.

3. Suppose the experimenter wishes to compute an estimate of the population variance of the baseline scores (that is, the Phase 1 data only). The experimenter claims that with the Phase 1 data only, he could compute such an estimate based on 17 degrees of freedom. Is this possible? Why or why not?

16. Gail Goody believes that chewing gum during a statistics examination will improve exam scores. Hence she recruits four of her sorority sisters, Mugsy, Suzie, Dopey, and Rosie, to take part in an experiment. In the first-hour exam they do not chew gum, and in the second-hour exam they do. Their scores are as follows:

Sister	Exam 1	Exam 2
Mugsy	91	92
Suzie	73	80
Dopey	41	41
Rosie	80	90

1. Test the hypothesis that chewing gum improves examination scores.
2. Compute a confidence interval that will include the population mean difference score with a probability of 0.95.
3. What is wrong with this experimental design?

17. A college dean is interested in whether student ratings from an Introductory Chemistry class (of 50 students) differ from the ratings of an Introductory Psychology Class (of 40 students).

Each student rates the quality of the class on the following scale:

1 = excellent
2 = good
3 = average
4 = poor
5 = abominable

Thus one rating is obtained from each student in each of the two classes. The data are as follows:

$$\underline{\text{Chemistry } (N_1 = 50)}$$

$$M_1 = \sum_{i=1}^{50} \frac{X_{i1}}{50} = 2.40$$

$$SS_1 = \sum_{i=1}^{50} (X_{i1} - M_1)^2 = 100$$

$$\underline{\text{Psychology } (N_2 = 40)}$$

$$M_2 = \sum_{i=1}^{40} \frac{X_{i2}}{40} = 1.80$$

$$SS_2 = \sum_{i=1}^{40} (X_{i2} - M_2)^2 = 76$$

1. Calculate *est* σ^2.
2. Calculate *est* $\sigma_{M_1}^2$ and *est* $\sigma_{M_2}^2$.
3. Test the hypothesis that the ratings for Intro Chemistry *differ* from those for Intro Psychology.
4. Test the hypothesis that the ratings for Intro Chemistry *differ* from an "average" rating of 3.

Analysis of Variance 11

So far in this book we have been concerned with reasonably simple experimental situations. We described experiments in which the mean of a single group is tested against a constant, and we also described two-group experiments in which the mean of one group is tested against the mean of a second group. In this chapter we will proceed a step further in experimental complexity and consider a situation involving more than two groups. Suppose for example that an experiment involved three conditions and therefore three groups. In such a situation we might want to test whether the means of these three groups differed from one another. One possible way of addressing such an issue would be to perform several t-tests. We could perform a t-test of Group 1 against Group 2, then a second t-test of Group 1 against Group 3, and finally a third t-test of Group 2 against Group 3. However, for reasons we'll take up in Chapter 15, such a multiple t-test procedure turns out to be somewhat problematic. In this chapter we will instead develop a technique for performing a single, overall test of the question of whether the various groups differ significantly from one another in any way.

This technique is known as *analysis of variance* (ANOVA for short). In Digression 10.2 of the last chapter we had a preview of what analysis of variance is all about: Specifically we saw that ANOVA involves a consideration of two different types of variance. The first type of variance arises from the fact that the scores within a given group vary from one another. We have run into this type of variance a good deal in previous situations and have referred to it as σ^2, the population variance. The second type of variance arises from *between* groups—that is, it arises from the fact that the means of the groups differ from one another. We have dealt with this second type of variance in a situation involving only two groups. If the difference between the two means is large, we say that we have a good deal of between-groups variance. Conversely, if the difference between the two means is small, we say that we have a relatively small amount of between-groups variance.

To the degree that we have a large amount of between-groups variance, we tend to think that our independent variable is having some effect on our dependent variable. We have seen, however, that the amount of between-groups variance is only "big" or "small" in relation to the amount of within-group variance. That is, the larger the between-group variance (say, the difference between the two means), the more we would believe that there is some real difference between the groups. However, the larger the within-group variance (σ^2), the more we tend to think that any observed differences between groups may simply be due to random fluctuation. We have seen that in order to evaluate the amount of between-group variance relative to the amount of within-group variance, we form a ratio of the two—a z-ratio

$$z = \frac{M_1 - M_2}{\sigma_{M_1 - M_2}}$$

if the population variance is known and a t-ratio

$$t = \frac{M_1 - M_2}{est \; \sigma_{M_1 - M_2}}$$

if the population variance is unknown.

When we have more than two groups, we do much the same thing. We form a sort of ratio of the amount of between-groups variance to the amount of within-groups variance. This ratio is called an F-ratio. If this obtained F-ratio is large enough, we reject the null hypothesis and conclude that there are real differences among the groups—that is, that the independent variable is having some effect on the dependent variable. Let's see how this is done using a specific example. But before you proceed through this example, you should read Appendix E, where we establish some new notation.

> 11.1 Three different kinds of tests—a z-test, a t-test, and an F-test—consist of forming some sort of ratio of the amount of variance between groups to the amount of variance within groups.

THREE EXPERIMENTAL GROUPS: DO BLONDES HAVE MORE FUN?

Suppose we wish to investigate the question of whether hair color influences adolescent dating frequency. To deal with this question, we select three groups of teenagers—a group of $n = 4$ blondes, a second group of $n = 4$ brunettes, and a third group of $n = 4$ redheads. Thus our independent variable is hair color. The experiment will consist of observing each of the four boys in each of the three groups to determine how many times he dated a girl during a given week. Hypothetical data from this experiment are shown in Table 11-1. As usual our first task is to calculate the means for the three groups. Examination of the

TABLE 11-1.   Hypothetical Data: Number of Dates Per Week for Each of Four Blondes, Brunettes, and Redheads

Group 1: Blondes		Group 2: Brunettes		Group 3: Redheads	
x_{i1}	x_{i1}^2	x_{i2}	x_{i2}^2	x_{i3}	x_{i3}^2
$x_{11} = 6$	36	$x_{12} = 2$	4	$x_{13} = 2$	4
$x_{21} = 6$	36	$x_{22} = 4$	16	$x_{23} = 3$	9
$x_{31} = 10$	100	$x_{32} = 1$	1	$x_{33} = 5$	25
$x_{41} = 6$	36	$x_{42} = 1$	1	$x_{43} = 2$	4

$$T_1 = 28 \quad \sum_{i=1}^{4} x_{i1}^2 = 208 \qquad T_2 = 8 \quad \sum_{i=1}^{4} x_{i2}^2 = 22 \qquad T_3 = 12 \quad \sum_{i=1}^{4} x_{i3}^2 = 42$$

$$M_1 = \frac{T_1}{n} = \frac{28}{4} = 7 \qquad\qquad M_2 = \frac{T_2}{n} = \frac{8}{4} = 2 \qquad\qquad M_3 = \frac{T_3}{n} = \frac{12}{4} = 3$$

$$SS_1 = \sum_{i=1}^{4} (x_{i1} - M_1)^2 \qquad SS_2 = \sum_{i=1}^{4} (x_{i2} - M_2)^2 \qquad SS_3 = \sum_{i=1}^{4} (x_{i3} - M_3)^2$$

$$= \sum_{i=1}^{4} x_{i1}^2 - T_1^2/4 \qquad\quad = \sum_{i=1}^{4} x_{i2}^2 - T_2^2/4 \qquad\quad = \sum_{i=1}^{4} x_{i2}^3 - T_3^2/4$$

$$= 208 - 28^2/4 \qquad\qquad - 22 - 8^2/4 \qquad\qquad = 42 - 12^2/4$$

$$- 208 - 196 - 12 \qquad\qquad = 22 - 16 = 6 \qquad\qquad - 42 - 36 - 10$$

$$df_1 = n - 1 = 4 - 1 = 3 \qquad df_2 = n - 1 = 4 - 1 = 3 \qquad df_3 = n - 1 = 4 - 1 = 3$$

$$est_1\sigma^2 = SS_1/df_1 \qquad\qquad est_2\sigma^2 = SS_2/df_2 \qquad\qquad est_3\sigma^2 = SS_3/df_3$$

$$= \frac{12}{3} \qquad\qquad\qquad = \frac{6}{3} \qquad\qquad\qquad = \frac{6}{3}$$

$$= 4 \qquad\qquad\qquad = 2 \qquad\qquad\qquad = 2$$

pattern of means will provide some rough indication of whether or not hair color is actually having an effect on number of dates. The means are $M_1 = 7$ dates per week for the blonde group, $M_2 = 2$ dates per week for the brunette group, and $M_3 = 3$ dates per week for the redhead group. These means would seem to indicate that there are differences among the groups—that hair color really is having an effect on number of dates. We realize, however, that we must be cautious about precipitously leaping to such a conclusion. Perhaps hair color really has no influence whatsoever on dating behavior, and the differences among the means of the three groups are simply due to random fluctuation. We may just have happened to pick a sample of dazzlingly handsome blondes, and/or a group of plain brunettes and/or a group of average redheads. It would therefore be wise to go through a set of hypothesis-testing steps to provide a precise assessment of whether or not the independent variable is actually having an effect on the dependent variable.

Estimation of the Population Variance

In this situation we will need, as it happens, an estimate of σ^2, the population variance, just as was the case in the t-test situations. Our first step, therefore, is to obtain such an estimate. This is not difficult and is based on the same principles as the process of estimating the population variance in a two-group t-test situation.

An estimate from each group. Since we have three groups, we are in a position to obtain three separate estimates of the population variance. These three estimates have been obtained from the data in Table 11-1. As can be seen, the estimate from the blonde group is $est_1 \sigma^2 = 4$; the estimate from the brunette group is $est_2 \sigma^2 = 2$; and the estimate from the redhead group also turns out to be $est_3 \sigma^2 = 2$.

Combining to obtain one best estimate. We would guess that since there are equal numbers of subjects in each of the three groups, our one best estimate of σ^2 might be obtained by simply averaging the three estimates from the three groups (just as we averaged estimates from two groups in the analogous t-test situation). This intuition would be entirely correct, and the one best estimate of σ^2 is

$$est \, \sigma^2 = \frac{est_1 \sigma^2 + est_2 \sigma^2 + est_3 \sigma^2}{3}$$

$$= \frac{4 + 2 + 2}{3} = \frac{8}{3} = 2.67$$

We should also point out that we can be somewhat more general in our estimation procedure. Recall that in our two-group t-test situation, the best estimate of the population variance could always be obtained using the formula

$$est \, \sigma^2 = \frac{SS_1 + SS_2}{df_1 + df_2}$$

where SS_j refers to the sum of squares $(x_{ij} - M_j)^2$ from group j, and df_j refers to degrees of freedom, $n_j - 1$, from group j. Similarly in the situation where we have any number J rather than just two groups ($J = 3$ in the present example), we can compute a best estimate of the population variance using the formula

$$est \, \sigma^2 = \frac{\sum_{j=1}^{J} SS_j}{\sum_{j=1}^{J} df_j}$$

Once again we see that the best estimate of the population variance is obtained by dividing a sum of squares (in this case the total sum of squares from the J, or in this example, 3 groups) by an associated degrees of freedom (the sum of the degrees of freedom from the J, or in this example, 3 groups).

New terminology.   We will now introduce a new name for our estimate of σ^2 obtained from within groups. This new name is *mean square within*, which is abbreviated *MSW*. The term mean square within derives from the fact that the mean square within is the mean of the squared deviations within the groups.

> 11.2 Mean square within, or *MSW*, is our best estimate of σ^2, the variance of the populations from which the groups of scores come. *MSW* is the average of the J individual estimates of σ^2, and can be computed by dividing the sum of the sums of squares from the J groups by the sum of the degrees of freedom from the J groups.

Hypothesis-Testing Steps

At this point we start through our hypothesis-testing steps. Since we are in a slightly different position than we have been in the past—we have three groups rather than two groups—we are eventually going to run into some stumbling blocks. We'll handle those stumbling blocks when we get to them.

Null and alternative hypotheses.   As is usually the case, H_0 maintains that in the population the independent variable has no effect on the dependent variable. The alternative hypothesis, H_1, on the other hand, claims that the independent variable does have an effect on the dependent variable. In the present example,

H_0: The independent variable (in this case hair color) has no effect on the dependent variable (in this case number of dates per week). Referring to μ_1, μ_2, and μ_3 as the means of the blonde, brunette, and redhead populations of weekly dating frequencies, the null hypothesis would state that $\mu_1 = \mu_2 = \mu_3$.

Similarly, the alternative hypothesis would be

H_1: Hair color does have an effect on the number of dates. In other words, it is not true that $\mu_1 = \mu_2 = \mu_3$.

Notice that this alternative hypothesis is extremely vague, stating only that there is *some* difference among the three population means. So for example the alternative hypothesis would be true if

$\mu_1 > \mu_2 = \mu_3$

or if

$$\mu_1 > \mu_2 > \mu_3$$

or if

$$\mu_1 < \mu_2 < \mu_3$$

or if

$$\mu_1 = \mu_2 > \mu_3$$

or if

$$\mu_1 = \mu_2 < \mu_3$$

and so on. (There are other possibilities that would also imply the alternative hypothesis being true.) This vagueness of the alternative hypothesis is a major disadvantage of the ANOVA technique. For if we end up deciding that the null hypothesis is false—that the alternative hypothesis is true—we know only slightly more about our data than we did before we started. There are ways to be more precise about deciding what our data actually mean, but we shall defer this topic until subsequent chapters. First things first.

> 11.3 The alternative hypothesis is extremely vague in an ANOVA situation, stating only that it is untrue that all population means equal one another.

A summary score: Mean square between. Our next step is to obtain some summary score that reflects the extent to which the alternative as opposed to the null hypothesis is true. In the past, selection of such a summary score has been fairly straightforward. When testing a single group mean against a constant, the appropriate summary score was simply the mean. Likewise, when testing two means against one another, the appropriate summary score was the difference between the two means. But what is the appropriate summary score in this situation that involves three groups?

Here is where the idea of between-group variance comes in. Recall that we have been trying slowly but surely to identify the difference between two means we used as our summary score in previous examples with what we have started referring to as between-group variance. Now we shall be quite explicit about between-group variance. Specifically we will consider our three sample means and compute how much variance is among them. More specifically we will assume that these three sample means come from some population of sample means, and we will use our three sample means to estimate the variance of this population. Table 11-2 shows how this is done. We have three scores: in this case the scores are sample means. Let's refer to the mean of these three sample means

TABLE 11-2. Estimation of the Variance of the Distribution from Which Three Sample Means Are Drawn

M_j	M_j^2
$M_1 = 7$	49
$M_2 = 2$	4
$M_3 = 3$	9

$$\sum_{j=1}^{3} M_j = 12 \qquad\qquad \sum_{j=1}^{3} M_j^2 = 62$$

$$SS(\text{means}) = \sum_{j=1}^{3} (M_j - M)^2$$

$$M = \frac{\sum_{j=1}^{3} M_j}{J} \qquad\qquad -\sum_{j=1}^{3} M_j^2 - \frac{\left(\sum_{j=1}^{3} M_j\right)^2}{J}$$

$$= \frac{12}{3} \qquad\qquad\qquad = 62 - 12^2/3$$

$$= 4 \qquad\qquad\qquad\quad = 62 - 48$$

$$\qquad\qquad\qquad\qquad\quad = 14$$

as M. Now our estimate of the variance of the population from which these three sample means are drawn is simply the sum of squares from these sample means divided by the degrees of freedom. As is evident in Table 11-2, the sum of squares is 14. How many degrees of freedom are associated with this sum of squares? There are three scores (the three means), so there are two degrees of freedom (the number of scores minus one). Therefore, the estimate of the population variance from which these means come is

$$\frac{SS}{df} = \frac{14}{2} = 7 \tag{11.1}$$

What is the world like if the null hypothesis is true?　Let's for the moment assume that the null hypothesis is true, that is there are actually no differences among the three population means. If such were the case, then the 12 scores shown in Table 11-1 would all come from the same population. Now let's consider the variance between groups—that is, the variance between the three sample means—which we have just finished computing. Here is the crucial thing. *If the null hypothesis is true, then these three means are simply three scores drawn from a sampling distribution of sample means.* Therefore, the variance we estimated in Equation 11.1 is a quite familiar variance—it is simply σ_M^2, the variance of the sampling distribution of sample means for samples of size

$n = 4$. We have, of course, arrived at an estimate of this variance via a somewhat unfamiliar path—through the back door, so to speak. Consider that in the past we began by estimating σ^2, the population variance, and then divided by n to obtain an estimate of σ_M^2. In the present case, however, we *started* with three means and estimated σ_M^2 directly.

Now we will do a very tricky thing. We know the equation relating *est* σ^2 to *est* σ_M^2, namely

$$est\ \sigma_M^2 = \frac{est\ \sigma^2}{n}$$

where n is the sample size. This formula can be rearranged so that we can calculate *est* σ^2 given *est* σ_M^2:

$$est\ \sigma^2 = (n)(est\ \sigma_M^2)$$

When we plug our data into this equation we find that our estimate of σ^2 is

$$est\ \sigma^2 = (4)(7) = 28$$

This estimate of σ^2, the population variance that is obtained by considering variance between groups—between the sample means—has a name. It is called mean square between or *MSB*.

H_0 true: three sample means are from the same distribution

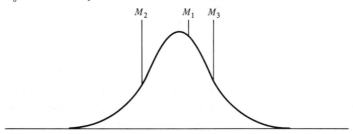

H_1 true: three sample means do not come from the same distribution

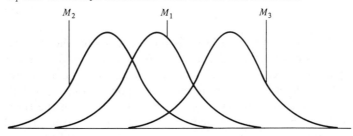

FIGURE 11-1. Origin of sample means if H_0 is true versus if H_1 is true. If H_0 is true (top panel), then the three sample means all come from the same sampling distribution of sample means whose variance is σ_M^2. If H_1 is true, then the three sample means come from different distributions and their variance is expected to be greater than σ_M^2.

We now find ourselves in an apparently puzzling situation. We began by obtaining our one best estimate of the population variance using fairly familiar techniques. This estimate, *MSW*, was 2.67. We then calculated another estimate of the population variance in a somewhat more indirect fashion—by first estimating σ_M^2 using the three sample means and then multiplying by n, the sample size. However, this estimate of σ^2 from between groups, *MSB*, turned out to be 28 which is more than 10 times the original estimate of 2.67 we obtained from within groups. Why the discrepancy?

There are three possible reasons: The first possibility is that σ^2, the population variance, really is somewhere around 28, but we happened to get an abnormally small estimate of the population variance from within groups. The second possibility is that σ^2 is actually somewhere around 2, but we happened to get an abnormally large estimate of the population variance from between groups—that is, we happened by chance to pick three means that were quite different from one another. The third possibility is the interesting one. This possibility is that the null hypothesis is actually false (recall that this entire train of logic is based on the assumption that the null hypothesis is true). Figure 11-1 illustrates why the null hypothesis being false would lead to an inflated estimate of the population variance from between groups. The top panel of Figure 11-1 depicts a situation in which the null hypothesis is true. In this case the three sample means are all plucked out of the same sampling distribution of sample means. In contrast, the lower panel of Figure 11-1 depicts a situation in which the null hypothesis is false. In this situation there are three different distributions of sample means; M_1 is drawn from one distribution; M_2 is drawn from another distribution; M_3 is drawn from yet a third distribution.[1] If the three sample means come from three separate distributions, they would tend to be more spread out than would be the case if they came from the same distribution. Therefore, an estimate of the population variance derived from these three spread-out sample means would be larger than it would be were the null hypothesis true and the three sample means all from the same distribution.

> 11.4 Mean square between or *MSB* is a reflection of how much variation there is among the sample means. In this sense it constitutes an appropriate summary score. To the degree that *MSB* is large, we tend to believe that H_1 is true, and to the extent that *MSB* is small we tend to believe that H_0 is true. Note that if H_0 *is* true, then *MSB* is an estimate of σ^2.

---

[1] It actually need not be the case that all three means come from three separate distributions. For the null hypothesis to be false, it is sufficient that all three means not come from the same distribution. But if for example M_1 and M_2 came from one distribution and M_3 came from a second distribution, the same arguments would hold—the variance among the three means would still be greater than would be the case if all three means came from the same distribution.

The *F* distribution.   This brings us back to the remarks that we made at the beginning of this chapter. We want to make our decision to reject or not reject the null hypothesis based on the relationship of variance between groups (mean square between) to variance within groups (mean square within). Furthermore, we have just discovered that if the null hypothesis is true, then mean square between and mean square within should both be estimates of the same variance (σ^2) and therefore should be about equal to one another. However, if the alternative hypothesis is true, then mean square between ought to be bigger than mean square within.

How are we to assess the extent to which mean square between is greater than mean square within? As suggested, the simplest way of making such an assessment is to take their ratio:

$$\text{ratio} = \frac{MSB}{MSW}$$

which, in the case of the present example, would be

$$\text{ratio} = \frac{28.0}{2.67} = 10.49$$

If this ratio is big enough, we will reject the null hypothesis in favor of the alternative hypothesis.

Fortunately, we can be quite specific about the question of what defines "big enough." The ratio of any two independent estimates of the population variance can be thought of as a random variable, and it turns out that the distribution of this random variable is well known. The distribution is called an *F* distribution (named after its originator, Sir Ronald Fisher).

Let's enumerate a few characteristics of the *F* distribution. First, as was the case with the *t* distribution, there is not a single unitary *F* distribution but rather there is a family of *F* distributions. The reason for this is as follows. Note that *MSW* can be based on any number of degrees of freedom, as can *MSB*. Now the

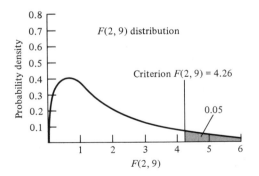
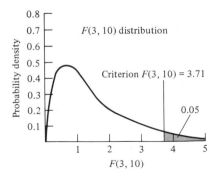

FIGURE 11-2.   Examples of two members of the *F* distribution family. Along with the distributions are the criteria that chop off the upper 5% of the distributions.

distribution of the F-ratio, MSB/MSW, depends on the number of degrees of freedom in both MSB and MSW. Therefore, there is one F distribution corresponding to each value of degrees of freedom in MSB and in MSW. The particular F distribution with which we are concerned in our example is an F distribution based on 2 and 9 degrees of freedom (2 being the number of degrees of freedom going into the variance estimate in the numerator and 9 being the number of degrees of freedom going into the estimate in the denominator). This F distribution is referred to as $F(2,9)$.

The second characteristic concerns the shape of the F distribution. We know that the value of an F-ratio can never be less than 0 (since no estimate of a population variance can be negative). However, an F-ratio can be arbitrarily large. Therefore, an F distribution should be skewed in a positive direction. Figure 11-2 shows two different members of the F distribution family.

11.5 The ratio of any two estimates of a population variance is distributed as an F. If the estimate in the numerator is based on df_1 degrees of freedom and the estimate in the denominator is based on df_2 degrees of freedom, then their ratio is distributed as an $F(df_1, df_2)$.

Digression 11.1

More on the F-Ratio

We wish to reemphasize the point that the ratio of *any two independent estimates* of some population variance is, by definition, distributed as an F. By and large F-ratios with which we will be concerned will be ratios of the two estimates MSB and MSW.

There are, however, other situations in which we might be interested in an F-ratio. Consider, for example, a t-test situation in which we have two groups and from each group, an estimate of σ^2, the population variance. Recall our assumption that these two estimates, $est_1\ \sigma^2$ and $est_2\ \sigma^2$, are estimates of the same population variance. Suppose we wanted to test this assumption. If the assumption is correct, then the ratio

$$\frac{est_1\ \sigma^2}{est_2\ \sigma^2}$$

would be drawn from an F distribution with $n_1 - 1$ and $n_2 - 1$ degrees of freedom. Alternatively, the ratio

$$\frac{est_2\ \sigma^2}{est_1\ \sigma^2}$$

would be distributed as an F with $n_2 - 1$ and $n_1 - 1$ degrees of freedom. The size of either ratio could be assessed for its extremeness. If the ratio were sufficiently extreme, then we might conclude that the assumption that $est_1 \, \sigma^2$ and $est_2 \, \sigma^2$ are estimating the same variance is false.

How extreme is our score?   Let's get back to the problem at hand. Recall that we have two estimates of σ^2, MSW and MSB. If the null hypothesis is true—that is, if hair color has no effect on number of dates—then mean square between and mean square within should be estimates of the same population variance, and therefore their ratio should be distributed as an $F(2,9)$. As noted above, we would expect this ratio to be generally somewhere around 1.[2] Additionally, we have demonstrated that if the alternative hypothesis is true, then mean square between ought to be bigger than mean square within. Therefore, our obtained $F(2,9)$ ought to be somewhat larger than 1. Now we can go through a familiar procedure. We will consider the distribution of our score (the F) assuming that the null hypothesis is true, and we will establish a criterion score that chops off the upper 5% of the distribution. Thus, if the null hypothesis is true, a score as extreme as our to-be-established criterion would be obtained with a probability of only 0.05.

***F* distribution tables.**   In order to determine what this criterion score should be, we make use of F distribution tables (painstakingly constructed by Sir Ronald Fisher). F distribution tables, like t distribution and z distribution tables, are found in many places. In particular, they are found in Appendix F of this book.

F distribution tables are even more abbreviated than are t distribution tables. Specifically, the F distribution tables in this book can only be used to establish criteria that chop off the upper 10%, 5%, 2.5%, and 1% of F distributions based on selected degrees of freedom. Each such table takes up an entire page and it is organized as follows. First, the rows correspond to degrees of freedom in the denominator (for purposes of the present discussion the degrees of freedom on which MSW is based). In our example we have an MSW based on 9 degrees of freedom, so we would go to the ninth row in the relevant table. The columns of the table correspond to the degrees of freedom in the numerator (the degrees of freedom on which MSB is based). In our present example we have 2 degrees of freedom between (since we have three groups) so we go to the second column. At the intersection of row 9 and column 2 we find an entry of 4.26 which is the criterion $F(2,9)$ that chops off the upper 5% of the $F(2,9)$ distribution. This criterion is depicted in Figure 11-2 (along with the criterion that chops off the upper 5% of the other F distribution depicted in Figure 11-2).

---

[2]Actually, the expected value of an F distribution is $df_2/(df_2 - 2)$ where df_2 refers to the number of degrees of freedom in the denominator of the F-ratio. This is a mysterious expression, and the authors are frankly unable to provide any intuitive rationale for it.

We see that our obtained $F(2,9)$ of 10.49 is considerably larger than the criterion $F(2,9)$ of 4.26. This conforms to our intuition that if MSB is an estimate of the same variance as is MSW, then it is very unlikely that the ratio of MSB to MSW would be as great as 10 to 1. In any case since our obtained $F(2,9)$ is greater than our criterion $F(2,9)$, we reject the null hypothesis and conclude that hair color really does have an effect on dating frequency. In other words we conclude that the three groups of scores shown in Table 11-2 are indeed drawn from different populations of scores.

In the next section we shall go through a somewhat more systematic way of carrying out an analysis of variance. At this point, however, let's summarize what we have done so far. We began with three groups of scores and a null hypothesis stating that these three groups of scores all came from a single population (that is, that hair color had no effect on dating frequency). We obtained an estimate of the population variance by computing an individual estimate from within each of the three groups and then averaging these three estimates to get a one best estimate. This estimate was based on 9 degrees of freedom—three from within each of the three groups. The technique of obtaining an estimate of the population variance from within groups was familiar to us since we did it routinely in our t-tests.

At this point our procedure departed somewhat from that used in the past, and we invoked the following rationale, "Suppose," we said, "that the null hypothesis is true. Should there in such an event be any differences among the means of the three groups?" The answer is, yes, there should be. Even if the three groups of scores came from the same population, we would expect their means to differ a bit, simply due to random fluctuation. However, we can be very specific about the relationship between the population variance and the variance among the three means if the null hypothesis is true. They are simply related by a factor of n, where n is the size of the samples on which the means are based. Using this logic we were able to calculate an estimate of the population variance by considering the variance between the three means. However, this estimate of the population variance turned out to be considerably greater than our "real" estimate of the population variance obtained from within groups, leading us to believe that the difference—the variance—among the three means was due to more than simply random fluctuation. Therefore, we rejected the null hypothesis and concluded that the independent variable of hair color really was causing some of the differences among the three groups.

COMPUTATIONAL FORMULAS FOR ANALYSIS OF VARIANCE

At this point we hope you have a reasonably intuitive understanding of what analysis of variance is all about, and now we'll derive some computational formulas. That is, we shall work out a "cookbook" method to make actually performing an ANOVA on any set of data relatively easy. Our strategy will be to first derive a formula for mean square within, then derive a formula for mean square between, and finally construct an ANOVA "summary table" that will make the performance of an analysis of variance neat and systematic. We'll begin by assuming equal numbers of subjects in our groups and then generalize to a situation in which unequal ns are permitted.

A Formula for Mean Square Within

We shall assume that we have J groups with n subjects per group. We have seen that

$$MSW = \frac{\sum_{j=1}^{J} SS_j}{\sum_{j=1}^{J} df_j}$$

where SS_j refers to the sum of squares from the jth group. **We shall refer to**

$$\sum_{j=1}^{J} SS_j$$

as "sum of squares within" or SSW. Likewise, we shall refer to

$$\sum_{j=1}^{J} df_j$$

as "degrees of freedom within" or dfW. We derive equations separately for dfW and SSW.

Degrees of freedom within. Establishment of how many degrees of freedom there are from within groups is fairly straightforward. First we observe that we have $n - 1$ degrees of freedom in each of our J groups. Therefore, the total number of degrees of freedom we have is

$$J(n - 1)$$

which in turn is equal to

$$Jn - J$$

Referring to Appendix E, we see that Jn is designated N, the total number of scores in the experiment. Therefore,

$$dfW = N - J$$

Sum of squares within. To obtain our formula for sum of squares within, we'll make heavy use of the notation established in Appendix E. We have already noted that sum of squares within is equal to the sum of all the sums of squares for each of the J groups, or

$$SSW = \sum_{j=1}^{J} SS_j \tag{11.2}$$

Now let's focus on one of those sums of squares—SS_1 or the sum of squares from Group 1. We know that this sum of squares is equal to

$$SS_1 = \sum_{i=1}^{n} (x_{i1} - M_1)^2$$

which in turn may be written using the computational formula

$$SS_1 = \sum_{i=1}^{n} x_{ij}^2 - \frac{T_1^2}{n}$$

(Note that T_1 is the total of the first group.) In general, for Group j

$$SS_j = \sum_{i=1}^{n} x_{ij}^2 - \frac{T_j^2}{n} \tag{11.3}$$

Now to obtain the sum of squares within, we substitute Equation 11.3 into Equation 11.2 to obtain

$$SSW = \sum_{j=1}^{J} \left[\sum_{i=1}^{n} x_{ij}^2 - T_j^2/n \right]$$

We can carry through the summation sign to the left of the brackets to obtain

$$SSW = \sum_{j=1}^{J} \sum_{i=1}^{n} x_{ij}^2 - \sum_{j=1}^{J} T_j^2/n \tag{11.4}$$

What this somewhat formidable looking formula says is really quite simple. The first term involves squaring every single score (every x_{ij}) and adding up all these squared x_{ij}s. The second term involves squaring each group total (T_j), dividing that squared group total by n, and adding up these T_j^2/ns.

To illustrate the use of this formula, let's apply it to the hair color example we described previously. First, as noted, we have to square each of the 12 scores in Table 11-1 and compute the sum of these 12 squared scores. That is,

$$\sum_{j=1}^{J} \sum_{i=1}^{n} x_{ij}^2 = 6^2 + 6^2 + 10^2 + 6^2$$
$$+ 2^2 + 4^2 + 1^2 + 1^2$$
$$+ 2^2 + 3^2 + 5^2 + 2^2 = 272$$

Next, we must square each group total, divide that squared group total by n (in this case 4, since there are four subjects per group), and add these all up. Thus,

$$\sum_{j=1}^{J} T_j^2/n = 28^2/4 + 8^2/4 + 12^2/4 = 248$$

So we see that the sum of squares within is equal to

$$SSW = \sum_{j=1}^{J} \sum_{i=1}^{n} x_{ij}^2 - \sum_{j=1}^{J} T_j^2/n$$

$$= 272 - 248 = 24$$

which conforms to the $SS_1 + SS_2 + SS_3 = 12 + b + b = 24$ that we could compute from Table 11.1.

Mean Square Between

Recall our general rule that any estimate of the population variance (to which we are now referring as a "mean square") is equal to a sum of squares divided by an associated degrees of freedom. Thus, MSW was equal to SSW divided by dfW.

The same is true of MSB; MSB is said to be equal to a *sum of squares between* (SSB) divided by an associated *degrees of freedom between* (dfB).

To derive the relevant formulas, we first obtain an estimate of σ_M^2 based on our J sample means. (Recall that this estimate is based on the assumption that the null hypothesis is true.) As mathematically proved in Digression 11.2, the mean of the J means (M) is simply equal to the grand mean of all the scores in the experiment, or T/N. Now let's write down a formula for estimating σ_M^2 from the J sample means.

$$est\ \sigma_M^2 = \sum_{j=1}^{J} \frac{(M_j - M)^2}{J - 1}$$

Digression 11.2

Proof that the Mean of the Group Means Is Equal to T/N

First, note that the mean of the M_js may be expressed as

$$M = \frac{\sum_{j=1}^{J} M_j}{J} \qquad (D.1)$$

Now we know that

$$M_j = \frac{T_j}{n} \qquad (D.2)$$

Substituting Equation D.2 into Equation D.1,

$$M = \frac{\sum_{j=1}^{J} T_j/n}{J} = \frac{\sum_{j=1}^{J} T_j}{nJ}$$

Since

$$\sum_{j=1}^{J} T_j = T \text{ and } nJ = N,$$

$$m = T/N$$

At this point we remind ourselves that what we really want is not an estimate of σ_M^2 but rather an estimate of σ^2. To obtain an estimate of σ^2, we must multiply our estimate of σ_M^2 by n, to obtain

$$MSB = \text{est } \sigma^2 = (n)\text{est } \sigma_M^2 = \frac{n \sum_{j=1}^{J} (M_j - M)^2}{J - 1} = \frac{SSB}{dfB} \qquad (11.5)$$

The expression in the numerator of Equation 11.5 is SSB, and likewise, the expression in the denominator is dfB.

> 11.6 Our estimate of σ^2 from between groups, MSB, is defined to be SSB/dfB. SSB is equal to:
>
> $$SSB = n \sum_{j=1}^{J} (M_j - M)^2$$
>
> whereas dfB is simply the number of groups minus one, or $J - 1$.

A raw-score formula for SSB.   Using our deviation-to-computational-score equation the sum of squares between may be expressed as

$$SSB = n \sum_{j=1}^{J} (M_j - M)^2 = n \left[\sum_{j=1}^{J} M_j^2 - \left(\sum_{j=1}^{J} M_j \right)^2 / J \right]$$

We can algebraically fiddle with this expression:

$$SSB = n \sum_{j=1}^{J} M_j^2 - \frac{n}{J} \left(\sum_{j=1}^{J} M_j \right)^2 \qquad (11.6)$$

Note that

$$M_j = \frac{T_j}{n} \qquad (11.7)$$

Substituting Equation 11.7 for M_j into Equation 11.6,

$$SSB = n \sum_{j=1}^{J} \frac{T_j^2}{n^2} - \frac{n}{J} \left(\frac{\sum_{j=1}^{J} T_j}{n} \right)^2 \qquad (11.8)$$

Now noting that

$$\sum_{j=1}^{J} T_j = T \tag{11.9}$$

and substituting Equation 11.9 into Equation 11.8

$$SSB = \sum_{j=1}^{J} \frac{T_j^2}{n} - \frac{n}{j}\left(\frac{T^2}{n^2}\right)$$

or

$$SSB = \sum_{j=1}^{J} \frac{T_j^2}{n} - \left(\frac{1}{Jn}\right)T^2$$

Finally, noting that $Jn = N$, the total number of scores in the experiment,

$$SSB = \sum_{j=1}^{J} \frac{T_j^2}{n} - \frac{T^2}{N} \tag{11.10}$$

Equation 11.10 says that to compute the sum of squares between we proceed as follows: First we square each group total, divide that squared total by n and then add up these T_j^2/ns. (Note that this term has made a prior appearance as the second term in the equation for sum of squares within.) The second term in Equation 11.10 simply involves squaring the T, grand total, and then dividing by N, the number of scores in the entire experiment. (T^2/N is often referred to as the "correction factor." It will pop up again and again in the future.)

An Analysis of Variance Summary Table

Now let's gather together these various formulas and summarize them in one big, tidy table. Such a table is referred to as *an analysis of variance summary table* and is depicted in Table 11-3. The summary table shown in Table 11-3 is relatively straightforward and contains four columns. The four different columns show (1) the source of variance (consisting in this case of variance from between groups and variance from within groups); (2) degrees of freedom associated with the sources of variance; (3) the sums of squares

TABLE 11-3. A One-Way ANOVA Summary Table

Source	df	SS	MS(est σ^2)
Between	$dfB = J - 1$	$SSB = \sum_{j=1}^{J} T_j^2/n - T^2/N$	$MSB = SSB/dfB$
Within	$dfW = N - J$	$SSW = \sum_{j=1}^{J}\sum_{i=1}^{n} x_{ij} - \sum_{j=1}^{J} T_j^2/n$	$MSW = SSW/dfW$

TABLE 11-4.   An ANOVA Summary Table for the Hair Color and Dating Example

Source	df	SS	MS	Obtained F	Criterion F
Between	2	56	28	10.49	4.26
Within	9	24	2.67		

associated with the sources of variance; and (4) the mean squares which are the sums of squares divided by their corresponding degrees of freedom.

Our hair color example using an ANOVA table.   Let's briefly run through our analysis of the hair color experiment using this analysis of variance summary table. The summary table for this example is shown in Table 11-4. We have already calculated all the entries that go into this table—degrees of freedom between and within. Additionally we calculated the relevant mean squares, and thus we have added a fifth and sixth column. The fifth column shows the $F(2,9)$ that results from dividing MSB by MSW. The sixth column lists the criterion $F(2,9)$ which is 4.26. As can be seen, the conclusion based on Table 11-4 is precisely the same as the conclusion that emerged in the last section. This, of course, *has* to be true since we performed algebraically equivalent calculations in both cases. In this section we just did those calculations somewhat more systematically.

The unequal ns case.   Now let's drop our requirement that there be equal numbers of observations in each of our J groups. Instead, we will make the more general assumption that Group j has n_j scores. This transition from equal to unequal is actually quite simple with respect to our computational formulas. We simply replace each n in our formulas with n_j. The resulting summary table is shown in Table 11-5. As can be seen, the changes take place in the term

$$\sum_{j=1}^{J} \frac{T_j^2}{n_j}$$

which is the second term in the formula for SSW and the first term in the

TABLE 11-5.   An ANOVA Summary Table: Unequal ns Are permitted; thus, Group j Is Assumed to Have n_j Scores

Source	df	SS
Between	$J - 1$	$SSB = \sum_{j=1}^{J} T_j^2/n_j - T^2/N$
Within	$N - J$	$SSW = \sum_{j=1}^{J} \sum_{j=1}^{n_j} x_{ij}^2 - T_j^2/n_j$

formula for *SSB*. To compute this term, we square each group total and then divide that squared total by the *number of scores that went into that total*. These T_j^2/n_js are then added up.

A START-TO-FINISH EXAMPLE: TOOTHPASTE AND CAVITY PREVENTION

Let's assume once again that we are dental researchers interested in comparing various brands of toothpaste for effectiveness in cavity prevention. Suppose that we have four different brands of toothpaste—Dazzle, White-O, Irrident, and No-Cav. We initially plan to have six subjects use each of these four types of toothpaste for one year. At the end of the year we intend to see how many cavities each of the subjects in the four groups has developed. Unfortunately we lose track of some of our subjects during the year, and we are left with $n_1 = 3$ subjects in the Dazzle group, $n_2 = 5$ subjects in the White-O group, $n_3 = 4$ subjects in the Irrident group, and $n_4 = 2$ subjects in the No-Cav group. Table 11-6 shows the data—the number of cavities—from the remaining subjects in the four groups.

TABLE 11-6. Hypothetical Data: Number of Cavities Obtained by Each Subject in Four Different Groups; a Group Is Defined by the Type of Toothpaste Used by Members of the Group

Group 1: Dazzle	Group 2: White-O	Group 3: Irrident	Group 4: No-Cav
$n_1 = 3$	$n_2 = 5$	$n_3 = 4$	$n_4 = 2$
x_{i1}	x_{i2}	x_{i3}	x_{i4}
	$x_{12} = 4$		
	$x_{22} = 1$	$x_{13} = 6$	
$x_{11} = 2$	$x_{32} = 3$	$x_{23} = 2$	
$x_{21} = 1$	$x_{42} = 3$	$x_{33} = 4$	$x_{14} = 3$
$x_{31} = 3$	$x_{52} = 4$	$x_{43} = 4$	$x_{24} = 1$
$T_1 = 6$	$T_2 = 15$	$T_3 = 16$	$T_4 = 4$
$M_1 = 6/3 = 2$	$M_2 = 15/5 = 3$	$M_3 = 16/4 = 4$	$M_4 = 4/2 = 2$

$$\sum_{everything} x_{ij}^2 = 2^2 + 1^2 + 3^2 + 4^2 + 1^2 + \cdots + 9^2 + 1^2 = 147$$

$$\sum_{groups} T_j^2/n_j = 6^2/3 + 15^2/5 + 16^2/4 + 10^2/2 = 129$$

$$T = 6 + 15 + 16 + 4 = 41$$

$$T^2/N = 41^2/14 = 120$$

Hypotheses

Let's set up the null and alternative hypotheses. The null hypothesis is that the independent variable—in this case toothpaste brand—has no effect on the dependent variable—in this case number of cavities developed in one year. Referring to μ_1, μ_2, μ_3, and μ_4 as the means of the populations of people who brush with Dazzle, White-O, Irrident, and No-Cav, respectively, the null hypothesis states that

$$H_0: \mu_1 = \mu_2 = \mu_3 = \mu_4$$

The alternative hypothesis, as usual claims, that the independent variable does have an effect on the dependent variable. The null hypothesis is very vague as to the exact nature of the differences among the various toothpaste brands, stating merely that

$$H_1: \text{It is not true that } \mu_1 = \mu_2 = \mu_3 = \mu_4$$

In Table 11-6 we calculated the means of the four groups, $M_1 = 2$, $M_2 = 3$, $M_3 = 4$, and $M_4 = 2$. There are certainly no monumental differences among these four means, and it is not at all obvious, looking at the data, whether or not we should reject the null hypothesis. Our subsequent hypothesis testing steps will consist of obtaining an estimate of σ^2 from within groups (MSW) and from between groups (MSB). A comparison of these two estimates will then constitute the test of whether or not the null hypothesis may be safely rejected.

Calculation of ANOVA Summary Table Entries

Now our next step is to calculate sums of squares and degrees of freedom. First we have seen that the formula for sum of squares within[3] is

$$SSW = \sum_{\text{everything}} x_{ij}^2 - \sum_{\text{groups}} \frac{T_j^2}{n_j}$$

---

[3]Once again we shall revert to informal notation in our summation signs. Thus

$$\sum_{\text{everything}}$$

refers to "sum over all N scores" in the experiment, that is, $\sum_{j=1}^{J} \sum_{i=1}^{n_j}$ and

$$\sum_{\text{groups}}$$

means "sum over all J groups," that is, $\sum_{j=1}^{J}$.

and that the formula for sum of squares between is

$$\sum_{\text{groups}} \frac{T_j^2}{n_j} - \frac{T^2}{N}$$

The three terms making up these two formulas have been computed in Table 11-6. They are:

$$\sum_{\text{everything}} x_{ij}^2 = 147$$

$$\sum_{\text{groups}} \frac{T_j^2}{n_j} = 129$$

and

$$T^2/N = 120$$

Therefore,

$$SSW = 147 - 129 = 18$$

and

$$SSB = 129 - 120 = 9$$

To calculate degrees of freedom is quite simple. Degrees of freedom within are $N - J$ or $14 - 4 = 10$. An alternate way of computing dfW is to note that there are 2 degrees of freedom from within Group 1, 4 degrees of freedom from within Group 2, 3, degrees of freedom from within Group 3, and 1 degree of freedom from within Group 4, for a total of 10 in all. Likewise, we know that there are $J - 1 = 4 - 1 = 3$ degrees of freedom between. Now we are in a position to set up our analysis of variance table (see Table 11-7).

We have obtained an $F(3,10)$ of 1.67. Figure 11-2 shows what the $F(3,10)$ distribution looks like (assuming that the null hypothesis is actually true and that MSW and MSB are therefore estimates of the same variance). Consulting our F distribution tables, we see that the criterion $F(3,10)$ that chops off the upper 5% of the distribution is 3.71. Our obtained $F(3,10)$ of 1.67 is not as extreme as our criterion $F(3,10)$ of 3.71; we therefore fail to reject the null hypothesis. We do not have enough evidence to conclude that brand of toothpaste really does have an effect on the number of cavities developed.

TABLE 11-7. ANOVA Summary Table for the Toothpaste Experiment

Source	df	SS	MS	Obtained F	Criterion F
Between	3	9	3	$F(3, 10) = 1.67$	$F(3, 10) = 3.71$
Within	10	18	1.8		

PARTITION OF THE TOTAL SUM OF SQUARES

Let's return to our prototypical analysis of variance situation, involving J groups with n_j subjects in Group j (see Appendix E). Suppose we wanted to treat all scores in the experiment as a single large set of scores. Suppose further that we compute one big sum of squares from the entire N scores. How would we obtain such a sum of squares? An obvious way to proceed would be to compute the sum over all xs of the $(x - M)^2$s, where M refers to the mean of all N scores or T/N. More precisely, the formula for obtaining this total sum of squares would be

$$\sum_{\text{everything}} (x_{ij} - M)^2$$

Now we can use our deviation-to-raw-score formula to obtain

$$\sum_{\text{everything}} (x_{ij} - M)^2 = \sum_{\text{everything}} x_{ij}^2 - \left(\frac{\sum_{\text{everything}} x_{ij}^2}{N}\right)^2$$

or since

$$\sum_{\text{everything}} x_{ij} = T$$

$$\sum_{\text{everything}} (x_{ij} - M)^2 = \sum_{\text{everything}} x_{ij}^2 - \frac{T^2}{N}$$

How many degrees of freedom would be associated with this sum of squares? Since the sum of squares is based on a total of N scores, the degrees of freedom would be $N - 1$.

Consider the analysis of variance summary table which is redepicted in Table 11-8. Let's do a seemingly odd thing and add up $df B$ and $df W$. Similarly

**TABLE 11-8.   A General ANOVA Summary Table
Illustrating Partition of the Total Sum of Squares**

Source	df	SS
Between	$J - 1$	$\sum_{\text{groups}} T_j^2/n_j - T^2/N$
+	+	+
Within	$N - J$	$\sum_{\text{everything}} x_{ij}^2 - \sum_{\text{groups}} T_j^2/n_j$
Total	$N - 1$	$\sum_{\text{everything}} x_{ij}^2 - T^2/N$

let's add up SSB and SSW. We can see the results in Table 11-8. The sum of the degrees of freedom is $N - 1$, and the sum of the sum of squares is

$$\sum_{\text{everything}} x_{ij}^2 - \frac{T^2}{N}$$

Notice, interestingly, that the sum of dfB and dfW is simply equal to $N - 1$, which is precisely what we decided the degrees of freedom would be if we were to consider all N scores in the experiment as just one large set of scores. Likewise, the total of SSB and SSW is equal to the sum of squares that we obtained when we considered all N scores in the experiment to be just one big set of scores.

Dividing the Pie

Let's now put together the pieces of this algebraic jigsaw puzzle. Consider first all N scores in a given experiment. Typically these scores differ from one another in some way or other. That is, there exists variance among all N scores, which we have dubbed sum of squares total (or SST). As Figure 11-3 shows, one can think of SST as simply one big mass of sum of squares, which has $N - 1$ degrees of freedom associated with it.

Now we recognize that this total mass of sum of squares—the total variance of the entire experiment—is actually divided into variance from two separate sources which are variance from within groups and variance from between groups. Therefore, as shown in Figure 11-4, we can conceive of dividing

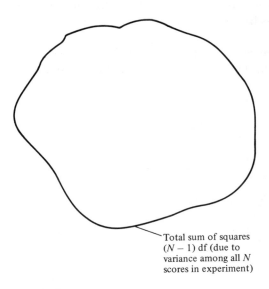

Total sum of squares
$(N - 1)$ df (due to
variance among all N
scores in experiment)

FIGURE 11-3.   A representation of the total sum of squares based on $N - 1$ degrees of freedom.

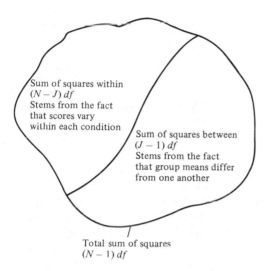

Sum of squares within
$(N - J)$ df
Stems from the fact
that scores vary
within each condition

Sum of squares between
$(J - 1)$ df
Stems from the fact
that group means differ
from one another

Total sum of squares
$(N - 1)$ df

FIGURE 11-4. Total sum of squares may be divided up into sum of squares from within groups (SSW) based on $N - J$ degrees of freedom and sum of squares from between groups (SSB) based on $J - 1$ degrees of freedom.

this total mass of sum of squares into one portion consisting of sum of squares within and another portion consisting of sum of squares between. Sum of squares between plus sum of squares within thereby makes up the entire sum of squares. Likewise, associated with sum of squares within are $N - J$ degrees of freedom and associated with sum of squares between are $J - 1$ degrees of freedom. We see, therefore, that the total $N - 1$ degrees of freedom we started with is made up of degrees of freedom within plus degrees of freedom between.

What we have just depicted is a quite simple way of dividing the "variance pie." In this relatively simple one-way analysis of the variance situation, the pie is sliced into just two pieces. In subsequent chapters we'll consider considerably more complex analysis of variance situations and see that the pie is divided and subdivided in fairly complex ways without changing the general idea. **There is a total amount of variance (SST) stemming from the fact that all the scores in the entire experiment vary from one another, and this total variance can be divided up into identifiable components.**

11.7 To the degree that all N scores in an experiment differ from one another, there is a total sum of squares, SST, which is based on $N - 1$ degrees of freedom. SST is divided into a sum of squares from between groups (SSB, based on $J - 1$ of the $N - 1$ degrees of freedom) and a sum of squares from within groups (SSW, based on the other $N - J$ degrees of freedom).

Digression 11.3

The Linear Model

A fairly common way of viewing an analysis-of-variance situation is in terms of what is referred to as a *linear model*. Here x_{ij}, the ith score in the jth group is represented by the equation,

$$x_{ij} = \mu + \alpha_j + \epsilon_{ij}$$

Here μ represents the population mean at all scores which could potentially emerge in the experiment (of which the scores that *do* emerge constitute a random sample). The second term in the equation, α_j, is an "effect"—that is, a change relative to μ caused by whatever treatment is applied to Group j. Finally, ϵ_{ij} is an *error* component that is distributed with variance σ^2 (the same σ^2 we have been talking about all along). Within this framework, it is stipulated that the $\alpha_j s$ sum to zero across groups that is,

$$\sum_{j=1}^{J} \alpha_j = 0$$

and that the $\epsilon_{ij} s$ sum to zero within each group—that is, for any group, j:

$$\sum_{i=1}^{n_j} \epsilon_{ij} = 0$$

The null and alternative hypotheses may now be formulated as:

$H_0: \alpha_1 = \alpha_2 = \cdots = \alpha_j = 0$

$H_1:$ not H_0

We will not be discussing the linear model framework in great detail in this book, as we prefer (somewhat arbitrarily) the partition-of-sum-of-squares framework. Hays (1973) provides an excellent description of the linear model.

CONFIDENCE INTERVALS IN AN ANOVA SITUATION

In the last chapter we delineated the virtues of computing confidence intervals around obtained sample means. In an ANOVA situation these virtues do not change. In general, presentation of confidence intervals provides a fairly good intuitive picture of how good the sample means as estimates of the population means—that is, how powerful the obtained data are.

Recall that for any sample mean, M_j, the formula for computing a confidence interval around M_j is

x percent confidence interval $= M_j \pm [\text{criterion } t(df)] \, [est \, \sigma_{M_j}]$

where the criterion $t(df)$ is the two-tailed criterion t corresponding to an α-level of 1 minus the percent confidence level we want (for example, the 0.05 criterion t for a 95% confidence interval). Note also that df is the degrees of freedom upon which $est \, \sigma_{M_j}$ is based.

We have seen that in an ANOVA situation, $est \, \sigma^2$ is relabeled MSW and is based on $N - J$ degrees of freedom. **Therefore, the formula for computing confidence intervals in an ANOVA situation is**

x percent confidence interval $= M_j \pm [\text{criterion } t(N - J)] \, est \, \sigma_{M_j}$

and note that

$$est \, \sigma_{M_j} = \sqrt{\frac{est \, \sigma^2}{n_j}} = \sqrt{\frac{MSW}{n_j}}$$

Table 11-9 illustrates the computation of confidence intervals for the hair color and the toothpaste examples described in this chapter. In Figure 11-5 we have shown the results graphically.

11.8 The confidence interval around M_j in an ANOVA situation is:

Confidence interval $= M_j \pm [\text{criterion} + t(dfW)^{2Q}_{\alpha=0.05}] \sqrt{\dfrac{MSW}{n_j}}$

TABLE 11-9. Computation of 95% Confidence Intervals for the Haircolor and Toothpaste Examples.

Hair color (see Table 11-1 for data)
$MSW = 2.67$
Criterion $t(9)^{2Q}_{\alpha=.05} = 2.262$

Mean	n_j	$\sqrt{MSW/n_j}$	Confidence interval $= M_j \pm [\text{Criterion } t] \, \sqrt{MSW/n_j}$
$M_1 = 7$	4	$\sqrt{2.67/4} = 0.817$	$7 \pm (2.262)(0.817) = 7 \pm 1.85$
$M_2 = 2$	4	$\sqrt{2.67/4} = 0.817$	$2 \pm (2.262)(0.817) = 2 \pm 1.85$
$M_3 = 3$	4	$\sqrt{2.67/4} = 0.817$	$3 \pm (2.262)(0.817) = 3 \pm 1.85$

Toothpaste (see Table 11-6 for data)
$MSW = 1.80$
Criterion $t(10)^{2Q}_{\alpha=.05} = 2.228$

Mean	n_j	$\sqrt{MSW/n_j}$	Confidence interval $= M_j \pm [\text{Criterion } t] \, \sqrt{MSW/n_j}$
$M_1 = 2$	3	$\sqrt{1.80/3} = 0.775$	$2 \pm (2.228)(0.775) = 2 \pm 1.727$
$M_2 = 3$	5	$\sqrt{1.80/5} = 0.600$	$3 \pm (2.228)(0.600) = 3 \pm 1.337$
$M_3 = 4$	4	$\sqrt{1.80/4} = 0.671$	$4 \pm (2.228)(0.671) = 4 \pm 1.495$
$M_4 = 2$	2	$\sqrt{1.80/2} = 0.949$	$2 \pm (2.228)(0.949) = 2 \pm 2.114$

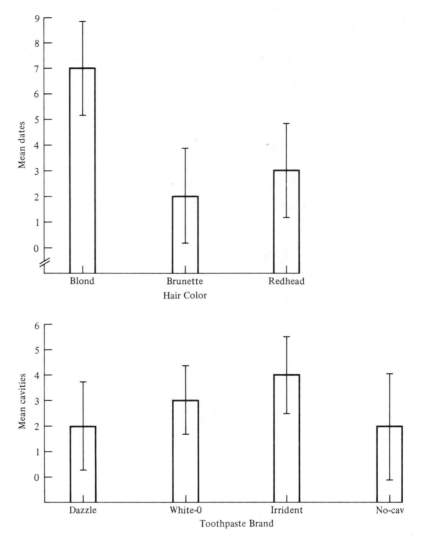

FIGURE 11-5. Graphs of (*a*) average number of dates as a function of hair color, and (*b*) average number of cavities as a function of toothpaste brand (95% confidence intervals are shown around all sample means).

AESTHETICS AND RULES

We have completed the major topics that we wanted to discuss with respect to one-way analysis of variance. At this point we want to indicate two aspects of the analysis of variance situation that (a) we hope will give you a feeling for some of the beauty and aesthetics of the mathematical structures involved in analysis of variance, and (b) will somewhat more pragmatically provide you with some mnemonics for the various formulas.

Relationship Between Degrees of Freedom and Sums of Squares

Let's consider a sum of squares with its associated degrees of freedom. We listed sum of squares total, sum of squares between, and sum of squares within along with their degrees of freedom in Table 11-8. When sums of squares and associated degrees of freedom are listed in this way, it's evident that there is a fairly close relationship between the structure of any given sum of squares and its degrees of freedom. First consider the row labeled "Total." Degrees of freedom are $N - 1$. Total sum of squares can be viewed as the sum of N things (the sum of all squared scores in the experiment) minus the sum of one thing (T^2/N). The same holds true for sums of squares and degrees of freedom between and within. Degrees of freedom within is equal to $N - J$ and correspondingly SSW is equal to the sum of N things (once again, the sum of all the N squared scores in the experiment) minus the sum of J things, the T^2/n_js. Finally degrees of freedom between is equal to $J - 1$, and SSB is the sum of J things minus the sum of one thing. We should point out that not only does this general correspondence between sums of squares and degrees of freedom hold for the sums of squares described in this chapter, but it will hold for all subsequent degrees of freedom and sums of squares as well.

The "Square and Divide" Rule

A convenient rule to keep in mind when formulating these equations is the following: Whenever you square some total, divide that squared total by the number of scores that went into the total. For example, sum of squares within is composed of the two terms:

$$\sum_{everything} x_{ij}^2 \quad \text{and} \quad \sum_{groups} T_j^2/n_j$$

Note that each x^2 is simply the sum of one thing—so that you divide by 1, which of course is equivalent to not dividing by anything at all. Likewise, T_j is the sum of n_j things. So whenever you square a T_j, divide by n_j. The third term that pops up in these formulas is T^2/N. Once again, T is the sum of N different things—therefore, whenever T is squared, it should be divided by N. Once again, this rule—which we have just demonstrated for these particular formulas—will hold up in any analysis of variance situation, no matter how complex. We hope this rule will facilitate remembering the various formulas.

SUMMARY

In this chapter we have progressed in terms of experimental complexity from a simple t-test situation involving two groups to an analysis of variance situation involving more than two groups. Although we do different tests in the two different situations—a t-test in a two-group situation and an ANOVA in a multiple-group situation—we are actually doing common things in the two situations. We are first getting a measure of the population variance (in the analysis of variance situation we referred to this measure as "mean square within"). In both the t-test and the ANOVA situations this estimate of the population variance is obtained by computing one estimate of the population variance from each group and then combining these estimates using a weighted average.

We then obtain some measure (summary score) of how much variance there is between groups. In the t-test situation this summary score was simply the difference between the means of the two groups. In an ANOVA situation the summary score consisted basically of the variance among the three means and was referred to as mean square between.

We then proceeded to determine how much variance we had between groups relative to the amount of variance within groups. We made this determination by taking the ratio of MSB to MSW. We called this ratio an F-ratio, and we learned that the distribution of the F-ratio is well known. Furthermore if the null hypothesis is correct, then any difference among the means must simply stem from fluctuation due to the population variance. In such a situation MSB and MSW are estimates of the same population variance, and their ratio (the F-ratio) ought to be about 1. In contrast, if the alternative hypothesis is true, then there ought to be more variance among the groups than we would expect from fluctuation due to the population variance alone. Therefore, the F-ratio that we get—mean square between divided by mean square within—ought to be large.

PROBLEMS

1. An experiment is done to test the effect of age on memory in a free-recall paradigm. Three groups of subjects are chosen—10-year-olds, 21-year-olds, and 75-year-olds—each with five subjects in it. Each of the 15 subjects is shown a list of 20 words and then asked to write down as many words as possible. The results are as follows:

Number of Words Recalled

10-year-olds	21-year-olds	75-year-olds
$x_{11} = 1$	$x_{12} = 9$	$x_{13} = 3$
$x_{21} = 4$	$x_{22} = 8$	$x_{23} = 5$
$x_{31} = 5$	$x_{32} = 7$	$x_{33} = 7$
$x_{41} = 6$	$x_{42} = 10$	$x_{43} = 7$
$x_{51} = 4$	$x_{52} = 6$	$x_{53} = 8$

1. Plot the three means.
2. State the null hypothesis and alternative hypothesis.
3. For each of the three groups, compute an estimate of σ^2.

4. Take the average of these three estimates to get MSW.
5. Now get MSW by determining the sum of squares within and dividing it by the degrees of freedom within. Do you get the same answer as you do for question 3?
6. Use the three means to estimate σ_M^2.
7. Now use your answer from question 5 plus the fact that $est\ \sigma_M^2 = est\ \sigma^2/n$ to get the mean square between.
8. What is your F-score? How many degrees of freedom? What is the criterion F for the 0.05 level? Does your obtained F exceed this? Should you reject H_0? What does it mean if you reject H_0?
9. Draw 95% confidence intervals around the three means you plotted.

2. Some years ago there was a controversy involving whether smoking banana skins made one high, and a test was made. Three randomly selected groups of 6 subjects per group were given a dried-up substance to smoke and told it was banana skins. In fact for the first group the substance was tea; for the second group the substance was marijuana; for the third group the substance was indeed banana. After smoking the substance, each subject rated how high he or she was on a scale from 7 (meaning "very high") to 1 (meaning "not high at all"). The following ratings were obtained:

Tea	Marijuana	Banana
5	2	3
6	3	4
3	6	4
5	6	5
4	4	5
4	5	1

1. Test the hypothesis that these three substances differ from each other in terms of how high they make the subjects feel.
2. Plot these three means along with 95% confidence intervals.
3. During the experiment above simple reaction times are measured. Average reaction times for the three groups are as follows:

Reaction Time (tenths of a second)

Tea	Marijuana	Bananas
1.0	1.5	0.8
1.1	2.0	1.0
0.9	1.8	0.9
1.2	1.8	1.2
1.0	1.9	1.1
1.3	1.7	1.0

4. Plot these three means along with 95% confidence intervals.
5. Test the hypothesis that the three substances differ with respect to their effects on reaction time.

3. The following two "conditions" each contain 7 random numbers between 0 and 10. (Random numbers were obtained from an SR-51A calculator.)

Condition 1	Condition 2
$x_{11} = 9$	$x_{12} = 3$
$x_{21} = 6$	$x_{22} = 4$
$x_{31} = 4$	$x_{32} = 7$
$x_{41} = 2$	$x_{42} = 4$
$x_{51} = 6$	$x_{52} = 2$
$x_{61} = 6$	$x_{62} = 0$
$x_{71} = 0$	$x_{72} = 3$

1. Compute an estimate of σ^2 by considering variation within each of the two groups (compute MSW).
2. Compute a second estimate of σ^2 by considering variation between the means of the two groups (compute MSB).
3. Do these two estimates of σ^2 differ significantly? Should they? Why or why not?

4. A psychologist was interested in the effects of different types of liquor on a mirror drawing task. He administered five types of liquor (beer, wine, vodka, gin, and scotch). Each of the five types of liquor was drunk by five subjects (that is, five Ss drank wine, five drank beer, and so on). Each group then performed the mirror drawing task and the number of errors was recorded in the following table:

Wine	Scotch	Gin	Beer	Vodka
5	9	8	4	7
7	13	11	8	10
6	10	12	10	9
8	11	7	6	8
4	7	7	7	6

1. Plot the five means along with 95% confidence intervals.
2. Test the hypothesis that the five liquors differ with respect to effect on mirror drawing.

5. Joe Shablotnik is interested in whether different brands of mouthwash lead to differences in breath freshness. He has three different groups of subjects. He gives Scope to one group, Listerine to another, and Lavoris to a third. Freshness of breath is then measured with a breathometer.

Higher breathometer scores indicate fresher breaths, and breathometer scores are known to be distributed in the population with a variance $\sigma^2 = 2.0$.

The data are as follows:

Group 1 (Scope)	Group 2 (Listerine)	Group 3 (Lavoris)
$n_1 = 4$	$n_2 = 2$	$n_3 = 3$
$x_{11} = 2$	$x_{12} = 7$	$x_{13} = 5$
$x_{21} = 4$	$x_{22} = 10$	$x_{23} = 6$
$x_{31} = 4$		$x_{33} = 4$
$x_{41} = 2$		

1. Plot the three means along with 95% confidence intervals.
2. Perform the appropriate test on these data. Make sure you make use of the fact that the population variance is known.

3. Suppose you wanted to test whether MSW obtained from your data is a faithful estimate of σ^2. Perform such a test. Use the 0.05 α-level.
4. Suppose you did not know σ^2. Perform a standard analysis of variance on these data.

6. In a verbal-learning task, nonsense syllables are presented for later recall. Three different groups of subjects see the nonsense syllables at a 1-sec, 5-sec, or 10-sec presentation rate. The data (number of errors) for the three groups are as follows:

1-sec Group	5-sec Group	10-sec Group
13	11	3
15	14	5
15	13	6
12	12	6
13	16	9
13	12	7
9	11	2
8	9	4
15	10	
	8	
	8	

1. Plot the three means along with 95% confidence intervals.
2. Perform an analysis of variance on the data.

7. Consider the data of Chapter 10, Problems 6, 7, 8, 10, and 11. Re-do these problems using an analysis of variance rather than a t-test. In each case compare the t you obtained with the F you now obtain. What is the relationship between the t and the F in each case?

8. For Problem 6 of this chapter, evaluate the question: Do the scores in Group 3 come from a population whose *variance* is greater than the variance of populations of Groups 1 and 2 scores? (Assume the populations of Groups 1 and 2 have the same variance.)

9. Use the information from Chapter 7, Problem 10, to solve the following:

1. Compute the 82% confidence intervals around the means of the two groups.
2. Compute the power corresponding to the alternative hypothesis:

$$\mu_1 - \mu_2 = 2$$

Parts 3–7 of this problem assume that you do not trust the known variance estimate, and you plan to estimate the variance instead.

3. Retest whether the groups differ.
4. Compute the 95% confidence interval around each mean.
5. Test whether your estimated variance differs from the "known" σ^2 of 4. (Use $\alpha = 0.10$.)
6. Test whether the variance estimate from Group 1 differs from the variance estimate of Group 2. (Use $\alpha = 0.10$.)

12 Two-Way Analysis of Variance (and Beyond)

We have hitherto been describing experimental situations designed to investigate the effect of one independent variable on one dependent variable. For example we described experiments investigating the effect of CAI program type on math ability; of hypnosis on memory; of brain lesions on sexual behavior; of toothpaste brand on whiteness of teeth, and so on. In each of these examples we have established a null hypothesis stating that the independent variable (drug type, hypnosis, lesion vs. no lesion, toothpaste brand, or whatever) has no effect on the dependent variable.

However, it is often the case that social scientists, adventurous and impetuous beings that they are, want to study the effects of more than one independent variable on some dependent variable. For example, someone may wish to study how *two* independent variables *jointly* affect a dependent variable. Such a situation is the topic of this chapter.

A TWO-FACTOR EXPERIMENT: EFFECTS OF INCENTIVE AND RETENTION INTERVAL ON MEMORY PERFORMANCE

Suppose we are interested in how memory works. As part of our research program we plan to investigate the effects of two separate independent variables on memory performance. The first independent variable is *incentive,* and the specific question we are addressing is whether people remember more if they are paid more. The second independent variable in which we imagine ourselves to be interested is *retention interval,* the period intervening between the time information is initially studied and the time the subject attempts to remember it. With regard to retention interval the question is whether people tend to remember less after longer periods of time.

One way of investigating these two independent variables would be to simply do two free-recall experiments. (We have described a free-recall experiment as one in which a subject is read a list of unrelated words and subsequently tries to remember as many of the words as possible.) In the first experiment we could vary incentive. Subjects in one group would receive 1¢ per word for each of the words they recalled, whereas other subjects in a second group would receive 25¢ per recalled word. The appropriate null hypothesis would then state that the independent variable (incentive) has no effect on the dependent variable of memory performance (number of words recalled). A between-groups t-test could then be performed to assess whether this null hypothesis should or should not be rejected.

We could then do a second experiment varying retention interval. One group of subjects would hear a list of words and would immediately be tested on this list. A second group would hear a list of words but would wait one hour before being tested. Finally a third group of subjects would hear the list of words but would not be tested until after an interval of five hours had elapsed. The appropriate null hypothesis in this experiment would be that the independent variable—now retention interval—has no effect on the dependent variable, which is still memory performance. We could then perform a one-way analysis of variance (with three conditions) to assess whether or not we should reject this null hypothesis.

Combining Independent Variables: Factorial Designs

Rather than doing two experiments such as we have just described, it is often better to do a single experiment in which the two independent variables are combined and studied together. The experimental design typically used to do this is called a *factorial design*. A representation of a factorial design is provided in Table 12-1.

New terminology. A factorial design is a bit more complicated than the designs we previously discussed, and we need some new terminology to describe it. First we introduce the term *factor*, which is synonymous with the term independent variable. (We shall be using the terms "factor" and "independent variable" pretty much interchangeably.) As Table 12-1 shows, our example

---

TABLE 12-1.   A Representation of a Factorial (Two-Factor) Design

		Factor 1: Retention Interval		
		Level 1 (0 hours)	Level 2 (1 hour)	Level 3 (5 hours)
Factor 2: Incentive	Level 1: 1¢			
	Level 2: 25¢			

design using two independent variables is referred to as a two-factor design. We are calling retention interval Factor 1 and incentive Factor 2.

Each factor has some number of *levels*. The term "levels" may be identified with that to which we have referred in the past as "conditions" or "groups." In our current example we have chosen to include three levels of the retention interval factor. Level 1 is set to 0 hours, Level 2 is set to 1 hour, and Level 3 is set to 5 hours. For the second factor—incentive—we have opted to include 2 levels. Level 1 of this factor is 1¢ per word and Level 2 is 25¢ per word. Note that the number of levels per factor, as well as the values of the levels, are arbitrary choices. For instance we arbitrarily chose three levels of retention interval. The values of these levels were arbitrarily chosen to be 0, 1, and 5 hours.

The design shown in Table 12-1 is a completely factorial (sometimes called a completely crossed) design, referring to the fact that for any given level of one factor, we include all levels of the other factor. Consider for example the 1¢ level of Factor 2. We have combined this level with all three levels—0, 1, and 5 hours—of Factor 1. In general a two-factor design with J levels of Factor 1 and K levels of Factor 2 is referred to as a $J \times K$ design. In our example J is equal to 3 and K is equal to 2, so we have a 3×2 design.

As can be seen, this 3×2 design produces six separate conditions—that is, six separate groups of subjects will be required to complete the experiment. Each condition consists of some *combination* of one particular level of Factor 1 and one particular level of Factor 2. For example, there is the 0 hour/1¢ condition, the 0 hour/25¢ condition, the 1 hour/1¢ condition, and so on. Notice that in any $J \times K$ design, we simply multiply J times K to compute the number of conditions or groups required for the experiment. In this case J times K is $3 \times 2 = 6$.

These conditions in a multifactor design are often referred to as *cells* because, as depicted in Table 12-1, each condition constitutes one cell of the overall structure of the design.

> 12.1 A $J \times K$ factorial design has two factors with J levels of Factor 1 and K levels of Factor 2. The total number of cells (or separate conditions) is equal to $J \times K$.

Representation of a particular score: Triple subscripts.   Recall that in a simple, one-factor design (such as a *t*-test or a one-way ANOVA), we represent a particular score by x_{ij}. In such a situation j refers to the condition (that is, the level of the one factor) from which the score comes and i refers to the particular subject within the condition. In a two-factor design, however, a score is represented by *three* subscripts—that is, as x_{ijk}. Here j represents the level of Factor 1. In our example j would be 1, 2, or 3, referring to Levels 1, 2, or 3 of the retention interval factor. Likewise, k represents the level of the second factor. In our example k would be 1 or 2 in reference to the 1¢ or the 25¢ level of Factor 2. Thus, j and k jointly specify which of six conditions or cells x_{ijk} comes from. As usual i, the first subscript, specifies the particular subject within a condition. Consider for example the score x_{231}. We would know that this score was obtained by a subject in the condition corresponding to the third level of Factor 1 (third column) and to the first level of Factor 2 (first row). Furthermore since the first subscript is 2, we would know that this was the score obtained by the second subject within this cell.

Hypothetical Data

With this new terminology under our belt, let's deal with some data. In Table 12-2 we have redrawn the design depicted in Table 12-1, and we have entered some hypothetical data. (The $3 \times 2 = 6$ conditions have been temporarily labeled 1–6 for reasons that will become evident below.) In Table 12-2 we have computed a number of things from our data: First the total and the mean of scores in each cell are shown. Second we have computed the totals for each column and for each row. (Here we have sneaked in a little more new terminology. The total for any given column j is referred to as T_{Cj}, referring to the fact that it is the total of column j. Likewise the total of any given row k is referred to as T_{Rk}. These row and column totals are referred to as *marginals*, a term which stems from the fact that data having to do with row and column

---

TABLE 12-2. Hypothetical Data from the Retention Interval/Incentive Experiment; Scores Represent Number of Words Recalled

		Factor 1 *Retention Interval*			
		0 hours	*1 hour*	*5 hours*	
		(Condition 1)	(Condition 2)	(Condition 3)	
		$x_{111} = 6$	$x_{121} = 4$	$x_{131} = 5$	
		$x_{211} = 4$	$x_{221} = 4$	$x_{231} = 4$	
	1¢	$x_{311} = 5$	$x_{321} = 4$	$x_{331} = 3$	$T_{R1} = 39$
		$T_{11} = 15$	$T_{21} = 12$	$T_{31} = 12$	
		$M_{11} = 15/3 = 5$	$M_{21} = 12/3 = 4$	$M_{31} = 12/3 = 4$	
Factor 2: *Incentive*		(Condition 4)	(Condition 5)	(Condition 6)	
		$x_{112} = 11$	$x_{122} = 6$	$x_{132} = 6$	
		$x_{212} = 8$	$x_{222} = 10$	$x_{232} = 4$	
	25¢	$x_{312} = 8$	$x_{322} = 8$	$x_{332} = 5$	$T_{R2} = 66$
		$T_{12} = 27$	$T_{22} = 24$	$T_{32} = 15$	
		$M_{12} = 27/3 = 9$	$M_{22} = 24/3 = 8$	$M_{32} = 15/3 = 5$	
		$T_{C1} = 42$	$T_{C2} = 36$	$T_{C3} = 27$	$T = 105$ $N = 18$

totals are written in the margins of the table.) Finally we have computed N which (as before) is the total *number* of scores in the experiment and T, the grand total, which (as before) is the sum of all scores in the experiment.

> 12.2 In a two-factor design the totals we must compute are: (a) the totals of each of the $J \times K$ cells, (b) the J column totals, (c) the K row totals, and (d) the grand total. Note that the grand total, T, is, as always, the sum of all of the N scores in the experiment.

Plotting the means: An intuitive feel for what's going on. Prior to performing any statistical analyses on the data, it is always a good idea to provide ourselves with a rough feeling for the results of the experiment. As we have suggested previously, the best way of generating such a rough idea is to plot the means on a graph. This has been done for our hypothetical data in Figure 12-1.

Notice that in Figure 12-1 we have introduced yet another new concept. In the past, results have been represented by a function relating the dependent variable (whose value we represented on the ordinate of the graph) to the independent variable (whose value we represented on the abscissa). However, in this two-factor situation, we have not one but two independent variables. Since we have only one abscissa, we are forced to make a choice about which independent variable is going to be represented on the abscissa. This choice is arbitrary, and in our present example we have chosen to represent Factor 1

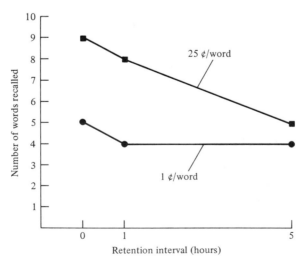

FIGURE 12-1.   A plot of data from a two-factor design. The dependent variable is plotted as a function of one of the factors. A separate curve is plotted for each level of the other factor.

(retention interval) on the abscissa. Now what are we going to do about the other independent variable—incentive, in our example? The answer is that we plot a separate curve for each level of the other independent variable. Thus, we have one curve representing memory performance as a function of retention interval for the 1¢ condition and a second curve representing memory performance as a function of retention interval for the 25¢ condition.

The curves shown in Figure 12-1 gives us a general gut-level idea of what is going on with our data. In general it appears as though both independent variables are having effects on the dependent variable. People seem to remember less after a longer retention interval (as indicated by the fact that the two curves are declining). Furthermore, people seem to remember more if they are paid more (as indicated by the fact that the 25¢ curve is higher than the 1¢ curve). As usual, however, we would like some check to assure ourselves that these apparent effects are not just due to random or chance fluctuation. In particular we would like to go through some sort of hypothesis-testing procedure to decide whether the observed differences among the means are in fact due to variation in the independent variable(s) or whether they might simply stem from chance factors. The mechanics of performing such tests constitute the next section.

STATISTICAL ANALYSES: A TWO-WAY ANOVA

The mechanics of doing these analyses are not really very different from the mechanics of doing the sort of one-way ANOVA discussed in the last chapter. In fact the first step in a two-way ANOVA is to actually perform a one-way ANOVA on the data.

A Six-Condition One-Way ANOVA

We have seen that our 3×2 design yields six separate cells or conditions. For the moment let's ignore the fact that the six conditions result from combining two independent variables and let's just consider them to be six separate conditions (indicated by the labeling of the conditions as 1–6, shown in Table 12-2). Considering our design this way, we can do a one-way ANOVA on the data to detect whether there is any statistically significant differences whatever among the six conditions.

The null and alternative hypotheses.   Since we are now considering ourselves to merely have six separate conditions, the mechanics of doing a one-way analysis of variance are fairly straightforward. First we set up a null hypothesis which states that the independent variable (in this case condition) has no effect on the dependent variable (in this case memory performance). Referring to the population means of the six conditions as μ_1, μ_2, μ_3, μ_4, μ_5, and μ_6, the null hypothesis states that

H_0: $\mu_1 = \mu_2 = \mu_3 = \mu_4 = \mu_5 = \mu_6$

Likewise, the alternative hypothesis states that the independent variable does have some effect on the dependent variable or in this case

H_1: It is not true that $\mu_1 = \mu_2 = \mu_3 = \mu_4 = \mu_5 = \mu_6$

Sums of squares and degrees of freedom.   Now let's compute the sum of squares between and the sum of squares within. Recall that **the equation for computing sum of squares between is**

$$SSB = \sum_{\text{conditions}} \frac{T_j^2}{n} - \frac{T^2}{N}$$

The first term in this expression, $\sum_{\text{conditions}} T_j^2/n$ is easily computed to be

$$\sum_{\text{conditions}} \frac{T_j^2}{n} = \frac{15^2}{3} + \frac{12^2}{3} + \frac{12^2}{3} + \frac{27^2}{3} + \frac{24^2}{3} + \frac{15^2}{3}$$
$$= 681$$

We have seen that T, the total of all the scores in the experiment, is equal to 105. N in this case is 18, since we have a total of 18 scores in the experiment. Therefore, T^2/N is equal to

$$\frac{T^2}{N} = \frac{105^2}{18} = 612.5$$

Thus, to compute SSB,

$$SSB = 681 - 612.5 = 68.5$$

The formula for sum of squares within is

$$SSW = \sum_{\text{everything}} x_{ij}^2 - \sum_{\text{conditions}} \frac{T_j^2}{n}$$

In this case, the first sum is over all N scores of x_{ij}s: thus

$$\sum_{\text{everything}} x_{ij}^2 = 6^2 + 4^2 + 5^2 + 4^2 + \cdots + 4^2 + 5^2 = 701$$

We have already computed the second term in the equation, which was 681. Thus

$$SSW = 701 - 681 = 20$$

What about degrees of freedom? Since we consider ourselves to have six conditions, $dfB = 6 - 1 = 5$. Likewise, dfW may be computed as $N - J$ or $18 - 6 = 12$. (Alternatively, we see that since we have two degrees of freedom within each of the six groups, we can also arrive at $2 \times 6 = 12$ degrees of freedom within.)

An ANOVA summary table.   Now with our sums of squares and degrees of freedom all computed, we are in a position to set up an ANOVA summary table as shown in Table 12-3. To compute mean squares between and within, we divide sums of squares by associated degrees of freedom. Mean square within is 1.67. Therefore, our best guess is that the scores within each of the six groups come from populations of scores whose variance, σ^2, is 1.67. Our mean square between is 13.7, which is considerably greater than the MSW of 1.67. In fact the ratio of MSB to MSW, the obtained F, turns out to be 8.20. Looking in our F tables, we find the criterion $F(5,12)$ to be 3.11. Therefore, we are in a position to soundly reject the null hypothesis of no differences among the six conditions. Somehow or other, one or both of the independent variables is having an effect on memory performance.

TABLE 12-3.   One-Way ANOVA Summary Table; We Are Treating Our 3 × 2 Design as a One-Way Design with Six Conditions

Source	df	SS	MS	F	Crit F
Total	17	88.5			
Between	5	68.5	13.7	8.20	3.11
Within	12	20.0	1.67		

12.3 By considering a $J \times K$ design to be simply a one-way design with $J \times K$ conditions, we can compute a sum of squares between the $J \times K$ cells, SSB as well as SSW the sum of squares from within the $J \times K$ cells.

Dividing the pie.   Note that the total sum of squares in this experiment—the SST which would result from considering the variation among all 18 scores in the experiment—is

$$SST = SSB + SSW = 68.5 + 20 = 88.5$$

and, likewise, the total number of degrees of freedom in the experiment is

$$dfT = dfB + dfW = 5 + 12 = 17$$

Figure 12-2 shows how this total "pie" of 88.5 sum of squares with its 17 degrees of freedom may be conceptualized as being divided into its components stemming from variance between groups and variance within groups. Below, we're going to complicate this pie somewhat.

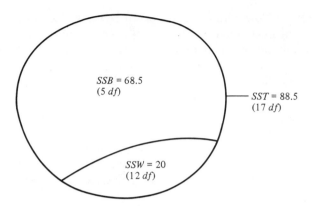

FIGURE 12-2. Initial "division of the pie" for a two-factor design. The total sum of squares and degrees of freedom are initially divided into components stemming from within and between groups.

A Two-Way Analysis of Variance

Now we would like to dig deeper into our data and assess the *independent* effects of each of our two factors in turn. That is, having discovered that something is having an effect, we want to determine what it is.

A sum of squares due to retention interval (columns). For the time being let us ignore the fact that we have included incentive as a factor in our experiment. In other words let us pretend that we have just run a one-way ANOVA-type experiment with retention interval as our only factor. Such a charade is depicted in Table 12-4, and there are considered to be $n_C = 6$ scores in each of the three columns. Let's proceed as if we were going to do a one-way ANOVA on these data.

Our null and alternative hypotheses are fairly straightforward. First the

TABLE 12-4. Retention Interval Is Assumed to Be the Only Factor in the Experiment; Thus, Each Condition (Column) Contains $n_C = 6$ Scores

	Retention Interval		
0 hours	*1 hour*	*5 hours*	
6	4	5	
4	4	4	
5	4	3	
11	6	6	
8	10	4	
8	8	5	
$T_{C1} = 42$	$T_{C2} = 36$	$T_{C3} = 27$	$T = 105$

null hypothesis would state that the independent variable—retention interval in this case—has no effect on the dependent variable of memory performance. Referring to μ_{C1}, μ_{C2}, and μ_{C3} as the population means of the three conditions, our "column null hypothesis" would state that

H_0: $\mu_{C1} = \mu_{C2} = \mu_{C3}$

Likewise, our alternative hypothesis would state that retention interval does have some effect on memory performance, or

H_1: It is not true that $\mu_{C1} = \mu_{C2} = \mu_{C3}$

Now let us proceed to get a sum of squares between these three conditions. The equation for this sum of squares (which we will refer to as SSC or sum of squares due to columns) is a slightly disguised form of our equation for SSB

$$SSC = \sum_{\text{columns}} \frac{T^2_{Cj}}{n_C} - \frac{T^2}{N}$$

Note that T and N are unchanged. We still have the same $N = 18$ scores in our experiment, and the total of these 18 scores is still $T = 105$. Thus, T^2/N is still equal to 612.5. We can now without any difficulty discover that the sum of squares due to columns is equal to

$SSC = (42^2/6 + 36^2/6 + 27^2/6) - 612.5$ or

$= 631.5 - 612.5 = 19$

Let's record the fact that sum of squares due to columns is equal to 19 and move on to our second independent variable.

A sum of squares due to incentive (rows).   Now we pretend that in our experiment we varied only incentive. That is, we temporarily ignore the fact that retention interval was also varied. Table 12-5 shows the situation that would

TABLE 12-5.   Incentive Is Assumed to Be the Only Factor in the Experiment; Thus, Each Condition (Row) Contains $n_R = 9$ Scores

Incentive					
	1¢	6	4	5	$T_{R1} = 39$
		4	4	4	
		5	4	3	
	25¢	11	6	6	$T_{R2} = 66$
		8	10	4	
		8	8	5	
					$T = 105$

result. Now we set up the null hypothesis that incentive has no effect on memory performance. Referring to μ_{R1} and μ_{R2} as the population means of the two rows, our "row null hypothesis" would state that

$$H_0: \mu_{R1} = \mu_{R2}$$

Likewise, our alternative hypothesis would state that

$$H_1: \text{It is not true that } \mu_{R1} = \mu_{R2}$$

Let's now compute a sum of squares due to rows. Once again, T and N are unchanged (they *never* change) and T^2/N is still equal to 612.5. The "n" with which we shall be concerned, to which we now refer as n_R, is equal to 9, since there are nine scores in each of our two incentive conditions. Thus, our **equation for sum of squares due to rows is**

$$SSR = \sum_{\text{rows}} T^2_{Rk}/n_R - T^2/N$$

which is

$$SSR = (39^2/9 + 66^2/9) - 612.5 \quad \text{or}$$
$$= 653 - 612.5 = 40.5$$

Thus, we record the fact that our sum of squares due to rows is 40.5.

Performing the tests.   The next thing we do is to set up a new ANOVA summary table, which we have done in Table 12-6. As we can see, we still have a total sum of squares that reflects the total variance among all 18 scores in the experiment. Additionally we have subdivided this total sum of squares into *SSB* (based on 5 degrees of freedom) and *SSW* (based on 12 degrees of freedom). Now, however, we consider *SSB*—the sum of squares between the six conditions—and subdivide it still further. In particular we have determined that of the original 68.5 sum of squares between the six conditions, 19.0 of that sum of squares is taken up by Factor 1 (columns) and 40.5 of *SSB* is taken up by Factor 2 (rows). How many degrees of freedom are associated with columns and rows?

TABLE 12-6.   Two-Way Analysis of Variance Summary Table; the Sum of Squares Between the 6 Conditions (*SSB*) Is Compartmentalized into Sum of Squares Due to Columns and Sum of Squares Due to Rows

Source	*df*	*SS*	*MS*	*F*	*Crit F*
Between	5	68.5			
Column	2	19	9.5	5.69	3.89
Rows	1	40.5	40.5	24.25	4.75
Within	12	20	1.67		
Total	17	88.5			

This is fairly straightforward: Since there are three columns, there are $3 - 1 = 2$ degrees of freedom due to columns. Likewise, since there are two rows, there are $2 - 1 = 1$ degree of freedom due to rows.

> 12.4 The sum of squares between (in this case between the six cells) can be broken down into that portion due to columns (Factor 1) and that portion due to rows (Factor 2).

We now divide the column and the row sums of squares by their respective degrees of freedom. These mean squares due to columns and rows of 9.5 and 40.5, respectively, are shown under the heading "Mean Square" (MS) in Table 12-6.

Now let's suppose that our "column null hypothesis" is true. That is, imagine that Factor 1, or retention interval, really has no effect on memory performance. *In this case, the mean square due to columns should be an estimate of σ^2, the population variance.* By comparing MSC to our "best estimate" of σ^2 our MSW of 1.67—we can determine whether or not we should reject the null hypothesis of no effect due to retention interval. More specifically the ratio of MSC to MSW will be distributed as an F with 2 and 12 degrees of freedom if the column null hypothesis is correct. In fact, the obtained F turns out to be 5.69. Looking at our F-tables, we discover that the criterion $F(2,12)$ is equal to 3.89. Since our obtained $F(2,12) = 5.69$ exceeds our criterion $F(2,12) = 3.89$, we reject the column null hypothesis and conclude that retention interval really does have some effect on memory performance. This conclusion stems from the fact that the mean square columns is much larger than would be expected merely by chance.

We now perform an analogous operation with Factor 2 (rows). Dividing SSR (40.5) by dfR (1), we obtain a mean square due to rows of 40.5. Once again if our "row null hypothesis" is correct—that is, if the 1¢ versus 25¢ manipulation is having no effect on memory performance—then this mean square due to rows will be an estimate of the population variance. We can therefore compute the ratio of MSR to MSW to obtain an $F(1,12)$ of 24.25. Since the criterion $F(1,12)$ equals 4.75, we are in a position to reject the row null hypothesis and conclude that incentive has some effect on memory performance just as does retention interval.

> 12.5 Sums of squares due to columns and rows can be divided by degrees of freedom due to columns $(J - 1)$ and rows $(K - 1)$ to produce mean squares due to columns and rows. These mean squares can then be tested against MSW to test for significant column or row effects.

So far a two-way analysis of variance appears to boil down to two one-way analyses of variance. Specifically we first computed a sum of squares due to columns by ignoring the fact that our design actually included two rows. We then did the converse, computing a sum of squares due to rows by ignoring the fact that the design actually involved three columns. We then tested two independent null hypotheses—first that there was no effect due to columns, and second that there was no effect due to rows.

The mechanics of testing each of these null hypotheses were identical to those used in a one-way analysis of variance situation. We divided each sum of squares by its associated degrees of freedom to obtain a mean square. It was then the case that if a particular null hypothesis were true, the corresponding mean square would be an estimate of σ^2, the population variance. Since we already had a "best" estimate of the population variance—mean square within which, in this case, was equal to 1.67—we formed appropriate F-ratios to assess whether each of the null hypotheses could safely be rejected. In our example both F-ratios—the F-ratio due to Factor 1 and the F-ratio due to Factor 2—were much bigger than we could expect by chance. Therefore, we rejected each of the two null hypotheses—the first due to columns and the second due to rows—successively. That is, we concluded that both our factors—retention interval and incentive—were having an effect on memory performance.

The missing sum of squares.   You've probably noticed something peculiar about the ANOVA summary table shown in Table 12-6. We started with a sum of squares between of 68.5. We then alleged that we subdivided this SSB into sums of squares due to columns and to rows. However, the sum of SSC and SSR is equal to $19.0 + 40.5 = 59.5$. Since we started with $SSB = 68.5$, there seem to be 9.0 sum of squares missing. Likewise, we began with 5 degrees of freedom between the six cells. Two of these degrees of freedom went to columns and one degree of freedom went to rows. Hence, only 3 of the original 5 degrees of freedom have been accounted for, and there seem to be 2 degrees of freedom missing. These missing sums of squares and degrees of freedom are due to a new component of variance known as an *interaction* and unique to multifactor designs. It is a very interesting component whose existence is the major reason for our use of multifactor (in this case two-factor) designs. Interactions are the topic of our next section.

INTERACTIONS

In order to simplify our discussion of interactions, let's modify our example somewhat and suppose that we have only two levels of Factor 1 (retention interval) which are: Level 1 = 0 hours; Level 2 = 1 hour. Let's continue to suppose we still have two levels of Factor 2 (incentive): 1¢ and 25¢. We now have a 2 × 2 design.

Three Hypothetical Outcomes

Figure 12-3 shows three ways in which the data of our experiment might turn out. In each of the three cases shown in Figure 12-3 the dependent variable—memory performance—is plotted as a function of Factor 1 (retention interval). Separate curves are plotted for each value of the other independent variable, incentive. As is quite evident, the three outcomes represented by the three cases are somewhat different from one another.

Case I: No interaction.   Let's consider the data shown in Case I. Without worrying for the moment about statistical analyses of the data, it seems apparent that both the independent variables are having an effect on the dependent variable. That is, the 25¢ conditions are doing better than the 1¢ conditions, and additionally performance seems to be poorer after 1 hour than after 0 hours.

Now let's confine our attention to the effect of incentive on memory performance. Suppose you're asked what the *magnitude* of the incentive effect is; that is, how *much* better do people remember when awarded 25¢ as opposed to 1¢ per remembered word? The answer to this question is fairly straightforward. People paid 25¢ recall an average of three more words than do people paid 1¢. After 0 hours' retention interval, nine versus six words are recalled. Similarly after an interval of 1 hour, the scores are four versus one word.

Now let's switch our attention to the other independent variable, that of retention interval and inquire about its effect—that is, how much forgetting takes place after an interval of 1 hour has elapsed? Again the response to this inquiry would be fairly straightforward: People appear to forget five words after a period of 1 hour. The drop is from nine to four words when 25¢ is being paid and from six words to one word when 1¢ is being paid.

Now we can summarize the state of affairs depicted in Case I by saying that *the effect of one factor is independent of the level of the other factor.* That is, to specify the magnitude of either factor, we need not concern ourselves with the other factor. As we shall see shortly, another way of stating this is to say that the factors *do not interact* with one another.

Case II: Some interaction.   Now let's look at the data shown in Case II of Figure 12-3. Once again, the 25¢ conditions seem to be doing better than the 1¢ conditions, and furthermore the 0-hour retention interval seems to be doing better than the 1-hour retention interval. However, the pattern of data shown in Case II is subtly—but importantly—different from the pattern of data shown in Case I. To illustrate this difference, suppose we once again inquire about the magnitude of the incentive effect. How much better, we ask, does a subject do when given 25¢ versus 1¢ per word? Here the response is not quite so simple as it was before. We must qualify our answer somewhat and take retention interval into account. After a 0-hour retention interval, people remember three more words in the 25¢ as in the 1¢ condition (the means are 9 versus 6). However, after an hour's retention interval, people only remember one more word in the

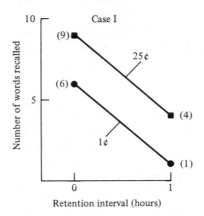

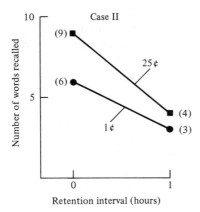

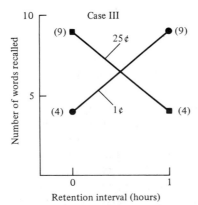

FIGURE 12-3. Three hypothetical outcomes of a retention interval/incentive experiment. The three outcomes depict successively greater amounts of interaction in the data.

25¢ condition than in the 1¢ condition (the means are 4 versus 3). This question can also be asked in the reverse way: How much forgetting has taken place after an interval of 1 hour? Once again, the answer to this question depends on the level of incentive. If people are paid 25¢ per word, then they forget five words after 1 hour. On the other hand if people are only paid 1¢ per word, then they forget three words after 1 hour. Considering the Case II data, we find that no matter how we ask the question, we end up with the conclusion that the effect of one independent variable depends on the level of the other independent variable.

Digression 12.1

Symmetry of Interactions

We have suggested that if the effect of Factor 1 is independent of the level of Factor 2 (that is, if there is no interaction), then the converse is also true—the effect of Factor 2 must be independent of the level of Factor 1. To prove this (for a 2 × 2 design), let's refer to μ_{jk} as the mean of the cell corresponding to the jth level of Factor 1 and the kth level of Factor 2. Asserting that the effect of Factor 1 is independent of the level of Factor 2 is to say that

$$\mu_{21} - \mu_{11} = \mu_{22} - \mu_{12} \tag{D.1}$$

We can rearrange the terms of Equation D.1 to obtain

$$\mu_{21} - \mu_{22} = \mu_{11} - \mu_{12} \tag{D.2}$$

Equation D.2 is a mathematical formulation of the statement that the effect of Factor 2 is independent of the level of Factor 1.

Likewise, if the effect of Factor 1 *does* depend on the level of Factor 2, then conversely, the effect of Factor 2 must depend on the level of Factor 1. The proof of this is quite analogous. To say that the effect of Factor 1 depends on the level of Factor 2 is to say that

$$\mu_{21} - \mu_{11} \neq \mu_{22} - \mu_{12} \tag{D.3}$$

Rearranging terms in equation D.3,

$$\mu_{21} - \mu_{22} \neq \mu_{11} - \mu_{12} \tag{D.4}$$

Equation D.4 says that the effect of Factor 2 depends on the level of Factor 1.

We are now in a position to define what an interaction is. *An interaction exists to the degree that the effect of one independent variable depends on the level of the other independent variable.* In Case I we saw that there was no interaction by this definition: rather, the two factors in Case I affected memory performance independently of one another. However, in Case II there is an interaction. In Case II the effect of one independent variable does depend on the level of the other independent variable. The two factors do not operate independently on the dependent variable of number of words recalled; they interact.

Case III: A complete interaction.   To dramatize further what we mean by an interaction, let's examine Case III, which depicts a somewhat unlikely outcome of the data, but it does illustrate a point. Here the effect of each factor depends *completely* on the level of the other factor. Suppose we inquire about the effect of incentive. We see that after a 0-hour retention interval people remember five more words in the 25¢ condition than in the 1¢ condition. However, after a retention interval of one hour, this state of affairs reverses—people remember five more words in the 1¢ condition than in the 25¢ condition!

The same general conclusion holds true if we inquire about the effect of retention interval. As Case III shows, if people are paid 25¢ per word then they forget five words after a retention interval of one hour. However, if people are paid 1¢ per word then they *gain* five words after a retention interval of one hour. No matter how the question is posed, the effect of one independent variable as depicted in Case III depends *completely* on the level of the other independent variable. Here we say that there is a complete interaction.

Interactions imply nonparallel lines.   You have probably noticed that a "quick and dirty" method of determining whether or not two factors interact is to assess the degree to which the curves on the relevant graph are parallel. Thus, in Case I where there is no interaction, the two curves are parallel. In Case II where there is some interaction, the lines are not quite parallel. And finally in Case III, where there is complete interaction the lines are obviously nonparallel.

> 12.6   Two factors interact to the degree that the magnitude of the effect of one factor depends on the level of the other factor. Graphically interactions imply and are implied by nonparallel lines.

Digression 12.2

Scaling Problems

In Chapter 10 we discussed assumptions underlying analysis of variance, and in the course of this discussion, we mentioned the *scaling assumption*. The scaling assumption sometimes comes back to haunt us when we interpret interactions, and some brief remarks on it are in order here. These remarks involve subtle issues and are concerned primarily with the relationship between theory and data rather than with the statistical treatment of data per se.

Underlying Variables Are Measured by Dependent Variables

In most instances we must distinguish between two variables in a psychology experiment. The first is some underlying psychological variable which plays a part in some theory, and the second is the dependent variable that we observe and use as a measure of the unobservable underlying variable. In the example at hand the underlying variable of concern might be termed *memory strength.* The dependent variable that we use to measure memory strength is, as we have seen, number of words recalled.

We assume that our dependent variable is at least *monotonically related* to the underlying variable. That is, if memory strength is greater in one situation than in another, we expect number of words recalled to be correspondingly greater, and vice versa. However, the exact function relating our underlying variable and our dependent variable is often unknown. The implication of this fact is that our dependent variable is generally measured on an *arbitrary scale,* and any other chosen dependent variable that is monotonically related to the dependent variable would be equally acceptable. For instance if instead of using number of words recalled as our dependent variable, we had chosen to use the *logarithm* of number of words recalled or *square root* of number of words recalled, these would have been equally acceptable as measures of memory strength.

Some Interactions Can Be Transformed Away

Now here is the crucial thing: Suppose we have an interaction such as the one depicted in Case II of Figure 12-3. Now suppose we change our minds and choose to use some other dependent variable that is a monotonic transform of number of words recalled. *It would be possible to choose this dependent variable such that the interaction would disappear.* That is, when the data were plotted in terms of our new dependent variable, the curves would be parallel. Since our choice of dependent variable is arbitrary to begin with, it would really not make sense to conclude that there is or is not an interaction in terms of memory strength since the interaction could be made to appear or disappear at will by an appropriate choice of a dependent variable.

This does not mean that it is hopeless to try to interpret interactions. There are classes of situations in which interactions may be interpreted meaningfully and unambiguously. The first such class occurs when the scale of our dependent variable is equivalent to the scale of our underlying variable of interest. *Time* is a good example of such a scale. For instance we may be interested in the time it takes for internal mental events to occur, and in some experiment we may use reaction time as the dependent variable to measure this "internal time." The second class of situations involves certain types of interactions which must hold up over all monotonic transformations of the dependent variable. One such type of interaction is called a "crossover interaction." The interaction shown in Case III of Figure 12-3 is an example of a crossover interaction.

This whole topic is actually somewhat beyond the scope of this book. But we raise it to give you a flavor for the issues and to warn you to be cautious about implications of interactions for theories and hypotheses. For more extensive discussions of the topic, see the papers listed under "Measurement" in the Bibliography.

Statistical Treatment of Interactions

In the above discussion of Cases I, II, and III we have ignored the old bugaboo of the possibility of chance fluctuation. That is, in Case I we claimed that we had no interaction. In Case II we claimed that there was some interaction. The *reason* that we claimed that there was an interaction in Case II was that the effect of each factor depended on the level of the other factor. However, this dependence may just be due to chance fluctuations. That is, the four means plotted in Case II are sample means. It may perfectly well be the case that in terms of *population* means, there is no interaction.

All of which is to say that the existence of an interaction must be tested statistically. What we must do is set up some "interaction null hypothesis" claiming that there is no interaction in terms of population means. We will likewise set up an alternative hypothesis that there *is* some interaction in terms of population means. We will then be able to test this interaction null hypothesis just as we test the effects of our two independent variables.[1]

Sum of squares and degrees of freedom due to interaction.   Recall that in the previous section we initially computed a sum of squares between our various groups. We then subdivided this *SSB* by computing a sum of squares due to each of the two factors (*SSC* and *SSR*). Likewise, we subdivided degrees of freedom due to each of the two factors. However, there appeared to be a missing sum of squares and correspondingly a missing degree of freedom. We now identify those missing sums of squares and degrees of freedom as due to the interaction. The more interaction in the data, the bigger will be the sum of squares due to interaction—just as the more effect there is due to any independent variable, the larger will be the sum of squares (the variance) due to that independent variable.

---

[1]The effects of the factors (that is, the effects of rows and columns) are referred to as *main effects*. Effects of interactions are referred to as interaction effects.

Accordingly considering Cases I, II, and III in Figure 12-3, we would guess that the sum of squares due to interaction would be very small (probably zero, in fact, since there is no interaction) for Case I, somewhat larger for Case II, and quite large for Case III.

Let's test out these intuitions. In Table 12-7 (which, although it contains a formidable amount of computation, would be worth your while to go through) we have computed sums of squares between, sums of squares due to columns, and sums of squares due to rows for each of the three cases. (We have arbitrarily assumed an $n = 4$ subjects per cell in these computations.) **We have also computed sums of squares due to interaction by simply subtracting sums of squares due to columns and due to rows from the sum of squares between with which we started.** As we can see, the results of these computations agree with our intuitions. The sum of squares due to interaction turns out to be 0 for Case I, 4 for Case II, and 100 for Case III. Thus, we see that the size of an interaction is reflected quite nicely by the size of the sum of squares that corresponds to that interaction.

A test of the "interaction null hypothesis." Let's return to the original data that we described early in this chapter. Table 12-8 re-creates the analysis of variance summary table from Table 12-6. Along with the sum of squares and degrees of freedom due to the two main effects, we have listed the sum of squares

---

TABLE 12-7.   Computation of Sums of Squares for Each of the 3 Cases. (It Is Assumed that There Are $n = 4$ Scores Per Cell.)

	Case 1 Retention Interval				Case 2 Retention Interval				Case 3 Retention Interval		
	0 hrs	*1 hr*			*0 hrs*	*1 hr*			*0 hrs*	*1 hr*	
25¢	$T_{11} = 36$	$T_{21} = 16$	$T_{R1} = 52$	25¢	$T_{11} = 36$	$T_{21} = 16$	$T_{R1} = 52$	25¢	$T_{11} = 36$	$T_{21} = 16$	$T_{R1} = 52$
1¢	$T_{12} = 24$	$T_{22} = 4$	$T_{R2} = 28$	1¢	$T_{12} = 24$	$T_{22} = 12$	$T_{R2} = 36$	1¢	$T_{12} = 16$	$T_{22} = 36$	$T_{R2} = 52$
	$T_{C1} = 60$	$T_{C2} = 20$	$T = 80$		$T_{C1} = 60$	$T_{C2} = 28$	$T = 88$		$T_{C1} = 52$	$T_{C2} = 52$	$T = 104$

Case 1:

$$SSB = \sum T_{jk}^2/n - T^2/N$$
$$= [36^2/4 + 16^2/4 + 24^2/4 + 4^2/4] - 80^2/16$$
$$= 536 - 400 = 136$$

$$SSC = \sum T_{Cj}^2/n_C - T^2/N$$
$$= [60^2/8 + 20^2/8] - 80^2/16$$
$$= 500 - 400 = 100$$

$$SSR = \sum T_{Rk}^2/n_R - T^2/N$$
$$= [52^2/8 + 28^2/8] - 80^2/16$$
$$= 436 - 400 = 36$$

$$SSI = SSB - SSC - SSR$$
$$= 136 - 100 - 36 = 0$$

Case 2:

$$SSB = \sum T_{jk}^2/n - T^2/N$$
$$= [36^2/4 + 16^2/4 + 24^2/4 + 12^2/4] - 88^2/16$$
$$= 568 - 484 = 84$$

$$SSC = \sum T_{Cj}^2/n_C - T^2/N$$
$$= [60^2/4 + 28^2/4] - 88^2/16$$
$$= 548 - 484 = 64$$

$$SSR = \sum T_{Rk}^2/n_R - T^2/N$$
$$= [52^2/4 + 36^2/4] - 88^2/16$$
$$= 500 - 484 = 16$$

$$SSI = SSB - SSC - SSR$$
$$= 84 - 64 - 16 = 4$$

Case 3:

$$SSB = \sum T_{jk}^2/n - T^2/N$$
$$= [36^2/4 + 16^2/4 + 16^2/4 + 36^2/4] - 104^2/16$$
$$= 776 - 676 = 100$$

$$SSC = \sum T_{Cj}^2/n_C - T^2/N$$
$$= [52^2 + 52^2] - 104^2/16$$
$$= 676 - 676 = 0$$

$$SSR = \sum T_{Rk}^2/n_R - T^2/N$$
$$= [52^2 + 52^2] - 104^2/16$$
$$= 676 - 676 = 0$$

$$SSI = SSB - SSC - SSR$$
$$= 100 - 0 - 0 = 100$$

TABLE 12-8.   Two-Way Analysis of Variance Summary Table; *SSB* Is Subdivided into Sums of Squares Due to Columns, Rows, and Interaction

Source	df	SS	MS	F	Crit F
Between	5	68.5			
Columns	2	19	9.5	5.69	3.89
Rows	1	40.5	40.5	24.25	4.75
Interaction	2	9	4.5	2.69	3.89
Within	12	20	1.67		
Total	17	88.5			

and degrees of freedom due to interaction. We have also divided the sum of squares due to interaction by its associated degrees of freedom to arrive at a mean square due to interaction.

Now as you may have guessed, if the "interaction null hypothesis" is true—that is, if there is actually no interaction in terms of population means—then the mean square due to interaction is an estimate of σ^2, the population variance. That is, suppose there is actually no interaction in terms of population means. Would we expect the two lines on the graph shown in Figure 12-1 to be *completely parallel?* The answer is no, we would not. Due to the population variance, we would expect some shifting of the sample means relative to the population means. So we would expect some interaction variance in the data.

We are therefore in a position to test statistically the presence of an interaction, just as we were in a position to test the presence of main effects. To test the interaction, we compute the ratio of mean square interaction (*MSI*) to *MSW*. If there is, in fact, no interaction in terms of population means, then this ratio should be distributed as an *F* with 2 and 12 degrees of freedom. As can be seen in Table 12-8, the value of this *F* is 2.69. Looking in our *F*-tables, we find the criterion $F(2,12)$ to be 3.89. Since the $F(2,12)$ due to interaction is not as extreme as the criterion $F(2,12)$, the interaction is not statistically significant. That is, the magnitude of the interaction is not enough to warrant the conclusion that there is a real interaction, an interaction in terms of population means.

12.7   A sum of squares due to interaction (*SSI*) is that left over after *SSC* and *SSR* have been removed from *SSB*—that is, $SSI = SSB - SSC - SSR$. Likewise $dfI = dfB - dfC - dfR$. A mean square due to interaction is computed by $MSI = SSI/dfI$. If there is no interaction in terms of population means, then *MSI* is an unbiased estimate of σ. Therefore, the significance of the interaction may be assessed by computing the *F* obtained by MSI/MSW.

A COMPLETE EXAMPLE: CINEMA PREFERENCES

The foregoing discussion has included quite a mishmash of different concepts. We have proceeded through several examples in bits and pieces with the goal of providing some intuitive concepts of what two-way ANOVA is all about. Now we'll go through a complete example all at once to provide a picture of how an ordinary two-way ANOVA is carried out.

Let's suppose that we are social psychologists interested in people's reactions to various types of movies. We plan to show subjects one of three types of movies. The first type of movie will consist of hardcore pornography (X-rated, such as *Wet Nostrils*). The second type of movie will consist of softcore films (R-rated, such as *From the Waist Up*), and the third type will consist of family films (G-rated, such as *Dumbo Meets the Little Engine that Could*). Thus, Factor 1 in this design is movie *type*.

Now we decide that various segments of the population might react differently to these various types of movies, so we include a second factor in our design, the *age* of the audience. This factor will have two levels. Level 1 will consist of young people (age range 20 to 30), whereas the second level will consist of old people (age range 70 to 80). The experiment will consist of showing subjects movies and asking them to rate how much they like the movies. Let's assume that the rating is on a 1 to 10 scale (where 1 means "can't stand it" and 10 means "love it").

Once again we have a 3×2 design, and let's suppose that we have $n = 2$ subjects in each cell. Table 12-9 shows hypothetical data from the experiment.

TABLE 12-9. Hypothetical Data Involving Ratings of Various Types of Movies by Viewers of Different Ages

Factor 2: Age		Factor 1: Movie Type			
		Level 1: Hardcore	Level 2: Softcore	Level 3: Family	
	20	$x_{111} = 7$ $x_{211} = 7$ $\overline{T_{11} = 14}$ $M_{11} = 7$	$x_{121} = 9$ $x_{221} = 7$ $\overline{T_{21} = 16}$ $M_{21} = 8$	$x_{131} = 10$ $x_{231} = 10$ $\overline{T_{31} = 20}$ $M_{31} = 10$	$T_{R1} = 50$
	70	$x_{112} = 2$ $x_{212} = 2$ $\overline{T_{12} = 4}$ $M_{12} = 2$	$x_{122} = 6$ $x_{222} = 4$ $\overline{T_{22} = 10}$ $M_{22} = 5$	$x_{132} = 8$ $x_{232} = 6$ $\overline{T_{32} = 14}$ $M_{32} = 7$	$T_{R2} = 28$
		$T_{C1} = 18$	$T_{C2} = 26$	$T_{C3} = 34$	$T = 78$

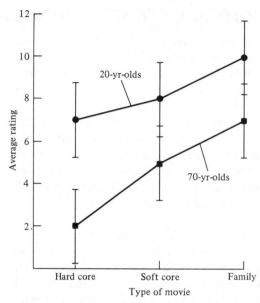

FIGURE 12-4.   Average rating as a function of movie type. Separate curves are plotted for each value of audience age (95% confidence intervals are plotted around the sample means)

We have computed the means for each of the six cells and plotted them in Figure 12-4 to provide a rough idea of what is going on in the data.[2] The graph in Figure 12-4 appears to indicate that (1) the harder the core, the less the movie is liked; (2) 20-year-olds like movies better than do 70-year-olds; and (3) there seems to be an interaction in the data. That is, 70-year-olds seem to be affected more by the type of movie than are 20-year-olds. Naturally, however, some or all these apparent effects may be due simply to random fluctuation, so we would like to perform the appropriate statistical checks.

Null and Alternative Hypotheses

We actually have three effects whose statistical significance we would like to assess. We have the two main effects of type of movie and age group; and in addition we have the interaction effect. Therefore, we have three null and alternative hypotheses to establish. In order that we not lose ourselves in the jumble of subscripts that will be necessary for stating hypotheses, Table 12-10 establishes notation for various population means. First μ_{C1}, μ_{C2}, and μ_{C3} refer to the population mean ratings for hardcore, softcore, and family movies, respectively. Likewise, μ_{R1} and μ_{R2} refer to population mean ratings for 20-year-olds and 70-year-olds, respectively. Finally μ_{jk} refers to the population mean of the jk^{th} cell in the design. (For instance μ_{12} would refer to the population mean rating of 70-year-olds watching hardcore movies.)

---

[2]We will shortly explain how the confidence intervals around the means were computed.

TABLE 12-10.   Notation for Various Population Means: Cell Population Means Are Denoted as μ_{jk}; Column Population Means Are Denoted as μ_{Cj}; and Row Population Means Are Denoted as μ_{Rk}

		Factor 1			
		Level 1	Level 2	Level 3	
Factor 2	Level 1	μ_{11}	μ_{21}	μ_{31}	μ_{R1}
	Level 2	μ_{12}	μ_{22}	μ_{32}	μ_{R2}
		μ_{C1}	μ_{C2}	μ_{C3}	

Type of movie.   Our "column null hypothesis" states that

   H_0 (columns): $\mu_{C1} = \mu_{C2} = \mu_{C3}$

That is, the column null hypothesis states that the type of movie—hardcore, softcore, or family—has no effect on how much people like the movie in general. Likewise, the alternative hypothesis would say that type of movie does have some effect on how much people like the movie, or

   H_1 (columns): It is not true that $\mu_{C1} = \mu_{C2} = \mu_{C3}$

Effects of age.   Our null hypothesis due to rows would state that age has no effect on movie enjoyment or

   H_0 (rows): $\mu_{R1} = \mu_{R2}$

Likewise, the alternative hypothesis for rows would state that age does make a difference in terms of how much people like movies, or

   H_1 (rows): It is not true that $\mu_{R1} = \mu_{R2}$

Interaction.   Finally we want to establish null and alternative hypotheses having to do with the interaction of age by movie type. The interaction null hypothesis would state that the difference between the two levels of Factor 2 (20- versus 70-year-olds) would not depend on the type of movie, or

   H_0 (interaction): $(\mu_{11} - \mu_{12}) = (\mu_{21} - \mu_{22}) = (\mu_{31} - \mu_{32})$

Another mathematically equivalent way of stating the null hypothesis is that the effect of movie type is not influenced by the age group. Mathematically this way of stating the null hypothesis would be

   H_0 (interactions): $(\mu_{31} - \mu_{21}) = (\mu_{32} - \mu_{22})$ *and* $(\mu_{21} - \mu_{11}) = (\mu_{22} - \mu_{12})$

The alternative hypothesis would simply state that the effect of each factor does in some way or another depend on the level of the other factor. The simplest way of stating the alternative hypothesis is

   H_1 (interaction): It is not true that $(\mu_{11} - \mu_{12}) = (\mu_{21} - \mu_{22}) = (\mu_{31} - \mu_{32})$

Computation of Sums of Squares

In Table 12-11 we have computed all the terms that are necessary for determining the relevant sums of squares. In Table 12-12 we have computed the sums of squares themselves. Sum of squares within has been computed in the usual way—by taking the sum over all scores of the x_{ijk}^2s and subtracting the sum of the T_{jk}^2s divided by n (in this case, n equals 2 since there are two scores per cell). We then computed SSB by considering our design as a simple one-way analysis of variance design with six separate conditions. We computed SSC and SSR by respectively ignoring the existence of rows and columns. And finally we computed the sums of squares due to interaction by considering SSB and subtracting those sums of squares that have been used up by columns and rows.

TABLE 12-11. **Various Terms Necessary for Computing Sums of Squares in a Two-Way ANOVA Design**

$$\sum_{\text{everything}} x_{ijk}^2 = 7^2 + 7^2 + 9^2 + 7^2 + \cdots + 8^2 + 6^2 = 588$$

$$\sum_{\text{cells}} T_{jk}^2/n = 14^2/2 + 16^2/2 + \cdots + 14^2/2 = 582$$

$$\sum_{\text{columns}} T_{cj}^2/n_C = 18^2/4 + 26^2/4 + 34^2/4 = 539$$

$$\sum_{\text{rows}} T_{Rk}^2/n_R = 50^2/6 + 28^2/6 = 547.3$$

$$\frac{T^2}{N} = 78^2/12 = 507$$

TABLE 12-12. **Computation of Sums of Squares in a Two-Way ANOVA Design**

$$SSW = \sum_{\text{everything}} x_{ijk}^2 - \sum_{\text{cells}} T_{jk}^2/n = 588 - 582 = 6$$

$$SSB = \sum_{\text{cells}} T_{jk}^2/n - T^2/N = 582 - 507 = 75$$

$$SSC = \sum_{\text{columns}} T_{Cj}^2/n_C - T^2/N = 539 - 507 = 32$$

$$SSR = \sum_{\text{rows}} T_{Rk}^2/n_R - T^2/N = 547.3 - 507.0 = 40.3$$

$$SSI = SSB - SSC - SSR = 75.0 - 32.0 - 40.3 = 2.7$$

A Two-Way Analysis of Variance Summary Table

Finally in Table 12-13 we have established an ANOVA summary table. In this summary table we have included all relevant sums of squares and degrees of freedom from our entire experiment. As usual we start with a total sum of squares (SST) deriving from the fact that all $N = 12$ scores in the entire experiment vary from one another. We then subdivide SST into sum of squares between the six groups (SSB) and sum of squares from within each of the six groups (SSW). Finally we divide SSB into its constituent components of rows, columns, and interaction.

Each of these sums of squares has associated with it a particular number of degrees of freedom. The total degrees of freedom are equal to $N - 1$ or, in this case, $12 - 1 = 11$. These total degrees of freedom break up into 5 degrees of freedom between (since there are six groups) and 6 degrees of freedom within. The 6 degrees of freedom within can be calculated as either $N - J = 12 - 6 = 6$, or by noting that there is 1 degree of freedom within from each of the six groups. Finally there are 2 degrees of freedom due to columns (since there are three columns), 1 degree due to rows (since there are 2 rows), and 2 degrees of freedom due to the interaction. These 2 degrees of freedom due to the interaction may be computed in either of two ways. First we know that we had 5 degrees of freedom between the six groups to begin with. We used up a total of 3 of these for rows and columns, so there are 2 left over for the interaction. Or the interaction degrees of freedom may be obtained by multiplying degrees of freedom due to columns (2) by degrees of freedom due to rows (1) to obtain $2 \times 1 = 2$ degrees of freedom for the interaction

TABLE 12-13. Two-Way ANOVA Summary Table for Movie Type/Age Experiment

Source	df	SS	MS	F	Crit F
Between	5	75			
Columns	2	32	16	16	5.14
Rows	1	40.3	40.3	40.3	5.99
Interaction	2	2.7	1.35	1.35	5.14
Within	6	6	1		
Total	11	81			

Mean squares: Estimates of σ^2. Now we obtain mean squares due to columns, rows, and interaction in the usual way—we divide the sum of squares by its associated degrees of freedom. We note that mean square within—which turns out to be 1.0—is still our best estimate of σ^2, the population variance. Although the formula does not make it entirely evident, we should remind ourselves that MSW is the estimate of the population variance that would result if we were to compute one estimate of σ^2 from within each of the six cells and then average these estimates. Thus, it is the estimate of the population variance to which all other estimates (mean squares) are to be compared.

The mean squares for columns, rows, and interactions turn out to be 16.0, 40.3, and 4.35, respectively. In the fifth column of Table 12-13 we have listed the relevant Fs obtained by dividing mean squares due to columns, rows, and interactions by MSW.

Statistical conclusions.   Finally in column six of Table 12-13 we have listed the relevant criterion Fs. As we can see, the obtained Fs for both columns (movie type) and rows (age) exceed the corresponding criterion Fs and therefore may be declared statistically significant. Thus, we conclude that the type of movie (hardcore, softcore, or family) has an effect on how much the movies are liked and, likewise, we conclude that age (70 versus 20) has an effect on how well movies are liked.

The obtained interaction $F(2,6)$ of 1.35 is less extreme than the criterion $F(2.6)$ of 5.14. Therefore, we are not in a position to declare the movie type by age interaction statistically significant. We do not have enough evidence to conclude that the difference between 70-year-olds and 20-year-olds depends on the type of movie being viewed.

> 12.8 In a two-way ANOVA situation we do three separate F-tests: one for Factor 1 (columns), one for Factor 2 (rows), and one for the interaction.

Dividing the Pie

Figure 12-5 shows how the "variance pie" is divided and subdivided in this two-way ANOVA situation. We initially have a total sum of squares equal to 81.

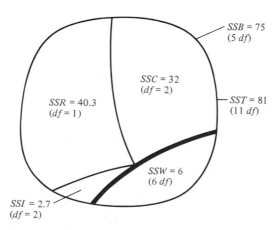

FIGURE 12-5.   Final division of the pie for a two-factor design. Sum of squares between is further subdivided into variance from columns, rows, and interaction.

This *SST* stems from the fact that all 12 scores in the experiment differ from one another and is thus based on 11 degrees of freedom.

SST is first subdivided into variance (sum of squares) from *within* cells (*SSW*) and variance from *between* cells (*SSB*). Five of the original 11 degrees of freedom are used up between groups and the other 6 go to within groups.

Now we focus on *SSB*. Variance between groups stems from three separate components. First, *SSC* results from variation among column means. Second, *SSR* results from variation among row means. And finally, *SSI* results from interaction between rows and columns.

THE EQUAL *ns* REQUIREMENT

In a *z*-test, a *t*-test, or a one-way ANOVA situation, it was perfectly reasonable to have unequal numbers of subjects in the various conditions. We saw that to the degree that *n*s were equal, violation of various assumptions became less problematic. But equal *n*s did not itself constitute an assumption necessary for the validity of the statistical tests.

To perform the sort of multifactor ANOVA we have just described, the situation is quite different. Here it is *required* that there be equal *n*s in all cells of the design. Unequal *n*s will cause severe problems if a normal multifactor ANOVA is attempted. If one does end up with unequal *n*s, there are two possible solutions: First, observations may be randomly discarded from the high-*n* cells. Second, one may perform what is known as an *unweighted means ANOVA*. A detailed description of an unweighted means ANOVA is beyond the scope of this book, but you can see the bibliography for appropriate references if you're interested.

> 12.9 In a multifactor design there must be equal numbers of observations in each cell, or else more complicated analysis is required.

CONFIDENCE INTERVALS

The necessity for equal *n*s does have the felicitous consequence that calculation of confidence intervals in a multifactor ANOVA situation is quite simple. Recall from Chapter 11 that the formula for computing a confidence interval around some sample mean, M_j in an ANOVA design is

$$\text{Confidence interval} = M_j \pm [\text{criterion } t] \sqrt{MSW/n_j}$$

The equal *n*s requirement dictates that all the n_js must be equal. Thus, in a two-way ANOVA, only one confidence interval need be computed, and this one confidence interval is placed around *all* sample means.

To compute the confidence intervals shown in Figure 12-4, we used the data from Table 12-3 as follows:

$MSW = 1.0$

criterion $t(6)_{\alpha=.05}^{2Q} = 2.447$

$n = 2$

Thus,

95% confidence interval $= M \pm (2.447) \sqrt{1/2} = M \pm 1.73$

LOOKING AT THE GRAPHS:
AN INTUITIVE VIEW OF WHAT'S HAPPENING

We have been constantly suggesting that it is useful to get an intuitive picture of what's happening with data by looking at appropriate graphs of the condition means. We saw that the behavior of the curves provides an idea of whether the two independent variables are having any effect. Futhermore, the extent to which the curves are parallel versus nonparallel provides an idea of whether or not there is any interaction. In this section we would like to consider this issue in a bit more depth. Consider Figure 12-6, which shows typical data from a 2×2 design. Factor 1 is shown as the abscissa variable, and separate curves are drawn for each level of Factor 2. What aspects of the curves indicate to us whether there is a main effect of Factor 1, a main effect of Factor 2, and an interaction?

Here are some rules of thumb. To determine whether there is an effect of Factor 1, we first take the average of the two points corresponding to Level 1 of Factor 1 and we then take the average of the two points corresponding to Level 2 of Factor 1. We then ask whether these two averages differ from one another. To the degree that they do differ, we say that there is an effect of Factor 1; and to the degree that they do not differ, we say that there is no effect of Factor 1.

Now let's concern ourselves with Factor 2. Again we take two averages. The first is the average of the curve corresponding to Level 1 of Factor 2, and the second is the average of the curve corresponding to Level 2 of Factor 2. We now ask whether *these* two averages differ. To the degree that they do, we say there is an effect of Factor 2, and to the degree that they do not, we say that there is no effect of Factor 2.

This technique is a systematic way of doing a relatively simple thing. To look at the effect of Factor 1, we just look at the *average* of each of the two levels of Factor 1, and we compare these average values. That is, we say, "Are the curves in general going up or going down?" Likewise, when considering Factor 2, we are comparing the average values of the two levels of Factor 2. We are saying, "Are the two curves different from one another on the average?"

Finally as noted earlier, to determine whether or not there is interaction we simply assess the degree to which the two curves are parallel. To the degree that

Is there an effect of Factor 1?

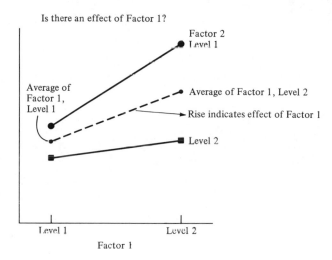

Is there an effect of Factor 2?

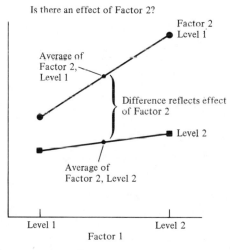

FIGURE 12-6.   An intuitive view of whether there are effects of Factor 1 and Factor 2. As shown in the top panel, an effect of Factor 1 is reflected by an increase or decrease of the average of the two curves. As shown in the bottom panel, an effect of Factor 2 is reflected by a difference between the two curves.

the two curves are parallel, we believe there is *no* interaction, and to the degree that the curves are nonparallel, we believe there *is* an interaction.

More Examples of 2 × 2 Designs

Figure 12-7 shows other possible outcomes of a 2 × 2 design. For each of these examples, we have provided a rough notion of whether there is an effect of

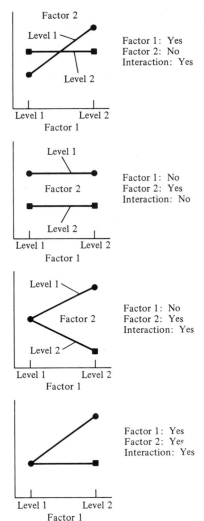

FIGURE 12-7. Various possible outcomes for 2×2 designs.

Factor 1, an effect of Factor 2, and an interaction. The purpose of giving these curves is simply to provide you with an intuitive feel for what different types of data mean.

More Than Two Levels

Figure 12-8 shows some examples of designs in which one or both of the factors involve more than two levels. Here the general idea of assessing effects of Factor 1, Factor 2, and the interaction are the same as we have described above. However, with more data points to be concerned with, you can see that things are somewhat more complicated.

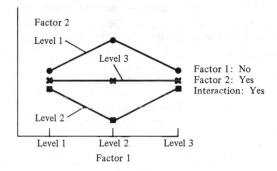

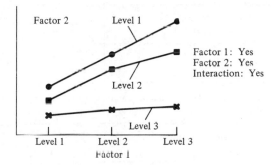

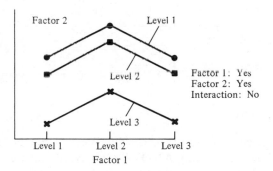

FIGURE 12-8.   Various possible outcomes for 3 × 3 designs.

HIGHER-ORDER ANOVA

As we suggested earlier in this chapter, there is no need to restrict ourselves to two independent variables. In fact we can have arbitrarily many factors in an experimental design. Consider, for example, the memory experiment we described earlier. Recall that this was a 3 × 2 design—three levels of retention interval by two levels of incentive. Suppose, however, we also had been interested in whether there were any sex differences with respect to memory performance.

We could then essentially have run the experiment twice—once for males and once for females, and we would have had the $3 \times 2 \times 2$ design—retention interval by incentive by sex—that is depicted in Table 12-14. Notice that this design has $3 \times 2 \times 2 = 12$ separate cells or conditions.

TABLE 12-14.   Representation of a Three-Way ANOVA Design

Sex: Level 1 (Females)

		Retention Interval		
		Level 1 (0 hours)	*Level 2 (1 hour)*	*Level 3 (5 hours)*
Incentive	*Level 1 (1¢)*			
	Level 2 (25¢)			

Sex: Level 2 (Males)

		Retention Interval		
		Level 1 (0 hours)	*Level 2 (1 hour)*	*Level 3 (5 hours)*
Incentive	*Level 1 (1¢)*			
	Level 2 (25¢)			

Analysis of a Three-Way ANOVA

This design is of course a three-way ANOVA design. We will not go through a detailed description of how a three-way ANOVA is analyzed, but we will describe it at a general level.

A three-way ANOVA table.   Table 12-15 shows an abbreviated summary table for the $3 \times 2 \times 2$ design we have been discussing (assuming $n = 3$ scores per cell). Additionally the right-hand side of Table 12-15 shows the general case of a $J \times K \times L$ design with n subjects per cell.

As always SST, the total sum of squares, reflects the fact that all $J \times K \times L \times n$ scores differ from one another. SST is divided into eight components. Initially it is divided into SSW (from within cells) and SSB (from between cells). SSB is then subdivided into seven components, six of which should strike familiar chords.

TABLE 12-15.   Summary Tables for a Three-Way ANOVA Design

Specific Case (from Table 12–14)		General Case ($J \times K \times L$ design)	
Source	df	Source	df
Between	11	Between	$(J \times K \times L) - 1$
Retention Interval ($F1$)	2	A	$J - 1$
Incentive ($F2$)	1	B	$K - 1$
Sex ($F3$)	1	C	$L - 1$
$F1 \times F2$	2	$A \times B$	$(J - 1) \times (K - 1)$
$F1 \times F3$	2	$A \times C$	$(J - 1) \times (L - 1)$
$F2 \times F3$	1	$B \times C$	$(K - 1) \times (L - 1)$
$F1 \times F2 \times F3$	2	$A \times B \times C$	$(J - 1) \times (K - 1)$ $\times (L - 1)$
Within	24	Within	$J \times K \times L \times (n - 1)$
Total	35	Total	$(J \times K \times L \times n) - 1$

The six familiar components are the main effects of the three factors and three two-way interaction terms. We have already discussed the meaning of the retention interval by incentive interaction. The other two-way interactions—sex by incentive and sex by retention interval—are exactly analogous, reflecting respectively the degree to which the effect of incentive differs for the two levels of sex (or vice versa) and the degree to which the effect of retention interval differs for the two levels of sex (or vice versa). Note that to compute degrees of freedom for any two-way interaction, one merely multiplies the degrees of freedom corresponding to the two factors involved in the interaction.

A three-way interaction.   The seventh component is unfamiliar. It is the three-way interaction—in this case the retention interval by incentive by sex interaction. It is based on $2 \times 1 \times 1$, or in general $(J - 1) \times (K - 1) \times (L - 1)$ degrees of freedom. It reflects *the degree to which any two-way interaction depends on the level of the third factor.* In the example at hand it represents either (a) the degree to which the nature of the retention interval $\times$ incentive interaction depends on the level of sex, or (b) the degree to which the nature of the retention interval $\times$ sex interaction depends on the level of incentive, or (c) the degree to which the nature of the incentive $\times$ sex interaction depends on the level of retention interval.

12.10 An $A \times B \times C$ interaction reflects the dependence of the $A \times B$ interaction on the level of C, or alternatively the dependence of the $A \times C$ interaction on the level of B, or alternatively the dependence of the $B \times C$ interaction on the level of A.

Performing a three-way ANOVA. Computing the sums of squares for a three-way ANOVA is relatively straightforward, but it is also long and tedious and therefore best left to a computer. (The BIOMED statistics package, available on most campus computers, is quite up to this task.) But once the sums of squares are computed, everything proceeds normally. All sums of squares are divided by their corresponding degrees of freedom to produce mean squares. Each of the seven components of variance between cells (mean squares due to three main effects, three two-way interactions, and one three-way interaction) are then tested against MSW.

> 12.11   Essentially there is nothing new about performing a three-way ANOVA. All relevant mean squares are tested against MSW (mean square from within cells).

Higher Still

The process of adding factors could continue ad infinitum. However, from a practical standpoint proliferation of factors within a single experiment is probably unwise for two reasons: First it becomes difficult to fill all the cells with subjects because running subjects is costly and time-consuming. Second and more importantly as the number of factors increases, the number of effects to be interpreted mushrooms, and comprehension of the results becomes exceedingly difficult. For example in a five-factor ($J \times K \times L \times M \times N$) design, there are five main effects, $\binom{5}{2}$ or 10 two-way interactions, $\binom{5}{3}$ or 10 three-way interactions, five four-way interactions, and one five-way interaction! What is one to make of the fact that the retention interval by incentive by sex interaction depends on the subjects' age? Or that the retention interval by incentive by sex by age interaction depends on the color of the experimenter's shirt? You probably had a difficult time understanding three-way interactions. More complicated ones get very bewildering very fast.

> 12.12   In a W-way design, there are $2^W - 1$ effects to be tested. When W is greater than three, the whole thing starts to become a sort of a mess.

SUMMARY

In this chapter we have added yet a new dimension of complexity to our experimental designs. Specifically we have examined the situation in which we look at the simultaneous effects of two independent variables on one dependent variable. We saw that the design typically used to accomplish this is called a factorial design. In a two-factor design there are two factors (independent variables). Each factor has two or more levels. In a factorial design all levels of the first factor are combined completely with all levels of the second factor. The number of conditions in this experiment is thus computed by the number of levels of the first factor times the number of levels of the second factor.

There are three effects that we wish to test in this kind of design. First there are two main effects of Factor 1 and Factor 2. Testing these main effects involves much the same technique that we used in the one-way ANOVA design. We first computed sums cf squares for each of the two factors. Dividing these sums of squares by the associated degrees of freedom provided us with mean squares for the two factors. A mean square for a particular factor is an estimate of σ^2, the population variance, if and only if that factor is actually having no effect on the independent variable (that is, if and only if the null hypothesis corresponding to that factor is true). Therefore, we were able to form F-ratios by dividing the mean square due to each of the factors by our "best estimate" of the population variance, which was MSW. These F-tests were essentially identical to the F-tests we did in a one-way ANOVA situation. By comparing our obtained F with some criterion F, we were able to decide whether or not to reject a null hypothesis of some factor's having no effect on the dependent variable.

The major novelty encountered in this chapter concerned a new source of variance called an interaction. We saw that in a two-factor design, an interaction exists to the extent that the effect of one factor depends on the level of the other factor. In contrast, if there is no interaction in the data, then the two factors are said to operate independently of one another. We saw that we were able to perform a statistical test for the presence or absence of an interaction. That is, we were able to obtain a sum of squares and corresponding degrees of freedom for the interaction just as we were able to obtain a sum of squares and corresponding degrees of freedom for the two main effects. We were then able to establish an interaction null hypothesis that there is really no interaction in terms of population means. If this is indeed the case, then the interaction mean square that we obtain in the data is an estimate of the population variance—just as for a given factor, if that factor has no effect on the dependent variable in terms of population means, then the mean square due to that factor should be an estimate of the population variance. We are therefore able to divide our interaction mean square by MSW to test for the presence of an interaction. This F-test was just like all the previous F-tests that we did.

Finally we discussed higher-order designs. In general we saw that they were just extensions of lower-order designs, the main new feature being that there were increasingly exotic, complicated and unwieldy higher-order interactions with which to deal.

PROBLEMS

1. A psychologist is interested in the effects of marijuana and motivation on memory. She uses a free-recall task in which a subject is presented a list of

20 words and is asked to recall as many of the words as possible. The dependent variable is thus the number of words recalled. The design is 2×2. Subjects are either given marijuana or not. In addition subjects have either high motivation (they're given 10¢ for each word they remember) or low motivation (they're given 1¢ for each word they remember). There are 10 subjects in each of the four conditions:

The data are as follows for the four conditions:

		Marijuana		
		Yes	No	
Motivation	Low	I $T_{11} = 20$ $M_{11} = 2$	II $T_{12} = 80$ $M_{12} = 8$	$T_{R1} = 100$ $n_{R1} = 20$
	High	III $T_{21} = 100$ $M_{21} = 10$	IV $T_{22} = 120$ $M_{22} = 12$	$T_{R2} = 220$ $n_{R2} = 20$
		$T_{C1} = 120$ $n_{C1} = 20$	$T_{C2} = 200$ $n_{C2} = 20$	$T = 320$ $n = 40$

For Questions 1 and 2 of this problem, you can view this design as a one-way design, that is

		Condition		
I. low motivation marijuana	II. low motivation no marijuana	III. high motivation marijuana	IV. high motivation no marijuana
$T_1 = 20$ $M_1 = 2$	$T_2 = 80$ $M_2 = 8$	$T_3 = 100$ $M_3 = 10$	$T_4 = 120$ $M_4 = 12$

For Questions 1 and 2, consider this to be a one-way design with 4 conditions and 10 subjects per condition. Assume $SSW = 3600$.

1. What is the value of SSB? What df is associated with SSB?
2. Compute MSB and MSW. What is F? Is it significant?

For parts 3–6 consider this as a two-way (2×2) design and refer to the data in the 2×2 table.

3. Compute the sum of squares corresponding to the motivation (row) variable. How many df does this sum of squares have?
4. What is the mean square for the motivation variable? What is the F? Is this F significant?
5. Compute the SS and MS for the marijuana (column) variable. What is the F for this variable? Is this F significant?

6. Compute the *SS* and *MS* for the interaction. What is the *F* for the interaction? Is it significant?

2. An experiment is done to test the effect of age and dress of a speaker on attitude of the audience.

Groups of college students listen to a speaker talk on why Shablotnik should be president. The speaker is either neatly dressed (tie and jacket) or grubbily dressed (blue jeans and t-shirt). Additionally the speaker is either in his 20s, 30s, or 40s. After the talk the subjects rate their attitude about whether Shablotnik should be elected, on a scale ranging from 0 (should not be elected) to 7 (should be elected).

There are two subjects per group. The data are as follows:

		Speaker's Age		
		20s	*30s*	*40s*
Speaker's Dress	grubby	5 5	3 4	2 1
	neat	0 1	1 1	2 1

1. Plot means and 95% confidence intervals for the 6 cells.
2. Test the effects of age, dress, and interaction on attitude scores.

3. An experiment is done to test whether marijuana has any effect on perceived time duration. To test this, marijuana is given to one group of subjects and a placebo to another. Additionally half the subjects in each group are males and the other half are females. The design is thus 2 × 2: male/female × drug/placebo. There are six subjects in each cell.

All subjects sit in a room and are asked to talk into a tape recorder until they think 10 minutes has elapsed. The dependent variable in the experiment is then the actual amount of time spent in the room. The data are as follows:

	Drug			Placebo		
Males	5 6	3 6	4 7	11 13	9 11	8 10
Females	8 6	7 8	9 9	11 13	14 12	9 10

1. Plot means and 95% confidence intervals for the four cells.
2. Perform the appropriate analysis to determine whether there are significant effects of sex, drug, and interaction.

4. A physiological psychologist is interested in the effects of three types of brain lesions on learning in rats. He is also interested in testing out the brain lesions on three types of rats: rats bred to be dumb, rats bred to be moderately intelligent, and rats bred to be smart. He has, therefore, a 3 × 3

design: three levels of lesion crossed with three levels of intelligence. Assume there are five observations per cell, and that $SST = 262$.

The dependent variable is number of trials needed to learn a maze, and the data are as follows (M_{ij} refers to the mean of a given cell):

		Type of Lesion		
		I	*II*	*III*
Intelligence	high	$M_{11} = 4$	$M_{12} = 2$	$M_{13} = 5$
	medium	$M_{21} = 6$	$M_{22} = 5$	$M_{23} = 6$
	low	$M_{31} = 6$	$M_{32} = 10$	$M_{33} = 7$

1. What are the cell and marginal and grand totals? Put them in a table like the one above.
2. Plot the data in two ways: with type of lesion on the abscissa and with intelligence on the abscissa. Compute 95% confidence intervals around all means.
3. What are SSB, SSR, SSC, SSI, and SSW? What are all the dfs and MSs? Put them all into an ANOVA table.
4. Compute the appropriate Fs. What is significant?

5. A new drug is invented which will presumably speed up learning in rats. A group of 40 rats is subdivided into two groups: one group injected with the drug, the other, with a placebo. Each of the two groups is further divided in half: ten rats in each group are rewarded with a tasty sugar cube in maze-learning, whereas the other ten are rewarded with a less-interesting rat pellet. The design may thus be viewed as 2 (drug/no drug) × 2 (sugar/pellet). The data in terms of trials to learn the maze are as follows:

	Drug		*No Drug*	
Sugar	5	7	13	15
	9	6	10	9
	4	5	10	8
	4	6	11	13
	8	5	13	12
Pellet	8	9	10	16
	10	9	9	15
	11	12	12	9
	7	10	15	10
	8	13	8	12

1. Plot the four means along with 95% confidence intervals.
2. Perform a two-way analysis of variance on the data.
3. Describe how you would interpret the results.

6. The Gazelle Motorcycle Corporation is trying a new type of braking system for their motorcycles. The experiment involves comparing the stopping

distance with the new versus the old braking system. Additionally, the two systems are compared for three sizes of motorcycles: 125 cc, 500 cc, and 100 cc. There are six motorcycles in each of the six cells, and the data are as follows:

Stopping Distance (feet)

	125 cc		500 cc		1000 cc	
Old Braking System	53	50	55	60	71	75
	48	50	61	60	72	61
	47	51	52	65	63	70
New Braking System	54	42	51	55	49	45
	48	50	50	49	50	48
	55	45	52	51	55	53

1. Plot the means along with the 95% confidence intervals for the six cells.
2. Are there effects of size, type of braking system? Do these factors interact?
3. How would you describe the effect of the new braking system?

7. An experiment is done to investigate the effects of verbal description on person perception. The experiment works as follows. While waiting in an antechamber, a subject is approached by a confederate who begins a conversation. In the course of the conversation the confederate casually describes the experimenter, using one of three adverbs ("somewhat," "very," or "extremely") factorially combined with three adjectives ("nice," "efficient," or "nasty"). After going through a dummy verbal learning experiment, the subject fills out a questionnaire. Of interest is the answer to the question, "How did you like the experimenter?" The scale ranges from 0 (couldn't stand him) to 10 (loved him).

Five subjects are assigned to each of the nine possible combinations of adjective and adverb. The data are as follows:

Rating

		Adjective					
		Nice		Efficient		Nasty	
	Somewhat	5	5	5	5	4	3
		6	5	6	4	3	6
		4		6		5	
Adverb	Very	6	6	5	6	3	1
		9	5	6	7	3	2
		7		4		2	
	Extremely	9	7	4	7	1	3
		8	8	5	8	2	1
		9		4		1	

1. Plot the nine cell means along with 95% confidence intervals.
2. Are there effects of adjective, adverb, and their interaction?
3. How would you interpret these results?

8. The city of Puyallup is trying to cut down on juvenile delinquency. To do this, it is planned to introduce a technique called "modeling therapy." Eight teenagers—four males and four females—are selected for an experiment to determine whether modeling therapy is effective in terms of reducing delinquent behavior. Half the males and half the females (the experimental groups) are provided with therapy for a year. The remaining subjects are not given therapy and constitute the control group. During the year the number of delinquent acts for each subject is recorded. The data are as follows:

| | | Factor 1 (treatment) | |
		Experimental (therapy)	Control (no therapy)
Factor 2 (sex)	Male	2 3	3 6
	Female	1 5	8 6

1. Are there significant effects of treatment, sex, and interaction?
2. What is the 90% confidence interval for the cell means?
3. This experiment is repeated in Seattle, where it is known that the variance of the population of delinquent acts is $\sigma^2 = 1.0$. Only one subject is in each of the four cells with the following data:

| | | Factor 1 (treatment) | |
		Experimental	Control
Factor 2 (sex)	Male	1	3
	Female	1	6

Are there significant effects of treatment, sex, and interaction?
4. How would you interpret all these results?

9. An experiment is done to test the effect of varying amounts of light on electrical activity in Merkin plant leaves. Three different light levels—no light, medium light, and bright light—are presented to Ralph and Irving, two different Merkin plants. Each plant goes through each condition twice. Due to error in procedure, however, Ralph goes through the bright light condition six times rather than twice.

The data recorded are voltage amplitudes on the leaves and are as follows (numbers represent millivolts):

	Light Condition		
	No Light	*Medium Light*	*Bright Light*
Ralph	11 9	10 10	11 10 9 11 10 9
Irving	1 3	2 2	1 3

1. Compute a mean for each column.
2. Compute a mean for each condition for each plant (thereby getting two numbers per plant for each condition). Compute a mean for each condition by taking the mean of these two numbers.
3. Looking at the data, do you think there is an effect of amount of light? Why or why not?
4. Compute sums of squares for between rows, columns, interaction, and within. Do not be alarmed if something strange happens.
5. Why do you think it is a bad thing to have unequal cell frequencies in this kind of design?
6. What would you do to correct this shortcoming?

10. The following graphs represent data from 2 × 2 designs. In each case, assume $SSB = 100$. Make rough guesses as to what the sums of squares are for Independent Variable 1, Independent Variable 2, and the interaction.

1.

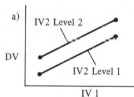

a)

Sums of squares
IV1:
IV2:
Interaction:

2.

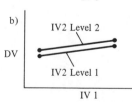

b)

Sums of squares
IV1:
IV2:
Interaction:

3.

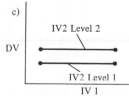

c)

Sums of squares
IV1:
IV2:
Interaction:

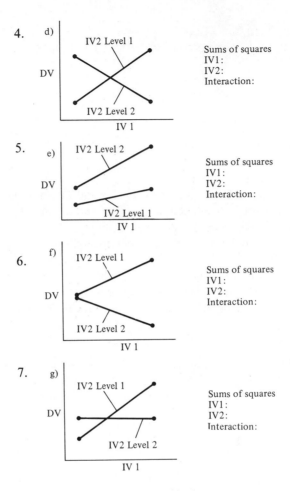

4. d)

Sums of squares
IV1:
IV2:
Interaction:

5. e)

Sums of squares
IV1:
IV2:
Interaction:

6. f)

Sums of squares
IV1:
IV2:
Interaction:

7. g)

Sums of squares
IV1:
IV2:
Interaction:

11. Here are data from a 3 × 3 design:

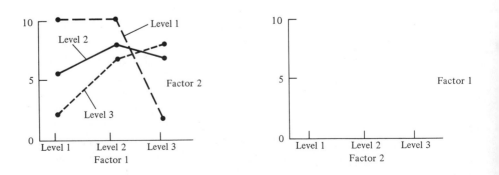

Redraw these data on the right-hand graph, with Factor 2 on the abscissa and Factor 1 as the curve parameter.

Within-Subjects (or Repeated-Measures) Designs

13

Remember within-subjects designs? We have sporadically discussed them in previous chapters. Recall that in a within-subject design a given subject participates in all experimental conditions. Within-subjects designs may be distinguished from between-subjects designs in which the separate experimental conditions involve separate groups of subjects and any given subject therefore participates in only one condition.

In this chapter we'll offer a systematic discussion of within-subjects designs and the mechanics by which they are analyzed. We shall see that the formulas and computations involved in analyzing within-subjects designs are very similar to those encountered in the previous two chapters. However, some of the concepts involved in within-subjects designs will be novel.

The reason we've deferred this long before returning to the topic of within-subjects designs is that there turns out to be a very strong relationship between within-subjects designs and the sort of two-factor design described in Chapter 12. In fact, as we shall see, a within-subjects design *is* a kind of two-factor design.

SLEEP DEPRIVATION AND REACTION TIME

Let's consider a situation in which we are interested in the effect of sleep deprivation on how long it takes a person to react to the onset of some stimulus. Our independent variable in this design will be the amount of time the subject has gone without sleep. Let's suppose that we have chosen three levels of this particular independent variable (that is, three conditions). In Condition 1, the subject is wide awake—it's only been 2 hours since he slept. In Condition 2, the subject has gone 16 hours without sleep, and in Condition 3 the subject has gone 24 hours without sleep. The dependent variable is the amount of time it takes to respond to the onset of a light. A typical task used for such a situation is called a *simple reaction-time task* and works as follows. Imagine Dierdre, a subject, seated at a table, her finger resting on a telegraph key. Before her is a light. Periodically the light is turned on for a short period of time and Dierdre's task is to press the key as rapidly as possible following the onset of the light. The amount of time intervening between the onset of the light and Dierdre's response is then referred to as simple reaction time. In normal circumstances, simple reaction time is roughly 200 milliseconds.

Hypothetical Results: Between Subjects

Let's first imagine that we use a between-subjects design to investigate the effect of sleep deprivation on reaction time. That is, one group of subjects participates in the 2-hour condition; a second group of subjects participates in the 16-hour condition; a third group of subjects participates in the 24-hour condition. Figure 13-1 illustrates some hypothetical data that might be obtained in such an experiment. In Figure 13-1 we have represented the mean reaction times by *x*s. Additionally Figure 13-1 depicts data points for individual subjects, which are represented by dashes scattered around the means. Thus, in Group 1 one subject had a score of 120 milliseconds, another subject had a score of 180 milliseconds, and so on.

Ponder now the data in Figure 13-1. Just intuitively does there seem to be any difference among the three groups? That is, if we were to do a statistical

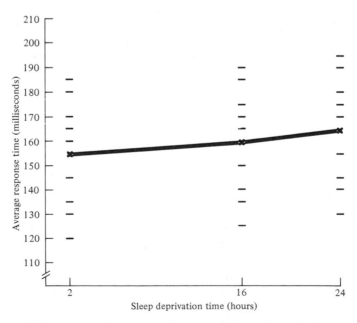

FIGURE 13-1.   Hypothetical data from a between-subjects design. The *X*s represent the condition means and the dashes represent individual data points around the means.

test—a one-way ANOVA—on these data, does it look as if it would produce a significant F? The answer seems to be no. There are, to be sure, small differences among the three means—however, the variance of individual subjects around each of the group means (σ^2) is so large that it would probably overwhelm the relatively meager amount of variance among the three means. It therefore would seem reasonable to conclude that the observed differences among the three means are probably due to random fluctuation.

A Within-Subjects Design

Now let's suppose that we have done the same experiment but used a within- rather than a between-subjects design. That is, suppose that the same subjects participated sometimes in the 2-hour condition, at other times in the 16-hour condition, and at still other times in the 24-hour condition. Imagine that this situation produced the data shown in Figure 13-2. Notice an interesting thing about Figure 13-2: all the individual data points as well as the three condition means are identical to those depicted in Figure 13-1, but since the data shown in Figure 13-2 are assumed to come from a within-subjects design, each subject produces his or her own little individual curve relating reaction time to amount of sleep deprivation. These individual subject curves are shown in Figure 13-2 along with the curve connecting the means. Now once again let us consider

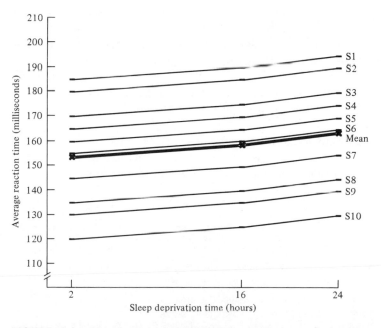

FIGURE 13-2.   Hypothetical data from a within-subjects design. Each subject's individual curve is plotted along with the curve connecting the means.

the question: Would we intuit from the data of Figure 13-2 that there is any real effect of sleep deprivation on reaction time? In contrast to the previous situation, our answer here would be yes.

Consistency over subjects.   The reason for this guess is that all the subjects show the *same pattern*—the same increase in reaction time across the sleep-deprivation conditions. The overall (average) effect of sleep deprivation is not particularly great—just as it was not particularly great considering the data shown in Figure 13-1. However, the crucial thing here is that the effect of the independent variable on the dependent variable is *consistent* across subjects. It is because of this consistency—because of the fact that *each subject* shows a similar rise in reaction time over conditions—that we tend to believe that the effect is "real" and not just due to chance fluctuation.

As we have noted, the data points themselves are identical in Figures 13-1 and 13-2. However, our interpretation of the data is quite different in the two situations. From the data shown in Figure 13-1, which is assumed to be a between-subjects design, we concluded that sleep deprivation probably had no effect on reaction time. However, from the data shown in Figure 13-2, which is assumed to come from a within-subjects design, we conclude that there is an effect of sleep deprivation on reaction time.   What is the crucial difference between these two situations? The answer is that in a between-subjects design, the variance among subjects—that is, the extent to which individual subjects differ from one another in reaction-time scores—is critical. We have identified this variance in the past as σ^2, the population variance. The more variance there is over subjects—the bigger is σ^2—the less is the power and the worse is the situation. That is, the bigger is σ^2, the more difficult it is to detect any effect of the independent variable. However, in a within-subjects design, variance among subjects is inconsequential. It is of no concern that one subject happens to be generally fast whereas another subject happens to be generally slow. All we care about is that subjects perform consistently in one condition relative to each other.

The foregoing discussion has been meant to provide an intuitive feeling for the major difference between a between-subjects and a within-subjects experimental design. Now let's make our example a bit more concrete and see if we can formalize this difference.

13.1 In a between-subjects design we reject the null hypothesis if the effect of conditions is large relative to the variance of subjects within conditions. But in a within-subjects design, we reject the null hypothesis if the effect of conditions is large enough with respect to the consistency of subjects across conditions.

TABLE 13-1.   **Representation of the Structure of a Within-Subjects Design**

	Sleep-deprivation condition		
	2 hours	*16 hours*	*24 hours*
Subject 1			
Subject 2			
Subject 3			
Subject 4			

Let's suppose we have four subjects in the experiment we just described. Since our design is within-subjects, each of the four subjects is assumed to participate in each of the three conditions. We therefore have the design depicted in Table 13-1. This design should look somewhat familiar.

Subjects as a factor.   In fact this design looks very much like the sort of two-factor design we described in the previous chapter. Viewing this design as a two-factor design, Factor 1 would be sleep-deprivation time. This factor might be viewed as a "real factor"—it is an independent variable that is conceptually quite similar to other independent variables we described before, and it has three levels.

Our other "factor," however, is subjects! Now "subjects" may not seem very much like a factor, but there is no reason not to consider it as such. It is a factor that, in this example, has four levels, each level corresponding to an individual subject. Therefore, we have a 3 × 4 factorial design. A 3 × 4 design, of course, produces the 12 separate cells shown in Table 13-1.

Actual Numbers

In Table 13-2 we have reproduced our 3 × 4 design and inserted some data. Each of the data points represents the mean reaction time of a given subject in a given sleep-deprivation condition. Thus, Subject 1 had an average reaction time of 100 milliseconds in the 2-hour condition, 200 milliseconds in the 16-hour condition, and 300 milliseconds in the 24-hour condition. From the data shown in Table 13-2, we have computed means and totals just as if this were a real two-factor design. Since each cell only contains one score, things are somewhat simplified in the sense that each cell total and mean is simply equal to the score contained in the cell. The column and row totals are obtained in the usual way; T_{Cj} is the total of the four scores in the jth column, and likewise, T_{Rk} is the total of the three scores in the kth row. Also, $T = 42$ is the total of all $N = 12$ scores in the experiment.

TABLE 13-2. Data from Hypothetical Experiment Involving Reaction Time as a Function of Amount of Sleep Deprivation. Cell Entries Represent Mean Reaction Time (in tenths of a second) for a Given Subject in a Given Condition

| | *Factor 1: Sleep-deprivation condition* | | | |
	Level 1: 2 hours	*Level 2: 16 hours*	*Level 3: 24 hours*	
Level 1: Subject 1	$x_{11} = T_{11} = M_{11} = 1$	$x_{21} = T_{21} = M_{21} = 2$	$x_{31} = T_{31} = M_{31} = 3$	$T_{R1} = 6$
Level 2: Subject 2	$x_{12} = T_{12} = M_{12} = 2$	$x_{22} = T_{22} = M_{22} = 2$	$x_{32} = T_{32} = M_{32} = 5$	$T_{R2} = 9$
Level 3: Subject 3	$x_{13} = T_{13} = M_{13} = 3$	$x_{23} = T_{23} = M_{23} = 6$	$x_{33} = T_{33} = M_{33} = 6$	$T_{R3} = 15$
Level 4: Subject 4	$x_{14} = T_{14} = M_{14} = 2$	$x_{24} = T_{24} = M_{24} = 4$	$x_{34} = T_{34} = M_{34} = 6$	$T_{R4} = 12$
	$T_{C1} = 8$ $M_{C1} = 2.0$	$T_{C2} = 14$ $M_{C2} = 3.5$	$T_{C3} = 20$ $M_{C3} = 5.0$	$T = 42$

Factor 2: Subjects (row label)

Graphing the data. As we stated earlier it's a good idea to initially graph the data in order to provide a rough, intuitive idea of what's going on. The present example is no exception. In Figure 13-3 we have graphed the data from Table 13-2 just as we would graph the data from any two-factor design. We have represented the independent variable (sleep-deprivation time) on the abscissa, and a separate curve has been plotted for each of the four subjects.

Notice that Figure 13-3 is basically the same as Figure 13-2. In each case we plotted individual subject curves relating the dependent variable (reaction time) to an independent variable. We now have simply formalized the procedure and shown it to be analogous to plotting data from a two-factor experiment.

> 13.2 A within-subjects design is formally similar to a two-factor design. Here one of the factors is the "actual" independent variable, and the other factor is subjects. Each subject constitutes one level of this factor.

Subject-by-condition interaction as error term. In our intuitive discussion of whether or not there was a real effect in a within-subjects design, we focused on the degree to which subjects were consistent across the experimental conditions. Looking at the data in Figure 13-2, we concluded that there probably was a real effect because all subjects showed the same sort of increase in reaction time across the sleep-deprivation condition. Now we can be a little more precise

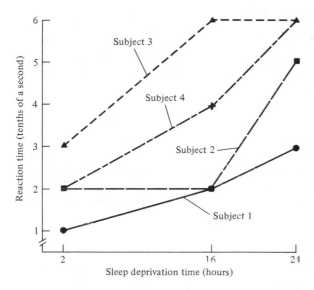

FIGURE 13-3.   The data from Table 13-2 shown in graphical form. Data from each of the four subjects is shown individually.

about what we mean by this. When we view a within-subject design as a two-factor design, the degree to which subjects are consistent across conditions is reflected by the amount of *interaction* between the two factors. To say that subjects are consistent across conditions is simply to say that the interaction between subjects and the independent variable is small.

The smaller the interaction, the more we tend to believe that any observed differences among the condition means are due to a real effect of the independent variable. Conversely, the larger is the interaction (the less consistent subjects are across conditions), the more we tend to believe that any differences among the condition means are just due to random fluctuation.

When analyzing a between-subjects design, we saw that the question of whether observed differences among means were real was assessed by forming an *F*-ratio of variance between groups to variance within groups. In such a situation variance within groups was referred to as the "error term." It is "garbage variance"—variance that we have not controlled. In a within-subjects design, however, the subject-by-condition interaction is our "garbage variance." And as we shall see, the mean square due to interaction is the error term going into the *F*-ratio, replacing mean square within.

> 13.3 In a within-subjects design, the error variance is the subject-by-condition interaction. Therefore, the appropriate *F* for testing the effect of conditions is *MSC/MSI.*

Analysis of a Within-Subjects Design as a Two-Factor Design

Let's continue treating this experiment as we would a two-way design. In a two-way design we are concerned with sums of squares due to columns, rows, and interaction. Let's proceed to compute such sums of squares.

Sum of squares between. This situation is actually a little bit different from two-way designs we have encountered previously in that there is now only one score per cell. This means we cannot compute a within sum of squares (we have 0 degrees of freedom and therefore zero variance within each of the 12 cells). However, there is nothing to prevent us from computing a sum of squares between the 12 cells using the same old equation that we know and love:

$$SSB = \sum_{\text{cells}} \frac{T_{jk}^2}{n} - \frac{T^2}{N}$$

Here n is 1 since there is only one observation per cell, and T_{jk}, the total of each cell, is simply equal to the single score that is found in each cell. Sum of squares between may therefore be computed as

$$SSB = (1^2/1 + 2^2/1 + 3^2/1 + 2^2/1 + \cdots + 6^2/1) - 42^2/12$$
$$= 184 - 147 = 37$$

Sums of squares due to rows and columns. The sums of squares due to rows and columns are computed in a comparable way. The equation for sums of squares due to columns is

$$SSC = \sum_{\text{columns}} \frac{T_{Cj}^2}{n_C} - \frac{T^2}{N}$$

just as it was in the last chapter. Here n_C is equal to 4 since there are four scores in each column, so SSC may be computed as

$$SSC = (8^2/4 + 14^2/4 + 20^2/4) - 42^2/12$$
$$= 165 - 147 = 18$$

Sum of squares due to rows may be computed analogously. Here the equation is

$$SSR = \sum_{\text{rows}} \frac{T_{Rk}^2}{n_R} - \frac{T^2}{N}$$

and n_R is equal to 3 since there are 3 scores in each row. Therefore,

$$SSR = (6^2/3 + 9^2/3 + 15^2/3 + 12^2/3) - 42^2/12$$
$$= 162 - 147 = 15$$

Sum of squares due to interaction. We also compute the sum of squares due to interaction in the same way we did in the previous chapter. We take the SSB

with which we started and subtract the sum of squares due to rows and columns that we have used up,

$$SSI = 37 - 18 - 15 = 4$$

13.4 Even with only one score per cell sums of squares due to rows, columns, and interaction can be computed just as they have always been.

A Summary Table

We can now make up an ANOVA summary table in the same way we have done before. Here the total sum of squares in the entire experiment is the sum of squares between all 12 scores in the experiment. Therefore, SSB is equal to SST. Sum of squares between is (as usual) divided into sum of squares due to rows, columns, and interaction as indicated in Table 13-3.

Now we go through the same procedure we have before: we divide each sum of squares by its associated degrees of freedom to produce mean squares, which are shown in the fourth column of Table 13-3.

Null and alternative hypotheses. We can establish null and alternative hypotheses in this situation just as we did in the past. The null hypothesis states that the independent variable (sleep deprivation) has no effect on the dependent variable (reaction time). In contrast, the alternative hypothesis maintains that sleep deprivation *does* affect reaction time.

An *F*-ratio. What do we want to do now? Note that we have a summary score which is the mean square due to conditions (columns). To the extent that this summary score is large, we tend to think that the independent variable is having an effect on the dependent variable. However, as we have discussed, to the extent

---

TABLE 13-3. ANOVA Summary Table for a Within-Subjects Design. Since There Is Only One Observation Per Cell, There Is No Within-Cell Variance

Source	df	SS	MS	Obtained F	Crit F
Between	11	37			
Columns	2	18	9	13.49	5.14
Rows	3	15	5		
Interaction	6	4	0.667		

that the interaction between subjects and conditions is large, we tend to think that this effect may be due simply to chance fluctuation. Intuitively then the appropriate F would be an $F(2,6)$ formed by dividing mean square due to conditions by mean square due to interaction. Later we shall be more specific about why this is the appropriate F. For the time being let's just state that, if indeed there is no effect of conditions, then the mean square due to conditions is an estimate of the same variance as is the mean square due to interaction. We can therefore test the obtained $F(2,6)$ in the normal way. The criterion ($\alpha = 0.05$) $F(2,6)$ is 5.14 and the obtained $F(2,6)$ is 13.49. We can therefore reject the null hypothesis and conclude that there exists a real effect on reaction time due to sleep-deprivation condition.

Sum of squares due to subjects.   So far we have said nothing about the variance due to rows (subjects). Notice that this sum of squares (and mean square) reflects the extent to which subjects differ from one another in the population. In the past we have been very concerned with this effect because in a between-subjects design the variance among subjects has been referred to as the population variance and has formed our error term. However, the situation is quite different in this within-subjects design. Here we do not care about the sum of squares due to subjects. As noted above, it makes no difference to us if one subject happens to be very slow and another subject happens to be very fast. What is of importance is whether or not subjects behave consistently across conditions. So for the time being we shall simply ignore the variance due to subjects.

> 13.5 Sum of squares due to subjects is irrelevant in a within-subjects design.

CONFIDENCE INTERVALS IN A WITHIN-SUBJECTS DESIGN

Recall that in a between-subjects ANOVA situation, a confidence interval around a sample mean, M, was computed using the formula,

Confidence interval = $M \pm$ [criterion t] $\sqrt{MSW/n}$

where n was the number of scores on which M was based. The formula for a roughly analogous confidence interval in a within-subjects design is much the same except that we substitute the interaction mean square—our error term in a within-subjects design—for MSW, our error term in a between-subjects design. Thus,

Confidence interval = $M \pm$ [criterion t] $\sqrt{MSI/n}$ (13.1)

Note that the criterion t is based on the number of degrees of freedom due to interaction. Also n is the number of observations upon which the sample mean is based—that is, the number of subjects. Digression 13.1 demonstrates where this formula comes from. For now, let's compute confidence intervals from the data shown in Tables 13.2 and 13.3 as an illustration.

Digression 13.1

Rationale Behind Within-Subjects Confidence Intervals

Derivation of within-subjects confidence intervals is rooted in the assertion made earlier that variation among subjects is of no consequence in a within-subjects design. Let's follow up on the implications of this assertion.

   The first implication is that if variation among subjects is of no consequence, then we should be able to *remove* that particular variation without causing any trouble or changing any conclusions. This removal has been accomplished in Table D-1 as follows. We first compute the

TABLE D-1.   Removal of Between-Subject Variance in a Within-Subjects Design

(a) *Raw data (from each subject in tenths of a second)*

Subject	2 hours	16 hours	24 hours	Mean	Deviation
1	1	2	3	2	−1.5
2	2	2	5	3	−0.5
3	3	6	6	5	+1.5
4	2	4	6	4	+0.5
	$T_{C1} = 8$	$T_{C2} = 14$	$T_{C3} = 20$	$M = 3.5$	

(b) *Data after removing between-subjects variance. (A constant equal to each subject's deviation has been subtracted from each of that subject's scores.)*

Subject	2 hours	16 hours	24 hours	Mean	Total
1	2.5	3.5	4.5	3.5	10.5
2	2.5	2.5	5.5	3.5	10.5
3	1.5	4.5	4.5	3.5	10.5
4	1.5	3.5	5.5	3.5	10.5
	$T_{C1} = 8$	$T_{C2} = 14$	$T_{C3} = 20$		$T = 42$

$$SS \text{ "Within"} = \sum_{\text{everything}} x_{ij}^2 - \sum_{\text{conditions}} T_j^2/4$$
$$= 169 - 165 = 4$$

mean score for each subject across the three conditions and list these means in column 5 of part *a* of Table D-1. Note that the mean of these four subject means is 3.5. Now we characterize each subject in terms of *deviation* from this overall mean of 3.5. For instance, Subject 1, the fastest subject, has an average score of 1.5 below the mean and therefore has a deviation score of -1.5. Removal of between-subject variance now consists of subtracting a given subject's deviation from each of that subject's scores. This process yields the adjusted data shown in part *b* of Table D-1. Note that the column totals which reflect the pattern of condition means are unchanged by this operation. Also the pattern of each subject's scores is unchanged. The only thing we have done is to create a common mean (of 3.5) for all four of the subjects. We have artificially prevented the subjects from varying from one another.

The data from part *b* of Table D-1 have been plotted in Figure 13-4. Here the means from the three conditions are presented as large circles. The critical thing is that *the variation of individual data points around each mean stem solely from subject-by-condition interaction variance.* That is, if there were no variance due to interaction, there would be no variation of individual scores around the means in Figure 13-4.

Now we assert that a confidence interval should appropriately reflect the variation of data points around the sample means in Figure 13-4. Let's therefore estimate ''σ^2'' where this σ^2 is now viewed as the variance of the scores around the sample means. Returning to part *b* of

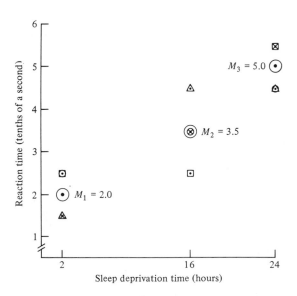

FIGURE 13-4.   The data from Figure 13-3 are replotted with variation among the subjects removed. Hence there is less variation of individual data points around the condition means. This residual variance is all due to subject-by-condition interaction.

Table D-1 we note that this procedure is somewhat analogous to computing *MSW* in a one-way between-subjects situation. And in fact when we compute a "sum of squares within," which we have done at the bottom of Table D-1 it turns out to be 4.0. This is exactly what the interaction sum of squares turned out to be in the original analysis (Table 13-3), which makes sense since this "*SSW*" *is* the interaction sum of squares.

If we were to continue our analogy to a between-subjects situation, we would compute an estimate of within-conditions variance as follows:

$$est \; \sigma^2 = \frac{"SSW"}{"dfW"} = \frac{4.0}{9.0} = 0.444$$

Note that "*dfW*" would be 9: 3 degrees of freedom from within each of the three groups. But now we see an apparent contradiction. The *interaction* mean square is computed to be

$$MSI = \frac{SSI}{dfI} = \frac{4.0}{6.0} = 0.667$$

as shown in Table 13-3. That is, we divide *SSI* by 6, the degrees of freedom due to interaction, and not by 9.

Which *is* the appropriate estimate of the variance of scores around the sample means shown in Figure 13-4? The answer, as suggested earlier, is that the appropriate estimate is *MSI* and is based on (in this case) 6 degrees of freedom. Thus, we estimate that the scores around the sample means in Figure 13-4 are drawn from populations with variances of 0.667.

The breakdown in our analogy to a between-subjects situation occurs for the following reason. The data shown in part *b* of Table D-1 have been constrained in a way that does not occur in a between-subjects design in that each row has been forced to add to the same thing, and this additional constraint causes us to lose degrees of freedom. If you hold column totals and row totals constant in a 3 × 4 table, you will find that you only can fiddle around with 6 of the 12 scores. That is, once you determine 6 of the scores, the other 6 are predetermined.

In any case to arrive at a confidence interval, we note that each of the three means in Figure 13-4 must come from a sampling distribution of sample means whose variance is estimated (based on *dfI* degrees of freedom) to be

$$est \; \sigma_M^2 = \frac{est \; \sigma^2}{n} = \frac{MSI}{n}$$

This implies the formula for confidence intervals that is represented as Equation 13.2 above.

At the bottom of Table 13.2 we have listed the means of the three conditions. Note now that

$$n = 4 \text{ subjects}$$

$$MSI = 0.667$$

$$dfI = 6$$

$$\text{criterion } t(6)^{2Q}_{\alpha-0.05} = 2.447$$

Therefore for each sample mean, M

$$\text{Confidence interval} = M \pm 2.447 \sqrt{0.667/4} = M \pm 1.00$$

Figure 13.5 plots the sample means along with their confidence intervals.

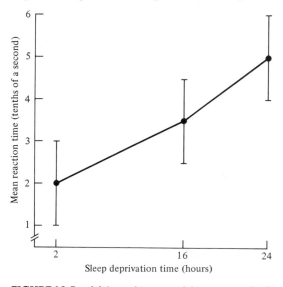

FIGURE 13-5.   Within-subjects confidence intervals plotted around the condition means.

13.6 A confidence interval in a within-subjects design may be computed by the equation:

$$\text{Confidence interval} = M_i \pm [\text{criterion } t\,(dfI)] \sqrt{MSI/n}$$

where n is the number of subjects.

MULTIPLE OBSERVATIONS PER SUBJECT PER CONDITION

In the experiment we have just described, each subject was assumed to have participated just once in each experimental condition. Therefore, each subject

TABLE 13-4.   Data from a Hypothetical Experiment Involving Reaction Time as a Function of Amount of Sleep Deprivation. Each Subject Contributes Two Observations to Each Condition.

	Factor 1: Sleep-deprivation condition			
	Level 1: 2 hours	Level 2: 16 hours	Level 3: 24 hours	
Level 1: Subject 1	$x_{111} = 1$ $T_{11} = 2$ $x_{211} = 1$	$x_{121} = 3$ $T_{21} = 4$ $x_{221} = 1$	$x_{131} = 2$ $T_{31} = 6$ $x_{231} = 4$	$T_{R1} = 12$
Level 2: Subject 2	$x_{112} = 1$ $T_{12} = 4$ $x_{212} = 3$	$x_{122} = 2$ $T_{22} = 4$ $x_{222} = 2$	$x_{132} = 6$ $T_{32} = 10$ $x_{232} = 4$	$T_{R2} = 18$
Level 3: Subject 3	$x_{113} = 3$ $T_{13} = 6$ $x_{213} = 3$	$x_{123} = 6$ $T_{23} = 12$ $x_{223} = 6$	$x_{133} = 5$ $T_{33} = 12$ $x_{233} = 7$	$T_{R3} = 30$
Level 4: Subject 4	$x_{114} = 1$ $T_{14} = 4$ $x_{214} = 3$	$x_{124} = 6$ $T_{24} = 8$ $x_{224} = 2$	$x_{134} = 6$ $T_{34} = 12$ $x_{234} = 6$	$T_{R4} = 24$
	$T_{C1} = 16$	$T_{C2} = 28$	$T_{C3} = 40$	$T = 84$

Factor 2: Subjects (row label on left side)

contributes just one score to each condition. However, there is no logical basis for this limitation. It would be perfectly possible, for example, to run the experiment this week, then bring back exactly the same subjects and run the same experiment again next week. In fact we could cycle through this procedure as many times as we wish, and it would therefore be perfectly possible to accumulate as many scores per subject per condition as we wanted (ignoring certain practical limitations such as mutiny on the part of the subjects).

Let's suppose in fact that we run the identical experiment twice for each subject. (Each rerunning of the experiment is referred to as a "replication.") Hypothetical data from such a procedure are shown in Table 13-4. Here the design appears structurally identical to the two-factor designs described in the previous chapter. And indeed it will turn out that computations of the various sums of squares will be identical to computations in a two-factor design. But this design is still a within-subjects design and is conceptually analogous to the within-subjects design described in the previous section. As such it is conceptually quite different from the two-factor designs described in Chapter 12. Let's see how we would analyze it.

Sums of Squares

In Table 13-4 we included all the relevant totals needed for computing all various sums of squares. As we did in Table 13-2, we included the column totals (T_{Cj}s) and the row totals (the T_{Rk}s). In addition we computed T, the total of the $N = 24$ scores in the entire experiment. Finally we computed the 12 cell totals (the T_{jk}s).

Sum of squares within. Since we now have more than one score per cell, we are once again in a position to compute a within-cell sum of squares. This SSW is computed by the same formula we used in the past, specifically

$$SSW = \sum_{\text{everything}} x_{ijk}^2 - \sum_{\text{cells}} \frac{T_{jk}^2}{n}$$

Plugging the data into this formula, we see that

$$SSW = (1^2 + 1^2 + 3^2 + 1^2 + \cdots + 6^2 + 6^2) - (2^2/2 + 4^2/2$$
$$+ 6^2 + \cdots + 4^2/2 + 8^2/2 + 12^2/2)$$
$$= 388 - 368 = 20$$

Before we proceed to computation of other sums of squares, let's reflect a bit on the SSW. Although we are calling it a sum of squares within, it refers to a conceptually different entity from that which it has in the past. In a two-way, between-subjects design each score from within a given cell came from a *different subject*. Therefore, the sum of squares within cells we computed in the past represented variance *across subjects*—that is, it represented the degree to which one subject differed from another subject who had been treated identically.

However, the sum of squares within that we have just computed is different. In the design that we are presently considering, the various scores within a given cell represent scores coming from the *same subject, treated identically at different times*. The SSW we have just computed therefore represents the degree to which a given subject differs from himself or herself when treated under identical conditions at different times. We'll have more to say about this point below.

Sums of squares due to between, columns, rows, and interaction. The rest of the sums of squares in the experiment are computed using familiar techniques and are almost boringly straightforward. The formula for sum of squares between cells is

$$SSB = \sum_{\text{cells}} \frac{T_{jk}^2}{n} - \frac{T^2}{N}$$

which is

$$SSB = (2^2/2 + 4^2/2 + 6^2/2 + \cdots + 4^2/2 + 8^2/2 + 12^2/2) - 84^2/24$$
$$= 368 - 294 = 74$$

As before, this SSB is divided into sums of squares from columns, rows, and interaction. The sum of squares due to columns may be computed as

$$SSC = \sum_{\text{columns}} \frac{T_{Cj}^2}{n_C} - \frac{T^2}{N}$$

$$= (16^2/8 + 28^2/8 + 40^2/8) - 84^2/24$$

$$= 330 - 294 = 36$$

Likewise, to compute the sum of squares due to rows,

$$SSR = \sum_{\text{rows}} \frac{T_{Rk}^2}{n_R} - \frac{T^2}{N}$$

$$= (12^2/6 + 18^2/6 + 30^2/6 + 24^2/6) - 84^2/24$$

$$= 324 - 294 = 30$$

As before the sum of squares due to interaction is part of *SSB*, the sum of squares between cells. To compute *SSI*, therefore, we take the original sum of squares between and subtract the sum of squares already used up by columns and rows to obtain

$$SSI = SSB - SSC - SSR$$

$$= 74 - 30 - 36 = 8$$

A Summary Table

We are now in a position to set up an analysis of variance summary table, as in Table 13-5. We see that the total amount of variance in the experiment—sum of squares between plus sum of squares within—is equal to 94. This sum of squares is based on 23 degrees of freedom (since there is a total of 24 scores in the entire experiment). This total sum of squares is broken up into sum of squares between the 12 cells and sum of squares within the 12 cells. There are 11 degrees of

TABLE 13-5. An ANOVA Summary Table for a Within-Subjects Design

Source	df	SS	MS	Obtained F	Crit F
Total	23	94			
Between	11	74			
Columns	2	36	18	13.5	5.14
Rows	3	30	10	6.00	3.49
Interaction	6	8	1.333	0.80	3.00
Within	12	20	1.667		

freedom between cells since there are 12 cells. Likewise, there are 12 degrees of freedom within cells—1 degree of freedom stemming from the 2 scores in each of the 12 cells. There are 3 degrees of freedom due to rows (since there are 4 rows), 2 degrees of freedom due to columns (since there are 3 columns) and finally $3 \times 2 = 6$ degrees of freedom due to the row by column interaction. All the sums of squares and degrees of freedom are shown in Table 13-5. In addition the relevant sums of squares have been divided by their appropriate degrees of freedom to produce corresponding mean squares.

Null and alternative hypotheses. The null and alternative hypotheses are the same as they were in the previous example. The null hypothesis states that sleep deprivation has no effect on reaction time, whereas the alternative hypothesis states that sleep deprivation does have some effect on reaction time.

An obtained F. Now how shall we test the null hypothesis? Since the ANOVA summary table (Table 13-5) looks very similar to the two-way ANOVA summary tables depicted in the previous chapter, we might be tempted to compute an F in the same way we did then. That is, our inclination might be to divide the mean square due to conditions (columns) by mean square within. However, such a procedure would be incorrect. As we have noted, this situation aligns itself conceptually with the within-subjects design in the previous example. That is, a test of the null hypothesis is still provided by an answer to the question: To what extent are subjects behaving consistently over conditions? Once again the appropriate error term that must emerge is the subject-by-condition (rows by columns) interaction. To the degree that this interaction is small, we tend to say that subjects are behaving consistently over conditions, and to the extent that this interaction is large, we say that subjects are not behaving consistently over conditions. Therefore, the interaction is still the entity that constitutes the denominator of the obtained F-ratio. The numerator, of course, remains the summary score of mean square due to conditions (columns). Mean square due to conditions is still the number representing the amount of variance among the three conditions.

Thus, the obtained $F(2,6)$ may be computed as

Obtained $F(2,6) = MSC/MSI = 18/1.33 = 13.53$

The criterion $F(2,6)$ is 5.14. Therefore, we reject the null hypothesis and conclude that the independent variable of sleep deprivation is really having an effect on the dependent variable of reaction time.

> 13.7 In a within-subjects design, the effect of conditions is *always* tested against the mean square due to interaction.

Other tests. We have previously alleged that the variance due to subjects (rows) is of typically no concern. Suppose, however, this issue *were* of interest for some reason. That is, suppose we wanted to test whether or not the population mean reaction times of the four subjects differed from one another. Is it possible to perform such a test? Indeed it is. The *F*-ratio for this test is formed by dividing the mean square due to subjects by the *mean square within cells*. Such a procedure yields an obtained $F(3,12)$ of 6.00 The criterion $F(3,12)$ equals 3.69; we can therefore declare that we have a significant effect of subjects. As usual such a conclusion means that if the four subject population means do not differ from one another, the probability of obtaining this extreme an $F(3,12)$ is less than 0.05.

In like fashion suppose we were interested in whether there is any "real" interaction between subjects and conditions. (Again, "real" refers to "in terms of populations means.") We test this question in an analogous way. That is, we divide the mean square due to interaction by the mean square within to get an obtained $F(6,12)$ of 0.80. Naturally, this is not a significant F (no F less than 1.0 can be significant), so we would conclude we do not have enough evidence to claim that there is any real interaction between subjects and conditions in the population.

> 13.8 Effects of subjects and interaction can be tested only when there are multiple observations per subject per condition. If such is the case, then these effects are tested against MSW (mean square within cells).

EXPECTED MEAN SQUARES

These last remarks have probably seemed mysterious, vague, and arbitrary. We have provided some intuitive rationalization for why we test conditions against interaction. However, we have provided no rationalization whatsoever for why we test subjects and subject-by-condition interaction against variance within cells. What determines what gets tested against what? The solution to this puzzle lies in what is referred to as *expected mean squares*.

Different Sources of Variance

Let's return to the experimental design shown in Table 13-4. Recall our discussion of how the mean square within a given cell should be conceptualized. This mean square represented the degree to which the given subject varied within himself or herself, which is perhaps the most basic form of variance that exists in an experiment. Since this within-a-given-subject variance is so basic, we will refer to it as σ_e^2 (or basic error variance).

Components of variance due to subjects (rows).

Now let's concern ourselves with variance across subjects, reflected by the degree to which the row totals differ from one another in Table 13-4. Let's consider what *causes* these row totals to differ from one another. There are actually two sources of variation in row totals. The first is obvious; it is the extent to which there are any inherent differences among the various subjects. That is, to the extent that I am different from you—because of our different genes or upbringing or experience or whatever—I would expect my score to differ from your score. However, any difference between my score and your score also stems in part from σ_e^2—the variance that exists due to the fact that you are not totally consistent within yourself, nor am I totally consistent within myself. To clarify this notion, imagine that there were actually no inherent difference whatsoever between you and me. Now if both our reaction times were measured in the same task, would we then expect our scores to be identical? No, of course we wouldn't. I might be having a slightly bad day and you might be having a slightly good day—therefore your score would be a bit better than mine, or vice versa. The point here is that the observed differences between our scores—which we would label as mean square due to subjects or *MSR*—would include a component of σ_e^2.

We can state this a bit more formally by saying that we *expect* the observed mean square due to subjects (*MSR*) to be composed of both inherent variance among subjects (which we shall call σ_s^2) and of variance within a subject (σ_e^2). Thus[1]

$$E(MSR) = \sigma_s^2 + \sigma_e^2 \tag{13.1}$$

Components of variance due to interaction.

The same argument can be made for the variance that we observe due to subject-by-condition interaction. Suppose there is no population interaction between you and me. That is, suppose a plot of our means across the various conditions were to produce parallel lines. Would we, in such an event, expect the *observed* interaction variance (*MSI*) to be zero? Once again the answer is no. Once again since we are not completely consistent within ourselves we would expect our observed means to jiggle around a little bit, which would be expected to produce some degree of interaction in the data. Therefore, the observed variance due to interaction, *MSI*, is composed

---

[1]Actually the expressions for the expected mean squares are not correct, because the various σ^2s are missing some coefficients. To be precise, the expected mean squares are:

$$E(MSW) = \sigma_e^2$$
$$E(MSR) = \sigma_e^2 + nJ\sigma_S^2$$
$$E(MSC) = \sigma_e^2 + n\sigma_I^2 + nK\sigma_C^2$$
$$E(MSI) = \sigma_e^2 + n\sigma_I^2$$

We leave out these coefficients in the text because they are of little interest and obscure the relevant points that we are trying to make.

both of any real interaction between subjects and conditions in the population (which we label as σ_I^2) plus a component of σ_e^2. Therefore,[1] the expectation of *MSI* is

$$E(MSI) = \sigma_I^2 + \sigma_e^2$$

Components of variance due to conditions.   Finally, we come to variance due to treatments or conditions, and the arguments are similar. We would expect the observed mean square due to conditions, *MSC*, to be composed both of any real variance caused by the independent variable (which we label as σ_C^2), plus a component of variance that is due to the subject by condition interaction, plus a component of variance due to σ_e^2. Thus,[1]

$$E(MSC) = \sigma_C^2 + \sigma_I^2 + \sigma_e^2$$

Expected Mean Squares as Part of a Summary Table

Much of the above may be summarized by including these expected mean squares as part of our analysis summary table, which has been done in Table 13-6. Looking at the expected mean squares in Table 13-6, we find it quite evident what should be tested against what. Let's begin by testing whether there is any effect due to conditions. We see that the expected mean square due to conditions is

$$E(MSC) = \sigma_C^2 + \sigma_I^2 + \sigma_e^2$$

The expected mean square due to subject-by-condition interaction is

$$E(MSI) = \sigma_I^2 + \sigma_e^2$$

Therefore—and here is the critical thing—*MSC* and *MSI* are estimates of the same population variance ($\sigma_I^2 + \sigma_e^2$) *if and only if* $\sigma_C^2 = 0$. Next of course we note

TABLE 13-6.   Expected Mean Squares as Part of an ANOVA Summary Table. (A within-subjects design with multiple observations per cell is assumed.)

Source	df	SS	MS	E(MS)
Total	$N - 1$	SST		
Between	$JK - 1$	SSB		
Columns	$J - 1$	SSC	MSC	$\sigma_e^2 + \sigma_I^2 + \sigma_C^2$
Rows	$K - 1$	SSR	MSR	$\sigma_e^2 + \sigma_S^2$
Interaction	$(J - 1)(K - 1)$	SSI	MSI	$\sigma_e^2 + \sigma_I^2$
Within	$N - J$	SSW	MSW	σ_e^2

[1]See footnote 1 page 406.

that $\sigma_C^2 = 0$ is simply another way of stating that the null hypothesis is true. Therefore, if H_0 is true, then MSC and MSI are estimates of the same variance, and MSC/MSI should be distributed as an F-ratio.

We can briefly run through the same arguments with respect to testing whether there are significant effects due to subjects and to subject-by-condition interaction. The expected mean square due to subject, MSR, is equal to $\sigma_S^2 + \sigma_e^2$. Therefore, if $\sigma_s^2 = 0$ (if there is actually no variance due to subjects in the population), then the expectation of MSR and MSW is the same thing (σ_e^2), and their ratio should be distributed as an F. As should be quite transparent at this point, the same argument holds true when testing mean square due to subject-by-condition interaction.

> 13.9 Any observed mean square is composed of variance from various sources. The expected mean square is essentially the sum of the variances from these sources.

FIXED VERSUS RANDOM EFFECTS

Many statisticians and statistics textbooks frame the issues we have just been describing in terms of what are referred to as fixed and random effects. "Effects" may be thought of simply as variance due to particular factors or independent variables. The difference between fixed and random effects lies in the degree to which the experimenter wishes to generalize about the effects. Let's consider an example of this.

The Effective Pain Reliever

To be more specific, let's suppose that we are doing an experiment to determine whether Anacin or Bufferin is more effective in relieving headaches. Notice that the independent variable in such an experiment would be "drugs," and the two levels of the independent variable would be Bufferin and Anacin. What sorts of conclusions would we want to make about the independent variable—drugs—as a whole? The answer is that we would want to make conclusions *only* about the two particular levels of the independent variable— Bufferin and Anacin— we have chosen. That is, an acceptable conclusion of our experiment would be "Anacin and Bufferin do (or do not) differ from one another."

In such a situation the effect of "drugs" would be treated as a *fixed effect*. A fixed effect is an effect that may be generalized only to the levels of the independent variable we have opted to use in our experiment. Most independent variables are treated as fixed effects.

In contrast, consider an independent variable that has many potential levels of which some have been chosen for use in the experiment. Any variation in the data due to such an independent variable would be treated as a *random effect*. The most common random effect is an effect due to *subjects,* which we described in previous sections. Note that we have treated subjects as a factor or an independent variable in our examples. Typically, we wish to be able to extend our conclusions beyond the particular levels of this factor (the particular subjects) that we happen to have chosen. When we treat a factor as a random effect, we wish to be able to generalize our results to *all possible levels of* the factor (that is, to all potential subjects). We would like to be able to say that any difference in pain relief between Anacin and Bufferin applies to all individuals in the population (that is, all potential levels of the "subject" factor). Likewise, in terms of the experiments we described earlier in this chapter, we would like to be able to conclude that all potential subjects show an increase in reaction time as a function of sleep deprivation.

> 13.10 The world is divided into fixed effects and random effects. An effect is fixed if we wish to generalize it only to the levels we have used in the experiment. An effect is random if we wish to generalize it to all possible levels in the population.

Other Random Effects

Although "subjects" is, as noted, one of the most common random effects used in experiments, it is by no means the *only* random effect ever used. To illustrate another random effect, suppose that the Polar Bear Publishing Company has just published a new spy novel entitled *The Tacoma Connection.* Because Polar Bear's marketing experts disagree as to whether the book should have a black cover or a white cover for optimal selling, the company decides to do an experiment. A random sample of 20 bookstores is selected from across the country. Half the copies of *The Tacoma Connection* in any given bookstore have black covers and the other half have white covers.

Notice now that *bookstores* is the random effect. That is, when analyzing the data, the Polar Bear Company will treat bookstores just as we have treated subjects in the past. Polar Bear will therefore obtain two scores—number of white books sold and number of black books sold—from each bookstore and do a "within-bookstores *t*-test" to assess whether or not there is any difference between black versus white covers in terms of the probability that a book will be sold.

Note that Polar Bear could have alternatively opted to use a "between-bookstores" design. Of the 20 bookstores, a random half might have been assigned all black copies of the new thriller and the other half might have been assigned all white copies. In this situation it would be appropriate to do a "between-bookstores" t-test. In either case, the independent variable with which we are concerned—cover color—would have but two levels—black and white. We would only be interested in generalizing the results of the experiment to these two particular levels—these two colors. Therefore, cover color would be treated as a fixed effect. On the other hand it is certainly desirable to be able to generalize the result to all bookstores. The 20 bookstores chosen for the experiment are simply a random sample of all possible bookstores in the country or in the world. Therefore, in this example, "bookstores" is treated as a random effect.

HIGHER-ORDER DESIGNS

In Chapter 12 we discussed the concepts of higher-order designs and their associated higher-order interactions. A quite common design incorporates two factors—that is, two fixed effects—that are completely within-subjects. This, of course, would translate into a three-way $J \times K \times n$ design where n is the number of subjects.

Analysis of this Kind of Design

Table 13-7 depicts an ANOVA summary table for this sort of design along with an enumeration of what gets tested against what. As can be seen, the rule is quite simple: any effect gets tested against the interaction of that effect with subjects. This should not be surprising. We have already seen that what is of interest in a within-subjects design is the consistency of an effect across subjects. Consistency, in turn, is reflected by the magnitude of the corresponding interaction. The only thing even slightly novel is that the two-way ($A \times B$) interaction is tested against the three-way ($A \times B \times S$) interaction. As we saw in the last chapter, this interaction may be seen as reflecting the degree to which the $A \times B$ interaction is consistent across subjects (levels of S). The more consistency, the smaller the interaction.

TABLE 13-7.   ANOVA Summary Table for a $J \times K \times n$ Design

Source	df	SS	MS	Test
A	$J - 1$	SSA	MSA	$MSA/MS(A \times S)$
B	$K - 1$	SSB	MSB	$MSB/MS(B \times S)$
S(ubjects)	$n - 1$	SSS	MSS	
$A \times S$	$(J - 1) \times (n - 1)$	$SS(A \times S)$	$MS(A \times S)$	
$B \times S$	$(K - 1) \times (n - 1)$	$SS(B \times S)$	$MS(B \times S)$	
$A \times B$	$(J - 1) \times (K - 1)$	$SS(A \times B)$	$MS(A \times B)$	$MS(A \times B)/MS(A \times B \times S)$
$A \times B \times S$	$(J - 1) \times (K - 1) \times (n - 1)$	$SS(A \times B \times S)$	$MS(A \times B \times S)$	

Still Higher

These same principles apply to any level within-subjects design. For instance a factorial design with three fixed factors would boil down to an $A \times B \times C \times n$ design. The effects of A, B, and C would be tested against $A \times S$, $B \times S$, and $C \times S$; the effects of $A \times B$, $A \times C$, and $B \times C$ would be tested against $A \times B \times S$, $A \times C \times S$, and $B \times C \times S$; and finally the effect of $A \times B \times C$ would be tested against $A \times B \times C \times S$. The same interpretational difficulties discussed in the last chapter apply here as well of course.

> 13.11 Any effect in a within-subjects design is tested against the interaction of that effect with subjects.

More Complicated Designs

Sometimes multifactor designs incorporate some between-subjects and some within-subjects factors. For example if one wished to incorporate sex as a factor, one would have no choice but to incorporate it as a between-subjects factor. The analysis of these sorts of designs is beyond the scope of this book, but the bibliography will provide appropriate references.

SUMMARY

In this chapter we have provided a systematic discussion of within-subjects designs, and the following issues were treated:

1. A within-subjects design is structurally quite similar to a two-way analysis of variance design. In both cases two factors are involved. In the case of a within-subjects design one of the factors is *subjects*, and each subject corresponds to one level of this factor. Computationally analysis of a within-subjects design is very similar to analysis of a two-factor between-subjects design. In fact computation of the various sums of squares is identical.

2. Here, however, the similarity ends. We have seen that hypothesis testing in a within-subjects design involves somewhat different concepts than does hypothesis testing in a between-subjects design. In both cases the null and alternative hypotheses are the same. As always the null hypothesis is that the independent variable has no effect on the dependent variable, whereas the alternative hypothesis states that the indepen-

dent variable does have an effect on the dependent variable. In a between-subjects design we answer this question by assessing the amount of variance between the various conditions relative to the amount of variance within a given condition—that is, relative to the amount of variance due to subjects. Therefore, in a between-subjects design the degree to which individual subjects differ from one another is very important. It is the error term.

In contrast, when considering a within-subjects design, we deal with the question of whether subjects behave consistently across the various conditions. In a within-subjects design the degree to which individual subjects differ from one another is of no concern to us. What *is* of concern is the *interaction* between subjects and the dependent variable. The greater this interaction, the less consistently subjects behave over conditions. Therefore, the interaction forms the error term and the mean square interaction becomes the denominator of the *F*-ratio that we used to test the null hypothesis.

3. We discussed two (actually quite similar) types of within-subjects designs. The first type of within-subjects design was relatively simple: each subject contributed only a single score to each condition (that is, to each cell). We saw, however, that it is perfectly reasonable for a subject to contribute multiple observations to a given cell. In such a multiple-observation situation the analysis of variance becomes a bit more complex. Specifically we now find ourselves with a sum of squares within which did not exist in the single-observation-per-cell situation. We emphasized—and we reemphasize here—that this mean square within—although computationally identical to mean squares within encountered before—is conceptually quite different. The mean square within that we described before represented variance across subjects—it reflected the degree to which individual subjects differed from one another. However, *this* mean square within represents the degree to which a given subject differs within himself, herself, or itself from one time to another. It represents, therefore, a quite different component of variance than it did in the past.

4. The notion of *expected mean square* may be used to assess what terms should be tested against what other terms. In particular any given mean square that is computed in an experiment is expected to be the sum of various components of variance. We saw that to test any component of variance, σ_x^2, it was necessary to find two mean squares that differed from one another only in that the expectations of their mean squares differed from one another by σ_x^2.

5. One way of looking at ANOVA situations is in terms of fixed versus random effects. Fixed effects constitute what we ordinarily think of as independent variables. We choose particular levels of those independent variables, and we are typically interested in generalizing only to the

particular levels of the independent variable that we have chosen. On the other hand random effects are independent variables from which we randomly choose a number of levels, but we wish to be able to generalize to all possible levels of that independent variable. "Subjects" is the most common random effect, but there are many other possible ones.

6. The concept of a within-subjects design can be expected to be the case in which there is more than one fixed factor. The general rule is that any effect is tested against its interaction with subjects.

PROBLEMS

1. A social psychologist is interested in the amount of stage fright people have as a function of the number of people in the audience. Four subjects are each asked to imagine that they will have to recite a poem in front of one, five, or 15 people. They are asked to rate how frightened they would be on a scale from 0 (not frightened at all) to 7 (scared to death).

    The data are as follows:

	Audience Size		
Subject	1	5	15
1	3	6	5
2	1	5	6
3	3	6	7
4	1	3	5

    1. Plot the means for the three conditions along with their 95% confidence intervals.
    2. Does the audience size have a significant effect on amount of stage fright?

2. A clinical psychologist has developed a new type of training to relieve anxiety in overanxious patients. To test the treatment, she performs two experiments. In the first experiment the 10 patients simply rate their anxiety (on a scale from 0 "not anxious to all" to 7 "extremely anxious") each day for five days. She then begins the new treatment (the second experiment), and the patients again rate their anxiety for five days running.

    The data are as follows:

Experiment 1 (no treatment)

Patient	Day 1	Day 2	Day 3	Day 4	Day 5
1	6	7	7	6	7
2	7	6	6	5	7
3	5	6	7	6	7
4	5	6	7	6	6
5	5	5	5	5	5
6	6	5	4	7	6
7	5	6	6	7	6
8	6	7	5	6	5
9	6	6	7	5	4
10	7	6	7	6	6

Experiment 2 (treatment)

Patient	Day 1	Day 2	Day 3	Day 4	Day 5
1	6	7	6	6	4
2	7	7	6	4	2
3	6	7	6	3	4
4	7	6	7	6	5
5	6	6	5	4	3
6	7	6	5	4	3
7	6	7	7	6	5
8	5	3	2	6	4
9	6	7	3	1	1
10	3	4	5	2	3

1. For each experiment, plot the mean anxiety rating over days with the 95% confidence intervals.
2. Perform an analysis of variance on each of the two sets of data.
3. Would you conclude that the treatment was effective?

3. A new technique called "sterilestud" is invented in an attempt to eliminate mosquitoes. The technique involves releasing sterile male mosquitoes who will mate with females but produce no offspring. For a test of the technique,

five different national forests are divided into three sections each. In the first section nothing is done. In the second section DDT is sprayed. In the third section the sterilestud technique is used. The sections are then examined for presence of mosquitoes. The data (mosquitoes per square yard) are as follows:

Forest	No Treatment	DDT	Sterilestud
1	614	512	123
2	320	300	250
3	502	500	313
4	750	600	430

   1. Plot the means for the three techniques along with 95% confidence intervals.

   2. Are there significant differences among the three techniques?

4. A drug called "remembrin" is invented to prevent forgetting. Remembrin is administered to five subjects, who are then presented with a 30-word list: Of the 30 words, 10 are countries, 10 are sports, and 10 are foods. Immediately following the list, the subjects are asked to remember as many countries as possible from the list. After 24 hours, subjects are asked to remember sports; and after a week, subjects are asked to remember foods. The data (number of words recalled) are as follows:

Subject	Immediate	24 hours	1 week
1	6	7	3
2	4	3	5
3	7	7	8
4	3	2	0
5	5	4	6

   1. Plot the three means along with their 95% confidence intervals.

   2. Perform an analysis of variance on the data.

   3. Would you conclude that remembrin works? Name at least two things wrong with this experiment.

5. A physician has a hypothesis that the probability of catching cold varies with season of the year. She keeps track of five patients in terms of how many colds they have during winter, spring, summer, and fall. She keeps records for three years. Thus for each patient she has a record of the number of colds caught in each of the four seasons for three years. The data are as follows:

Number of Colds

Patient	Winter			Spring			Summer			Fall		
Bill	3	1	2	1	0	1	3	3	2	0	0	0
Sam	1	1	2	0	1	1	3	4	3	1	1	2
Fred	1	1	1	0	3	0	2	1	2	1	0	0
Ralph	3	4	3	2	2	1	5	2	2	1	2	3
Irving	2	2	3	1	2	1	2	1	0	1	0	1

(*Note:* in each cell, three numbers represent the number of colds for that patient in that season for the three separate years.)

1. Compute a mean for each cell.
2. Plot the mean number of colds for the four seasons along with 95% confidence intervals.
3. Perform a complete analysis of variance on the data. Is there indeed an effect of season? Additionally determine whether there is an effect of patient and season by patient interaction.

6. Consider Chapter 5, Problems 3, 4, 5, and 8. Reevaluate these problems using analysis of variance rather than a sign test.

7. Consider Chapter 10, Problem 6. Redo this problem using an analysis of variance rather than a *t*-test.

8. Consider the date from Problem 5 in this chapter. For each patient graph the mean number of colds for the four seasons along with the 95% confidence interval.

9. An experimenter is interested in how long it takes to learn various types of verbal material. He chooses lists of three types of verbal material: English words, Russian words, and nonsense syllables. Each of the four subjects gets all three types of lists, and the number of trials taken to learn each type of list is measured for each subject. The following data are obtained:

Type of list

Subject	English	Russian	Nonsense
1	4	9	8
2	5	5	8
3	7	8	9
4	8	10	15

1. What is the between sum of squares?
2. What are *SS* (subjects), *SS* (conditions), and *SS* (interaction)?
3. Arrange these in an ANOVA table.

4. Is there a significant effect of condition?
5. Suppose that you got rid of subject variance in the above. That is, you carried out the following procedure:

Subject 1: Add 3 points to each of the scores
Subject 2: Add 6 points to each of the scores.
Subject 3: Leave the scores alone.
Subject 4: Subtract 9 points from each of the scores.

6. What would be the *SS* (subjects), *SS* (conditions), and *SS* (interaction) for this new set of data?
7. Make an ANOVA table for these data. Is there a significant effect of condition?

10. Joe Shablotnik is interested in whether the typing rates of his two secretaries, Ralph and Shirley, differ for three kinds of to-be-typed material: letters, memos, and book chapters. Hence he gets a sample of typing for each of the three kinds of material from each secretary. The data are as follows (cell entries are typing rates in words per minute):

	Memos	Letters	Chapters
Ralph	60	75	50
Shirley	30	10	30

1. Can Joe conclude that different types of material are typed at different rates?
2. Suppose within-person variance (σ^2) is known to be 25. Can you conclude that there is a difference between Ralph and Shirley in terms of typing rate?

11. A sociologist is studying suicide rates in various U.S. cities. *Out of all cities,* she randomly selects Seattle and Boston (just as a psychologist might randomly select two subjects). For each city she determines the number of suicides for each season of the year (winter, spring, summer, fall) for the two years 1976 and 1977. The data are as follows.

Number of Suicides
Season

	Winter		Spring		Summer		Fall		
Seattle	18 16	$T_{11} = 34$	4 4	$T_{21} = 8$	5 7	$T_{31} = 12$	10 14	$T_{41} = 24$	$T_{R1} = 78$
Boston	10 8	$T_{12} = 18$	7 7	$T_{22} = 14$	5 9	$T_{32} = 14$	11 11	$T_{42} = 22$	$T_{R2} = 68$
	$T_{C1} = 52$		$T_{C2} = 22$		$T_{C3} = 26$		$T_{C4} = 46$		$T = 146$

1. Test whether season of the year affects number of suicides in U.S.

cities. Additionally test whether there are differences among U.S. cities in terms of suicide rates and whether there is an interaction between cities and seasons.

2. Suppose the sociologist decides that she'll be satisfied if her conclusions extend *only* to Seattle and Boston. Test whether there is an effect of season, whether Boston and Seattle differ from one another, and whether there is an interaction between season and city.

12. An experiment is done to test the effects of alcohol consumption on reaction time. The experiment involves three conditions. A subject in condition 1 is given a shot of water; a subject in condition 2 is given a shot of beer; and a subject in condition 3 is given a shot of gin. Reaction times (in tenths of a second) are then measured. Three subjects partake twice in each of the three conditions. The data are as follows:

	water	beer	gin	
Subject 1	1 2	2 1	3 4	$T_{R1} = 13$
Subject 2	2 2	1 3	4 4	$T_{R2} = 16$
Subject 3	10 12	10 14	14 15	$T_{R3} = 75$
	$T_{c1} = 29$	$T_{c2} = 31$	$T_{c3} = 44$	$T = 104$

1. Are there significant effects of alcohol condition, subject and subject by condition interaction? Arrange things in ANOVA table.

2. Do subjects 1 and 2 differ from one another?

Correlation, Prediction, and Linear Relationships

14

In the past chapters we have been dealing with increasingly complex experimental designs that have shared the characteristic of involving examination of relationships among variables. Specifically we have consistently been addressing the question of whether changes in one or more independent variables lead to associated changes in some dependent variable.

We noted that there is an asymmetry in the relationship between independent and dependent variables, in the sense that we *control* independent variables but *measure* dependent variables. We have also been working under an implicit assumption regarding the direction of causality, which is that changes in the independent variable cause changes in the dependent variable, and not the other way around.

Actually, however, this sort of independent variable/dependent variable arrangement is a special case of a much larger class of research designs in which the general *association* or *correlation* between two variables is the issue of interest. You probably have some intuitive idea of what it means for two variables to be correlated. Height is correlated with weight, poverty is correlated with crime, athletic ability is correlated with physical fitness, and so on. Formally the idea of correlation can best be considered in terms of *prediction*. Let's examine the idea of an experiment as an exercise in prediction.

THE CONCEPT OF PREDICTION

To provide a flavor for statistical analysis as predictive ability, consider a simple experiment (similar to one that we have described in the past) in which we wish to investigate the effects of alcohol consumption on simple reaction time. Let's suppose that $N = 50$; subjects are randomly divided into 5 groups of $n = 10$ subjects per group. The five groups are given varying amounts of alcohol so as to bring their blood alcohol content to certain levels. Subjects in Group 1 are given no alcohol at all; thus subjects in Group 1 have a blood alcohol level of zero. Subjects in Groups 2–5 are given sufficient amounts of alcohol to raise their blood alcohol levels to 0.01%, 0.02%, 0.03%, and 0.04%, respectively. (In most states 0.10% blood alcohol constitutes legal drunkenness.) In the experiment each of the 50 subjects is given a series of reaction time trials in which his or her mean reaction time to the onset of a light is measured.

Suppose that this experiment produces the data that are shown in Table 14-1. We have also calculated that MSW is 30.81. This of course means that our best estimate of σ, the standard deviation of the population distributions that the scores in each of the 5 groups comes from is $\sqrt{MSW} = \sqrt{30.81} = 5.55$. For reasons that will become apparent presently, the standard deviation of the entire set of 50 scores has also been computed. This standard deviation (actually our best estimate of the standard deviation of the population that all $N = 50$ scores comes from) is 15.21. Finally, the grand mean, M, of all 50 scores is 213.44.

Let's now return to the issue of prediction. Suppose that you are interested in Ralph, a randomly selected individual from the population. I tell you that Ralph will be drinking such that his blood alcohol level will be somewhere between 0 and 0.04%. However, you don't know the exact value. What would you expect Ralph's reaction time to be?

In this situation you have very little information, and your best guess would be that Ralph's reaction time will be 213.44, the grand mean of the $N = 50$ scores in the experiment. Having made this best guess, how could you assess your confidence in it? A moment's thought should convince you that your confidence should depend on the rather large standard deviation of 15.21 computed over all

TABLE 14-1. Reaction Time Scores (in milliseconds) for the 50 Subjects in the Alcohol Study

Group 1 (0%)	Group 2 (0.01%)	Group 3 (0.02%)	Group 4 (0.03%)	Group 5 (0.04%)
192	205	208	231	235
194	198	209	228	230
189	201	220	216	233
178	208	216	220	228
193	216	221	225	237
201	203	210	226	230
199	207	215	220	241
198	200	217	218	242
196	198	208	223	225
190	205	210	229	230
$M_1 = 193.0$	$M_2 = 204.1$	$M_3 = 213.4$	$M_4 = 223.6$	$M_5 = 233.1$

$$SSW = 1386.6$$
$$MSW = 1386.6/45 = 30.81$$

Standard deviation over all $N = 50$ scores is:
$$\sqrt{(SST/dfT)} = \sqrt{11334.42/49} = \sqrt{231.31} = 15.21$$

50 scores. Since you don't know how much alcohol Ralph is going to have, his reaction time could be quite short (if he happened to be a fast subject and had no alcohol) or quite long (if he happened to be a slow subject and had a large amount of alcohol) or anything between.

Suppose on the other hand that you were provided the additional information that Ralph's blood alcohol level was 0.2%. You would then be able to use this additional information quite profitably because you know that under such circumstances subjects have a mean reaction time estimated to be $M_2 = 204.1$. Thus, 204.1 would be your best estimate of what Ralph's reaction time would be under these circumstances. Furthermore, your confidence in this prediction would be based not on the larger standard deviation of all $N = 50$ scores, but rather on *est* $\sigma = 5.50$, the estimate of the standard deviation of scores within a given condition. In short having information about the value of the independent variable (Ralph's blood alcohol level in this example) substantially increases your ability to predict the dependent variable which, in this example, is Ralph's reaction time.

> 14.1 If there is some relationship between an independent and a dependent variable, and we know what that relationship is, this increases our ability to predict the state of the dependent variable assuming that we know the state of the independent variable.

Prediction and Conditional Probability

We hope this line of reasoning strikes a familiar chord. Many chapters ago we raised similar considerations with respect to the relationship between unconditional and conditional probability. Take a moment to go back and reread the section in Chapter 1 on these issues. Remember the example involving the relationship between doing the reading for an exam and subsequently passing the exam. We saw that in the absence of additional information the unconditional probability of passing an exam was 0.75. However, if we had knowledge about whether an individual had or had not done the reading for the exam, we could be more precise in our evaluation of the person's chance of passing. In particular if the person had done the reading, then the probability that the person would pass was $p(P|R) = 0.83$, whereas if the person had not done the reading, then that individual's probability of passing was reduced to $p(P|\overline{R}) = 0.67$. We saw in this situation that having information about the state of one variable (having done versus not having done the reading) improved our predictive ability with respect to another variable—namely the probability of passing the exam.

Independence and the Null Hypothesis

Extending this analogy just a bit further, we were also concerned in Chapter 1 with the issue of independence. Two variables, A and B, were said to be independent if the state of one had no bearing on the state of the other—that is, if $p(A|B) = p(A|\overline{B}) = p(A)$. Such a state of affairs is equivalent to there being no effect of A on B—equivalent that is, to the null hypothesis being true with respect to A and B. Thus, only if A and B are not independent are we in a position to use one to predict the other—just as, only is there is an effect of an independent variable on a dependent variable are we in a position to use the state of the independent variable to predict the state of the dependent variable.

CORRELATIONAL VERSUS EXPERIMENTAL STUDIES

We now introduce a major departure from the sort of study that has constituted the focus of the book thus far. This departure has to do with the way independent and dependent variables are defined. As noted above, we have seen that an experimenter selects—that is, manipulates—the value of the independent variable. The dependent variable on the other hand is not manipulated, but simply measured. This sort of study is referred to as an *experimental study*.

In a *correlational study,* in contrast, no variable is manipulated; rather all variables are measured. As we shall see, this difference between the two types of studies has rather strong consequences in terms of how the data are to be interpreted. To acquire a flavor for the differences between a correlational and an experimental study, consider the following pair of examples. Imagine that we are educational psychologists, and we want to study the relationship between the number of hours per week that a student spends reading for a course and that student's average hour exam score in the class. Our expectation is that more hours spent reading will lead to higher exam scores.

An Experimental Study

To investigate this issue, we randomly assign each student in the class to one of three conditions. Students in Condition 1 are required to spend one hour a week reading for the course. Students in Conditions 2 and 3 are required to spend, respectively, 2 and 3 hours a week on the reading. At the end of the course the mean hour exam score is computed for each of the three conditions. Let's suppose that this experiment produces the data shown in Figure 14-1. As indicated, there seems to be a fairly substantial increase in exam score as a function of number of hours spent reading. Let's suppose that the null hypothesis of no difference among the means is rejected. We would conclude from the results of this study that increasing amounts of reading time causes a student to obtain higher exam scores.

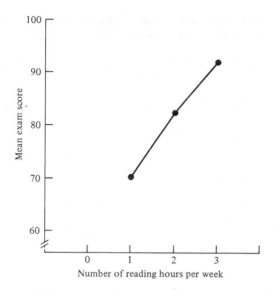

FIGURE 14-1. Hypothetical results from office hours experimental study. Average exam score increases as a function of number of reading hours.

A Correlational Study

Now suppose we were to investigate the relationship between reading time and exam scores in a somewhat different way. Rather than assigning students to conditions, we wait until the course is over and then conduct an interview with each student. In each interview we ask two questions. First we ask how many hours per week the student spent reading for the course, and second we ask what the student's average exam score was. Let's suppose that there are 15 students in this class, and the results of these 15 interviews are shown in Table 14-2. As can be seen, the first student, Annabelle Asgaard, spent an average of 1.5 hours per week reading and her average exam score was 82. Likewise, student number two, Brentano Bravado, only did about a half-hour of reading per week and ended up with an average exam score of 68. And so on.

Scatterplots.   We have taken the data from Table 14-2 and transformed them into the graph in Figure 14-2. Notice that Figure 14-2 is similar to Figure 14-1 in the sense that the ordinate represents exam score and the abscissa represents number of reading hours per week. In Figure 14-2, however, a single point has been plotted for each individual student. This sort of plot is called a *scatterplot* because of the manner in which the individual data points are scattered across the graph.

What would we conclude from the data shown in Table 14-1 and Figure 14-2? Once again there certainly seems to be a positive relationship between amount of reading time and exam scores. It seems evident from the scatterplot

TABLE 14-2. Fifteen Students in a Course (For each student is listed the number of hours per week spent reading for the course and average exam score.)

Student	Number of Hours Spent Reading	Average Exam Score
Annabelle Asgaard	1.5	82
Brentano Bravado	0.5	68
Engelburt Eech	1.25	80
Francine Fox	3.00	92
Gilda Gofer	2.0	76
Henry Haldaman	1.25	78
Geoffrey R. Loftus	1.00	68
Myrna Mezzanine	1.25	85
Nutly Notly	2.75	91
Quasar Quigley	2.25	79
Sophia Spacek	1.75	75
Ravenna Ravine	3.00	88
Teriffa Trapeze	2.5	85
"Wailin'" Willy Whiskey	0.25	67
Zerma Zaminsky	2.0	80

that the students who tend to do a lot of reading also tend to get high exam scores. In contrast, students who do relatively little reading seem to end up with relatively low exam scores. So do we get the same information from this correlational study as we did from the experimental study described earlier? The answer is that we do not.

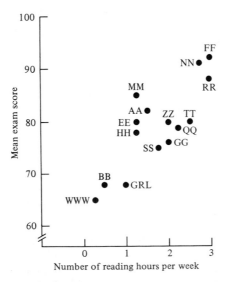

FIGURE 14-2. Hypothetical results from reading-time correlational study. Each point represents the average exam score and the amount of reading time for a particular student. The student's initials identify his or her data point.

Correlation Is Not Causality

Recall that following our experimental study we were able to conclude that more reading time caused higher exam scores. However, this sort of causal conclusion would not be justified following a correlational study. Rather, we could conclude that more reading time *may* cause higher exam scores, but there would be at least two other explanations for the data.

First it may be that causality goes in the opposite direction—earning higher exam scores may prompt the student to read more. That is, perhaps some students tended to do quite well on the exams simply because they were smart or lucky or whatever. Reinforced by and delighted with their high grades, these individuals were motivated to plunge into the reading with great gusto. On the other hand students who did poorly on exams tended in their disgust to avoid the textbook as much as possible and did relatively little reading. This is one possible alternative explanation for the data shown in Figure 14-2.

The second alternative explanation is that some external factor is causing the variation across students in both amount of reading and exam grades. For example consider the student's motivation. It is quite possible that highly motivated students tend to read more and also tend to study harder so as to get higher exam scores. Conversely, low-motivated students may neither have the incentive to read the book nor have the incentive to do anything else that is required to get a good exam score.

Digression 14.1

Other Examples of Correlated Variables

To amplify on the point that correlation does not imply causation, let's enumerate a few other pairs of variables that are correlated but which are (probably) not causally related.

Are Ice Cream Consumers Guilty of Genocide?

Over days of the year the number of ice cream cones eaten in New York City turns out to be correlated with the number of deaths in Bombay. That is, on days when many ice cream cones are eaten in New York there tend to be more deaths in Bombay than on days when fewer ice cream cones are eaten in New York. But does this mean that one should feel guilty each time one eats an ice cream cone in New York? Of course not. Eating ice cream in New York does not *cause* the demise of some poor individual in Bombay. Rather, there is a third factor—temperature—causing both. That is, days that tend to be hot in New York (days in the middle of summer) tend also to be hot in Bombay. People tend to eat ice cream cones on hot days in New York, and people tend to die during hot days in Bombay.

Are Call Girls Seeking Salvation?

Over cities the number of ministers is correlated with the number of prostitutes—that is, cities with many prostitutes also tend to have many ministers. Once again, however, there is probably not a causal relationship between these two variables. Rather, there is some third variable such as city size mediating both the number of ministers and the number of prostitutes. For example a very big city like Chicago would have a large number of ministers and also a large number of prostitutes. In contrast, a relatively small city such as Tacoma would have a smaller number of ministers as well as a smaller number of prostitutes.

Control versus measurement.   As we have noted, an experimental study is one in which we control one variable (the independent variable) and measure the other (the dependent variable). In a correlational study on the other hand we do not control anything. We just measure both the variables in which we are interested. The critical difference between an experimental and a correlational study is encapsulated in the idea of *control* in conjunction with the idea of *random assignment*. In an experimental study we (that is, the experimenters) choose (that is, control)—what condition any given subject is going to be in. In our example this means that we assign a given student to the one-hour, the two-hour, or the three-hour reading condition, and the student is *constrained* to participate in that particular condition. Furthermore, since students are randomly assigned to the various conditions, there is presumably no other variable that distinguishes the students in the various conditions aside from the amount of reading that they do. For instance there is no reason to expect that students in the three-hour condition would be systematically more motivated or more intelligent or in any other way distinguishable from students in the one-hour condition. Following an experimental study, we are therefore able to conclude that any variation in the dependent variable is caused by variation in the independent variable.

As we have seen, however, nothing is controlled in a correlational study. Therefore, it is not necessarily true that students who do a lot of reading differ from students who do relatively little reading *only* in terms of amount of reading. Since the students themselves have the ability to choose the amount of reading they do, it is perfectly possible that variation in reading time is linked to other factors such as variation in motivation or intelligence. Thus when we find a relationship in a correlational study, we are not in a position to determine what is causing what.

14.2 In an experimental study we control the independent variable and measure the dependent variable. In a correlational study we measure all variables. It is possible to infer causality from the results of an experimental study, but not from the results of a correlational study.

Correlational studies are quite commonly performed because it is often not possible to control the variables in which we are interested. For example, the hypothetical reading time experiment described above would probably never actually be done, because it is not very ethical or polite to control the amount of weekly reading that a student does. Similar situations come readily to mind. Suppose, for example, that we were sociologists interested in the relationship between the burglary rate in a given city and the number of door locks purchased in that city. We are not really in a position to control either the burglary rate or the number of locks purchased, so the best that we can do is to look at the correlation between the two variables. The same kind of constraint applies to a physician interested in studying the relationship between temperature on a given day and the number of people having heart attacks on that day. The physician cannot randomly assign some days to a high temperature condition and other days to a low temperature condition. In fact there are so many uncontrollable variables running amok in the world that experimental studies are probably the exception rather than the rule in most research.

LINEAR RELATIONSHIPS

Let's now return to examining the nature of correlational studies. As we have seen, two variables are correlated if and only if the state of each variable is in some way predictable from the state of the other. However, the fact that two variables are correlated does not imply anything about the specific nature of the function that relates them.

It is often the case, though, that a *linear* relationship is postulated between two variables, X and Y. The existence of a linear relationship implies two things.

1. Both X and Y are variables whose levels can be represented by numbers that form at least an interval scale. Thus for example the variable "sex" would *not* be such a variable since its two levels, "male" and "female," are not numbers, but rather qualitative attributes. Variables such as "blood level of alcohol" and "reaction time" would qualify, since their levels can be represented by numbers such as "0.02" and "205," respectively.

2. The underlying relationship between X and Y may be described by the

equation

$$Y = bX + a$$

where b and a are constants representing the slope and the Y-intercept, respectively, of the function. If we believed, for example, that reaction time (RT) and blood alcohol level (L) were linearly related, then they would similarly be related by the equation

$$RT = bL + a \tag{14.1}$$

> 14.3 If a linear relationship exists between two variables, then predicting the value of one given the value of the other consists of plugging an X-value into a linear equation of the form
>
> $$Y = bX + a$$
>
> and noting the resulting—that is, the predicted—Y-value.

There are two reasons why a linear relationship is often of interest. First due to the physical structure of the world, many pairs of variables *are* linearly related. For instance, the time it takes to travel between two points at a constant speed is a linear function of the distance to be traversed. Second in attempting to construct a simple and elegant theory of some phenomenon, it often makes sense, at least as a first attempt, to *assume* the simplest possible relationship among variables, and a linear relationship is indeed a simple one.

In this chapter we shall be concentrating on linear relationships. Assuming that a relationship between two variables is linear if it exists, then we can enumerate three possible subtypes of relationships that two variables, X and Y, can bear on one another.

Positive Relationship

Two variables are said to be positively related when big values of Y go with big values of X and small values of Y go with small values of X. For example, height and weight are positively related over people. People who are tall tend to weigh a lot and people who are short tend to weigh relatively little, as indicated in part *a* of Figure 14-3.

Negative Relationship

Two variables are negatively correlated when big values of Y are associated with small values of X, and vice versa. For example, a golf professional's average golf score is negatively related to the amount of money the professional

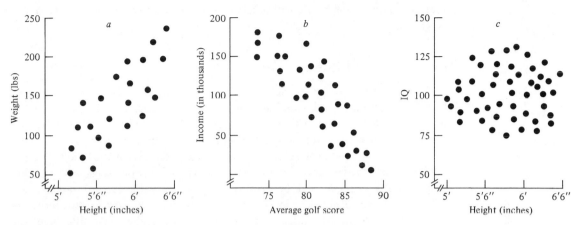

FIGURE 14-3.   Examples of pairs of variables that are (a) positively correlated, (b) negatively correlated, or (c) uncorrelated.

makes—those with low golf scores make more money golfing than do those with high scores, as indicated in part *b* of Figure 14-3.

No Relationship

Finally if there is no relationship between two variables, they are said to be uncorrelated. For example, height and IQ are uncorrelated across people. Tall people tend to be neither smarter nor dumber than short people, as depicted in part *c* of Figure 14-3.

> 14.4 When we are concerned with linear prediction, we view two variables, X and Y, as being either positively correlated, negatively correlated, or uncorrelated.

MEASURES OF LINEAR RELATIONSHIPS: REGRESSION LINES AND THE CORRELATION COEFFICIENT

So far we have dealt with correlation and linear relationships in a general sense. In this section we will discuss them from a quantitative point of view and deal with two major issues: how to construct a best-fitting straight line (called a *regression line*) through a scatterplot and how to assess the degree to which the underlying population relationship between X and Y may be characterized as a straight line.

Monday Night Rituals

Suppose that we are sociologists interested in the relationship between leisure-time activities and drinking behavior among American males. As part of our investigation we would like to examine the question of whether there is a relationship between watching football games and drinking beer. To deal with this question, we randomly select six males from the U.S. population. We ask each individual in the sample how many football games per season he watches and how many beers a week he drinks. Suppose that this study produces data depicted as a scatterplot in Figure 14-4. A glance at this scatterplot indicates that the two variables do indeed seem to be related in the sense that men who tend to watch large numbers of football games also tend to drink large quantities of beer. Let's see how we can generate some quantitative descriptors of this relationship.

A Regression Line

Recall our earlier discussion of linear relationships. There we pointed out that a linear relationship between two variables, X and Y, is characterized by an

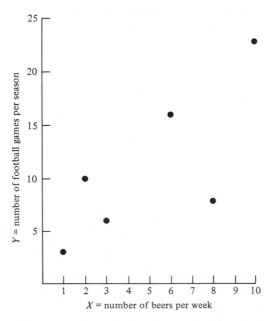

FIGURE 14-4. Scatterplot over people representing the correlation between number of football games watched per season and number of beers drunk per week.

equation of the form:

$$Y = bX + a$$

where a and b are parameters of the equation.

Note that any unique pair of values of the parameters a and b will produce a unique straight line. We now make the following assertion: given any scatterplot relating X and Y (such as the one in Figure 14-4) there is *one* best line—that is, one particular a and one particular b—that best fits that scatterplot. In a moment we shall defend that assertion and describe how the best a and b are arrived at. For now, examine Figure 14-5, which shows what will turn out to be the best-fitting straight line through the scatterplot of Figure 14-4.

Regression as prediction.   Before describing where the best-fitting line of Figure 14-5 comes from, we should emphasize that this line may be viewed as the line for most accurately predicting Y from X—in this case, number of football games from number of beers. That is, suppose a new man were to be selected at random who is known to drink an average of nine beers per week. How many football games would he be predicted to watch? To answer this question, we would simply plug the value of $X = 9$ into the best-fitting regression equation to come out with the value of 17 football games, as depicted in Figure 14-6.

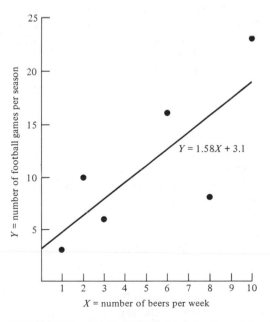

FIGURE 14-5.   The best-fitting regression line relating football games to beers.

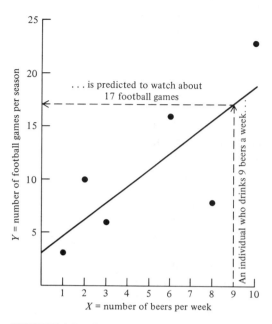

FIGURE 14-6. Regression as prediction. From our best-fitting regression line we can predict a Y-value given any arbitrary X-value.

Prediction and prediction errors. We will now deal with the questions: Where does the best-fitting regression line come from? How is it computed? By what criteria is it "best"? We begin by creating some more precise notation, which is illustrated in Figure 14-7. Each data point in Figure 14-7 is labeled as (X_i, Y_i). Now consider the regression line. Each of the X_is has associated with it some *predicted* Y_i score which we label Y_i'. Finally, for each data point there is an *error* which is the difference between the actual observed point and the predicted point. This error is referred to as $(Y_i - Y_i')$.

Good and bad regression lines. Now let's return to the question of where the regression line comes from. Intuitively there are some regression lines that seem to be *good* regression lines and other regression lines that seem to be *bad* regression lines. Figure 14-8 shows some examples of regression lines that vary in their goodness. For instance, panel *a* represents a relatively good regression line, whereas panels *b* through *e* represent relatively bad regression lines. Panel *f* represents a horrible regression line.

What distinguishes good regression lines from bad regression lines? The answer is that the errors—the $(Y_i - Y_i')$ scores—are in general big around the bad regression lines and smaller around the good regression lines. This observation provides a clue for how to generate a *best* regression line. We want to find that regression line that will in some way minimize the errors—the $(Y_i - Y_i')$ scores.

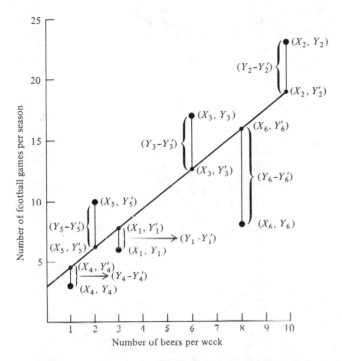

FIGURE 14-7. Notation for dealing with regression lines. Each observed data point is labeled as (X_i, Y_i). Each *predicted* data point is labeled (X_i, Y'_i). The error in prediction is the difference between the observed and predicted data points, that is, $(Y_i - Y'_i)$.

To do this, we first need some sort of overall score to reflect the size of the $(Y-Y')$s. One possible candidate for such a score is simply the sum of the $(Y_i - Y'_i)$s. Perhaps we should try to find the regression line that minimizes this sum.

Appealing though it seems, such a procedure has a flaw that we have encountered before, which is that some of the $(Y_i - Y'_i)$s will be positive and others will be negative. Since the negative errors will tend to cancel out the positive errors, the sum of the $(Y-Y'_i)$s would not be a very good measure of total error.

Another possibility would be to sum up the *absolute* values of the error scores. This would be an improvement. But it turns out that the *best* thing to do is to square each $(Y_i - Y'_i)$ and add up these $(Y_i - Y'_i)^2$s. Thus, for any given regression line, a "total squared error" can be defined by the expression

$$\text{Total squared error} = \sum_{\text{data points}} (Y_i - Y'_i)^2$$

Table 14-3 lists the total squared errors associated with each of the regression lines shown in Figure 14-8. A glance at Table 14-3 should confirm our intuitions. Those regression lines which seem like bad regression lines produce big total

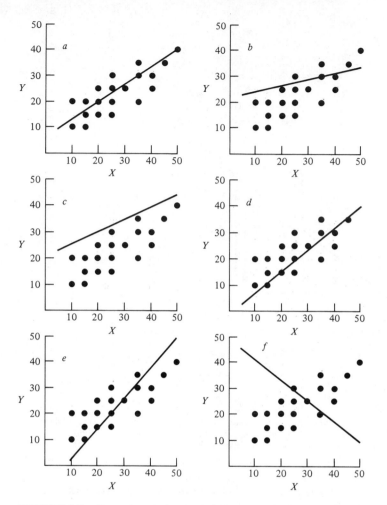

FIGURE 14-8. Examples of what are intuitively "good" or "bad" regression lines.

TABLE 14-3. Error Scores Associated with Each of the Regression Lines Shown in Figure 14-9.

Regression Line	$\Sigma (Y_i - Y_i')^2$
a	490
b	1,184
c	2,549
d	789
e	1,340
f	5,711

For each regression line, the error score is computed by the expression

$$E = \Sigma (Y_i - Y_i')^2$$

where Y_i is the observed data point and Y_i' is the predicted data point (that is, the vertical projection of Y_i on to the regression line).

squared errors, and those regression lines which seem like better regression lines produce smaller total squared errors. Now what we would like to do is calculate that regression line that *minimizes* the total squared error.

Digression 14.2 derives the best-fitting regression line for any given scatterplot. As can be seen, the slope, *b*, of the best-fitting regression line is computed from the equation

$$b = \frac{n\Sigma X_i Y_i - (\Sigma X_i)(\Sigma Y_i)}{n\Sigma X_i^2 - (\Sigma X_i)^2}$$ (14-2)

Once we have computed *b*, we can simply plug it into Equation D.4 of Digression 14.2

$$a = \frac{\Sigma Y_i - b\Sigma X_i}{n_1}$$ (14.3)

to determine *a*. Thus, Equations 14.2 and 14.3 together allow us to calculate the best-fitting regression line.[1]

Digression 14.2

Derivation of the Regression Line that Will Minimize $\Sigma(Y_i - Y_i')^2$

As noted, a regression equation takes the form,

$$Y_i' = bX_i + a$$

Our goal is to select the values of *a* and *b* that will yield the line minimizing $\Sigma(Y_i - Y_i')^2$.

Let us refer to $\Sigma(Y_i - Y_i')$ as "*E*" (for error). Thus,

$$E = \Sigma(Y_i - Y_i')^2$$ (D.1)

Since

$$Y_i' = bX_i + a$$ (D.2)

---

[1]Note that from Equation 14.3, we can derive

$$a = \frac{\Sigma Y_i}{n} - \frac{b\Sigma X_i}{n}$$

Now note that Y_i/n and X_i/n are the means of the *Y*-scores and *X*-scores, respectively. Designating these means as M_Y and M_X,

$$a = M_Y - bM_X$$

or

$$M_Y = nM_X + a$$

This means that (M_Y, M_X) is always one point on the best-fitting regression line.

We can substitute Equation D.2 into Equation D.1 to obtain

$$E = \Sigma[Y_i - (bX_i + a)]^2 \tag{D.3}$$

Expanding out Equation D.3,

$$E = \Sigma[Y_i^2 - 2Y_i(bX_i + a) + (bX_i + a)^2]$$
$$= \Sigma[Y_i^2 - 2bX_iY_i - 2aY_i + b^2X_i^2 + 2abX_i + a^2]$$
$$= \Sigma Y_i^2 - 2b\Sigma X_iY_i - 2a\Sigma Y_i + b^2\Sigma X_i^2 + 2ab\Sigma X_i + na^2$$

(where n is the number of data points we have).

Our strategy will now be as follows. Since we wish to find those values of a and b that minimize E, we will take partial derivatives of E first with respect to a and then with respect to b. Setting these partial derivatives to zero will provide two simultaneous equations relating a to b. Solving these sumultaneous equations will give expressions for a and b. Thus

$$\frac{\partial E}{\partial a} = -2\Sigma Y_i + 2b\Sigma X_i + 2na$$

Setting $\partial E/\partial a$ to zero,

$$\Sigma Y_i - b\Sigma X_i = na$$

or

$$a = \frac{\Sigma Y_i}{n} - b\frac{\Sigma X_i}{n} \tag{D.4}$$

Now,

$$\frac{\partial E}{\partial b} = -2\Sigma X_iY_i + 2b\Sigma X_i^2 + 2a\Sigma X_i$$

Setting $\partial E/\partial b$ to zero,

$$b\Sigma X_i^2 = \Sigma X_iY_i - a\Sigma X_i$$

or

$$b = \frac{\Sigma X_iY_i - a\Sigma X_i}{\Sigma X_i^2} \tag{D.5}$$

Now substituting Equation D.4 into Equation D.5,

$$b = \frac{\Sigma X_iY_i - \left[\dfrac{\Sigma Y_i - b\Sigma X_i}{n}\right]\Sigma X_i}{\Sigma X_i^2}$$

$$= \frac{\Sigma X_iY_i}{\Sigma X_i^2} - \frac{\Sigma X_i\Sigma Y_i}{n\Sigma X_i^2} + \frac{b(\Sigma X_i)^2}{n\Sigma X_i^2}$$

or

$$b - \frac{b(\Sigma X_i)^2}{n\Sigma X_i^2} = \frac{\Sigma X_i Y_i}{\Sigma X_i^2} - \frac{\Sigma X_i \Sigma Y_i}{n\Sigma X_i^2} \tag{D.6}$$

Factoring out b from the left side of Equation D.6,

$$b\left[1 - \frac{(\Sigma x_i)^2}{n\Sigma x_i^2}\right] = \frac{\Sigma X_i Y_i}{\Sigma X_i^2} - \frac{\Sigma X_i \Sigma Y_i}{n\Sigma X_i^2}$$

or

$$b = \frac{\dfrac{\Sigma X_i Y_i}{\Sigma X_i^2} - \dfrac{\Sigma X_i \Sigma Y_i}{n\Sigma X_i^2}}{1 - \dfrac{(\Sigma X_i^2)}{n\Sigma X_i^2}}$$

or finally, multiplying numerator and denominator by $n\Sigma X_i^2$,

$$b = \frac{n\Sigma X_i Y_i - \Sigma X_i \Sigma X_i}{n\Sigma X_i^2 - (\Sigma X_i)^2}$$

Now let's compute what this regression line is for our example. In Table 14-4 we list all the terms we need to compute the best-fitting regression line. (We have also included some terms that we do not need, but that we will need later.) Using Equation 14.2, we can compute b

$$b = \frac{2586 - 1980}{1284 - 900} = \frac{606}{384} = 1.58$$

TABLE 14-4. Various Sums, Sums of Squares, and Sums of Cross Products Needed to Compute Regression Measures

Person	Beers (X)	X^2	Football Games (Y)	Y^2	XY
1	3	9	6	36	18
2	10	100	23	529	230
3	6	36	16	256	96
4	1	1	3	9	3
5	2	4	10	100	20
6	8	64	8	64	64
	$\Sigma X = 30$	$\Sigma X^2 = 214$	$\Sigma Y = 66$	$\Sigma Y^2 = 994$	$\Sigma XY = 431$

$$\Sigma X = 30 \qquad\qquad\qquad\qquad\qquad\qquad \Sigma Y = 66$$
$$(\Sigma X)^2 = 900 \qquad (\Sigma X)(\Sigma Y) = 1{,}980 \qquad (\Sigma Y)^2 = 4{,}356$$
$$\Sigma X^2 = 214 \qquad\qquad \Sigma XY = 431 \qquad\qquad \Sigma Y^2 = 994$$
$$n\Sigma X^2 = 1{,}284 \qquad\quad n\Sigma XY = 2{,}586 \qquad\quad n\Sigma Y^2 = 5{,}964$$

Likewise, using Equation 14.3, we can compute a, the intercept of the best-fitting line, to be

$$a = \frac{66 - (1.58)(30)}{6}$$

$$= \frac{18.6}{6} = 3.1$$

And that's how the best-fitting regression line shown in Figure 14-5 was arrived at.

14.5 A best-fitting regression line for predicting Y from X is the one that minimizes the sum of the squared ($Y-Y$) errors. To determine what this best-fitting line is, we calculate b, the slope parameter by plugging into Equation 14.2 and then calculate a, the intercept parameter by plugging into Equation 14.3.

How Good Is the Fit?

Now that we have determined how to compute the best-fitting regression line, let's turn our attention to some measure of how good the fit is—that is, how well the data points are fit by a straight line. To get a feeling for what we mean by goodness of fit, consider the various scatterplots depicted in Figure 14-9.

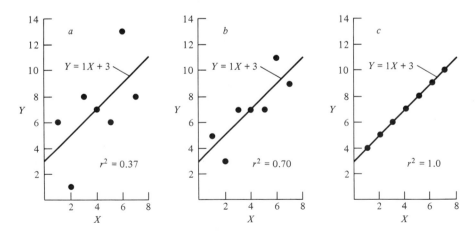

FIGURE 14-9. Different scatterplots which produce identical best-fitting regression lines vary in how good is the fit of the data points to the line. In general the further the data points tend to be from the regression line, the poorer is the fit. Goodness of fit is reflected by Pearson r^2.

Each of these scatterplots is fit best by the identical regression line—but as can be seen, some of the fits seem to be "good fits" and others seem to be "bad fits." That is, the total squared error is rather large in the poor-fit cases and relatively small in the good-fit cases.

Pearson r^2.   Fortunately there is a very good measure of how well the data points are fit by a straight line—that is, of how good the correlation is. This measure is called a Pearson r^2 after the statistician Karl Pearson. A Pearson r^2 is a number that varies between 0 and 1 (like probability). To the extent that the fit of the data to a straight line is good, the associated Pearson r^2 tends toward 1.0. Conversely, to the extent that the fit of the data to a straight line is poor, the associated Pearson r^2 tends toward 0. The Pearson r^2s are listed for each of the scatterplots in Figure 14-9. As can be seen, the better the fit, the bigger the r^2. In the extreme (part c of Figure 14-9) the data are fit *perfectly* by a straight line, and $r^2 = 1.0$, which is as big as it can be.

We should point out that variables that are negatively correlated are also related more or less well by a straight line. Figure 14-10 shows a series of scatterplots that reflect negatively correlated variables. Again good fits lead to smaller error terms and big Pearson r^2s. Poor fits lead to big error terms and small Pearson r^2s.

The formula for the Pearson r^2 is

$$r^2 = \frac{[n\Sigma X_i Y_i - (\Sigma X_i)(\Sigma Y_i)]^2}{[n\Sigma X_i^2 - (\Sigma X_i)^2][n\Sigma Y_i^2 - (\Sigma Y_i)^2]}$$

This formula looks somewhat formidable, but there is actually a nice symmetry to it. Let's reflect a bit on this symmetry. First we note that there are three

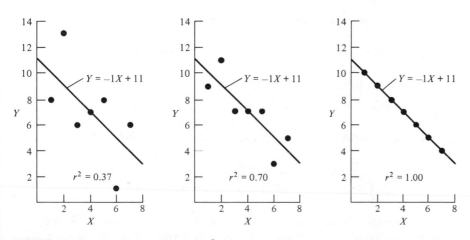

FIGURE 14-10.   The value of a Pearson r^2 reflects *only* the goodness of fit and is unaffected by whether the correlation is positive or negative.

factors in the formula—one factor in the numerator (which is squared) and two factors in the denominator. These three factors bear a good deal of similarity to one another. The first term in each of the three factors is n times the sum of two things multiplied together. In the case of the numerator the two things that are multiplied together are X_i and Y_i—thus, the first term in the numerator is $n \Sigma X_i Y_i$. In the first factor of the denominator the first term is $n \Sigma X_i X_i$ or $n \Sigma X_i^2$; and likewise, in the second factor of the denominator the first term is $n \Sigma Y_i Y_i$ or $n \Sigma Y_i^2$.

The second term of each factor is the product of two sums. In the numerator the two sums are ΣX_i and ΣY_i. In the first term of the denominator, the two sums are both ΣX_i—so we get a $(\Sigma X_i)^2$; and the second factor of the denominator involves two ΣY_is—so we get a $(\Sigma Y_i)^2$. See the symmetry?

The "guts" of the Pearson r^2 formula lie in the numerator, which is related to what is called the "covariance" of X and Y. Covariance is a measure of how much X and Y are related. To the extent that X and Y are positively related, their covariance will be large and positive. To the extent that X and Y are negatively related, their covariance will be large and negative. Note that if X and Y are *either* positively or negatively related, their squared covariance (which is in the numerator of the Pearson r^2 formula) will be large and positive.

The denominator of the Pearson r^2 is a "normalizing factor" that corrects for the scales used for measuring X and Y. It is because of this normalizing factor that r^2 falls between 0 and 1. As Digression 14.3 shows, the two factors in the denominator of the r^2 formula are very closely related to the variance of the Xs and the variance of the Ys.

Digression 14.3

Variance, Covariance, and r^2

As we have already noted, the equation for the Pearson r^2 is

$$r^2 = \frac{[n \Sigma XY - (\Sigma X)(\Sigma Y)]^2}{[n \Sigma Y^2 - (\Sigma Y)^2][n \Sigma X^2 - (\Sigma X)^2]} \tag{D.1}$$

Dividing numerator and denominator of Equation D.1 by n^4 yields

$$r^2 = \frac{[\Sigma XY/n - (\Sigma X)(\Sigma Y)/n^2]^2}{[\Sigma X^2/n - (\Sigma X/n^2)][\Sigma Y^2/n - (\Sigma Y/n^2)]} \tag{D.2}$$

The numerator of Equation D.2 is the squared covariance of X and Y which we label $[S_{XY}^2]$.[2] Note that the two terms in the numerator of Equation D.2 are the variances of the Y-scores and X-scores, S_Y^2 and S_X^2.

Thus, another way of expressing the Pearson r^2 is

$$r^2 = \frac{[S_{XY}^2]^2}{S_Y^2 S_X^2}$$

In Table 14-4 we have listed all the terms needed to compute a Pearson r^2 for our example data. The r^2 is:

$$r^2 = \frac{(2586 - 1980)^2}{(1284 - 900)(5964 - 4356)}$$

$$= 0.60$$

> 14.6 The Pearson r is a measure of how well the relationship between X and Y is characterized by a straight line. The r measure can vary from 0 (for the worst possible fit) to 1.0 (for the perfect fit).

The meaning of a Pearson r^2.   We have seen that the Pearson r^2 is a measure of goodness of fit. To the degree that two variables are correlated, the resulting Pearson r^2 will be close to 1 and to the degree that the two variables are not correlated, the resulting Pearson r^2 will be close to 0. It turns out, however, that the Pearson r^2 has a much more specific meaning. To see why this is so, let us first compute the total variance of the Y-scores (number of football games) via the usual equation for variance. (Note that this variance is the variance of the Y-scores themselves and *not* an estimate of any population variance.) So the equation for the variance of the Y-scores is

$$S_Y^2 = \frac{\Sigma Y_i^2}{n} - \left(\frac{\Sigma Y_i}{n}\right)^2$$

We have already listed the various terms used in this equation (Table 14-4), so the variance is easily computed to be

$$S_Y^2 = \frac{994}{6} - \left(\frac{66}{6}\right)^2$$

$$= 165.7 - 121.0 = 44.7$$

Now let us consider the Y' scores—the six scores that are *predicted* from the regression equation and the X_is. We have listed these Y'-scores in Table 14-5. Just as we computed the variance in the original Y-scores, we can also

TABLE 14-5.   Listing of Y'_i and $(Y_i - Y'_i)$ scores; the Y' Scores Are Computed Using the Equation $Y'_i = 1.58 X_i + 3.1$

X_i	Y_i	Y'_i	$(Y'_i)^2$	$(Y_i - Y'_i)$	$(Y_i - Y'_i)^2$
3	6	7.84	61.47	−1.84	3.39
10	23	18.90	357.21	4.10	16.81
6	16	12.58	158.26	3.42	11.70
1	3	4.68	21.90	−1.68	2.82
2	10	6.26	39.19	3.74	13.99
8	8	15.74	247.75	−7.74	59.91
		$\Sigma Y'_i = 66$	$\Sigma (Y'_i)^2 = 885.77$	$\Sigma (Y_i - Y'_i) = 0.00$	$\Sigma (Y_i - Y'_i)^2 = 108.61$

compute the variance in these Y'-scores. This variance is

$$S^2_{Y'} = \frac{\Sigma(Y')^2}{n} - \left(\frac{\Sigma Y'}{n}\right)^2$$

$$= \frac{885.77}{6} - \left(\frac{66}{6}\right)^2$$

$$= 147.6 - 121.0$$

$$= 26.6$$

Now let's divide the variance of the Y'-scores by the variance of the original Y-scores. This ratio is

$$\frac{S^2_{Y'}}{S^2_Y} = \frac{26.6}{44.7} = 0.60$$

which is exactly the same as the r^2 that we computed above. This is no coincidence. It turns out that a Pearson r^2 may be viewed as the *percentage of the total variance in the Y-scores that is accounted for by a straight line.* We shall have a bit more to say about this shortly, but first let's compute one other interesting variance.

Just as we computed the variance of the Y-scores and of the Y'-scores, we can also compute the variance of the error scores—of the $(Y-Y')$s. In Table 14-5 we listed these error terms—these $(Y_i-Y'_i)$s. The six $(Y_i-Y'_i)$s have a variance just like any other set of scores, which is

$$S^2_{Y-Y'} = \frac{\Sigma(Y_i-Y'_i)^2}{n} - \left[\frac{\Sigma(Y_i-Y'_i)}{n}\right]^2$$

$$= \frac{108.61}{6} - 0$$

$$= 18.1$$

Now let's add the variance due to the Y' scores (26.6) to the variance of the $(Y-Y')$ scores which we have just computed to be 18.1. This sum is 44.7, which is exactly equal to the variance of the original Y-scores.

Another variance pie.   Just as we did in ANOVA situations, we can take a variance in a correlational situation and subdivide it into its components. That is, we have some total variance of the "dependent variable"—of the Y-scores. This total variance is divided into two components—that variance due to regression (the variance of the Y'-scores) and that variance due to error (the variance of the $[Y-Y']$ scores).

> 14.7 The Pearson r^2 is equal to the variance of the Y'-scores divided by the variance of the Y-scores. It can also be viewed as the percentage of the variance in the Y-scores accounted for by variation in the X-scores, assuming a linear relationship between X and Y. The variance of the Y-scores may be viewed as the variance due to X (variance of the Y'-scores) plus variance not due to X [variance of the $(Y-Y')$ scores.]

Digression 14.4

Prediction of X from Y

So far our discussion of correlation has revolved around the prediction of Y-scores from X-scores. For instance, in our football/beer example, we concentrated on predicting number of football games watched (which we had dubbed as "Y") from the number of beers drunk (which we had dubbed "X").

However, this was completely arbitrary. Since both of our variables are in a sense "equal"—they are both measured—it makes no difference, from a mathematical point of view which we label Y and which we label X. Likewise, we can predict X from Y just as we can predict Y from X.

Referring to Figure 14-4, suppose that we wished to predict the X-scores from the Y-scores. That is, suppose that we were given the number of football games some individual watches and, based on that information, we wanted to predict the number of beers that individual will drink. Our first inclination might be to simply use the same regression line that we have already calculated (Figure 14-6), but such an inclination would lead us astray. Recall that the X-to-Y regression line is computed so as to minimize the squared error terms between the Y and Y's—in terms of Figure 14-7, we are concerned with minimizing the "vertical" errors. However, if we wish to predict X from Y, we want to minimize the errors between the observed and the predicted Xs—that is, we are concerned with minimizing the *horizontal* distances between the observed Xs and the regression line. It turns out that this criterion (usually) requires a different regression line.

Just as the formulas for obtaining the Y-from-X regression are

$$b = \frac{\Sigma XY - (\Sigma X)(\Sigma Y)}{\Sigma X^2 - (\Sigma X)^2}$$

and

$$a = \frac{\Sigma Y - b\Sigma X}{n}$$

the analogous formulas for obtaining the X-from-Y regression equation are

$$b = \frac{\Sigma XY - (\Sigma X)(\Sigma Y)}{\Sigma Y^2 - (\Sigma Y)^2}$$

and

$$a = \frac{\Sigma X - b\Sigma Y}{n}$$

The Pearson r^2, however, is identical whether predicting Y from X or predicting X from Y.

Given two variables, practical factors often dictate which should be the predictor and which should be the predicted variable. For example suppose we are interested in the relationship between a person's IQ score as a teenager and the person's yearly income as an adult. Since teenage IQ occurs prior to adult income, it makes sense to want to predict the latter from the former, rather than vice versa. By convention, the predictor variable is usually labeled as X and the predicted variable, Y.

The standard error of estimate. Since r^2 is the percentage of variance accounted for by a straight line, we can see that $(1 - r^2)$ is the percentage of variance that is *not* accounted for by a straight line—that is, the percentage of variance attributable to the $(Y-Y')$ or error scores. The square root of this expression, or

$$\sqrt{1 - r^2}$$

times the standard deviation of the Y-scores, or

$$S_Y\sqrt{1 - r^2}$$

is referred to as the standard error of estimate. It is somewhat similar to a confidence interval.

The Pearson r. We have just seen that the Pearson r^2 has the convenient property of being equal to the percentage of variance in the Y-scores that is accounted for by regression—that is, the amount of variance in the Y-scores that is predicted by the X-scores. Thus, r^2 is a very meaningful measure. It does, however, have the shortcoming that it does not reflect whether a given relationship is positive or negative. A measure that is often used to circumvent this problem is a Pearson r, which is the square root of a Pearson r^2. Accordingly, the Pearson r is computed by the formula

$$r = \frac{n\Sigma X_i Y_i - (\Sigma X_i)^2}{\sqrt{[n\Sigma X_i^2 - (\Sigma X_i)^2][n\Sigma Y_i^2 - (\Sigma Y_i)^2]}}$$

A Pearson r, which can vary from $+1.0$ to -1.0, is positive if the relationship between two variables is positive and negative if the relationship between two variables is negative. Figure 14-11 illustrates this state of affairs by showing

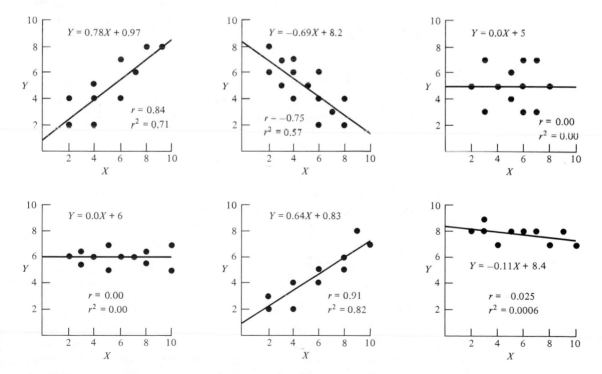

FIGURE 14-11.   Various scatterplots along with their associated regression lines, rs and r^2s. Note that r reflects the goodness of fit as well as whether the correlation is positive or negative. However, r^2 reflects goodness of fit only.

some examples of various scatterplots along with their associated Pearson rs and Pearson r^2s, and best-fitting regression lines.

TESTING THE SIGNIFICANCE OF A RELATIONSHIP

So far we have talked about what correlation is and how one measures it. In this section we'll describe how to test the *statistical significance* of a correlation.

A Null Hypothesis for Correlation

Let's continue with our current example. Suppose we were to set up a "null hypothesis" that in the American population there is actually *no* correlation between number of beers drunk and number of football games watched. That is, if we were able to compute a Pearson r^2 for the entire population (which, since it is a population parameter, is referred to as ρ^2) it would be zero. If such a null hypothesis were true, would we expect the sample correlation to be 0? The answer is no, because as usual there is error variance in the data. We may just happen to have picked a sample of six individuals for whom a spurious correlation exists. Therefore, we need some way of testing whether our observed correlation is significantly different from zero.

The r to t Transformation

Fortunately this is very easy to do. Without going into the mathematics of the situation, it turns out that we can compute the following statistic.

$$t(n - 2) = \frac{r \sqrt{n - 2}}{\sqrt{1 - r^2}}$$

This t is just like any other t; notice that it is based on $n - 2$ degrees of freedom (where of course n is the number of observations—the number of pairs of scores that we have). In our particular example, this t would be

$$t(4) = \frac{0.77 \sqrt{(4)}}{\sqrt{1 - 0.59}} = 2.41$$

One-tailed versus two-tailed tests.   As has always been the case, we must be concerned with whether we wish to do a one-tailed test or a two-tailed test on our data. In particular two situations may exist in terms of a priori notions about which way the relationship (if it exists) should go. On the one hand we might have some a priori notion that two variables will be positively correlated—or that two variables will be negatively correlated. In the example at hand we might well have had an a priori notion that if there exists any correlation at all between number of football games watched and number of beers drunk, then that correlation should be positive. Thus, we would be in a position to do a one-tailed test on the data. Just as is the case with any one-tailed test, we would only reject the null hypothesis if the observed relationship turned out to be in the expected direction (which it has in this particular instance).

On the other hand if we did not have any a priori notion as to whether the relationship is positive or negative, we would use a two-tailed criterion-t and we would reject the null hypothesis of zero-correlation if the observed correlation were sufficiently large in either a positive or a negative direction.

Power and number of observations.   Recall that in z-test and t-test situations the power of the test reflected the test's ability to detect (reveal as significant) any difference between the conditions that actually existed. The higher the power of the test, the smaller the probability of making a Type II error. Recall further that one of the main ingredients of power was the number of subjects in the experiment. The more subjects in the experiment, the greater the power of the associated test.

A correlational situation is much the same. The power of the r-to-t test reflects the test's ability to declare an observed correlation to be significant given that there actually is some correlation in the population. The formula for the r-to-t transformation makes this relationship between n and power exceedingly evident. Since the term $\sqrt{n-2}$ appears in the numerator of the r-to-t equation, it must be the case that, holding the observed r constant, the larger the number of observations on which that r is based, the greater will be the t that emerges. Thus, the moral in a correlational situation is the same as it is in an experimental situation. The more observations, the higher the probability of picking up any relationship that actually exists in the population.

> 14.8 Hypothesis testing in a correlational situation involves establishing a null hypothesis that no relationship exists between X and Y in the population and an alternative hypothesis that a relationship does exist. The test itself is carried out via the r-to-t transformation.

OTHER TYPES OF RELATIONSHIPS

The topics of correlation and regression are actually vast ones about which numerous books have been written. Because they are not the major topic of this book, we have gone into detail about only the very simplest type of correlation. In closing we should just like to briefly mention other, more complicated forms of correlation which one often encounters in the literature.

Curvilinear Correlation

We have described correlation as reflecting a relationship between two variables, X and Y. Moreover, we have so far constrained our discussion to relationships that are *linear*. Suppose, however, we are interested in the relationship between a person's age and the time it takes that person to run the hundred-yard dash. If we were to observe a large number of individuals and for each individual record both the age and the time to run the hundred-yard dash, we might get a relationship such as the one shown in Figure 14-12. As we can see, this relationship is very systematic and reasonably easy to describe. However, it is not a linear relationship, but is rather a *curvilinear* relationship.

When we were testing a linear relationship between two variables, X and Y, we were essentially seeing how well the observed data fit the hypothetical relationship

$$Y = bX + a$$

When, however, we suspect that there may be a curvilinear relationship between X and Y, we can similarly compute how well the two variables may be related by the quadratic relationship

$$Y = b_1X + b_2X^2 + a$$

Using techniques similar to the one described above, we can select the parameters b_1, b_2, and a so as to once again minimize the sum of the squared errors

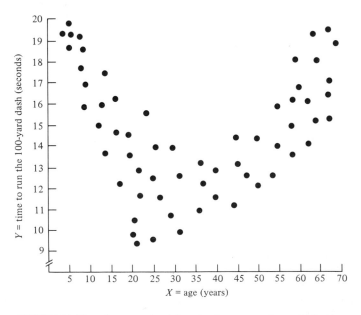

FIGURE 14-12. An example of data that form a curvilinear pattern.

between the predicted curvilinear regression line (which, of course, would be a *parabola*) and the observed data points. Once again, we would get an r^2 that is a measure of how good this fit is. Such an r^2 would be equal to the proportion of variance of the original Y-scores accounted for by the best-fitting curvilinear regression line.

> 14.9 Curvilinear correlation is a relationship between X and Y in which Y is assumed to be a quadratic function of X.

Multiple Regression

Recall that one of the ways of discussing correlation is in terms of prediction. That is, a regression line provides a theoretical relationship between X and Y. Given any particular X-score, we can therefore use the theoretical relationship (the regression equation) to predict what the value of the corresponding Y should be.

A situation in which the notion of prediction demonstrates itself somewhat more cogently is when there are several predictor variables—that is, several types of Xs. Suppose, for example, that an admissions committee is attempting to evaluate candidates who are applying to medical school. On what does such a committee base its evaluations? There are several pieces of information about each candidate that the admissions committee has at its disposal. For example, they have candidates' undergraduate grade point average (GPA) and the candidates' MCAT scores. (They undoubtedly have other pieces of information as well, but for simplicity let us confine our attention to these two.) How much weight should the admissions committee give to each of these two pieces of information?

Predictor variables predict a criterion variable.   Suppose that there exists some *criterion variable* such as grade point average in medical school, and let's suppose that the goal of the admissions committee is to choose those students who will eventually achieve the best medical school grade point averages.

Suppose that we consider all the students who are *already attending* medical school. For each such student we list three things—the value of that student's criterion variable (the student's grade point average in medical school), the student's undergraduate GPA, and the student's MCAT score, as is demonstrated in Table 14-6. The criterion variable (medical school GPA) has been labeled Y and the two predictor variables have been labeled X_1 and X_2.

A multiple regression equation.   Now let's suppose we have a linear equation that relates these three variables:

TABLE 14-6. Example of a Situation Involving One Criterion Variable (Y) and Two Predictor Variables (X_1 and X_2)

| | Predictor Variables | | Criterion Variable |
| | | | |
Student	X_1 = Undergrad GPA	X_2 = MCAT Score	Y = Med School GPA
1	3.25	620	3.12
2	3.95	630	3.43
3	3.50	710	3.87
4	2.95	790	3.32
5	3.57	690	3.00
.	.	.	.
.	.	.	.
.	.	.	.

$$Y = b_1X_1 + b_2X_2 + a \tag{14.4}$$

Once again using appropriate techniques, we can judiciously select the values of b_1, b_2, and a such that we get a best fit of the equation to the data. And this is our multiple regression equation. Along with the multiple regression equation we may also compute a multiple r^2 (which is actually referred to as R^2) that represents the percentage of variance in the criterion variable—the Y-scores—that is accounted for by multiple regression—that is, predicted by the two Xs.

Now let's go back to our original question of which students the admissions committee will choose for medical school. Since Equation 14.4 above provides the best prediction of the criterion variable (medical school GPA) from the predictor variables (GPA and MCAT) we can take each of our applicants' MCAT scores and GPA scores, plug them into Equation 14.4, and arrive at a prediction of medical school GPA. Then we can simply select those students whose predicted medical school GPAs are predicted to be the highest.[2]

In general if we have L predictor variables and 1 criterion variable, we can compute a regression equation of the form

$$Y = b_1X_1 + b_2X_2 + \ldots + b_LX_L + a$$

and we can always find the values of the coefficients—the bs and the a—that will give us an equation of best fit.

> 14.10 Multiple regression is a situation akin to multifactor ANOVA in which there is one to-be-predicted variable (Y) and more than one predictor variable (X).

---

[2]The University of Oregon Psychology Department uses precisely this technique for screening potential graduate students.

In this chapter we have very briefly gone through the concepts of correlation, prediction, and linear regression—what they are and how to deal with them quantitatively.

1. Both a correlational and an experimental study are designed to investigate relationships among variables. In an experimental study we control one (or more) independent variables and measure a dependent variable. From the results of an experimental study we can make causal inferences about the effect of one variable on another. However, in a correlational study we do not control anything. Rather, we just measure variables in an after-the-fact sort of way and look at the relationship between them. Following a correlational study we can infer the existence of relationships among variables, but we *cannot* infer that changes in one variable cause changes in a second variable.

2. To quantify a linear relationship, we can compute a regression line which is a linear function relating variable Y to variable X. A best-fitting regression line has the property that it minimizes the sum of the squared differences between the actual (observed) Y-scores and what these Y-scores are predicted to be (the Y'-scores). This regression line is also referred to as the line of best prediction from X to Y.

3. As a measure of how good the fit is, a Pearson r^2 is used. A Pearson r^2 is the ratio of the variance of the predicted (Y') scores to the total variance of the original Y-scores. As such it is interpreted as the percentage of variance in the Y-scores that is accounted for by linear regression—or the percentage of variance in the Y-scores that may be predicted by the X-scores. The Pearson r is nothing more than the square root of the Pearson r^2. A Pearson r may be positive or negative depending on whether the correlation between the two variables is positive or negative.

4. There are many other more complicated forms of regression. Curvilinear regression may be used when there is a suspected nonmonotonic or curvilinear relationship between two variables. Likewise, multiple regression may be used in a situation where there are several predictor variables (that is, several Xs) that are used to predict a single Y.

1. Draw scatterplots that would roughly correspond to the following situations (stated in terms of regression lines and Pearson rs).

   1. $Y = 0.65X + 2.5$ $r = 0.25$
   2. $Y = -0.51X + 1.5$ $r = 0.75$
   3. $Y = X + 3$ $r = 1.00$
   4. $Y = 4$ $r = 0$

5. $Y = 3X + 14$ $r = 0.95$
6. $Y = 1.6X + 2$ $r = -0.05$
7. $Y = 1.6X + 3$ $r = -0.99$

2. A sports psychologist is interested in the relationship between players' ages and their batting averages. Twelve players, chosen at random, yielded the following data:

	Player											
	1	2	3	4	5	6	7	8	9	10	11	12
Age	18	17	31	25	22	24	28	21	21	18	35	41
Average	0.225	0.350	0.150	0.275	0.269	0.200	0.320	0.315	0.195	0.200	0.310	0.275

1. Compute the regression line predicting average from age.
2. From this regression line compute each player's predicted average and error (difference between predicted and actual average).
3. Compute the variance of the actual averages, predicted averages, and errors. How do these variances relate to one another?
4. What is the ratio of the variance of the predicted averages to the variance of the actual averages?
5. Compute a Pearson r^2 for these data. How does the value of the r^2 compare to your answer to question 4?

3. The same sports psychologist is interested in the relationship between weight and discus-throwing ability. He selects a random sample of 15 people from the population and measures both the person's weight and the distance they are able to throw the discus. The data are as follows:

Person	Weight (lbs)	Discus distance (ft)
Maggie	120	125
Fred	165	215
Elaine	105	145
Suzie	128	129
Robert	220	175
Geoff	170	209
Beth	115	141
Earl	156	223
Linda	125	130
Tom	190	200
Betsy	160	132
George	130	250
John	200	180
Alinda	100	150
Sylvia	130	135

1. Draw a scatterplot of these data.
2. Compute the regression line predicting discus distance from weight.
3. Compute the Pearson r^2 relating these two variables. Is this Pearson r significantly different from 0?
4. Now repeat questions 2 and 3 for men and women separately.
5. What conclusion do you draw about one potential danger of computing correlations?

4. A sociologist is interested in the relationship between the size of a city and its per capita murder rate. Hence she chooses 10 cities in the United States. The sizes and per capita murder rates are shown below:

					City					
	1	2	3	4	5	6	7	8	9	10
Size (in 1000s)	150	990	75	520	610	304	190	100	60	120
Murder rate (per 100,000 per year)	1.2	11.4	3.1	4.0	3.1	4.2	2.0	0.3	1.1	0.9

    1. Draw a scatterplot depicting these data.
    2. Compute regression equations to predict per capita murder rate from city size, and vice versa.
    3. What is the correlation (Pearson r) between the two variables?
    4. Is the correlation significantly different from zero?

5. Consider the data of Chapter 5, Problem 8.

    1. Find the regression equation to predict the IQ of the high-SES twin from the IQ of the low-SES twin.
    2. Find the regression equation to predict the IQ of the low-SES twin from the IQ of the high-SES twin.
    3. What is the Pearson r^2 between IQ scores of high- and low-SES twins?
    4. Is this correlation significantly different from zero?

6. A developmental psychologist is interested in whether there is a relationship between sex and handedness. He selects a random sample of 24 children. To each child, he assigns a 1 or a 0 corresponding to whether the child is male or female and also a 1 or a 0 corresponding to whether the child is right- or left-handed. The data are as follows:

Child	Sex	Handed	Child	Sex	Handed	Child	Sex	Handed
1	1	1	9	0	1	17	1	0
2	0	1	10	1	1	18	1	0
3	0	0	11	0	1	19	0	1
4	1	0	12	1	1	20	1	0
5	0	1	13	0	0	21	0	1
6	1	1	14	0	0	22	1	1
7	1	0	15	1	1	23	1	1
8	1	1	16	1	0	24	0	0

Compute the correlation (Pearson r^2) between sex and handedness. Is this correlation significant?

7. A study is done by the telephone company to determine whether there is any relationship between sex (male or female) and type of job (operator or supervisor).

    Ten phone company employees are selected at random and assigned a 1 or a 0 corresponding to whether they are female or male and another 1 or 0 corresponding to whether they are an operator or a supervisor, respectively. The data are as follows:

Person	Female/Male (x)	Operator/Supervisor (y)
1	1	1
2	0	1
3	1	0
4	1	0
5	0	0
6	1	1
7	0	0
8	1	1
9	1	1
10	0	0

1. Determine r^2 for these data.
2. Test the significance of this r^2.

8. A dental researcher is interested in the relationship between the number of times teeth are brushed per day and number of cavities. The researcher asks six people how many times per day (1, 2, or 3) they brush their teeth and how many cavities they have. The data are as follows, showing the number of cavities for people who brush 1, 2, or 3 times a day:

Number of Brushings (x)		
1	2	3
8	1	2
6	5	
7		

Number of cavities

1. Compute the regression line relating number of brushings and number of cavities.
2. Compute r^2 for these data.
3. Test whether a significant nonzero relationship exists between the two variables.

After or Instead of the ANOVA

15

In chapters 11–13, we made our way through various types of analysis of variance situations. Throughout these chapters we were sidestepping a pervasive problem involving the extremely vague nature of the alternative hypothesis in any design that includes more than two conditions. In general given J conditions, the null hypothesis states that the population means for the J conditions are all equal to one another, that is,

$$H_0: \mu_1 = \mu_2 = \ldots = \mu_J \tag{15.1}$$

where μ_j is the population mean score for the jth condition.

Suppose we perform a one-way ANOVA following some experiment, and the obtained F is significant. What do we then conclude? Only that the null hypothesis is false—or equivalently that the alternative hypothesis is true. Unfortunately, however, as we have noted in previous chapters, a statement that a null hypothesis such as the one stated in Equation 15.1 is false really provides very little information. Such a statement merely says that somehow or other at least one of the population means differs from at least one other population mean. But this conclusion in no way isolates *where* the difference or differences actually occur, nor does it tell us anything about how large or important the differences are.

Conclusions stemming from a two-way ANOVA are typically plagued with the same sorts of problems. Recall that a two-factor design involves three separate F-tests—an F-test for each of the two main effects, and an F-test for the interaction effect. Here the general problem of isolating specific effects still remains—it's just slightly disguised. Suppose, for example, that we have a 3×4 design, and a two-way analysis of variance reveals both main effects and the interaction to be significant. Such a revelation is a start, but there remain many unanswered questions. What causes the difference among the four levels of Factor 1? What causes the differences among the three levels of Factor 2? And likewise there are many different patterns of means in a 3×4 design that constitute an interaction. (If you don't believe this, try plotting some hypothetical data for a 3×4 design and see how many different ways you can produce an interaction.)

In conjunction with descriptions of statistical tests, we have also discussed confidence intervals. We saw that plotting data along with, say, 95% confidence intervals provided us with a rough intuitive idea of where the population means fall relative to one another. Having a rough idea of where population means fall gives us, in turn, a rough idea of where the differences among means lie. But a rough idea is only a start. In this chapter we are going to deal with the problem of how to statistically isolate differences among means. This problem is not a simple one. It raises very sticky issues. Let's begin by describing the sticky issue of planned versus post-hoc comparisons.

> 15.1 Just knowing H_0 is false provides relatively little information in any ANOVA situation. After the ANOVA we need to acquire more information about the pattern of population means.

PLANNED AND POST-HOC COMPARISONS

Let's go back to the issue we raised long ago concerning one-tailed versus two-tailed t-tests (or z-tests). Recall that in a t-test or a z-test situation we have a relatively simple alternative hypothesis which in its most general form states that two population means are not equal to one another, or

$$H_1: \mu_1 \neq \mu_2$$

Sometimes, as we have pointed out, this alternative hypothesis is stated more precisely, because sometimes we have an a priori notion of which group will produce the higher mean given that a difference between the means really exists.[1] Therefore, this more precise (directional) form of the alternative hypothesis is

$$H_1: \mu_1 > \mu_2$$

Now recall that these two ways of stating the null hypothesis imply somewhat different tests. With an a priori notion of how the means should lie relative to one another, a one-tailed test is performed, and the null hypothesis is rejected only if the difference between the obtained sample means is in the expected direction. Lacking an a priori notion about which of the two means would be greater, a two-tailed test must be performed. With a two-tailed test the null hypothesis is rejected *either* if M_1 is sufficiently greater than M_2 or if M_2 is sufficiently greater than M_1. However, when a two-tailed test is performed, the required (criterion) difference between the two means necessary to reject the null hypothesis must be larger in order to maintain a 0.05 α-level. (If you are the least bit foggy on any of these points, we recommend a review of Chapter 6.)

---

[1]For our purposes a priori means "formulated prior to seeing the data."

When we move to an ANOVA situation with three or more conditions, an analogous although somewhat more complicated issue arises. That is, in some instances we may do an ANOVA-type experiment with no a priori idea about how the condition population means might fall relative to one another. Such an experiment might be termed "exploratory"; it would generally not be an experiment carried out with the purpose of testing any specific theory. To analyze this kind of experiment, we would be constrained to do what are referred to as *post-hoc* or *unplanned comparisons* among means. In contrast, we might in another experiment have some very good a priori notion of what the pattern of means should be. To analyze this kind of experiment, we would be able to do specific *planned comparisons*.

A post-hoc test is roughly analogous to a two-tailed *t*-test. And, as was the case with a two-tailed *t*-test, we have to be concerned with the pesky problem that the null hypothesis can be erroneously rejected in many different ways. A planned comparison, on the other hand, is more like a one-tailed test—that is, to test specific hypotheses, we perform a relatively small number of previously formulated tests that either pass or fail. In the next two sections we shall describe the properties of these two methods of data analysis.

> 15.2 If we have no a priori hypotheses as to the pattern of means, we perform post-hoc tests. If we do have a priori hypotheses, we perform planned comparisons.

POST-HOC TESTS (UNPLANNED COMPARISONS)

As noted above, post-hoc tests are exploratory. They are not tests done to test any specific hypothesis that has been formulated in advance of the experiment; rather, they provide an after-the-fact evaluation of potentially interesting effects which may emerge from the data. Let's consider a situation in which post-hoc tests might appropriately be applied.

Educational Techniques

Imagine we're educational psychologists interested in the effect of various types of study methods on amount of learning (as measured by final exam scores). To investigate this question, we plan to use 50 students registered for a particular course. At the beginning of the course we randomly divide these 50 students into groups of 10 students per group. Students in Group 1 (Reading Group) are required to do the course reading, but not required to attend lectures. Conversely, students in Group 2 (Lecture Group) are required to attend lectures but not to do any reading. Students in Group 3 (Freedom Group) have complete

TABLE 15-1. Hypothetical Data: Mean Final Exam Score as a Function of Learning Technique; 95% Confidence Intervals Around the Means Are Provided

$n = 10$ Group 1 (Reading)	$n = 10$ Group 2 (Lecture)	$n = 10$ Group 3 (Freedom)	$n = 10$ Group 4 (Tutor-1)	$n = 10$ Group 5 (Tutor-2)	$N = 50$
$M_1 = 82 \pm 6.36$ $T_1 = 820$	$M_2 = 95 \pm 6.36$ $T_2 = 950$	$M_3 = 71 \pm 6.36$ $T_3 = 710$	$M_4 = 83 \pm 6.36$ $T_4 = 830$	$M_5 = 85 \pm 6.36$ $T_5 = 850$	$T = 8,160$

freedom to do anything they want. Students in Group 4 (Tutor-1 Group) learn by the tutorial method; they are required to come to the instructor's office for one hour of tutoring a week. Students in Group 5 (Tutor-2 Group) are treated the same as are students in the Tutor-1 Group but are required to come for two hours of tutoring a week. At the end of the course all 50 students take a final exam, and the mean exam score of each of the five groups is computed. These five means[2] as well as the corresponding totals are shown in Table 15-1.

A one-way analysis of variance. A glance at Table 15-1 reveals that there seem to be differences among the means. The Lecture Group, for example, has an average score of 95% on the final exam, whereas the Freedom Group is performing only at the 71% level. It is of course conceivable that the differences among the means are due merely to chance fluctuations. Thus, the first step is to perform an overall one-way analysis of variance.

Let's suppose that the pattern of individual scores within groups is such that the sum of squares within turns out to be 4500. We compute the sum of squares between in the usual way,

$$SSB = \sum_{\text{groups}} \frac{T_j^2}{n} - \frac{T^2}{N}$$

or

$$SSB = (820^2/10 + \cdots + 830^2/10) - 8160^2/50$$

$$= 2928$$

At this point we can set up an ANOVA summary table as depicted in Table 15-2. There are 4 degrees of freedom between the five groups and 45 degrees of freedom within—9 degrees of freedom within each of the five groups. Since the obtained $F_{(4,45)}$ is 7.32, and the criterion $F_{(4,45)}$ is 2.61, we can easily reject

---

[2]In presenting these and other data in this chapter we will include 95% confidence intervals around the means. Since all our examples involve between-subjects designs with equal ns, the confidence intervals are computed using the formula

$$M \pm \text{criterion } t(dfw) \sqrt{MSW/n}$$

TABLE 15-2.   ANOVA Summary Table for Educational Technique Experiment

Source	df	SS	MS	Obt F	Crit F
Total	49	7,428			
Between	4	2,928	732	7.32	2.61
Within	45	4,500	100		

the null hypothesis of no differences among the five groups. Somewhere, somehow, differences may be assumed to exist.

Now what?   As we have noted previously, however, the fact that this overall F (sometimes referred to as an "omnibus F") is significant does not really provide very much information. From a practical standpoint we would like to know *which* of the differences among the groups are reliable ones. For example, is there any real difference between students who just go to the lectures and students who just do the reading? Are students reliably better when they are tutored for two hours a week versus one hour a week? And so on. In short once we have discovered the omnibus F to be significant, we would like to dig further into our data in order to unearth the specific differences among the means contributing to this overall effect. Post-hoc tests are indicated.

Problems with Post-Hoc Tests

Before illustrating the application of particular post-hoc tests to these data, we will discuss two pitfalls, both of which revolve around the fact that the more tests we do, the higher is the probability of our erroneously rejecting at least one null hypothesis by chance. To see why this is so, let's first compute the total number of tests of pairs of means (pairwise tests) that can be done. A listing of all such pairwise tests is provided in Table 15-3. To be a little more precise, we can reach back in our memory to what we know about counting rules. Counting the number of pairs of means is equivalent to computing the number of unordered combinations of 5 things taken 2 at a time. Thus,

$$\text{Number of comparisons} = \binom{5}{2} = \frac{5!}{(2!)(3!)} = 10$$

TABLE 15-3.   All Pairwise Comparisons that Can Be Done Among M_1, M_2, M_3, M_4, and M_5

M_1 versus M_2	M_2 versus M_3	M_3 versus M_4	M_4 versus M_5
M_1 versus M_3	M_2 versus M_4	M_3 versus M_5	
M_1 versus M_4	M_2 versus M_5		
M_1 versus M_5			

or 10 tests in all. In general of course when we have J means, the number of pairwise comparisons is

$$\binom{J}{2} = \frac{J!}{(2!)(J-2)!} = \frac{(J)(J-1)(J-2)\cdots(1)}{(2)(J-2)\cdots(1)} = \frac{J(J-1)}{2}$$

Now imagine that all these 10 tests are *independent* of one another (which they are not, as we shall see below) and imagine that the significance level of each test is set at the traditional value of 0.05. Suppose further that in fact *none* of the groups actually differs from the others in terms of population means. What would be the probability of erroneously rejecting a null hypothesis at least once? Note that on each test the probability of making a correct decision given the null hypothesis is true is 0.95. The probability of making at least one incorrect decision is one minus the probability of making all correct decisions, or

$$p(\text{at least one incorrect direction}) = 1 - (0.95)^{10} = 0.401$$

Thus, the chance of making at least one incorrect decision among our 10 tests would be about 40%, assuming that all 10 pairwise H_0s are true.

If this were all there were to the problem, things wouldn't really be all that bad, because although 40% is a fairly large probability of making an error, we would be obtaining quite a bit of information from the 10 tests. (Also bear in mind that out of any given 10 experiments reporting a 0.05 α-level, the probability is 40% that at least one of them will be wrong, assuming of course that the null hypothesis is actually true in all 10 experiments.)

Unfortunately, however, life is not quite this simple because, as we have noted, these 10 tests are actually not independent. To get a feeling for why they are not independent, suppose that by chance we happen to have assigned 10 not-very-bright students to the Freedom Group of our hypothetical experiment, and it is because of this happenstance that the mean score of the Freedom Group is so low. Now this would affect not one, but *four* of the 10 tests—it would affect the comparison of the Freedom Group with each of the other four groups. Thus, these tests would not be independent, and we cannot really apply the reasoning above to obtain the error probability. This means that not only is the error probability rather higher than we would like it to be, but we are placed in the uncomfortable position of not really knowing what the error probability is. From the point of view of trying to produce reasonably precise scientific conclusions, this is not a very acceptable state of affairs.

Two Types of Post-Hoc Tests

There are two extreme methods for dealing with these problems. The first is to basically ignore them and do a series of post-hoc tests anyway. The second is to require a stricter criterion for declaring two means to be significant and increase the strictness as the number of potential tests increases. It will turn out that we can construct a criterion such that the *overall* probability of making at least one Type I error over all 10 [or in general $J(J-1)/2$] tests is less than 0.05. Naturally such constraining of the Type I error probability can be carried out only at the cost of dramatically increasing β, the probability of a Type II error. The stricter the criterion, the more probable it is that some actual population difference between a pair of means will go undetected. (As noted in previous chapters, you can't get something for nothing.) These two techniques are known, respectively, as the least significant difference (LSD) method and the Scheffé method.

The LSD method.   The LSD method involves nothing more than simply doing one or more post-hoc t-tests. These t-tests are identical in most respects to the t-tests described in earlier chapters. The only real difference involves estimating the population variance (σ^2). In the past when we have done a between-groups t-test, the estimate of σ^2 comes from averaging the individual estimates of σ^2 from within each of the two groups. However, in an ANOVA situation we have an estimate of σ^2 obtained by averaging the individual estimates of σ^2 obtained from within each of the J groups. This estimate, of course, is MSW.

To illustrate the LSD method, supppose we want to test whether there is any significant difference between the Reading Group and the Lecture Group. The formula for the appropriate t-test would be

$$t = \frac{M_1 - M_2}{est\ \sigma_{M_1 - M_2}}$$

Now what is $est\ \sigma_{M_1 - M_2}$? Recall that the formula is

$$est\ \sigma_{M_1 - M_2} = \sqrt{est\ \sigma^2_{M_1 - M_2}} = \sqrt{\frac{est\ \sigma^2}{n_1} + \frac{est\ \sigma^2}{n_2}}$$

Since, as we have noted, our best estimate of σ^2 is MSW, and $n_1 = n_2 = 10$,

$$est\ \sigma_{M_1 - M_2} = \sqrt{\frac{MSW}{10} + \frac{MSW}{10}} = \sqrt{\frac{100}{10} + \frac{100}{10}} = \sqrt{20} = 4.47$$

Our t-test will now be based on 45 degrees of freedom because that is the number of degrees of freedom on which our best estimate of the population variance (MSW) is based. To complete our test then

$$t(45) = \frac{M_1 - M_2}{est\ \sigma_{M_1 - M_2}} = \frac{82.0 - 95.0}{4.47} = -2.91$$

This obtained $t(45)$ of -2.91 is more extreme than the criteria (two-tailed) $t(45)$ which are ± 2.01. Therefore, we are in a position to reject the null hypothesis and declare the difference between the Lecture Group and the Reading Group to be significant.

We could, if we wished, perform each of the other nine possible t-tests in a similar way, but there is a shortcut method, which is simply to determine the *smallest amount* by which two means can differ and still be significantly different. This shortcut is carried out as follows: First we know that any obtained $t(45)$ must be more extreme than ± 2.01 (the criterion t) in order to be declared significant. Thus, for any two means M_i and M_j to differ significantly, it must be true that their absolute difference divided by $\sigma_{M_i-M_j}$ must be greater than 2.01—that is, that

$$\frac{|M_i - M_j|}{est\ \sigma_{M_1-M_2}} \geq 2.01$$

Now we have already determined that estimate of $\sigma_{M_1-M_2} = 4.47$. Therefore,

$$\frac{|M_i - M_j|}{4.47} \geq 2.01$$

Multiplying both sides of this inequality by 4.47, we discover that

$$|M_i - M_j| \geq (2.01)(4.47) = 8.99$$

That is, two means must differ by 8.99 or more in order to be declared significantly different. This critical difference is our *least significant difference,* our LSD. It is now a very simple process to go through all possible comparisons and determine which pairs of means differ significantly by this criterion. For example, the Reading and the Lecture Groups are significantly different; as we have already determined, the means differ by 13. Likewise, the Reading and the Freedom Groups differ significantly; the means differ by 11. However, the Reading Group does not differ from the Tutor-1 or Tutor-2 Group, and so on.

> 15.3 In an LSD test the minimum difference between two means necessary to declare the two means to be significantly different is computed. Using this sort of procedure, the α-probability *on each test* is less than 0.05.

The Scheffé method. We have already discussed a number of problems inherent in the LSD method—we are doing a large number of tests, and the tests are not independent. Therefore, although the probability of erroneously rejecting the null hypothesis on *each individual test* is the traditional 0.05, the probability that a Type I error will be made at least once in the *entire post-hoc testing process* is (a) larger than 0.05 and (b) unknown. The Scheffé method eliminates these problems by creating a stricter criterion such that the *overall* probability of making a Type I error *over all tests* is less than 0.05. To illustrate use of the Scheffé method, let's once again do a t-test between the Reading and the Lecture Groups. As we have seen, this t was

$$t(45) = \frac{M_1 - M_2}{est\ \sigma_{M_1-M_2}} = \frac{82.0 - 95.0}{4.47} = -2.91$$

just as it did before. Now (strange though it may seem) we *square* this obtained t

to get

$$t^2 = (-2.91)^2 = 8.47$$

As shown in Digression 15.1, it turns out that any squared t is actually an F.

Digression 15.1

Proof that $t^2 = F$

The formula for t is

$$t = \frac{M_1 - M_2}{est\ \sigma_{M_1 - M_2}}$$

or

$$= \frac{M_1 - M_2}{\sqrt{\dfrac{2\ est\ \sigma^2}{n}}}$$

We square this t to obtain

$$t^2 = \frac{(M_1 - M_2)^2}{\dfrac{2\ est\ \sigma^2}{n}}$$

$$= \frac{n(M_1 - M_2)^2}{2\ est\ \sigma^2}$$

Note that $est\ \sigma^2$ is simply MSW. Thus,

$$t^2 = \frac{n(M_1 - M_2)^2/2}{MSW}$$

Since $F = MSB/MSW$, we now merely have to prove that

$$\frac{n(M_1 - M_2)^2}{2} = MSB$$

To do this, first note that $M_j = T_j/n$. Thus

$$MSB = \frac{n(M_1 - M_2)^2}{2} = \frac{n[T_1/n + T_2/n]^2}{2}$$

$$= \frac{n}{2}\left[\frac{T_1^2}{n^2} - \frac{2\ T_1 T_2}{n^2} + \frac{T_2^2}{n^2}\right]$$

$$= \frac{1}{2}\left[\frac{T_1^2}{n} + \frac{T_2^2}{n} - \frac{T_1 T_2}{n}\right] \qquad\qquad (D.1)$$

Now note that

$$\frac{1}{2}\left[\frac{T_1^2}{n} + \frac{T_2^2}{n}\right] = \frac{T_1^2}{n} + \frac{T_2^2}{n} - \frac{1}{2}\left[\frac{T_1^2}{n} + \frac{T_2^2}{n}\right] \tag{D.2}$$

Substituting Equation D.2 into Equation D.1,

$$MSB \stackrel{?}{=} \frac{1}{2}\left[\frac{T_1^2}{n} + \frac{T_2^2}{n} - \frac{T_1 T_2}{n}\right] = \left[\frac{T_1^2}{n} + \frac{T_2^2}{n}\right] - \frac{1}{2}\left[\frac{T_1^2}{n} + \frac{T_2^2}{n} + \frac{T_1 T_2}{n}\right]$$

Since

$$\frac{T_1^2}{n} + \frac{T_2^2}{n} = \sum_{\text{groups}} \frac{T_j^2}{n}$$

$$MSB \stackrel{?}{=} \sum_{\text{groups}} \frac{T_j^2}{n} - \frac{1}{2}\left[\frac{T_1^2}{n} + \frac{2 T_1 T_2}{n} + \frac{T_2^2}{n}\right]^2$$

Now since

$$\frac{T_1^2}{n} + \frac{2 T_1 T_2}{n} + \frac{T_2^2}{n} = \frac{(T_1 + T_2)^2}{n}$$

$$MSB \stackrel{?}{=} \sum_{\text{groups}} \frac{T_j^2}{n} - \frac{1}{2}\frac{(T_1 + T_2)^2}{n}$$

Note that

$$T_1 + T_2 = T$$

and

$$2n = N$$

Thus

$$MSB \stackrel{?}{=} \sum_{\text{groups}} \frac{T_j^2}{n} - \frac{T^2}{N} \tag{D.3}$$

Equation D.3 is the expression for *SSB*. But since there are only two groups, *dfB* = 1. Thus, the right side of Equation D.3 is also *SSB/dfB* = *SSB*/1 = *MSB*.

So now we have not a *t*, but an *F*, to use as a test statistic. But what do we use for the criterion *F*? We first look up the criterion *F* corresponding to degrees of freedom between and degrees of freedom within for the entire experiment. As we have already seen (Table 15-2) this criterion $F(4,45)$ is 2.61. **We now multiply this *F* of 2.61 by degrees of freedom between to obtain our** criterion for the Scheffé test

Criterion $F = 4(2.61) = 10.44$

and that is the criterion F we use.[3] In the present example the obtained F of 8.47 does not exceed the criterion F of 10.44. Accordingly we fail to reject the null hypothesis and conclude that we do not have enough evidence to declare the Lecture and the Reading Groups to be significantly different.

As we did with the LSD test, we can also make life a little easier by simply calculating the least significant difference necessary to declare two means to be significantly different by a Scheffé test. To do this, we again work backward from the criterion. Recall that the criterion F was 10.44, which in turn was a t^2. Therefore, when doing a Scheffé t-test, our criterion t must be

Criterion $t = \sqrt{\text{criterion } t^2} = \sqrt{10.44} = \pm 3.23$

Once again to declare any two means, M_i and M_j, to be significantly different requires that

$$\frac{|M_i - M_j|}{\text{est } \sigma_{M_i - M_j}} = \frac{|M_i - M_j|}{4.47} \geq 3.23$$

And again cross-multiplying by 4.47 we find that

$$|M_i - M_j| \geq 14.44$$

That is, two means must now differ by 14.44 or more in order to be declared different by a Scheffé test. In line with our notion that the criterion must be stricter for this Scheffé test than it was for the LSD test, we note that this criterion mean difference of 14.44 is rather larger than the criterion mean difference of 8.99 required for the LSD test. In fact with a Scheffé test the only two groups significantly different from one another are the two extreme groups—the Lecture Group and the Freedom Group.

> 15.4 In a Sheffé test, the criterion minimum difference is increased to the point where the probability of making any α-error over all possible comparisons is less than 0.05.

---

[3]Note that the greater the number of groups in the experiment, the more potential pairwise comparisons there are. With more pairwise comparisons there is more potential for error. Thus, to keep the overall error probability constant, it is necessary to have stricter criteria with more groups. Since the criterion F is obtained by multiplying by dfB—and since dfB increases as the number of groups is increased—this requirement is satisfied.

Digression 15.2

Other Types of Post-Hoc Tests

We have described two extreme types of post-hoc tests. As noted, the LSD test maintains a small β-probability at the expense of a large α-probability; and conversely, the Scheffé test maintains a small α-probability at the expense of a large β-probability.

Other tests fall between the LSD and the Scheffé. A complete description of all of them is beyond the scope of this book. We would like to point out, however, that several of these tests use a "variable-criterion" technique running roughly as follows. First, the group means are *ranked* from highest to lowest. Then a test is made of the highest versus the lowest mean using a particular criterion, C_1. If this test fails, the process is stopped. If the test passes, then comparisons are made of the lowest with the second-highest and the highest with the second-lowest mean, using a smaller criterion, C_2. This process continues using successively smaller criteria until the test fails.

PLANNED COMPARISONS

When post-hoc tests are performed, there simply is no way to avoid some kind of aggravation. With a fairly liberal method such as the LSD method, we are reasonably proficient at detecting existing population differences, but we spuriously increase our chances of making a Type I error—of declaring two groups to differ significantly when, in terms of population means, they actually do not. Conversely, the Scheffé test excels at guarding against this problem, but at the expense of increasing the probability of a Type II error. With a Scheffé test, there may be a number of interesting differences among groups, which will not be detected and will fall by the wayside.

As suggested above, a means of circumventing these problems is not to do post-hoc tests at all, but rather to formulate some hypothesis prior to doing the experiment (or at least prior to looking at the data). Such a hypothesis is similar to alternative hypotheses we've seen in the past in that it claims that some differences among the population means do exist. But it is better than previous alternative hypotheses in that it specifies or predicts exactly what the pattern of means ought to be.

Political Preferences: An Example of Planned Comparisons

Suppose we're sociologists interested in the factors contributing to a person's political preference. As part of a theory, we derive the hypothesis that

TABLE 15-4.   Hypothetical Data from the Political Preferences Experiment; Mean Conservatism Score; 95% Confidence Intervals Around the Means Are Provided

$n = 6$ Group 1 20–29 yrs.	$n = 6$ Group 2 30–39 yrs.	$n = 6$ Group 3 40–49 yrs.	$n = 6$ Group 4 50–59 yrs.	$n = 6$ Group 5 60–69 yrs.	$N = 30$
$M_1 = 2 \pm 2.91$ $T_1 = 12$	$M_2 = 4 \pm 2.91$ $T_2 = 24$	$M_3 = 4 \pm 2.91$ $T_3 = 24$	$M_4 = 8 \pm 2.91$ $T_4 = 48$	$M_5 = 12 \pm 2.91$ $T_5 = 72$	$T = 180$

older people will tend to be more conservative. To test this hypothesis, we administer the "Conslib Battery," a test designed to assess the conservativeness of a person's political beliefs. Let's imagine that we administer the Conslib to five groups, each group containing people drawn from a different age range. Assume that there are $n = 6$ people in each of the five groups.

Table 15-4 presents some hypothetical data from this experiment. We note a couple of things in perusing these data. First there do seem to be differences among the five means. Further—and more important—the differences among the means appear to be fairly systematic. In fact the pattern of means—the rise in conservatism over age group—is in accord with our original hypothesis that older people tend to be more conservative. As usual, however, we must be concerned with the problem that the observed differences among the means may be due merely to chance. (Since there not only seem to be differences among means but these differences also seem to be in the direction that we expected on an a priori basis, it would not really seem very likely that the differences are due to chance. Nonetheless, they may be, and we still have a guard against the possibility.)

Ideally we would like some sort of test not only to assess whether there are significant differences among the means but simultaneously to assess the degree to which the pattern among the means is the one anticipated by our hypothesis. Such a test is called a planned comparison. Let's now see how a planned comparison works and what it tells us.

Magic numbers.   The first step in doing a planned comparison is to make up a set of numbers (often called "weights" or w_js) with the following properties. First there must be one weight corresponding to each condition, or sample mean (thus, in our example there should be five weights corresponding to the five means). Second—and this is crucial—the pattern of the weights should correspond to the pattern of means predicted by the hypothesis. Since in the current example we expect the mean conservatism score to increase[4] across age group, one set of weights that would seem to do the trick is

$$w_j s = \{1, 2, 3, 4, 5\}$$

---

[4]We are actually testing the hypothesis that the means increase *linearly*. We do this for simplicity of discourse. Digression 15.4 takes up this issue in greater detail.

However, there is one additional constraint on the weights—they must add to zero. This constraint is not really very constraining, for to make these five numbers add to zero, we can simply subtract three from each of them to obtain

$$w_j s = \{-2, -1, 0, 1, 2\}$$

(Thus, $w_1 = -2$, $w_2 = -1$, and so on.) Note that subtracting (or adding) a constant to a set of numbers does not change the *pattern* of the numbers—that is, the relationship of the numbers to one another. But adding or subtracting an appropriate constant is a technique that can always make the sum of any set of numbers be anything we wish.

> 15.5 To test a hypothesis that a specific pattern of population means exists, we make up a set of numbers whose pattern corresponds to the hypothesized pattern. The only constraint on this set is that the numbers must add to zero.

The question we now address is: To what degree is the pattern of population means that correspond to our conditions similar to the pattern of weights we've chosen? Another way of asking this question is: How well do the weights correlate with the population means? If the correlation is high, then the hypothesis is a good hypothesis. If the correlation on the other hand is not very high, then the hypothesis is not so good. In the extreme the weights may have zero correlation with the population means. In that event the hypothesis would be viewed as having nothing to do with reality.

Naturally we do not know the population means, $\mu_1, \ldots, \mu_J$. But we do know the sample means, $M_1, \ldots, M_J$ (or in this case, M_1 through M_5) which are *unbiased estimates* of the population means. Thus, by finding the correlation between the sample means and the weights, we obtain an unbiased estimate of the correlation between the population means and the weights corresponding to the hypothesis.

A sum of squares for the hypothesis.   Recalling what we learned in the last chapter, we can compute a Pearson r^2 between the J pairs of weights and means as follows:

$$r^2 = \frac{\left[J \sum_{j=1}^{J} w_j M_j - \left(\sum_{j=1}^{J} w_j \right) \left(\sum_{j=1}^{J} M_j \right) \right]^2}{\left[J \sum_{j=1}^{J} w_j^2 - \left(\sum_{j=1}^{J} w_j \right)^2 \right] \left[J \sum_{j=1}^{J} M_j^2 - \left(\sum_{j=1}^{J} M_j \right)^2 \right]}$$

Digression 15.3 shows that this equation is algebraically equivalent to

$$r^2 = \frac{n\left(\sum_{j=1}^{J} w_j M_j\right)^2}{\left(\sum_{j=1}^{J} W_j^2\right)(SSB)}$$

where n is the number of scores in each group and SSB is, as usual, the sum of squares between groups.

Digression 15.3

The Pearson r^2 Between w_js and M_js

The r^2 is

$$r^2 = \frac{[J\Sigma M_j w_j - (\Sigma M_j)(\Sigma w_j)]^2}{[J\Sigma w_j^2 - (\Sigma w_j)^2][J\Sigma M_j^2 - (\Sigma M_j)^2]} \tag{D.1}$$

Note that since this correlation is over J pairs of means and weights, J is the "n" in the equation, and all summations are from $j = 1$ to J.

Since Σw_j is stipulated to be zero, Equation D.1 can be simplified considerably.

$$r^2 = \frac{(J\Sigma M_j w_j)^2}{[J\Sigma w_j^2][J\Sigma M_j^2 - (\Sigma M_j)^2]}$$

Multiplying numerator and denominator of Equation D.2 by n/J^2 and noting that $M_j = T_j/n$

$$r^2 = \frac{n(\Sigma M_j w_j)^2}{[\Sigma w_j^2][\Sigma(T_j^2/n) - (\Sigma T_j)^2/Jn]}$$

Since $\Sigma T_j = T$ and $Jn = N$,

$$\frac{\Sigma T_j^2}{n} - \frac{(\Sigma T_j)^2}{Jn} = \frac{\Sigma T_j^2}{n} - \frac{T^2}{N} = SSB$$

And thus:

$$r^2 = \frac{n(\Sigma M_j w_j)^2}{(\Sigma w_j^2)(SSB)}$$

Now recall that a Pearson r^2 relating X and Y is in general the percentage of variance among the Y-scores that is accounted for by variation in the X-scores. **If we consider SSB to represent variation among the group** means then

$$SSH = r^2(SSB) = \frac{n\left(\sum_{j=1}^{J} W_j M_j\right)^2}{\sum_{j=1}^{J} W_j^2} \tag{15.2}$$

is part of SSB and can be considered to be a sum of squares due to the hypothesis. That is, to the extent that the pattern of means faithfully reflects what is predicted by the hypothesis, the sum of squares will be big; and to the extent that the observed pattern of means does not fit the hypothesis, this sum of squares will be small. This sum of squares is based on 1 degree of freedom because it represents one specific pattern of means.

Applying Equation 15.2 to our present data, we see that a sum of squares due to our hypothesis may be computed as

$$SSH = \frac{6\,[(2)(-2) + (4)(-1) + (4)(0) + (8)(1) + (12)(2)]^2}{(-2)^2 + (-1)^2 + (0)^2 + (1)^2 + (2)^2}$$

$$= 345.6$$

Testing our hypothesis: A new kind of summary table.   Now let's construct a sort of ANOVA summary table. Although we have not computed it, let's assume that sum of squares within (which, of course, is based on 25 degrees of freedom) is equal to 300. We compute SSB using our familiar equation

$$SSB = \sum_{groups} \frac{T_j^2}{n} - \frac{T^2}{N}$$

which in this case is

$$SSB = (12^2/6 + \cdots + 72^2/6) - 180^2/30$$

$$= 1464 - 1080 = 384$$

Recall now that we have a sum of squares due to our hypothesis which turned out to be 345.6 and which, we have said, is based on 1 degree of freedom. As always we can divide the sum of squares due to the hypothesis (SSH) by the associated degrees of freedom (dfH) to arrive at a mean square due to the hypothesis of 345.6. All this information is summarized in Table 15-5.

Now consider the null hypothesis that our hypothesized pattern of means is uncorrelated with the actual populations means. If such were the case, then the mean square due to the hypothesis *will be an estimate of the population variance* (σ^2). Therefore, to test the significance of our hypothesis, we merely perform an F-test of mean square due to hypothesis divided by our best estimate of σ^2, which is MSW. The obtained $F_{(1,25)}$ that emerges from this procedure is 28.5. The criterion $F_{(1,25)}$ is 4.24, and we can therefore conclude that our

TABLE 15-5.   An **ANOVA** Summary Table for the Political Preference Experiment (this table contains information about our hypothesis).

Source	df	SS	MS	Obt F	Crit F
Total	29	684			
Between	4	384			
Hypothesis	1	345.6	345.6	28.8	4.35
Within	25	300	12		

hypothesis is resoundingly significant. Clearly our hypothesized pattern of means has something to do with reality.

The Pearson r^2.   So far we have skipped over the Pearson r^2 between the weights and the means and have gone directly to testing the null hypothesis of no correlation. As you might guess, however, the correlation itself—the estimated percentage of variance among the population means that is accounted for by the hypothesis—is of substantial interest. This r^2 may be easily computed by dividing SSH, the sum of squares due to the hypothesis by SSB the sum of the squares we began with. Applying this procedure to the data at hand, we find that the percentage of variance accounted for is

$$\% \text{ variance} = r^2 = \frac{SSH}{SSB} = \frac{345.6}{384.0} = .90$$

Thus, 90% of the sum of squares between groups is accounted for by the hypothesis.

The residual variance.   We now address the question of whether we can reject the hypothesis that the population means are *perfectly* correlated with the weights we have chosen (that is, whether r^2 between the $\mu_j s$ and $M_j s$ is equal to 1–0). Note that if we are *unable* to reject this hypothesis, then we would be in quite good shape in the sense that the hypothesis would in some sense be sufficient to account for all the data.

The logic by which we address this question involves examining the portion of SSB that is *not* accounted for by our weights. This variance is referred to as the *residual variance* and is obtained simply by subtracting SSH from SSB. This residual variance has associated with it the J-2 or in this case 3 degrees of freedom between left over after we have deducted from dFB the 1 degree of freedom due to the hypothesis itself.

In Table 15-6 we have created a more complete version of Table 15.5 including this residual information. We now have a new sum of squares (due to residual) and an associated degrees of freedom. As always we can divide the sum

TABLE 15-6.   A More Complete Summary Table; Information About the Residual Is Included

Source	df	SS	MS	Obt F	Crit F
Total	29	684			
Between	4	384			
Hypothesis	1	345.6	345.6	28.8	4.35
Residual	3	38.4	12.8	1.07	3.10
Within	25	300	12		

of squares by degrees of freedom to obtain a mean square due to residual, which in this case is 12.8.

Now suppose that the population means of our five conditions are reflected perfectly by the weights that we have chosen. In this case would we expect the sum of squares due to the hypothesis to take up the *entire* sum of squares between, leaving zero residual? The answer is no. Because of population variance (σ^2) we would expect the observed sample means to be somewhat displaced relative to the population means. Therefore we would not expect the sample means correlate perfectly with our weights, and in turn we would expect the residual variance to be nonzero. However, in such a situation the residual mean square would be an estimate of the population variance. We can therefore perform an *F*-test of mean square residual against mean square within to obtain an $F(3,20) = 1.07$. Since the criterion $F(3,20) = 3.10$, this is not significant. This means that our hypothesis is sufficient[5] to explain all the variation among the means in our experiment. The residual variance may be assumed to arise simply as a consequence of σ^2.

So where are we? Let's summarize a bit here. Prior to running our experiment, we had a specific hypothesis about what the pattern of means should be. We then did the experiment. Rather than simply being interested in whether there were *any* differences among the five sample means, however, we were primarily interested in the degree to which our pattern of observed sample means corresponded to the hypothesized pattern. To investigate this issue, we carried out the following procedures.

1. We first computed a sum of squares due to the hypothesis, *SSH*. This sum of squares was a portion of *SSB*, the sum of squares due to conditions. *SSH* was essentially obtained by computing the correlation (Pearson r^2) between the weights constituting the hypothesis and the sample means, and multiplying this r^2 by *SSB*.

---

[5]Actually "sufficient" may be too strong a word, because making this conclusion carries with it all the problems of accepting a null hypothesis. A more precise (although long-winded) way of stating the conclusion would be: "The data do not permit us to reject the idea that our hypothesis is sufficient to account for the observed pattern of means."

2. The sum of squares due to the hypothesis was based on 1 degree of freedom. By computing a mean square due to the hypothesis and testing this mean square against MSW, we evaluated the null hypothesis that the condition population means are uncorrelated with the hypothesis weights.

3. We then considered the sum of squares *not* accounted for by the hypothesis. We divided this residual sum of squares by its associated J-2 degrees of freedom to obtain a residual mean square. The residual mean square was in turn tested against MSW to evaluate the null hypothesis that the condition population means are correlated *perfectly* with the hypothesis weights.

15.6 We obtain a sum of squares due to our hypothesis. This sum of squares, SSH, comes out of SSB. Thus, it's also possible to obtain a residual sum of squares, $SSR = SSB - SSH$ that is *not* accounted for by the hypothesis. We are thus in a position to (a) test the significance of the hypothesis ($F = MSH/MSW$), (b) test the significance of the residual ($F = MSR/MSW$), and (c) compute what percentage of between-groups variance is accounted for by the hypothesis (percentage of variance = SSH/SSB).

Notice that we never did an omnibus F-test on our data. One of the fringe benefits of an a priori hypothesis is that there is no reason to do an omnibus F-test. Our interest does not lie in whether there are merely differences among the groups; rather, it lies in the degree to which the observed pattern of means corresponds to the hypothesized pattern of means. So generally when we do a planned comparison, we do *only* that planned comparison—we substitute it for the omnibus F-test.

MULTIPLE PLANNED COMPARISONS

The example we just described was a very simple one. We had one a priori notion about how the means should come out—that conservatism should increase with age group—and that was that. Often, however, things are not this simple. Sometimes we have two or more hypotheses that we would like to test simultaneously. Let's see how we'd do this.

Drug-Induced Time Distortion

A common phenomenon reported by users of marijuana is that of time distortion. That is, some given period of time (say, five minutes) often seems much longer. Let's suppose that we wish to test this reported phenomenon experimentally. To do so, we randomly assign 40 subjects to five groups, eight subjects per group. The five groups are defined as follows. Each subject in the first two groups is given marijuana in the form of a marijuana cigarette (hereafter referred to as a "joint"). Subjects in Group 1 are given one joint to smoke whereas subjects in Group 2 are given two joints to smoke. Subjects in Group 3 are given nothing. Subjects in Groups 4 and 5 are given "placebo joints" consisting of cigarettes containing oregano. Subjects in Group 4 are given one placebo joint to smoke, and subjects in Group 5 are given two placebo joints.

For the experimental task subjects are put in a comfortable room where they listen to soft rock music for exactly five minutes. After the five minutes are up, each subject is asked how much time he or she thinks has elapsed. (Notice that, to the extent that there is time distortion, the subjective time-duration report should be longer than the actual time of five minutes spent in the room.)

Hypotheses. Suppose there are two separate hypotheses we wish to test in this experiment. The first hypothesis is that smoking marijuana will increase time distortion relative to not smoking marijuana. An appropriate set of weights corresponding to this hypothesis would indicate Groups 1 and 2 showing higher time-distortion scores than do Groups 3, 4, or 5. The simplest set of weights that would fulfill these requirements (and also sum to zero) would seem to be $\{3, 3, -2, -2, -2\}$.

Our second hypothesis is that the more marijuana smoked, the more time distortion will occur. According to this hypothesis, subjects in Group 2 (the two-joint group) should show more time distortion than subjects in Group 1 (the one-joint group). An appropriate set of weights to reflect this hypothesis would be $\{-1, 1, 0, 0, 0\}$. Notice that this hypothesis does not make any claims about the relationships among Groups 3, 4, and 5, because it is concerned only with the two marijuana groups.

The data. With our hypotheses and the corresponding weights all formulated, let's look at some hypothetical data from this experiment, which are shown in Table 15-7. Table 15-7 lists the five groups along with their means and the totals. Glancing at the data, we see that both our hypotheses seem to be getting

---

TABLE 15-7. Hypothetical Data from the Marijuana/Time-Distortion Experiment: Mean Reported Time in Minutes (actual time is 5 minutes); 95% Confidence Intervals Around the Means Are Provided

$n = 8$ Group 1 (Marijuana— 1 joint)	$n = 8$ Group 2 (Marijuana— 2 joints)	$n = 8$ Group 3 (Nothing)	$n = 8$ Group 4 (Oregano— 1 joint)	$n = 8$ Group 5 (Oregano— 2 joints)	$N = 40$
$M_1 = 7 \pm 1.05$ $T_1 = 56$	$M_2 = 10 \pm 1.05$ $T_2 = 80$	$M_3 = 5 \pm 1.05$ $T_3 = 40$	$M_4 = 5 \pm 1.05$ $T_5 = 40$	$M_5 = 6 \pm 1.05$ $T_5 = 48$	$T = 264$

some support. The three control groups (Groups 3, 4, and 5) seem to show very little, if any, time distortion (that is, their mean reported times are very close to the actual time of five minutes). The two marijuana groups, in contrast, do seem to show time distortion, thereby supporting the first hypothesis. Also, the two-joint group seems to show greater time distortion than does the one-joint group in support of the second hypothesis. Let's see how we would do the appropriate tests to check out these notions.

Sums of squares. Initially we must compute a sum of squares between the five groups. Computation of SSB is necessary because it is the sum of squares out of which sums of squares due to our hypotheses will come. We compute SSB in the normal way and find that

$$SSB = \sum_{\text{groups}} \frac{T_j^2}{n} - \frac{T^2}{N}$$

$$= (56^2/8 + \cdots + 48^2/8) - (264)^2/40$$

$$= 1880.0 - 1742.4 = 137.6$$

Now let's compute sums of squares due to our hypotheses. We can compute a sum of squares due to each of our two hypotheses using Equation 15.2, introduced in the last section. For Hypothesis 1 the sum of squares is

$$SSH1 = \frac{8[(7)(3) + (10)(3) + 5(-2) + 5(-2) + 6(-2)]^2}{(3)^2 + (3)^2 + (-2)^2 + (-2)^2 + (-2)^2}$$

$$= 96.26$$

and, likewise, for Hypothesis 2 the sum of squares is

$$SSH2 = \frac{8[7(1) + 10(-1) + 5(0) + 5(0) + 6(0)]^2}{(1)^2 + (-1)^2 + (0)^2 + (0)^2 + (0)^2}$$

$$= 36$$

A summary table. We can now construct a table to summarize various sums of squares. Let's assume that when we computed the sum of squares within the five groups, this SSW was 75. (Note that this sum of squares within is based on 35 degrees of freedom—7 degrees of freedom from within each of the 5 groups.) We have seen that SSB was 137.6. Of this 137.6, 96.3 came from the first hypothesis and 36 from the second hypothesis. Therefore, what we have left in SSB is $137.6 - 96.3 - 36 = 5.3$. This 5.3 is *residual*.

Each of the two hypotheses is based on 1 degree of freedom. (As noted above, any given hypothesis is based on 1 degree of freedom because it represents a single possible pattern of means.) Since our two hypotheses together use up 2 degrees of freedom out of the original 4 degrees of freedom between, there are 2 degrees of freedom left over for the residual.

In Table 15-8 we have included a new column for "percentage of variance

TABLE 15-8.   An ANOVA Summary Table for the Marijuana/Time Distortion Experiment

Source	df	SS	% Var	MS	Obt F	Crit F
Total	39	212.6				
Between	4	137.6				
H1	1	96.3	70	96.3	44.9	4.17
H2	1	36	26	36	16.8	4.17
Residual	2	5.3	4	2.65	1.24	3.32
Within	35	75		2.14		

accounted for." As noted earlier, percentage of variance for a given hypothesis is simply the correlation (r^2) between the weights corresponding to that hypothesis and the sample means. We see that 70% of the variance among the means (that is, 70% of SSB) is accounted for by Hypothesis 1 and an additional 26% is accounted for by Hypothesis 2. Thus, only 4% of the variance among conditions is attributed to the residual.

We can test the significance of the two hypotheses by computing appropriate mean squares which are then F-tested against the mean square within. As is evident in Table 15-8, both these obtained Fs—44.9 for the first hypothesis and 16.8 for the second hypothesis—are wildly significant. We are therefore in a position to conclude that both of our hypotheses are accounting for significant proportions of the variance between the five groups. Both of them are good hypotheses; both are meaningful; both appear to reflect reality.

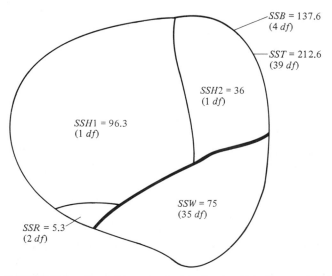

FIGURE 15-1.   Breakdown of variance when two hypotheses are used. SSB is subdivided into three components: sums of squares due to the first hypothesis, the second hypothesis, and the residual.

Furthermore, we can see that when we F-test the residual mean square against MSW, the resulting F is only 1.24, which is resoundingly nonsignificant. We are therefore in good shape. We conclude that our two hypotheses are sufficient to explain all the variance among groups. The residual variance can be assumed to be a mere by-product of σ^2, the population variance.

Dividing the pie. Figure 15-1 shows how the total "variance pie" in this situation may be envisioned. Once again we start with the total variance in the entire situation—variance that arises from the fact that all 40 scores in the experiment differ from one another. This variance pie is initially divided into *SSW* and *SSB*. Now as has been true in the past, we are interested in subdividing *SSB* into its various components. In this particular situation *SSB* is partitioned into three portions—variance stemming from two hypotheses and a residual.

15.7 Two separate sets of weights corresponding to two separate hypotheses can be constructed. Each can be tested, and each represents some percentage of *SSB*. The removal from *SSB* of the sums of squares due to the two hypotheses leaves a residual which can also be tested.

Independence of Hypotheses

Let's alter our hypotheses about the time-distortion experiment. Imagine that Hypothesis 1 is still the same as it was before—that smoking marijuana leads to more time distortion than not smoking marijuana. But suppose that Hypothesis 2 is slightly different. Suppose that, instead of saying more marijuana leads to more time distortion, Hypothesis 2 now says that it is necessary to smoke two or more joints in order to experience any time distortion at all. This hypothesis would therefore claim that subjects in Group 2, who had smoked two joints, should experience time distortion, and the other four groups should experience no time distortion. An appropriate set of weights for this new Hypothesis 2 would be $\{-1, 4, -1, -1, -1\}$. Computing the sum of squares due to this hypothesis, we find that

$$SSH2 = \frac{8[(7)(-1) + (10)(4) + (5)(-1) + (5)(-1) + (6)(-1)]^2}{(-1)^2 + (4)^2 + (-1)^2 + (-1)^2 + (-1)^2}$$

$$= 115.6$$

Now there are still 137.6 sums of squares between groups and furthermore 96.26 of that sum of squares is used up by Hypothesis 1. With this new second hypothesis we appear to use up an additional 115.6 sum of squares. But wait! Between the first hypothesis and our new second hypothesis we have used up

115.6 + 96.26 = 211.86 sum of squares, which is more than the 137.6 that we started with. What's wrong?

Let's reconsider the "variance pie" depicted in Figure 15-1. When we divided *SSB* into the three components—sum of squares due to Hypothesis 1, Hypothesis 2, and residual—we made the assumption that these three components used nonoverlapping sums of squares. This seems like an intuitively reasonable assumption. After all we have separate hypotheses—why shouldn't they use separate sums of squares?

However, this assumption supposes that the two hypotheses are *independent* of one another. Independent hypotheses are hypotheses that have nothing to do with one another—that is, discovering something about the truth of one hypothesis tells us nothing about the truth of another, independent hypothesis. When we think about our original two hypotheses, they do pretty much seem to be independent of one another. The first hypothesis simply said that smoking marijuana would cause some time distortion relative to not smoking marijuana—that is, that Groups 1 and 2 should show more distortion than Groups 3, 4, and 5. But the second hypothesis was quite different—it said that the more marijuana smoked, the more time distortion there is—that is, that Group 2 should show more distortion than Group 1. Intuitively, these hypotheses are independent in the sense that knowing something about the relationship of Groups 1 and 2 to Groups 3, 4, and 5 does not allow us to predict anything about the relationship of Group 1 to Group 2.

However, let's consider the relationship between Hypothesis 1 and our new Hypothesis 2. As before, Hypothesis 1 says that Groups 1 and 2 show more distortion than Groups 3, 4, and 5. However, the new Hypothesis 2 says that Group 2 shows more distortion than Groups 1, 3, 4, and 5. Now these hypotheses do not seem very independent. That is, if the second hypothesis were true—if Group 2 does show more distortion than Groups 1, 3, 4, and 5—then we would tend to believe that Groups 1 and 2 show more distortion than Groups 3, 4, and 5—that is, that the first hypothesis is true also.

Overlapping sums of squares. When two hypotheses are nonindependent, they *share* sums of squares. Thus, with our first hypothesis and our alternative second hypothesis, we would end up with a situation such as that depicted in Figure 15-2. Figure 15-2 resolves the seeming paradox that sums of squares stemming from two hypotheses add up to more than the original sum of squares between. The answer lies in the fact that part of the sums of squares are shared by the two hypotheses.

A test for independence. This is a fine kettle of fish," you say. We find that sometimes two hypotheses are independent of one another and other times hypotheses are not independent of one another. We have provided an intuitive rationale for deciding whether two hypotheses are independent, but this is not really very satisfying.

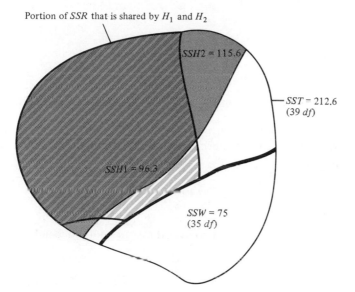

Portion of *SSR* that is shared by H_1 and H_2

SSH2 = 115.6

SST = 212.6
(39 *df*)

SSH1 = 96.3

SSW = 75
(35 *df*)

FIGURE 15-2. Nonindependent hypotheses share sums of squares. The shared area is depicted by the double-cross-hatched area.

Fortunately there is a very clear-cut test of whether two hypotheses are independent of one another. Two hypotheses are independent when the weights corresponding to the two hypotheses are uncorrelated with one another. Determining whether two sets of weights are uncorrelated is, in turn, very simple. Suppose we have two hypotheses whose corresponding weights we may depict as follows.

$$H1: \{w_{11}, w_{12}, \ldots, w_{1J}\}$$

$$H2: \{w_{21}, w_{22}, \ldots, w_{2J}\}$$

Now we compute a number which is a sum of the cross-products of the weights from the two hypotheses. Let's call this number x_{12}, that is

$$x_{12} = \sum_{j=1}^{J} w_{1j}w_{2j}$$

It is easy to show that x_{12} is essentially the numerator of the correlation coefficient (Pearson r) between the two sets of weights. So if x is *zero*, then the two hypotheses are uncorrelated and hence independent. If x is nonzero, the two hypotheses are correlated and hence not independent. To illustrate, consider our original two hypotheses which are

$$H1: \{3, 3, -2, -2, -2\}$$

and

$$H2: \{-1, 1, 0, 0, 0\}$$

The sum of the cross-products for these two hypotheses turns out to be

$$x_{12} = (3)(-1) + (3)(1) + (-2)(0) + (-2)(0) + (-2)(0)$$
$$= 0$$

Thus, the two hypotheses are independent. But now let's consider our first hypothesis and our alternative second hypothesis. In this case the sum of the cross-products turns out to be

$$x_{12} = (3)(-1) + (3)(4) + (-2)(-1) + (-2)(-1) + (-2)(-1)$$
$$= 15$$

Since 15 is not zero, we conclude that these two hypotheses are not independent.

15.8 Two hypotheses are independent if the sum of their cross-products is zero. Two hypotheses take up nonoverlapping sums of squares if and only if the two hypotheses are independent.

A third independent hypothesis. Consider our two original hypotheses, which, as we have seen, are independent of one another. Could we generate a third hypothesis that is independent of each of the first two? A moment's thought will reveal that the answer is yes. Suppose we have a hypothesis stating that the more *placebo* joints that are smoked, the more time distortion should occur. According to this hypothesis subjects in Group 5 (who smoke two placebo joints) should experience more time distortion that subjects in Group 4 (who smoke one placebo joint). The appropriate weights for this hypothesis would be

$$H3: \{0, 0, 0, -1, 1\}$$

It is quite easy to see that this hypothesis is independent of each of the other two. Comparing Hypothesis 1 and Hypothesis 3,

$$x_{13} = (3)(0) + (3)(0) + (-2)0 + (-2)(-1) + (-2)(1) = 0$$

Comparing Hypothesis 2 with Hypothesis 3,

$$x_{23} = (-1)(0) + (1)(0) + (0)(0) + (0)(-1) + (0)(1) = 0$$

These three hypotheses are said to be *mutually independent*. Each hypothesis is independent of each of the other hypotheses.

Let's compute a sum of squares due to this third hypothesis. To do this, we simply plug into Equation 15.2 and discover that

$$SSH3 = \frac{8\,[(7)(0) + (10)(0) + (5)(0) + (-1)(5) + (1)6]^2}{(0)^2 + (0)^2 + (0)^2 + (-1)^2 + (1)^2}$$

$$= 4$$

How many possible mutually independent hypotheses are there? It is possible to generate *as many independent hypotheses as there are degrees of freedom between groups*. This makes sense. Degrees of freedom between groups represents the number of ways that the groups are free to vary from one another. Each independent hypothesis accounts for one of these possible ways in which the groups can vary.

Since in our present situation $dfB = 4$, it would be possible to have four hypotheses all independent of one another. We have already come up with three mutually independent hypotheses, so it should be possible to come up with one more. And in fact a fourth hypothesis might be that smoking any number of placebo joints causes more time distortion than not smoking anything. Thus, according to this Hypothesis 4, Groups 4 and 5 should show more distortion than Group 3. The weights for this hypothesis would be

$$H4: \{0, 0, -2, 1, 1\}$$

Is this hypothesis independent of the other three? We can perform the appropriate tests and discover that

$$x_{14} = (3)(0) + (3)(0) + (-2)(-2) + (-2)(1) + (-2)(1) = 0$$

$$x_{24} = (-1)(0) + (1)(0) + (0)(-2) + (0)(1) + (0)(1) = 0$$

$$x_{34} = (0)(0) + (0)(0) + (0)(-2) + (-1)(1) + (1)(1) = 0$$

So it is indeed true that $H4$ is independent of $H1$, $H2$, and $H3$. We have already shown that $H1$, $H2$, and $H3$ are mutually independent. We now see that all four hypotheses are mutually independent.

Finally let's compute the sum of squares due to this fourth hypothesis

$$SSH4 = \frac{8\,[7(0) + (10)(0) + 5(-2) + (5)(1) + (6)(1)]^2}{(0)^2 + (0)^2 + (-2)^2 + (1)^2 + (1)^2}$$

$$= 1.3$$

Table 15-9 shows the sums of squares that we have computed from all four of our hypotheses. Notice what happens when we add these four sums of squares together—they add up to sum of squares between! This makes sense. Again using our pie representation, Figure 15-3 shows the variance pie and how it is divided among these four independent hypotheses. Notice that our four independent hypotheses *must* use up all of sums of squares between. This is a direct implication of our assertions that (a) independent hypotheses take up nonoverlapping sums of squares, and (b) in this example there can be only four independent hypotheses. If the sums of squares due to hypotheses did *not* take up

TABLE 15-9.   Sums of Squares from Four Independent Hypotheses in the Marijuana/Time Distortion Experiment

Hypotheses	Weights	df	Sum of Squares
1	$\{\ \ 3, 3, -2, -2, -2\}$	1	36
2	$\{-1, 1,\ \ \ 0,\ \ \ 0,\ \ \ 0\}$	1	96.3
3	$\{\ \ 0, 0,\ \ \ 0, -1,\ \ \ 1\}$	1	4.0
4	$\{\ \ 0, 0, -2,\ \ \ 1,\ \ \ 1\}$	1	1.3
		$4 = dfB$	$137.6 = SSB$

the entire sum of squares between, then there would be a little bit of sums of squares left over. But by our logic that little leftover sums of squares would correspond to another independent hypothesis. Since we cannot have another independent hypothesis, there can be no sums of squares left over. The sums of squares due to the four independent hypotheses completely partition the total sums of squares between.

In practice one generally does not use up all possible independent hypotheses. It is more typically the case that, preceding an experiment, there are only one or two hypotheses one wishes to test. So one typically goes through the procedure sketched in the initial portion of this section, testing one or perhaps two hypotheses and then looking at the residual. If the residual is nonsignificant, then one concludes that the original hypothesis or hypotheses are sufficient to explain all the variance between the various groups.

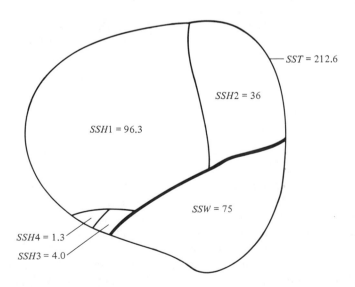

FIGURE 15-3.   Four independent hypotheses completely partition *SSB*.

SOME TECHNIQUES FOR MAKING UP WEIGHTS

In the last few sections we have tried to outline some of the major concepts necessary for working with planned comparisons. In this section we'll be a little bit more practically oriented and discuss some useful techniques for assigning actual sets of weights in various situations. In the examples we have been discussing, selection of the appropriate weights has been fairly straightforward, but not all situations are quite this simple. Let's first state two steps that are sometimes handy when selecting weights.

1. First choose any weights whatsoever that seem to be appropriate.
2. Then worry about making sure these weights add to zero.

These steps will prove convenient when selection of weights is at all complicated.

A Test for Linear Trend

A simple hypothesis that is often the first to be tested is that some dependent variable is a *linear* function of some independent variable. Accordingly one particular set of weights that is often applied to the data from some experiment is called a test for linear trend. We have actually already illustrated a test for linear trend in our age group/political preference example. Another such example is the following.

Suppose we are physiological psychologists interested in the effects of amphetamines on problem-solving behavior. To investigate this issue, we plan to give several dosages of amphetamines to several different groups of people. (Thus, the independent variable in this experiment is dosage.) We will then have the subjects in each group solve a difficult problem. The dependent variable will be the amount of time to solve the problem, and our hypothesis is that the amount of time to solve the problem will decrease linearly as a function of amphetamine dosage. Table 15-10 shows three variations of an experiment to test this question.

TABLE 15-10.   Three Possible Designs to Test the Effect of Amphetamine Dosage on Problem-Solving Time

Design	# Levels of Drug Dosage	Dosages				
		Level 1	Level 2	Level 3	Level 4	Level 5
1	5	0	10	20	30	40
2	4	0	10	20	30	
3	5	0	5	10	20	40

An odd number of groups.   In the first design we assume that we have five groups and the amphetamine levels are 0, 10, 20, 30, and 40 mg. To test the hypothesis that problem-solving time will decrease linearly as a function of amphetamine dosage, we simply use the weights $\{2, 1, 0, -1, -2\}$. Note that this set of weights reflects a linear hypothesis in that the differences between adjacent pairs are equal (they are all equal to 1). This reflects the notion that each equal increment in amphetamine dosage will produce an equal decrement in problem-solving time.

An even number of groups.   Now suppose that instead of having five groups, we had only four groups with dosages of 0, 10, 20, and 30 mg. What would the appropriate weights be in this situation? Well, one set of weights that might come to mind is $\{2, 1, -1, -2\}$. There is a problem with this particular set of weights, however, in that they are not linear. The decrement from 2 to 1 is 1, but then the weights drop from 1 to -1, a decrement of 2.

One appropriate set of weights to use in this situation would be $\{1\frac{1}{2}, \frac{1}{2}, -\frac{1}{2}, -1\frac{1}{2}\}$. These weights satisfy both the requirements of being linear and of adding to zero. Working with fractions is often tedious, however, so it would probably be more convenient to multiply this particular set of weights by 2 to arrive at $\{3, 1, -1, -3\}$.

In general with an odd number of levels, the appropriate weights for testing linear trend are $\{\ldots -3, -2, -1, 0, 1, 2, 3 \ldots\}$. With an even number of levels, the appropriate weights are $\{\ldots -5, -3, -1, 1, 3, 5 \ldots\}$.

Digression 15.4

Linear Trend with Uneven Spacing

The examples that we have just gone through share a common but important property—this is that there is equal spacing between adjacent levels of the independent variable. For example the difference between Level 1 and Level 2 is 10 mg; the difference between Level 2 and Level 3 is also 10 mg; and so on. Note that, as shown in Figure 15-4, if we plot our weights (which, recall, are supposed to reflect our hypothesized pattern of means) as a function of our independent variable, we get a straight line, just as we should. This is the meaning of the idea that the weights are linear weights.

Now suppose that the spacing between the levels of the independent variable were not even. For example, suppose that the amphetamine dosage levels that we chose were 0, 5, 10, 20, and 40 mg, as is the case with Design 3 in Table 15-10. Suppose now that we just cavalierly went ahead and used the weights $\{2, 1, 0, -1, -2\}$ as a test for linear trend. Would this really represent a test for linear trend? The answer is

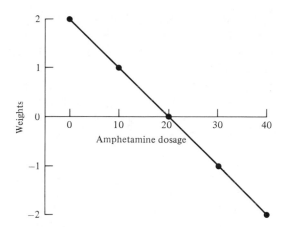

FIGURE D-1. Representation of the fact that the weights {2, 1, 0, −1, −2} reflect a linear function of the dosage levels 0, 10, 20, 30, 40.

no. Suppose we plot these weights as a function of the level of the independent variable, as has been done in Figure D-2. It's easy to see that this function is not linear. These weights assume, for example, that the drop in the dependent variable (amount of time to solve the problem) that occurs when dosage is increased from 0 to 5 mg is the same as the drop that occurs when dosage is increased from 20 to 40 mg. This does not produce a linear function—if the relationship between time to solve the problem and amphetamine dosage really were linear, then the drop in problem-solving time from increasing the dosage from 20 to 40 mg should be quite a bit greater (in fact four times as great) as the drop in time from increasing the level from 0 to 5 mg. Arriving at a solution to this

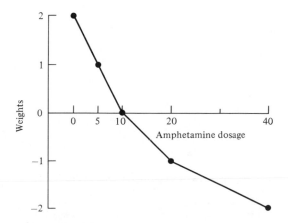

FIGURE D-2. The weights 2, −1, 0, −1, −2 do *not* reflect a linear function of the dosage levels, 0, 5, 10, 20, 40.

problem is quite simple. We use the two steps for choosing weights outlined at the beginning of this section. First we make up any set of weights that would be appropriate. The easiest set of weights to reflect linearity in this situation would simply be to let the weights equal the values we have chosen for the independent variable; that is, we choose our weights to be $\{0, -5, -10, -20, -40\}$.[6] Now of course these particular weights do not add to zero, but it is easy to make them add to zero. As we illustrated previously, we can simply compute the mean of the weights themselves, which is

$$\frac{-0 -5 -10 -20 -40}{5} = -15$$

Subtracting -15 from each of the weights, we come up with the weights of $\{15, 10, 5, -5, -25\}$. Do these weights represent a linear function of our dependent variable given the levels we have chosen? As illustrated in Figure D-3 they do indeed. So these weights would be perfectly reasonable to use, but we may as well make the weights as small as possible so that we will have relatively small numbers with which to work. So for convenience we can divide this set of weights by 5 to obtain our final set, which is $\{3, 2, 1, -1, -5\}$.

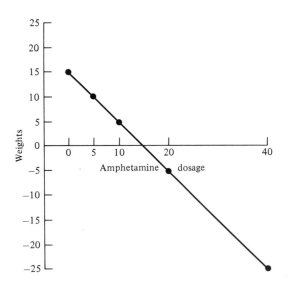

FIGURE D-3.   One appropriate set of weights to reflect a linear function of the dosage levels 0, 5, 10, 20, 40 is $\{15, 10, 5, -5, -25\}$.

---

[6]Because the weights will be decreasing as a function of amphetamine level, we make the weights increasingly more negative.

Digression 15.5

Testing for Monotonic Trend

Occasionally we find ourselves in a position of simply wanting to test the hypothesis that some dependent variable bears a *monotonic relationship* to our independent variable. Suppose for example we are social psychologists interested in people's preference for cars as influenced by the "prestige value" of the car. Suppose from rating data we order five types of cars in terms of their prestige value. The ordering (from lowest to highest prestige) turns out to be:

Ford, Volkswagen, Cadillac, Mercedes, Rolls-Royce

We now do an experiment in which subjects are asked to rate on a scale from 1–7 how much they would like to have the car, where 1 means "couldn't stand it" and 7 means "would love to have it." Suppose that an independent group of subjects rates each of the five cars.

This experiment would produce five means—a mean preference rating for each of the five cars. Now suppose we wish to test the hypothesis that preference rating increases monotonically with prestige value. That is, denoting the population mean preference ratings for the five cars as $\mu_F, \mu_V, \mu_C, \mu_M, \mu_R$, our hypothesis would be:

$$\mu_F < \mu_V < \mu_C < \mu_M < \mu_R$$

What would be the appropriate set of weights to test this hypothesis? One possibility would be to use linear weights, that is,

$$\{-2, -1, 0, 1, 2\}$$

However, another equally appropriate set would be

$$\{-3, -2, -1, 2, 4\}$$

In short, *any* set of monotonically increasing weights would be appropriate. Which of the infinite possible sets of weights should we use?

It turns out that the best possible set to be used is called "ends-doubled linear" weights. To form ends-doubled linear weights, we begin with the appropriate linear weights, in this case,

$$\{-2, -1, 0, 1, 2\}$$

and double the ends to obtain

$$\{-4, -1, 0, 1, 4\}$$

which are the weights we use. The reasons why these weights are best for testing monotonic trend are exceedingly complex, and we will not go into them here (although the Bibliography contains appropriate references).

Choosing Weights in Two-Way Designs

Recall that when we have a two-factor design, we are typically interested in the effect of Factor 1, the effect of Factor 2, and the effect of the interaction. However, suppose that we have some a priori hypothesis about what the pattern of means should be. In such a situation we can bypass the typical two-way ANOVA analysis, simply treat our design as a one-way design with $J \times K$ conditions, and do a planned comparison. Let's illustrate how we would do this.

Contented pets.   Suppose that we are interested in how happiness of dogs and cats is affected by the number of bones these creatures are given. To investigate this question, we arrive at a two-factor design. Factor 1 is type of animal, with the two levels of "dog" and "cat." Factor 2 is the number of bones that are provided for an animal, which can be either 0, 1, 2, 3, or 4. The dependent variable is the animal's happiness level, which we measure with a happinessometer.

Suppose we have a specific hypothesis that the pattern of happiness population means is that depicted in Figure 15-4. When no bones are given to an animal, the happiness level of dogs and cats should be the same. With increasing numbers of bones, dogs should get happier and happier. However, cats don't really care one way or another about bones, so their happiness shouldn't be affected by the number of bones they're given.

How would we analyze this design? One possibility would be to simply do a two-way ANOVA on the data once we have collected it. In this case we would probably expect both main effects and the interaction to be significant. But such a result really wouldn't tell us all that much. Given our specific hypothesis, it would make more sense to create a set of weights corresponding to this

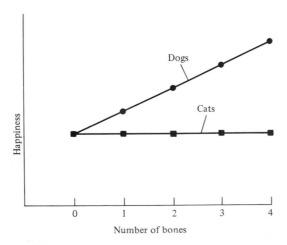

FIGURE 15-4.   Graphical representation of our hypothesis about how dogs and cats are affected by bones.

TABLE 15-11.   Construction of Appropriate Weights for the Dog/Cat Experiment

Step 1: Make up any set of weights

	Number of Bones				
	0	1	2	3	4
Dogs	1	2	3	4	5
Cats	1	1	1	1	1

Step 2: Subtract the mean of these weights (2 in this case) from each weight so that weights will sum to 0

	Number of Bones				
	0	1	2	3	4
Dogs	−1	0	1	2	3
Cats	−1	−1	−1	−1	−1

hypothesis and then do a planned comparison. To make up such weights we can once again use our two-step process. The first step is to make up any convenient set of weights reflecting the hypothesis without worrying about whether they add to zero. Such a potential set of weights is illustrated in the top part of Table 15-11. Now once again to make these weights add to zero, we simply calculate the mean of the weights themselves, which is

$$\frac{1 + 2 + \cdots + 1 + 1}{10} = \frac{20}{10} = 2$$

Now we subtract this mean from each of our original weights—that is, we change to deviation scores—to obtain the final set of weights shown in the bottom part of Table 15-11.

Notice that we actually are treating this design as a one-way design with ten separate conditions, and we have retained the 2×5 structure of the design merely for convenience. Notice further that treating this design as a one-way design means simply that the weights *over all 10 conditions* must add to 0. That is, there are no annoying requirements that the weights for each column or for each row add to zero.

We would analyze these data using an analysis of variance such as the one depicted in the ANOVA table (Table 15-12). We would have 9 degrees of freedom between the 10 conditions and 1 degree of freedom for our hypothesis. The remaining 8 degrees of freedom would be for the residual. Assuming n subjects per cell, **we would compute a sum of squares due to the**

TABLE 15-12.   ANOVA Summary Table for the Dog/Cat Experiment
(assume $n = 5$ animals in each of the 10 cells).

Source	df	MS
Total	49	
Between	9	
Hypothesis	1	MSH
Residual	8	MSR
Within	40	MSW

hypothesis using the equation

$$SSH = \frac{n \left(\sum_{\text{cells}} (w_{jk} M_{jk}) \right)^2}{\sum_{\text{cells}} w_{jk}^2}$$

That is, we would sum the cross-products of weights and means over all 10
conditions—again treating this just like a one-way design. We would then do
everything we would ordinarily do with a single planned comparison. We would
look at the significance level and the amount of variance between groups that is
captured by our 1 degree of freedom hypothesis. We would then see if the
residual were significant. Ideally the hypothesis would be significant and would
account for a fairly substantial proportion of the variance between groups—and
the residual would be nonsignificant.

> 15.9 If we have a specific hypothesis corresponding to a $J \times K$ design,
> we essentially just treat it as a one-way design with $J \times K$ conditions
> and proceed accordingly.

PLANNED COMPARISONS IN A WITHIN-SUBJECTS DESIGN

So far in this chapter we have been confining our discussion to between-subjects
designs. However, the transition to within-subjects designs involves no new
concepts, and the computational mechanics of performing a planned comparison
using a within-subjects design are extremely straightforward. Hence we shall go
through the topic rather briefly. An ANOVA summary table for a typical
within-subjects design is depicted in Table 15-13. Recall that in a within-
subjects design, we have a sum of squares due to conditions (SSC) which
represents the total amount of variance between the various conditions. In a

TABLE 15-13. Prototypical ANOVA Summary Table for a Within-Subjects Design (assume J conditions and n subjects).

Source	df	
Between	$Jn - 1$	
Conditions	$J - 1$	
Subjects	$n - 1$	Conditions tested against interaction.
$S \times C$	$(J - 1)(n - 1)$	

typical within-subjects design we simply perform an F-test on the mean square due to the conditions (MSC) against the mean square due to the subject by condition interaction (MSI).

Now suppose that we had some a priori hypothesis about how the pattern of means should come out. We can make up a set of weights corresponding to this hypothesis exactly as we have described in the preceding sections. Likewise, we use the weights to obtain a sum of squares due to the hypothesis using Equation 15.2 (where n represents the number of subjects), and we obtain the residual by subtracting SSH from SSC. Having a sum of squares due to conditions that is broken up into a sum of squares due to hypothesis and sum of squares due to residual, we are once again in a position to compute the percentage of the sum of squares due to conditions that is accounted for by the hypothesis. The only difference between a between-subjects design and a within-subjects design lies in the error term that we use to test our variance due to the hypothesis and the variance due to the residual. As in any within-subject analysis, the error term is the subject-by-condition interaction. We test both the hypothesis and its residual against this interaction error term. This situation is summarized in Table 15-14.

> 15.10 Use of planned comparisons in a within-subjects design is exactly analogous to use of planned comparisons in a between-subjects design. The only difference is that everything is tested against the subject-by-condition interaction.

TABLE 15-14. Prototypical ANOVA Summary Table for a Within-Subjects Design and a Planned Comparison

Source	df	
Between	$Jn - 1$	
Conditions	$J - 1$	
Hypothesis	1	
Residual	$J - 2$	Both hypothesis and residual are
Subjects	$n - 1$	tested against interaction.
$S \times C$	$(J - 1)(n - 1)$	

PERCENTAGE OF TOTAL VARIANCE ACCOUNTED FOR

In our discussion of planned comparisons we have noted that in addition to testing for significance, we can also pay attention to percentage of variance accounted for. In particular we described how we compute the percentage of between-group variance that is accounted for by a specific, a priori hypothesis.

In many cases, however, we are interested in how much of the *total* variance in the situation (reflected by SST) is accounted for by the independent variable. That issue is the topic of this section.

A Very Powerful Experiment

Let's suppose that we are educational psychologists interested in whether a student's exam score is affected by how many times a week the student attends office hours. We perform the following experiment to investigate this question. At the beginning of the course we randomly assign each of the 3000 students to one of three groups (thereby producing 1000 students in each group). Students in Group 1 are required to go to one office hour a week, students in Group 2 are required to go to two office hours a week, and students in Group 3 are required to go to three office hours a week. Suppose that this study produces the data shown in Figure 15-5. In Figure 15-5 the means of the three groups are represented by

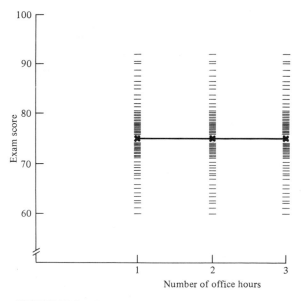

FIGURE 15-5.   Exam score as a function of number of office hours. The Xs represent the means of the 3 groups and individual scores are represented by dashes.

*X*s and the individual scores around the three means are represented by dashes. The mean exam scores for the three groups turn out to be 75.1, 75.4, and 75.7.

A very powerful ANOVA.   Looking at the data shown in Figure 15-5, would we guess that the differences among the three means are significant? Well on the one hand there seems to be a good deal of variance within groups as compared to the relatively meager amount of variance between groups. However, there are so many observations per group that the observed means are extraordinarily stable—that is, they must be very good estimates of the three population means. Let's do a one-way ANOVA on these data and see what happens.

Since there are 1000 students in each of the three groups, the group totals are 75,100, 75,400, and 75,700. We can compute our sum of squares between using the equation

$$SSB = \sum_{\text{groups}} \frac{T_j^2}{n} - \frac{T^2}{N}$$

With our present data, this sum of squares turns out to be

$$SSB = \left(\frac{75,100^2}{1000} + \frac{75,400^2}{1000} + \frac{75,700^2}{1000} \right) - \frac{226,200^2}{3000} - 90$$

Now suppose that the sum of squares within turned out to be 35,964. This gives us enough information to set up an ANOVA summary table as shown in Table 15-15. Note that in Table 15-15, mean square within—our best estimate of the population variance—turns out to be 12. The observed F (2,2997) turns out to be 3.75. Since the criterion F (2,2997) is 3.0, this F is significant.

Confidence intervals.   Let's think about our data just a little more. As noted, our best estimate of the population variance, *est* σ^2 (MSW), is 12. This of course means that within a given group, the scores are distributed with a standard deviation, *est* $\sigma = \sqrt{12} = 3.46$. A glance at the data in Figure 15-5 seems to indicate that this is quite reasonable.

Now let's compute the 95% confidence interval around each of the three means shown in Figure 15-5. Recall that the formula for the 95% confidence

TABLE 15-15.   ANOVA Summary Table for Office Hours Experiment

Source	df	SS	MS	F
Total	2,999	36,054		
Between	2	90	45	3.75
Within	2,997	35,964	12	

interval is

$$M \pm 1.96 \; est \; \sigma_M$$

Our estimate of σ is, as we have seen, 3.46. To obtain $est \; \sigma_M$, we use the familiar formula,

$$est \; \sigma_M = \frac{est \; \sigma}{\sqrt{n}} = \frac{3.46}{\sqrt{1000}} = 0.11$$

Therefore, the three confidence intervals are

$$M_1 \pm (1.96) \, (0.11) = 75.1 \pm 0.21$$

$$M_2 \pm (1.96) \, (0.11) = 75.4 \pm 0.21$$

$$M_3 \pm (1.96) \, (0.11) = 75.7 \pm 0.21$$

In Figure 15-6 we have replotted the three means along with the confidence intervals that we have just computed. An examination of Figure 15-6 provides a nice intuitive picture for why the relatively small differences among the means are significant. As we have noted, the sample means are very good estimates of the corresponding population means—that is, the confidence intervals around the sample means are quite small. The smaller the confidence intervals around the observed means, the less the means need to differ in order to be declared significantly different.

> 15.11 When a very powerful experiment is run, it is possible to detect a very small difference among population means.

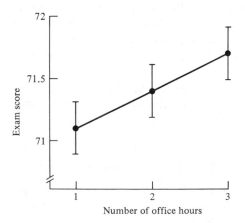

FIGURE 15-6. Mean exam scores as a function of office hours. For clarity, the scale has been changed relative to the scale in Figure 15.5. The vertical bars represent the 95% confidence interval for the population means of the three groups.

Should students go to office hours?   Based on the results of the study we have just described, we have concluded that there is a significant relationship between number of office hours and average exam score. That is, we can infer that going to more office hours leads to higher average exam scores. But so what? Does this mean that we should counsel students to go to more office hours in the hopes of raising their exam scores? Probably we should not, because of the fact that the relationship, although it exists, is small and inconsequential. It would not make sense to counsel students to spend an extra few hours a week at office hours when they would only expect all this effort to raise their exam scores by half a point or so relative to their not going to office hours.

Despite the significant result, we would note that there are probably many other factors—such as amount of studying, intelligence, motivation, and so on—that are much more important than office hours in determining the value of the student's average exam score. We would like to have something in this situation that reflects the fact that the independent variable of number of office hours really only accounts for a very tiny proportion of the variance in the average exam scores—even though we are quite certain that the relationship between the two variables really does exist.

> 15.12 Being certain that a relationship exists between two variables does not imply anything about the size or importance of the relationship.

Omega Squared

It turns out that there is such a measure. This measure is called "omega squared" and is represented by the symbol ω^2. Basically,

$$\omega^2 = \frac{\text{variance in the dependent variable predicted by the independent variable}}{\text{total variance in the dependent variable}}$$

The formula for ω^2 is sort of like

$$\omega^2 = \frac{SSB}{SST}$$

except that it's a little more complicated. It is actually

$$\omega^2 = \frac{SSB - (J-1)\,MSW}{SST + MSW} \tag{15.3}$$

The reason that one has to subtract $(J - 1)\,MSW$ in the numerator is due to the fact that sum of squares between is actually composed of a little more than just variance due to conditions. Recall that SSB actually has a little component of population variance—that is, even if there were actually zero effect due to conditions, we would expect SSB to be nonzero because the population variance will cause a little bit of displacement of the observed means relative to the population means. The reason that we must add MSW to SST in the denominator is more mysterious, and we will not go into it here.

To get a bit more feeling for ω^2, let us see what happens to Equation 15.3 in extremes. If we have a perfect relationship between our independent variable and our dependent variable—then $\omega^2 = 1.0$. In order for ω^2 to be equal to 1.0 it must be true that

$$SSB - (J - 1)\,(MSW) = SST + MSW$$

or

$$SSB - \frac{(J - 1)\,(SSW)}{N - J} = SSB + SSW + \frac{SSW}{N - J}$$

or

$$SSW + \left(\frac{1}{N - J}\right)SSW + \left(\frac{J - 1}{N - J}\right)SSW = 0 \qquad (15.4)$$

The only time Equation 15.4 is satisfied is when SSW equals zero. Thus, $\omega^2 = 1$ only when the entire variance in the dependent variable is accounted for by variation in the independent variable.

Now suppose there is *no* relationship between the independent variable and the dependent variable. In such a situation, $\omega^2 = 0$. Under what circumstances will ω^2 be equal to zero? This can only be true when the numerator of Equation 15.3 is equal to zero; that is, when

$$SSB = (J - 1)MSW$$

or

$$\frac{SSB}{J - 1} = MSW$$

or

$$MSB = MSW$$

In such an instance we say that variation among conditions (MSB) results only as a consequence of the population variance.

We should like to point out one slight anomaly of ω^2. Note that when MSB is less than MSW—in other words when the observed F is less than 1—then ω^2, according to Equation 15.3, will be negative. This particular property of ω^2 is a little bit irritating because conceptually a proportion (of variance or anything else) can never be negative. So if ω^2 is negative, we then just assume that it is equal to zero.

Let's just run through and compute ω^2 for our office hours problem. Sum of squares total turns out to be

$$SST = SSB + SSW = 90 + 35{,}964 = 36{,}054$$

Thus, ω^2 may be calculated as

$$\omega^2 = \frac{90 - (2)(12)}{36{,}054 + 12} = 0.0018$$

which, of course, is exceedingly tiny. Less than two-tenths of 1% of the total variation in exam scores is accounted for by number of office hours attended.

> 15.13 ω^2 is essentially the proportion of total variance in the dependent variable that is accounted for by the dependent variable.

ω^2 in Other ANOVA Situations

We have gone into some detail in describing the workings of ω^2 in a relatively simple one-way ANOVA situation. In t-tests, as well as in more complex types of ANOVA situations—multifactor designs, within-subjects designs—we can also obtain ω^2s. The relevant equations for these ω^2s are provided in Table 15-16.

Two-way ANOVA.   When we have more than one factor to worry about, we can get an ω^2 for each of the factors in the experiment, plus an ω^2 for the interaction. For example in a two-factor design we can get an ω^2 due to Factor 1, an ω^2 due to Factor 2, and an ω^2 due to the interaction. That is, we can get an

---

TABLE 15-16.   Formulas for ω^2 in the Various Situations

I. Two-way ANOVA with J levels of Factor 1 (columns) and K levels of Factor 2 (rows).

$$\omega^2 \text{(Factor 1)} = \frac{SSC - (J - 1)\,MSW}{SST + MSW}$$

$$\omega^2 \text{(Factor 2)} = \frac{SSR - (K - 1)\,MSW}{SST + MSW}$$

$$\omega^2 \text{(Factor 3)} = \frac{SSI - (K - 1)(J - 1)\,MSW}{SST + MSW}$$

II. Within subjects with J conditions (columns) and K subjects (rows).

$$\omega^2 = \frac{(J - 1)(MSC - MSI)}{(J - 1)(MSC - MSI) + K(MSR) + JK(MSI)}$$

indication of what percentage of the total variance in the entire experiment is accounted for by Factor 1, Factor 2, and the interaction.

Within-subjects design.   In a within-subjects design we can get an ω^2 due to conditions—again we can find what percentage of the total variance in the experiment is accounted for by the effect of conditions.

ω^2 and r^2

The astute reader has probably noticed that ω^2 bears a rather close relationship to the Pearson r^2 that we described in the previous chapter. In a correlational situation, r^2 represented the proportion of variance in some variable, Y, that was accounted for by a linear relationship with another variable, X. Likewise, ω^2 is the proportion of total variance in some dependent variable that is accounted for by any relationship of that dependent variable with some independent variable.

THREE MEASURES IN AN EXPERIMENTAL SITUATION

When an experiment is run, there are actually three measures of concern. The first is the significance level, which has been our focus throughout this book. To reiterate, the significance level of an experiment reflects our certainty that there is some relationship between the independent variable and the dependent variable. The second measure of concern is ω^2. As we have just seen, ω^2 is a measure of *how big* is the effect of the independent variable on the dependent variable relative to the total amount of variance that exists in the dependent variable. It is the proportion of the total variance in the dependent variable predictable from the independent variable. Finally we have to worry about the percentage of variance *between conditions* that is accounted for by specific hypotheses. This is where our notion of planned comparisons comes in. We have seen that significance level and ω^2 are unconcerned with the exact nature of the relationship between the dependent variable and the independent variable. A planned comparison, in contrast, reflects a specific hypothesized relationship among the means in some particular experiment. In most experiments the particular theory being tested predicts not only the existence of some relationship between two variables, but also predicts the nature of the relationship.

Relative Importance of These Three Measures

It is a matter of some debate as to which of these three measures we should pay the most attention to in an experiment. It's our opinion that finding a "significant effect" really provides very little information, because it's almost certainly true that *some* relationship (however small) exists between *any* two variables. And in general *finding* a significant effect simply means that enough observations have been collected in the experiment to make the statistical test of the experiment powerful enough to detect whatever effect there is. The smaller the effect, the more powerful the experiment needs to be of course, but no matter how small the effect, it's always possible in principle to design an experiment sufficiently powerful to detect it. We saw a striking example of this principle in the office hours experiment described above. In this experiment there was a relationship between the two variables—and since there were so many subjects in the experiment (that is, since the test was so powerful), this relationship was revealed in the statistical analysis. But was it anything to write home about? Certainly not. In any sort of practical context the size of the effect, although nonzero, is so small that it can almost be ignored.

It is our judgment that accounting for variance is really much more meaningful than testing for significance. So we arrive at the question of which is more important—total variance accounted for in the experiment (ω^2) or percentage of between-group variance accounted for by some specific hypothesis (planned comparisons)? The answer to this question depends on the purpose of the experiment. If the experiment has been carried out primarily for practical reasons, then ω^2 should probably be of primary concern. If, for example, we carried out our office hours experiment with the purpose of determining how to counsel students, then the ω^2 would be a very important measure. Due to the smallness of ω^2 we would conclude that for all practical purposes it really doesn't matter how much students go to office hours. Although there is some kind of an effect, it's so small that students should probably spend their time at home studying.

On the other hand if the experiment has been carried out to test some theory, then the result of the planned comparison is generally fairly important. Suppose that the office hours experiment were carried out in the context of some educational theory. In this case the fact that a positive relationship exists—however small—between office hours and exam grade might be of some theoretical significance.

In general, you just have to use your common sense about what to emphasize.

SUMMARY

This chapter has been concerned with techniques for identifying where interesting differences among means lie—or what the interesting patterns among means are—in an ANOVA-type design.

1. We can divide experiments into two basic types. The first type involves no a priori notion about what the pattern of means should look like. In this type of design we must do post-hoc tests. The second type of experiment does involve one or more a priori hypotheses about what the

pattern of means should be. We can analyze this type of experiment by doing one or more planned comparisons.

2. Post-hoc tests are appropriate only after an omnibus F-test has revealed some significant difference among the means somewhere. Post-hoc tests may then be used to determine *where* the significant differences lie. We saw that the major problem with post-hoc tests is that the more of them that can be done, the more likelihood there is that one or more of them will turn out to be significant just by chance. We saw that this problem is exacerbated by the fact that the tests are not independent. This means we do not have a good idea of what our error probabilities are.

3. In any case there are various techniques for doing post-hoc tests. These techniques vary mostly in terms of the degree to which they control the Type I error probability—the probability of erroneously declaring one or more differences among means to be significant. Two extreme types of tests were described in some detail. The first type—the LSD method—ensured that any individual comparison between two means had a Type I error probability of 0.05. The other type of test—the Scheffé test—required a stricter criterion and in fact was constructed such that the *overall* probability of making one or more Type I errors is held at 0.05. The Scheffé test eliminates the two principal disadvantages of the LSD method. First α is not allowed to balloon upward with a greater number of groups in the experimental design. Second α is known. However, the Scheffé test introduces the other very serious Scheffé problem that the probability of a Type II error—the probability of failing to detect some interesting difference—becomes large.

4. A planned comparison may be done when there is some a priori hypothesis about what the pattern of the means should look like. To do a planned comparison, one first generates a set of weights that reflects the hypothesized pattern of the means. These weights are then used to generate a sum of squares due to the hypothesis which is part of the sum of squares between conditions. This sum of squares due to the hypothesis is based on 1 degree of freedom and is therefore a mean square as well as a sum of squares. This mean square can be F-tested against the appropriate error term (sum of squares within in a between-subjects design or the sum of squares due to subject-by-condition interaction in a within-subjects design). Thus, the significance of the particular hypothesis may be assessed.

5. Furthermore the sum of squares due to the hypothesis may be divided by the sum of squares from whence it came (sum of squares between conditions) to determine the percentage of variance among the various conditions that is accounted for by the hypothesis. This percentage is equivalent to the Pearson r^2 between the weights and the sample means. The rest of the sum of squares between conditions—that sum of squares that is *not* accounted for by the hypothesis—is called the

residual. The residual is based on the number of degrees of freedom between minus one (the degree of freedom on which the hypothesis was based). Thus, a mean square due to the residual may also be computed, and this mean square may be F-tested against the appropriate error term. Ideally this residual will come out very nonsignificant, which means that the hypothesis is sufficient to explain the observed pattern of variation among the means. In the event that the residual does turn out significant, we know that the hypothesis is not sufficient to explain the observed variation among the means, and an addition to or modification of the original hypothesis would be called for.

6. Sometimes there is more than one a priori hypothesis to be tested. In this event we make up appropriate sets of weights and generate sums of squares for however many hypotheses have been generated. The significance of each of these hypotheses may then be tested. Typically there is some residual after the variance due to all the various hypotheses has been removed. This residual is based on the number of degrees of freedom between minus the number of hypotheses (since each hypothesis uses 1 degree of freedom). Thus, a mean square due to the residual can always be calculated, and this mean square can always be F-tested against the appropriate error term.

7. A pair of hypotheses may be distinguished in terms of whether or not they are independent. Two independent hypotheses take up nonoverlapping sums of squares, and the tests performed on them are independent tests. However, two dependent hypotheses take up overlapping sums of squares. It is generally better to work with independent as opposed to dependent hypotheses, because that way one gets the maximum amount of information out of two tests.

8. To make up weights in complicated situations, a two-step procedure is recommended. First we choose whatever set of weights seems to be appropriate for the hypothesis. Second the weights are adjusted so they add to zero. (An additional optional step is to multiply the weights by a constant to make them as simple as possible.)

9. Omega squared is the measure of the percentage of variance in the dependent variable.

10. In an experimental study there are actually three measures of concern—the significance level, the total variance accounted for by the independent variable (ω^2), and the percentage of variance between conditions accounted for by some planned comparison. The relative emphasis that should be placed on these three measures depends on the purpose of the experiment.

PROBLEMS

An experiment is done with four groups of 10 subjects per group. The following means are obtained:

Group 1	Group 2	Group 3	Group 4
$M_1 = 3$	$M_2 = 2$	$M_3 = 5$	$M_4 = 10$

Assume $SSW = 120$.

1. Plot the four means along with 95% confidence intervals.
2. By how much must two means differ to be significant at the 0.01 level using an LSD test?
3. By how much must two means differ to be significant at the 0.01 level using a Scheffé test?
4. Test the hypothesis that the means increase linearly. What percentage of SSB is accounted for by this hypothesis? Is the residual significant?

2. An experiment is done to test the hypothesis that Miracugrow causes increases in the heights of Merkin plants. Four groups of 10 plants per group are given varying amounts of Miracugrow, and the heights of plants after a year are measured. Here are the data for the four groups:

Group 1: 0 gm	Group 2: 4 gm	Group 3: 8 gm	Group 4: 12 gm
$n_1 = 10$	$n_2 = 10$	$n_3 = 10$	$n_4 = 10$
$M_1 = 2$	$M_2 = 5$	$M_3 = 5$	$M_4 = 8$

Assume that $MSW = 10$.

1. Do an ANOVA on these data. Plot the means with 95% confidence intervals.
2. Test the hypothesis that the means increase linearly with amount of Miracugrow. How much of the variance between groups does this hypothesis account for?
3. Test the residual from this hypothesis.
4. Determine the weights corresponding to a second hypothesis orthogonal to that of question 2.
5. Test the significance of this hypothesis.
6. Determine the weights corresponding to one hypothesis that will account for the *entire* sum of squares between.
7. Determine which groups are significantly different from one another, using the Scheffé method.
8. Determine which groups are significantly different from one another, using the LSD method.

3. Assume an experiment is run with three conditions (that is, three levels of the independent variable). There are four observations in each condition, and the following data are obtained:

	Condition	
I	*II*	*III*
2	4	1
8	10	8
14	16	15
10	12	9

Consider this to be a between-subjects design.

1. Graph the means, showing the dispersion of data points around them. Include the 95% confidence intervals. Does it look as if there's an effect of condition?
2. Do a one-way ANOVA. Put your findings in an ANOVA table. Is there a significant effect due to condition?
3. What are the weights corresponding to the hypothesis: Condition II is better than the other two?
4. What is the sum of squares corresponding to the hypothesis? What is the residual SS?
5. What percentage of the SSB is accounted for by the hypothesis?
6. Does the hypothesis account for a significant proportion of the variance?

4. Suppose you do an experiment with $n = 10$ subjects in each of four conditions. The four means are as follows:

$$M_1 = 5 \qquad M_2 = 8 \qquad M_3 = 6 \qquad M_4 = 10$$

Assume that $SSW = 180$.

1. Do an ANOVA on these data. Plot the means along with 95% confidence intervals.
2. By how much does a pair of means have to differ to be significantly different by an LSD test?
3. By how much does a pair of means have to differ to be significantly different by a Scheffé test?
4. Test the hypothesis that these means increase linearly. What percentage of variance does this hypothesis account for?
5. Is the residual from the hypothesis tested in question 4 significant?

5. An experiment is done by the University of Washington Physical Education Department to test the effect of altitude on the amount of time it takes to run the 100-yard dash. Four cities are selected: Seattle (sea level), Boise (1000 feet above sea level), Denver (5000 feet above sea level), and Taos (10,000 feet above sea level). Twelve runners are randomly divided into four groups of three runners per group, with one group running in each of the four cities. The times it takes (in seconds) are as follows:

		City	
Group 1	Group 2	Group 3	Group 4
Seattle (0 ft)	Boise (1000 ft)	Denver (5000 ft)	Taos (10,000 ft)
9	14	14	16
11	16	13	16
10	12	15	16

1. Is there a significant difference among the four groups? Plot the means along with 95% confidence intervals.

 2. What is the smallest difference needed to declare two means to be significantly different using an LSD test? (Use a 0.01 α-level.)

 3. What is the smallest difference needed to declare two means to be significantly different using a Scheffé test? (Use a 0.01 α-level.)

 4. Test the hypothesis that running time increases *linearly* with altitude.

 5. Test the residual from this hypothesis.

 6. Is the variance in running time in Denver significantly different from the variance in running time in Boise?

6. Consider the data from Chapter 11, Problem 4. Which of the groups differ significantly by a Sheffé test? By an LSD test?

7. Consider the data from Chapter 11, Problem 6.

 1. Which groups differ significantly by a Scheffé test? By an LSD test?

 2. Test the hypothesis that number of errors decreases linearly with presentation rate. What percentage of *SSB* does this hypothesis account for? Test the residual from the hypothesis.

8. Consider the data from Chapter 11, Problem 1.

 1. Which groups differ significantly by a Scheffé test? By an LSD test?

 2. Test the hypothesis that memory ability increases with age. Test the residual from this hypothesis.

 3. What do you think would be a more reasonable hypothesis?

9. Consider both sets of data shown in Chapter 11, Problem 2. For both sets, answer the following questions:

 1. Which groups are different from one another by a Scheffé test? By an LSD test?

 2. Test the hypothesis that marijuana makes people higher (leads to longer reaction times) than bananas or tea, which do not differ from one another.

 3. Test the hypothesis (orthogonal to the one in question 2) that bananas made you higher (lead to higher reaction times) than tea.

10. Consider the data from Chapter 13, Problem 4.
 Test the hypothesis that number of words recalled decreases linearly with retention interval.

11. Consider the data from Chapter 13, Problem 3.

 1. Test the hypothesis that Sterilestud and DDT are both more effective in mosquito control than nothing at all.

 2. Test the hypothesis (orthogonal to that of question 1) that Sterilestud is more effective than DDT.

12. Consider the data from Chapter 13, Problem 1.

    1. Test the hypothesis that stage fright increases linearly with audience size. What proportion of the sum of squares due to condition is accounted for by this hypothesis?

    2. Test the residual from this hypothesis.

13. Use the information from Problem 2 of Chapter 12 to solve the following:

    1. Which of the cells differ significantly from one another by an LSD test?

    2. Suppose that you have an a priori theory that, with grubbily dressed speakers, attitudes should decrease with age; however, with well-dressed speakers, attitudes should remain constant over age. Furthermore for a given age level attitudes should always be higher with grubbily dressed than with well-dressed speakers. That is, the data should look like this:

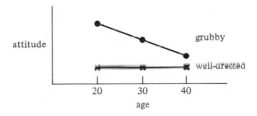

    Generate the weights corresponding to this hypothesis.

    3. Test the significance of the hypothesis. How much of the between-cell variance does it account for? Test the significance of the residual.

14. Use the information from Problem 1 of Chapter 12 to solve the following:

    (Assume it to be a 1-way design with 4 conditions and 10 subjects/condition.)

Assume $SSW = 3600$.

    1. What are the weights corresponding to the hypothesis: The high-motivation groups do better than the low-motivation groups?

    2. Compute a sum of squares corresponding to this hypothesis. What are the dfs corresponding to this sum of squares?

    3. Compute an F corresponding to the hypothesis. What are the dfs for this F? Is it significant?

    4. What are the residual sum of squares and mean squares? Compute an F corresponding to the residual. What are the dfs for this F? Is the F significant?

15. Consider a 3 × 3 design.

    1. Make up a set of weights to reflect this hypothesis:

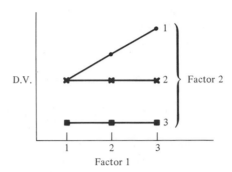

    2. Suppose there are 7 scores per cell and the means are:

Factor 1

	1	2	3
1	2	2	7
2	3	2	2
3	1	2	1

    Test the hypothesis and the residual (assume $MSW = 2$).

16. Here are two hypotheses for a four-group experiment:

$$H_1: \quad \{3 \quad 1 \quad -1 \quad -3\}$$
$$H_2: \quad \{1 \quad -1 \quad -1 \quad 1\}$$

Make up a third hypothesis that is orthogonal to these two.

17. Here is a hypothesis for a four-group experiment:

$$H_1: \quad \{1 \quad 1 \quad 1 \quad -3\}$$

Make up two more hypotheses such that all three are mutually independent.

18. Consider the data from Chapter 11, Problem 5 (note that σ^2 is known). Which groups differ by a Scheffé test? By an LSD test?

19. Consider the data from Chapter 12, Problem 6. Suppose Gazelle, Inc. has the following hypothesis about their braking systems. The new system is such that it will stop any size motorcycle in the same distance. However, with the old system stopping distance increases *linearly* with engine size.

1. Generate the weights corresponding to this hypothesis (*Caution!*). Note that 125, 500, and 1000 do not form a linear sequence.
2. Is this hypothesis significant? What proportion of SSB does it account for?
3. Test the residual from this hypothesis.

20. Suppose an experiment produces the following four means, each mean based on $n = 10$ scores.

$$M_1 = 2 \qquad M_2 = 4 \qquad M_3 = 2 \qquad M_4 = 0$$

Note that $SSB = 80$. Assume that $SSW = 360$.

1. Do an ANOVA on these data.
2. What percent of variance (ω^2) is accounted for by conditions?

21. A sociologist has the hypothesis that the average height of males living west of the Mississippi River is different from the average height of males living east of the Mississippi River. He takes random samples of 10,000 Easterners and 9500 Westerners, and gets the following data: (Height is in feet)

Easterners	Westerners
Mean height = 5.75	Mean height = 5.74
$\Sigma(X - M)^2 = 599$	$\Sigma(X - M)^2 = 632$

Test the hypothesis that the average height of Easterners differs from that of Westerners. What is ω^2 in this problem?

22. Consider the data from Chapter 10, Problems 5–8, 10–14, and 16–17. Compute ω^2 in each case.

23. Consider the data from Chapter 11, Problems 1, 2 (both sets of data), and 3–6. Compute ω^2 in each case.

24. Consider the data from Chapter 12, Problems 1–7. Compute ω^2 for rows, columns, and interaction in each case.

25. Consider the data from Chapter 13, Problems 1–5 and 9. Compute ω^2 for conditions for each set of data.

26. Consider the data from this chapter, Problems 1–5. Compute ω^2 for each set of data.

16 Chi-Square

As we have seen in past chapters the heart of the hypothesis-testing procedure consists of determining how some summary score is distributed assuming that some null hypothesis is true. Our strategy has been to identify a particular experimental situation and then to deal with the distribution of the summary score that, in some sense, naturally emerges from that situation.

A NEW DISTRIBUTION

In this chapter we are going to adopt a slightly different strategy, which is to first describe a theoretical distribution and then discuss what the distribution is used for. This distribution is the chi-square (χ^2) distribution, and summary scores arising out of several quite diverse experimental situations are distributed as χ^2.

Mathematical Properties of a χ^2 Distribution

Consider a normal distribution with a mean of μ and a variance of σ^2. Now imagine that we pluck a random score (x) out of this distribution, and with this x, we form a z-score in the usual way:

$$z = \frac{x - \mu}{\sigma}$$

Now suppose that we square this z to obtain

$$z^2 = \frac{(x - \mu)^2}{\sigma^2}$$

How would this novel variable—this z^2—be distributed? We can certainly formulate some intuitions. First since a z^2 is the square of something, it cannot be less than zero. But, a z^2 could go up as high as positive infinity. Therefore, like an F-distribution this z^2 should be skewed positively.

Actually we can be much more precise about some of the properties of this distribution. We know for example that 67% of all z-scores fall between -1 and $+1$. Note that any z between -1 and $+1$ will turn into a z^2 that falls between 0 and $+1$. Therefore, we know that about 67% of the z^2s should fall between 0 and $+1$. Using the other such properties of the z distribution (that we can easily determine from the z distribution tables), we can be as precise as we like about formulating this distribution of z^2-scores. In point of fact the distribution of this z^2 would look like the one in Figure 16-1. A z^2 has a name, which is a χ^2 *with 1 degree of freedom*. A χ^2 with 1 degree of freedom is represented as $\chi^2(1)$.

The fact that we have specified this χ^2 distribution as having 1 degree of freedom suggests that there might be χ^2 distributions with more than

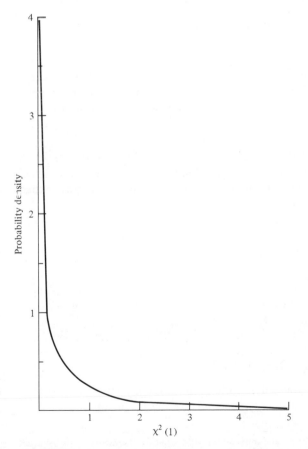

FIGURE 16-1.   Probability density function of a $\chi^2(1)$

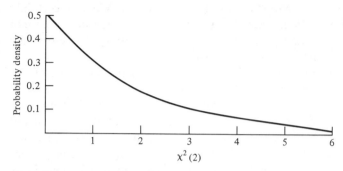

FIGURE 16-2.   Probability density function of a $\chi^2(2)$

1 degree of freedom, and in fact that's quite true. Suppose once again we pick out a random score, x_1, from our same normal distribution. From this x_1, we again form a z-score, z_1, and a corresponding z_1^2. We then repeat the operation, plucking out another random score x_2 and compute z_2^2. Now we compute the sum, $z_1^2 + z_2^2$. How is this sum distributed? The answer is that the sum is distributed as a χ^2 with 2 degrees of freedom—that is,

$$\chi^2(2) = z_1^2 + z_2^2$$

What would be the nature of this χ^2 distribution with 2 degrees of freedom? In general a sum of two z^2s ought to be greater than a sum of one z^2. So we would expect the mean of the $\chi^2(2)$ distribution to be higher than the mean of the $\chi^2(1)$ distribution. And in fact a $\chi^2(2)$ distribution would have the form that is shown in Figure 16-2.[1]

By now the idea of what a χ^2 distribution is should be fairly clear. In general suppose that we take n scores (x_is) out of a normal distribution, form a z-score from each of these n x_is, and then square each z-score. The sum of these z_i^2s will be distributed as a χ^2 with n degrees of freedom

$$\chi^2(n) = z_1^2 + z_2^2 + \ldots + z_n^2$$

Examples of other χ^2 distributions are shown in Figure 16-3.

16.1   A χ^2 with n degrees of freedom is represented as $\chi^2(n)$ and is the distribution of the sum of n z^2s.

Characteristics of the χ^2 Distribution

Before we go into the applications of the χ^2 distribution, let's discuss several of its important properties. These properties involve (a) the mean and variance

---

[1] Note that the scales of both the ordinate and the abscissa will change for the various χ^2 distributions we will be depicting in Figures 16-1 to 16-4.

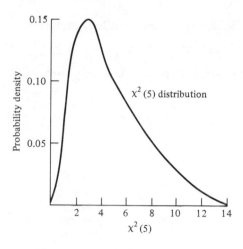

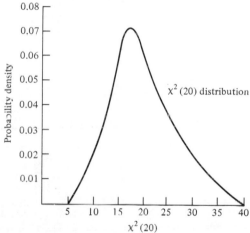

FIGURE 16-3. Probability density functions for χ^2 distributions with 5 and with 20 degrees of freedom.

of a χ^2 distribution, (b) additivity of χ^2 distributions, and (c) tables of the χ^2 distribution.

Mean and variance of a χ^2 distribution. Recall that a χ^2 with n degrees of freedom is defined as

$$\chi^2(n) = z_1^2 + z_2^2 + \ldots + z_n^2$$

The mean (expectation) of the $\chi^2(n)$ distribution may therefore be defined as

$$E[\chi^2(n)] = E(z_1^2 + z_2^2 + \ldots + z_n^2)$$

or

$$E[\chi^2(n)] = E(z_1^2) + E(z_2^2) + \ldots + E(z_n^2)$$

As Digression 16.1 shows, the expectation of any z^2 is equal to 1. Therefore

$$E[\chi^2(n)] = 1 + 1 + \ldots + 1 = n$$

The mean of the χ^2 distribution with n degrees of freedom is simply equal to n. This makes life quite simple.

Digression 16.1

The Expectation of z^2

Note that

$$z = \frac{x - \mu}{\sigma}$$

and therefore

$$z^2 = \frac{(x - \mu)^2}{\sigma^2}$$

Thus,

$$E(z^2) = \frac{E[(x - \mu)^2]}{\sigma^2}$$

Since

$$E[(x - \mu)^2] = \sigma^2$$

it follows that

$$E(z^2) = \frac{\sigma^2}{\sigma^2} = 1$$

Computation of the variance of a χ^2 distribution is somewhat less straightforward, so we will not provide an explicit derivation. Very simply the variance of a $\chi^2(n)$ distribution is equal to $2n$.

16.2 The mean of a $\chi^2(n)$ is n. The variance of a $\chi^2(n)$ is $2n$.

Additivity.   Consider two χ^2 distributions, one with n_1 degrees of freedom and the second with n_2 degrees of freedom. Suppose we take a χ^2 from the first distribution and a χ^2 from the second distribution and add them together to obtain

$$\chi^2(n_1) + \chi^2(n_2)$$

How will this sum of two χ^2s be distributed? Again the answer is quite simple:

$$\chi^2(n_1) + \chi^2(n_2) = \chi^2(n_1 + n_2)$$

that is, the sum of two χ^2s is also distributed as a χ^2. The reason for this is very straightforward. We consider first that

$$\chi^2(n_1) = z_{11}^2 + z_{21}^2 + \ldots + z_{n_1 1}^2$$

and, second, that

$$\chi^2(n_2) = z_{12}^2 + z_{22}^2 + \ldots + z_{n_2 2}^2$$

Therefore $\chi^2(n_1) + \chi^2(n_2)$ is the sum of $(n_1 + n_2)$ z^2s which, by definition, is distributed as a χ^2 with $n_1 + n_2$ degrees of freedom.

Tables of the χ^2 distribution.   The process of hypothesis testing using a χ^2 distribution will turn out to involve the same steps as hypothesis testing with any other distribution. As such it will be necessary to establish criterion χ^2s that chop off various areas of the χ^2 distribution. And as with other distributions tables have been worked out to permit us to do this quite simply.

Consider for example, a χ^2 distribution with 10 degrees of freedom, as in Figure 16-4. Suppose that we wish to find the criterion χ^2 that chops off the

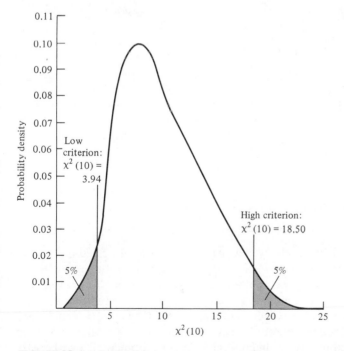

FIGURE 16-4.   Probability density function of a $\chi^2(10)$. Criteria chopping off the upper and lower 5% tails are shown.

upper 5% of this distribution. To do this, we would look at χ^2-tables which, like z-tables, t-tables, and F-tables, are found in Appendix F-5 of this book. The χ^2-table is organized very much like the t-table. First each row corresponds to some particular value of degrees of freedom. Since we are dealing in this example with a $\chi^2(10)$, we would direct our attention to the row corresponding to 10 degrees of freedom.

The columns in the χ^2-table represent the particular (upper) portions of the distribution that we wish to chop off. So if we wish to compute the criterion $\chi^2(10)$ that chops off the upper 5% of the distribution, we concern ourselves with the column labeled 0.05. At the intersection of this column and the row corresponding to 10 degrees of freedom, we would find the number 18.5, which is the criterion $\chi^2(10)$.

Chopping off other portions of the χ^2 distribution is also quite straightforward. For example, suppose that we wanted the criterion χ^2 that chopped off the lower 5% of the distribution. Since our χ^2-table is organized in terms of chopping off *upper* parts of the distribution, we would seek the criterion χ^2 that chops off the upper 95% of the distribution. Therefore, we would look in the column labeled 0.95, where we would find a criterion χ^2 equal to 3.94. Both of these criteria are shown in Figure 16-4.

> 16.3 As is the case with other distributions, tables have been constructed which allow us to determine criteria that chop off various portions of a χ^2 distribution.

USES OF THE χ^2 DISTRIBUTION

With this brief description of what a χ^2 distribution is and how it works, we now shall describe three uses for a χ^2: (a) testing an obtained variance against a constant, (b) testing the fit of a set of data to some theoretical distribution, and (c) testing for independence in a contingency table.

Computerized Teaching:
Testing Against an Exact Variance

Suppose we are educational psychologists once again interested in creative methods for teaching arithmetic to elementary school children. In particular, we plan to teach arithmetic using a computer-assisted instruction (CAI) system in which each child learns to use a computer terminal and is tutored in basic arithmetic via a computer. The expected advantage of a CAI system is that since each child is accorded a good deal of individual attention, the amount the child learns will be increased relative to being taught by a traditional classroom-style teacher.

However, use of a CAI system also involves a potential problem. It may be that the relatively smart students will catch on to the CAI system—the use of the terminal and so on—and will use it to great advantage in terms of increasing their learning. However, it may be that the not-so-bright students will have difficulty grasping the use of the system and therefore will not do as well as they might have in an ordinary classroom system (this is the "rich get richer and poor get poorer" problem).

How can we assess whether or not this hypothesized problem is really of concern? Suppose that at the end of the course each child is given a standard arithmetic exam to assess how much the child has learned in the course. Imagine that the scores on this exam range from 0 to 100, and it is known from a vast amount of prior experience that the population variance of this exam (σ^2) is equal to 25. Now suppose that we look at the exam score variance of children who have participated in the CAI-based course. If indeed our potential problem is a real one—that is, if it's really true that the brighter students learn more, but the poorer students learn less—then the variance of the CAI children should be greater than the population variance of 25.

Data.   Suppose there are 10 students in this program. The data (the standard exam scores) for each child are shown in Table 16-1. The sum of squares for this set of data is 846.4, and our best estimate of the population variance, *est* σ^2 is 94.04. Now this estimate of σ^2 of 94.04 certainly seems to be greater than the known population variance of 25. We might therefore be inclined to assume that the CAI children do indeed end up being more variable than children who are treated normally. As always, however, we have to be cautious about precipi-

TABLE 16-1
Hypothetical Data:
Arithmetic Scores of 10
Children Who Have
Participated in a CAI
Program for Learning
Arithmetic

Child	Score (x)
1	$x_1 = 68$
2	$x_2 = 90$
3	$x_3 = 70$
4	$x_4 = 91$
5	$x_5 = 72$
6	$x_6 = 80$
7	$x_7 = 85$
8	$x_8 = 82$
9	$x_9 = 91$
10	$x_{10} = 95$

tously leaping to such a conclusion. Perhaps just by chance we have selected a group of highly variable children. We would like some way of carrying out our traditional hypothesis-testing steps in order to determine whether or not we should really take this large variance seriously.

Hypothesis-testing steps.   The first thing we need is a null and an alternative hypothesis. As usual the null hypothesis says that the independent variable (whether the children were in normal or CAI arithmetic instruction) has no effect on the dependent variable. In this case the dependent variable is not the mean, as it has been in the past, but the *variability* of the scores. This null hypothesis may thus be stated as:

> H_0: The population of children who receive CAI instruction has an exam score variance (σ^2) equal to σ_0^2, which is the exam score variance of children who receive normal instruction. In this case $\sigma_0^2 = 25$.

Likewise, the alternative hypothesis maintains that the independent variable does have an effect on the dependent variable. Specifically we have the directional hypothesis that the variance of children who take part in the CAI program should *increase* relative to $\sigma_0^2 = 25$.

Next we need a summary score that will reflect the degree to which our null hypothesis or our alternative hypothesis is true. In this case it might seem that a reasonable summary score would be our estimate of the population variance, 94.04, obtained from the data. A better summary score to use, however, is something related to *est* σ^2:

$$\frac{\Sigma(x_i - M)^2}{\sigma_0^2}$$

This score is better because of the next step of our hypothesis-testing procedure, which is to determine how the summary score is distributed if the null hypothesis is true. **It turns out that the particular summary score we have chosen is distributed as a χ^2 with $n - 1$ degrees of freedom if the null hypothesis is true.** Digression 16.2 demonstrates why this is so. In our particular example we have $n = 10$ children and we can therefore obtain an observed $\chi^2(9)$ of

$$\text{Observed } \chi^2(9) = \frac{\Sigma(x - M)^2}{\sigma_0^2} = \frac{846.4}{25} = 33.86$$

What is our criterion $\chi^2(9)$? We consult our χ^2-tables, going to the row corresponding to 9 degrees of freedom and the column corresponding to 0.05. There we find a criterion χ^2 of 16.92. Since our obtained $\chi^2(9)$ of 33.9 is considerably more extreme than this criterion χ^2, we are in a position to reject the null hypothesis. We say that if the null hypothesis were true—if exam scores obtained by the children in our sample really did come from a distribution whose population variance was 25—then we would obtain a χ^2 as extreme as 33.9 less than 5% of the time. Therefore our conclusion would be that the CAI program does (unfortunately) seem to increase the variance of children who participate in it.

16.4 Consider some estimate of a population variance,

$$est\ \sigma^2 = \frac{\Sigma(x - M)^2}{n - 1}$$

If the variance, σ^2, that gives rise to this estimate equals σ_0^2, then

$$\frac{\Sigma(x - M)^2}{\sigma_0^2}$$

is distributed as $\chi^2(n - 1)$.

Digression 16.2

Distribution of $\dfrac{\Sigma(x - M)^2}{\sigma_0^2}$

We start with our definition of χ^2. If we draw n scores, $x_1, \ldots, x_n$ from a normal distribution with mean μ and variance σ_0^2, then we know that

$$\frac{\Sigma(x_i - \mu)^2}{\sigma_0^2}$$

is distributed as $\chi^2(n)$.

However, we do not know the population mean; rather, we are dealing with the sample mean, M. Let's inquire about the distribution of

$$\frac{\Sigma(x_i - M)^2}{\sigma_0^2}$$

We know that

$$\Sigma(x_i - \mu)^2 = \Sigma[(x_i - M) + (M - \mu)]^2$$
$$= \Sigma(x_i - M)^2 + 2\Sigma(x_i - M)(M - \mu) + \Sigma(M - \mu)^2 \qquad \text{(D.1)}$$

The middle term of Equation D.1 is zero since

$$2\Sigma(x_i - M)(M - \mu) = 2(M - \mu)\ \Sigma(x_i - M) = 2(M - \mu)(0) = 0$$

Therefore,

$$\Sigma(x_i - \mu)^2 = \Sigma(x_i - M)^2 + \Sigma(M - \mu)^2$$

Since $(M - \mu)^2$ is a constant,

$$\Sigma(x_i - \mu)^2 = \Sigma(x_i - M)^2 + n(M - \mu)^2$$

or, dividing by σ_0^2

$$\frac{\Sigma(x_i - \mu)^2}{\sigma_0^2} = \frac{\Sigma(x_i - M)^2}{\sigma_0^2} + \frac{(M - \mu)^2}{(\sigma_0^2/n)}$$

Note that

$$\frac{(M - \mu)^2}{(\sigma_0^2/n)}$$

is a z^2 drawn from a sampling distribution of sample means with parameters μ, and σ_0^2/n and therefore distributed as a $\chi^2(1)$. Likewise,

$$\frac{\Sigma(x_i - \mu)^2}{\sigma_0^2}$$

is distributed as a $\chi^2(n)$. Therefore,

$$\chi^2(n) = \frac{\Sigma(x_i - M)^2}{\sigma_0^2} + \chi^2(1)$$

or

$$\chi^2(n) - \chi^2(1) = \frac{\Sigma(x_i - M)^2}{\sigma_0^2}$$

and finally by additivity,

$$\frac{\Sigma(x_i - M)^2}{\sigma_0^2} = \chi^2(n - 1)$$

Notice that in this particular situation we used a one-tailed χ^2 test. Here, "one-tailed" means exactly what it has in the past. That is, we only concerned ourselves with the criterion that chopped off the *upper* 5% of the distribution. Suppose, however, that our initial alternative hypothesis had been nondirectional rather than directional. That is, suppose that, whereas we thought the CAI program might increase the variance of the children, it seemed equally reasonable that CAI might decrease the variance. In this situation we would perform a two-tailed χ^2-test. As with any two-tailed test we would want to find the criteria that chop off the upper and lower 0.025 of the distribution. These criteria would not be difficult to establish. Once again, we'd go to the row in our χ^2-tables corresponding to 9 degrees of freedom. To find the upper criterion, we would look in the column labeled "0.025," and there we would find a criterion $\chi^2(9)$ of 19.02. Similarly the lower 0.025 would correspond to the criterion $\chi^2(9)$ that chops off the upper 0.975 of the distribution. In this column we find a criterion of 2.70. If we were doing a two-tailed test, these would be our criteria. That is, any obtained $\chi^2(9)$ that was either lower than 2.70 or higher than 19.02 would permit us to reject the null hypothesis.

Fitting an Exact Distribution:
Are Babies Like Coins?

Would-be fathers often dream of siring a son. Sometimes this desire is sufficiently intense that the father of an all-daughter family will continue to demand more children in hopes that a baby boy will eventually emerge. "After all," the statistically knowledgeable father will say, "even though I now have four daughters, the probability is still 0.5 that my next kid will turn out to be a boy. So why not give it a go?"

This line of reasoning makes sense, but it does involve the hidden assumption that babies behave like coins. That is, it involves the assumption that the sex of each child cannot be in any way predicted by the sexes of the preceding children—just as the outcome of a coin toss cannot be predicted by the results of previous tosses.

In the case of babies this assumption may not be true. Another possibility is that a couple may be predisposed to have children of one sex or the other. If this were the case, then a couple with, say, four daughters would have a higher probability of producing a girl than a boy should they opt for a fifth child.

How might we investigate this issue? We could consult the Census Bureau records and consider all families who have exactly five children. Suppose that we find 1024 such families and plot the distribution of the number of boys in these 1024 families. This observed frequency distribution is shown in Table 16-2. So for instance 40 of the 1024 have no boys (all daughters), 184 of the families have one boy and four girls, and so on. Now suppose that the sex of each child is in fact independent of the sexes of all other children. How then should the number of boys be distributed over families? The answer is that the distribution should be binomial. That is, the probability of having r boys out of n children should be the same as the probability of obtaining r heads out of n flips of a coin.

If indeed number of boys is distributed binomially, then we should be able to compute an *expected frequency distribution* for the 1024 families. That is, we ought to be able to compute how many of the families we would expect to have no boys, how many of them we expect to have one boy, and so on. Table 16-3

TABLE 16-2.   Obtained Frequency Distribution of
Number of Boys (0–5) across 1024 Five-Child Families

Number of Families With . . .	*Observed Frequency (f_0)*
0 boys	40
1 boy	184
2 boys	300
3 boys	268
4 boys	196
5 boys	36

TABLE 16-3.   Expected Probability and Frequency
Distributions for Number of Boys in a Five-Child Family;
Assume that These Distributions Are Binomial with
$p = q = 0.5$ and $N = 1024$

Number of Boys	Probability	Expected Frequency (f_e = 1024 × Probability)
0	1/32	32
1	5/32	160
2	10/32	320
3	10/32	320
4	5/32	160
5	1/32	32

shows these expected frequencies, which we obtained by first constructing the
binomial probability distribution for $N = 5$ and $p = 0.5$. We then took the total
number in our sample—1024—and multiplied it by these probabilities to obtain
corresponding expected frequencies.

To the degree that the number of boys in a family of five really is
binomially distributed, the frequency distribution we observe in our sample
should resemble the expected binomial frequency distribution we constructed.
How are we going to assess the degree to which the expected and the observed
frequency distribution coincide with one another? Table 16-4 shows the two
distributions—the expected and the observed—lined up against one another. It's
evident that there is some discrepancy between the two distributions. We
observed 40 all-girl families, whereas we expected 32. We observed 184
four-girl/one-boy families, whereas we expected 160, and so on.

Suppose now that babies really do behave like coins. Due to chance
fluctuation it wouldn't have to be true that the expected and observed distribu-

TABLE 16-4.   Construction of a χ^2 from Expected and Observed Frequency
Distributions; Notice that the Greater the Disparity Between Expected and
Observed Distributions, the Greater the χ^2

Number of Boys	Observed Frequency (f_o)	Expected Frequency (f_e)	($f_e - f_o$)	($f_e - f_o$)^2	($f_e - f_o$)^2/f_e
0	40	32	−8	64	2.00
1	184	160	−24	576	3.60
2	300	320	20	400	1.25
3	268	320	52	2,704	8.45
4	196	160	−36	1,296	8.10
5	36	32	−4	16	0.50

$$\Sigma[(f_e - f_o)^2 / f_e] = 23.90$$

tions match *exactly*. But on the other hand there ought to be some degree of resemblance. Clearly we need some kind of summary score that will reflect the degree of correspondence between the expected and observed distribution. One possibility for constructing such a summary score would be to simply sum the differences between the observed and the expected frequencies for each of the six categories. However, this sum involves the familiar problem that some of the differences are positive and others are negative. And the problem is solved in a familiar way. We *square* the observed differences to obtain a set of $(f_e - f_o)^2$s. Actually we then do one more thing; we divide each of these $(f_e - f_o)^2$s by the corresponding f_e. These $(f_e - f_o)^2/f_e$s are found in the sixth column of Table 16-4. Now finally the sum of the $(f_e - f_o)^2/f_e$s constitutes a summary score indicating the degree to which the observed and the expected distributions coincide with one another.

Hypothesis-testing steps.   We are now in a position to carry out a set of fairly familiar hypothesis-testing steps to test the question of interest. In this case our null hypothesis will essentially be that having a baby is like flipping a coin; the sex of any child in a family will be male with a 0.5 probability and will be independent of the sex of all the other children in the family. As we have noted, this null hypothesis implies that over families the number of boys in the family should be distributed binomially. The alternative hypothesis as usual is simply that the null hypothesis is not true—in this case that the number of boys in a family should not be distributed binomially.

We have already determined a summary score to reflect the extent to which the null as opposed to the alternative hypothesis is true. This summary score is simply

$$\sum_{\text{categories}} \frac{(f_e - f_o)^2}{f_e}$$

This summary score appropriately represents the degree to which the null as opposed to the alternative hypothesis is true, because if the null hypothesis is true we would expect the summary score to be small, whereas if the alternative hypothesis is true, we would expect the summary score to be large.

Now as usual we must ask the question of how our summary score is distributed if the null hypothesis is true. Without going into the mathematics of the situation, if the null hypothesis is true, then this summary score is distributed as a χ^2 with $n - 1$ degrees of freedom. It is important to note that "n" in this case refers to the number of *categories making up our distribution*. In our particular example there are six categories, since we are concerned with the frequencies of families with 0, 1, 2, 3, 4, or 5 boys. (Do *not* fall into the trap of thinking n is equal to the number of observations—1024 in this example.) So the summary score we have obtained—23.90—should, if the null hypothesis is true, be distributed as a $\chi^2(5)$.

16.5 If an observed set of data comes from a particular underlying probability distribution, then we can compute a summary score:

$$\sum_{\text{categories}} \frac{\Sigma(f_e - f_o)^2}{f_e}$$

Where the f_e s are the expected frequencies and the f_o s are the observed frequencies, calculated from the underlying distribution. If the scores do come from the underlying distribution, then the summary score is distributed as $\chi^2(n - 1)$ where n refers to the number of categories.

The remainder of our hypothesis-testing procedure is quite straightforward. First note that we should be doing a one-tailed test, because if the null hypothesis is false, we would expect our summary score to be large. Under *no* circumstances would we expect our summary score to be too small. We therefore need a criterion χ^2 that chops off the upper 5% of the distribution. Looking at out χ^2 tables, we find this criterion $\chi^2(5)$ to be 11.07. Since our obtained $\chi^2(5)$ of 23.90 is considerably greater than our criterion $\chi^2(5)$ of 11.07, we can soundly reject the null hypothesis. We are in a position to say that the number of boys in a five-person family is not distributed binomially. Babies do not act like coins.

Let's just reiterate what we have done in terms of fitting a distribution. We had a particular hypothesis (in this case that babies act like coins), and we had some data (in this case the distribution of number of boys in five-person families). The hypothesis implies a particular type of distribution (in this case a binomial distribution). Given that we expect a particular type of distribution, we can determine a set of expected frequencies. We then compare the expected frequency distribution with the observed frequency distribution. We obtain a summary score that reflects the degree to which these two distributions coincide. If our hypothesis is true, then this summary score turns out to be distributed as a χ^2. Knowing the distribution of the summary score given our hypothesis, we can do what we have always done—we can see how extreme our summary score, our χ^2—is. If it is extreme enough so that we would only expect it to occur less than 5% of the time, then we are in a position to conclude that our hypothesis is false.

Testing for Independence in a Contingency Table

The final use of the χ^2 distribution we will describe—testing for independence in a contingency table—is very similar to the distribution-fitting procedure that we have just illustrated.

We discussed contingency tables in Chapter 1. To reinstate in your mind the idea of a contingency table, consider the following situation. Suppose we're political scientists interested in assessing whether there are political differences

TABLE 16-5. A 3 × 2 Contingency Table Depicting the Relationship Between Political Preference and Attitude Toward Legalization of Marijuana; Data Are Given Both in Terms of Frequencies and of Probabilities; Probabilities Are Frequencies Divided by 200 and Are Given in Parentheses.

| | | Attitude Toward Marijuana | | | |
		Should be Legalized	Should be a Misdemeanor	Should be a Felony	
Political Inclination	Democrats	40 (0.200)	70 (0.350)	40 (0.200)	150 (0.750)
	Republicans	35 (0.175)	5 (0.025)	10 (0.050)	50 (0.250)
		75 (0.375)	75 (0.375)	50 (0.250)	200 (1.000)

in attitudes toward marijuana. In particular we wonder whether Democrats and Republicans differ in their feelings about this controversial weed. To investigate the issue, we telephone 200 people drawn at random from the U.S. population. We first ask each person to identify him- or herself as a Democrat or a Republican. We then ask the person to classify his or her attitude toward marijuana as one of the following: (a) Marijuana should be legalized. (b) Marijuana should be a misdemeanor. (c) Marijuana should be a felony. Let's represent the results of this survey as a 3 × 2 contingency table shown in Table 16-5.

The numbers in the right-hand margin of Table 16-5 indicate that of our sample of 200, 150 were Democrats and 50 were Republicans. Likewise, the numbers in the bottom margin indicate that of our sample of 200 people 75 thought that marijuana should be legalized, 75 thought it should be a misdemeanor, and the remaining 50 believed that it should be a felony. For reasons that will become clear shortly, we have also represented these frequencies as probabilities.

Now let's ask whether one's attitude toward marijuana depends on whether one is a Democrat or a Republican. Or stated another way, do Democrats have a different pattern of attitudes toward marijuana legalization than do Republicans? We can attack this problem as follows. Table 16-5 shows what our data should look like if attitude toward marijuana and political inclination are indeed independent. Let's explain from where the numbers in Table 16-5 came. Recall that if any two events A and B are independent, then their joint probability, $A \cap B$, should be equal to $p(A) \times p(B)$. This reasoning may be applied to the data of Table 16-5. Suppose we define the following probabilities:

$p(R)$ = probability that person is a Republican

$p(D)$ = probability that person is a Democrat

$p(L)$ = probability that person thinks marijuana should be legalized

$p(M)$ = probability that person thinks marijuana should be a misdemeanor

$p(F)$ = probability that person thinks marijuana should be a felony

Then if political inclination and attitude toward marijuana are truly independent, it must be that in the population,

$$p(D \cap L) = p(D)p(L) = (0.750)(0.375) = 0.281$$

and

$$p(R \cap L) = p(R)p(L) = (0.250)(0.375) = 0.094$$

and so on. In short the joint probability in each cell of Table 16-5 is simply the product of the corresponding row and column probabilities.

So this is what our expected probability distribution over this contingency table is. As we did in the last section, let's use this expected probability distribution to create an expected frequency distribution. To do this, we simply multiply each of the cell probabilities by our sample size of $n = 200$ to produce the expected frequency distribution shown in Table 16-6. Now we are in much the same situation as we were in the last section. We have an expected frequency distribution and an observed frequency distribution. We can now argue that if our two factors are indeed independent, then the observed and the expected frequency distributions should resemble one another, and any discrepancy between the two should be due simply to chance. Once again we can derive a summary score which is equal to

$$\sum_{\text{cells}} \frac{(f_e - f_o)^2}{f_e}$$

Again if it is true that our two factors are independent and discrepancies between the observed and expected frequencies are due merely to chance, then this summary score should be distributed as a χ^2. The χ^2 is based on $(J - 1)$ $(K - 1)$ degrees of freedom where J is the number of columns and K is the number of rows. So in this case the χ^2 will be based on $(3 - 1)(2 - 1) = (2)(1)$ = 2 degrees of freedom.

We can compute our observed χ^2 to be

$$\chi^2(2) = (56.2 - 40)^2/56.2 + (56.2 - 70)^2/56.2 + (37.6 - 40)^2/37.6$$
$$+ (18.8 - 35)^2/18.8 + (18.8 - 5)^2/18.8 + (12.4 - 10)^2/12.4$$
$$= 4.67 + 3.39 + 0.15 + 13.96 + 10.13 + 0.46$$
$$= 32.76$$

Our criterion $\chi^2(2)$ is 5.99. Obviously our observed χ^2 of 32.76 is considerably greater than this criterion. Therefore we conclude that political inclination and attitude toward marijuana are not independent of one another. Looking at

TABLE 16-6.   Expected Frequencies and Probabilities for the Contingency Table Shown in Table 16.5, Assuming that Attitude Toward Marijuana and Political Preference Are Independent of One Another. The Expected Cell Probabilities Were Obtained by Multiplying the Corresponding Row and Column (Marginal) Probabilities. The Expected Frequencies Were Obtained by Multiplying Each Expected Probability by 200.

		Attitude Toward Marijuana			
		Should be Legalized	Should be a Misdemeanor	Should be a Felony	
Political Inclination	Democrats	56.2 (0.281)	56.2 (0.281)	37.6 (0.188)	150 (0.750)
	Republicans	18.8 (0.094)	18.8 (0.094)	12.4 (0.062)	50 (0.250)
		75.0 (0.375)	75.0 (0.375)	50.0 (0.250)	200 (1.000)

Table 6-4 should show fairly clearly what aspect of the data leads to the nonindependence. We know that 25% of the total sample are Republicans. But suppose we consider the first column only—those 75 people who believe that marijuana should be legalized. Of those 75 people 35, or 46%, are Republicans. So Republicans seem to be in favor of marijuana legalization more than they should be just by chance. Similarly of the 75 people who believe that marijuana should be a misdemeanor, 70, or more than 93%, are Democrats. So it is apparently the case that, of those people taking the stand that marijuana should be legalized, more than we would expect by chance are Republicans. Similarly of those taking the stand that marijuana should be a misdemeanor, more than we would expect are Democrats.

16.6 If the rows and columns of a $J \times K$ contingency table are independent, then the probability of a given cell should be the product of the corresponding marginal row and column probabilities. Thus, we can compute the expected frequency for each cell (by multiplying the expected probability by the total N). A summary score can then be computed as

$$\sum_{\text{cells}} \frac{(f_e - f_o)^2}{f_e}$$

If the rows and columns are independent then this summary score is distributed as χ^2 with $(J - 1)(K - 1)$ degrees of freedom.

RELATIONSHIPS AMONG DISTRIBUTIONS

We'd like to take this opportunity to point out some rather intimate relationships among the various distributions discussed throughout this book. We hope knowledge of these relationships will help provide a feeling for the considerable unity that underlies much of probability and statistics.

As we have emphasized, a t distribution is similar to a z distribution, but with an unknown rather than a known population variance. The estimate of the population variance is in turn dependent on the number of degrees of freedom. With infinite degrees of freedom, the population variance is known; thus,

$$t(\infty) = z$$

If one squares a $t(n)$, one has an $F(1,n)$, that is,

$$t^2(n) = F(1,n)$$

If there are infinite degrees of freedom, then since $t(\infty) = z$,

$$t^2(\infty) = z^2 = F(1,\infty)$$

However, as we have seen, a z^2 is actually a χ^2 with 1 degree of freedom. Therefore,

$$t^2(\infty) = z^2 = F(1,\infty) = \chi^2(1)$$

We can actually be more general about the relationship between the χ^2 and the F distributions. Recall that

$$F = \frac{\text{est}_1\sigma^2}{\text{est}_2\sigma^2}$$

where $\text{est}_1\sigma^2$ and $\text{est}_2\sigma^2$ are based on $n_1 - 1$ and $n_2 - 1$ degrees of freedom. Note that

$$\text{est}_1\sigma^2 = \frac{\Sigma(x_{i1} - M_1)^2}{n_1 - 1}$$

and

$$\text{est}_2\sigma^2 = \frac{\Sigma(x_{i2} - M_2)^2}{n_2 - 1}$$

Now since

$$\frac{\Sigma(x_{ii} - M_1)^2}{\sigma^2}$$

and

$$\frac{\Sigma(x_{i2} - M_2)^2}{\sigma^2}$$

are distributed as χ^2s with $n_1 - 1$ and $n_2 - 1$ degrees of freedom, respectively, it follows that

$$F[(n_1 - 1), (n_2 - 1)] = \frac{\chi^2(n_1 - 1)/(n_1 - 1)}{\chi^2(N_1 - 1)/(n_2 - 1)}$$

(Recall that we are assuming $est_1\sigma^2$ and $est_2\sigma^2$ are estimates of the same variance, σ^2).

A final note: we have seen in this chapter that we can use a χ^2 to test an obtained variance against some exact variance. The formula that we use for this is

$$\chi^2(n - 1) = \frac{\Sigma(x - M)^2}{\sigma^2}$$

Since σ_o^2 can be viewed as an estimate of σ_o^2 based on infinite degrees of freedom, we could also have performed the F-test

$$F(n - 1, \infty) = \frac{est\ \sigma^2}{v_o^2}$$

where $est\ \sigma^2$ is, of course,

$$\frac{(x - M)^2}{n - 1}$$

SUMMARY

The purpose of this chapter has been to introduce another useful probability distribution, the χ^2 distribution. A $\chi^2(n)$ is defined to be the sum of n z^2 scores, each z having been randomly drawn from a z distribution. It turns out that many variables in the world are distributed as a χ^2, and that a χ^2 is therefore used in a number of different hypothesis-testing situations.

1. A χ^2 may be used to test an obtained variance against a constant. In particular if a set of scores, $x_1, x_2, \ldots, x_n$, is drawn randomly from a population whose variance is σ_0^2, then

$$\frac{\Sigma(x_i - M)^2}{\sigma_0^2}$$

   is distributed as a χ^2 with $n - 1$ degrees of freedom.

2. A χ^2 may be used to test an obtained frequency distribution against a theoretical frequency distribution. In particular, with n observed frequencies (f_os) and their n corresponding expected frequencies (f_es),

the score

$$\frac{\Sigma \, (f_e - f_o)^2}{f_e}$$

is distributed as a χ^2 with $n - 1$ degrees of freedom.

3. Finally, a χ^2 may be used to test for independence in a $J \times K$ contingency table. Expected frequencies (f_e) in each cell are obtained by assuming the joint probability in each cell to be the product of the marginal probabilities. In this situation, the score

$$\frac{\Sigma \, (f_e - f_o)^2}{f_e}$$

is distributed as a χ^2 with $(J - 1)(K - 1)$ degrees of freedom.

PROBLEMS

1. Heights of U.S. males are known to be distributed with a mean, $\mu = 0.70''$ and a standard deviation, σ of $2''$. An anthropologist is inquiring as to whether heights of Bora-Bora males are distributed with a variance different from that of U.S. males. A sample of seven Bora-Bora males yields the following heights:

69″   68″   68″   67″   70″   71″   69″

Can the anthropologist reject the hypothesis that heights of Bora-Bora males is distributed with the same variance as the heights of U.S. males?

2. Boopsie-Cola sales are distributed over days with a mean of 82,000 bottles per day and a standard deviation of 1500 bottles. It is in Boopsie-Cola's interest to reduce this standard deviation because that would make marketing more efficient. Hence a new advertising technique is introduced with this goal. The first ten days of using the technique yield the following sales:

Day	# Sales	Day	# Sales
1	81,752	6	82,033
2	83,812	7	81,925
3	82,104	8	81,599
4	82,529	9	82,730
5	82,620	10	81,885

1. Has the new advertising technique been effective in reducing the standard deviation of sales over days?
2. Does the variance over the first five days using the technique differ from the variance over the last five days?

3. Joe Shablotnik is working for the U.S. Mint. His job is to ensure that new

coins are fair—that is, that they have an equal chance of coming up heads or tails when tossed.

The government has come out with a new coin known as the "Nixon nickel," with a portrait of Richard Nixon on the head side and a horse's tail on the tail side. To test the fairness of the Nixon nickel, Joe flips 100 Nixon nickels, and 61 of them turn up tails.

Test the following hypotheses:

H_0: The probability of a Nixon nickel coming up heads is 0.50.
H_1: The probability of a Nixon nickel coming up heads is not 0.50.

Do *not* use a binomial or a *z*-test.

4. An anthropologist has a hypothesis that 80% of the inhabitants of Clodovia will be left-handed. From a sample of 200 Clodovians, 140 turn out to be left-handed.

    1. Perform a *z*-test to test the hypotheses:

    H_0: 80% of Clodovians are left-handed.
    H_1: The percentage of left-handed Clodovians is something other than 80%.

    2. Use a χ^2 test to test the same hypotheses.

5. Two-thirds of the books in Ralph's science fiction shop supposedly involve time travel, whereas the other third are robot-oriented. Joe Shablotnik goes in and buys 180 books, which he selects randomly off the shelf. Upon arriving home he sorts them, to discover that 50 are about robots and the other 130 are about time travel.

    Joe suspects that Ralph has altered the ratio of time travel to robot books in the store. Is this suspicion justified?

6. Using the data from Chapter 5, Problem 11, evaluate the occultist's hypothesis using a χ^2 test.

7. Baskin-Sunshine has introduced four new flavors of ice cream: plasti-freeze; bubble gum; antiestablish-mint; and zirconium. To see if there are any differences in preferences for the four flavors, 40 people at random are given a choice of one of the four. The preferences are as follows:

        5 people preferred plasti-freeze
        20 people preferred bubble gum
        15 people preferred antiestablish-mint
        0 people preferred zirconium

    Test the hypothesis that there are no differences in preference among the population.

8. Joe Shablotnik keeps track of all dice roll totals during a marathon game of

Monopoly. (Each roll involves the throw of two dice.) He finds the following frequencies of dice totals.

Total	Frequency	Total	Frequency
2	122	7	580
3	204	8	560
4	319	9	401
5	399	10	321
6	502	11	210
		12	98

1. Compute an expected probability distribution and an expected frequency distribution for these data.
2. Would you conclude that the dice are fair dice? Or are they biased in some way?

9. IQ scores in the United States are normally distributed with a mean $\mu = 100$ and a variance $\sigma^2 = 225$. A random sample of 242 people from New York showed the following distribution of IQ scores:

IQ	Frequency of People
<55	20
55–70	17
71–85	29
86–100	52
101–115	63
116–130	42
131–145	13
>145	14

Would you say that New Yorkers are representative of the U.S. population in terms of their IQ scores?

Nonparametric Tests 17

In earlier chapters we alluded to the fact that many of the tests we discussed, particularly z-tests, t-tests, and F-tests, rest on certain assumptions about the populations from which sample scores were drawn. Specifically these tests assume that the scores from different conditions come from populations that (a) are distributed normally and (b) have equal variances. When we use one of these tests—say, the F-test—we are in effect saying, "*If the assumptions of the F-test have been met,* then our observed F of (say) 5.45 will permit us to reject the null hypothesis. Given that the null hypothesis is really true, the probability that we would get an F this large is quite small (less than 0.05)."

VIOLENT ASSUMPTION VIOLATIONS

Occasionally the data we collect flatly and flagrantly violate one or both of these assumptions. Suppose for example that a group of people takes a very easy exam in which scores can range from 0 to 100. In such a situation, there may be very many scores of 100 and very few scores of 60. We would have a markedly skewed distribution that is definitely and decidedly not normal. When such a clear violation occurs, it is not prudent to perform a t- or an F-test; instead, we must select an appropriate *nonparametric* (or *distribution-free*) test, which makes no assumptions about the parameters of the populations from which our samples were drawn.

> 17.1 When assumptions underlying z-tests, t-tests, and F-tests are flagrantly violated, it is prudent to use one of a class of tests known as nonparametric tests.

Computer Arithmetic Revisited

Suppose once again that we are interested in computer-assisted instruction (CAI) and are wondering whether children who learn via a computer behave differently from children who learn via regular instruction. In particular we hypothesize that there may be a difference between these two groups in terms of how quickly they can solve ordinary arithmetic word problems. To deal with this question, we select a random group of children and divide the sample randomly in half; one group learns arithmetic by computer and the other group learns by a live human teacher. We then give all 24 children a series of arithmetic problems to solve. We note how long in minutes each child takes to complete the problems, and we record the average time for each child. Suppose our observations produce the data listed in Table 17-1.

We now wish to go through our hypothesis-testing steps in order to determine whether the computer group takes significantly less time than the teacher group. Before doing hypothesis testing, however, we must obtain an estimate of the variance of the populations that have yielded the data in Table 17-1. This estimate is obtained by

$$est \ \sigma^2 = \frac{SS_1 + SS_2}{df_1 + df_2}$$

TABLE 17-1. Scores of Two Groups of Children on a Test; Each Score Represents the Number of Minutes Taken to Solve an Arithmetic Word Problem

Group 1 Regular Instruction X_{i1}	Group 2 Computer-Assisted Instruction X_{i2}
43	3
12	13
21	1
41	5
39	11
23	10
27	9
37	8
35	6
31	2
33	4
29	7
$\Sigma X_1 = 371$	$\Sigma X_2 = 79$
$n_1 = 12$	$n_2 = 12$
$M_1 = 30.91$	$M_2 = 6.58$
$\Sigma X_{i1}^2 = 12{,}379$	$\Sigma X_{i2}^2 = 675$

$SS_1 = \Sigma X_{i1}^2 - (\Sigma X_{i1})^2/n_1$
  $= 12{,}379 - (371)^2/12$
  $= 908.92$

$SS_2 = \Sigma X_{i2}^2 - (\Sigma X_{i2})^2/n_2$
  $= 675 - (79)^2/12$
  $= 154.92$

Note that

$$SS_1 = \Sigma X_{i1}^2 - \frac{(\Sigma X_{i1})^2}{n_1}$$

$$= 12{,}379.0 - \frac{(371)^2}{12} = 908.92$$

and

$$SS_2 = \Sigma X_{i2}^2 - \frac{(\Sigma X_{i2})^2}{n_2}$$

$$= 675.0 - \frac{(6.58)^2}{12} = 154.92$$

Therefore

$$est\ \sigma^2 = \frac{908.92 + 154.92}{11 + 11} = 48.36$$

Now that we have *est* σ^2 tucked away, we begin our hypothesis-testing steps. In terms of a null and an alternative hypothesis,

H_0: There is no difference between the regular instruction group and the computer-assisted instruction group in terms of how long they take to finish an arithmetic test, or $\mu_1 = \mu_2$, where μ_1 is the mean of the teacher group and μ_2 is the mean of the computer group.

H_1: There is a difference between the regular instruction group and the computer-assisted instruction group in terms of how long they take to finish an arithmetic test, or $\mu_1 \neq \mu_2$.

It seems appropriate to perform a *t*-test, with $df = n_1 + n_2 - 2$, or $12 + 12 - 2 = 22$. If we conduct a two-tailed test, the rejection region will consist of all values of $t \leq -2.09$ and $t \geq 2.09$. Our observed summary score in this situation is the difference between the sample means $(M_1 - M_2)$, which in this case is $(30.91 - 6.58) = 24.33$. If H_0 is true, this sample score is distributed with a mean $\mu_{M_1 - M_2}$ of zero and an estimated variance *est* $\sigma_{M_1 - M_2}^2$ equal to *est* $\sigma_{M_1}^2 +$ *est* $\sigma_{M_2}^2$. (We recall all this, of course, from Chapter 10.) The variance is thus computed as

$$est\ \sigma_{M_1 - M_2}^2 = est\ \sigma_{M_1}^2 + est\ \sigma_{M_2}^2 = \frac{est\ \sigma^2}{n_1} + \frac{est\ \sigma^2}{n_2}$$

$$= \frac{48.36}{12} + \frac{48.36}{12} = 4.03 + 4.03 = 8.06$$

And finally,

$$est\ \sigma_{M_1-M_2} = \sqrt{est\ \sigma^2_{M_1-M_2}} = \sqrt{8.06} = 2.84$$

Our obtained t may now be computed as:

$$t(22) = \frac{(M_1 - M_2) - \mu_{M_1-M_2}}{\sigma_{M_1-M_2}}$$

$$= \frac{24.33 - 0}{2.84} = 8.57$$

Since our observed $t(22)$ of 8.57 exceeds the criterion $t(22)$ of 2.09, we reject H_0.

When computing our estimate of σ^2, we noted that SS_1 and SS_2 were quite different (908.92 versus 154.92). This aroused our suspicion that perhaps the scores in Group 1 come from a population with a greater variance than do the scores in Group 2. And indeed, when we obtain a separate estimate of σ^2 from each group,

$$est_1\ \sigma^2 = \frac{SS_1}{df_1} = \frac{908.92}{11.0} = 82.63$$

and

$$est_2\ \sigma^2 = \frac{SS_2}{df_2} = \frac{154.92}{11.0} = 14.08$$

And we see that $est_1\ \sigma^2$ is almost six times as great as $est_2\ \sigma^2$. A t-test can stand a reasonable amount of mismatch, but a six to one ratio is really an intolerable violation of the homogeneity of variance assumption.

> 17.2 Gross violation of assumptions can generally be detected by examination of the data. (This is a good reason to examine data in close detail.)

Retreat to Nonparametrics: The Mann-Whitney U Test

What are we to do? Our t-test is not really legitimate, so it would be unwise to place complete faith in its conclusions. As suggested, the solution is to perform a nonparametric test that does not depend on assumptions. Nonparametric tests make primary use of the *ordering* of the data points (and therefore are sometimes referred to as *order tests*). There are various nonparametric tests corresponding to the tests we discussed in previous chapters. The sign test (Chapter 5) and the χ^2-test (Chapter 16) are two such examples. Several more illustrations will be provided in this chapter.

For the most part we can usually find some nonparametric test as a substitute for a parametric test. And for the most part it is safer to use a nonparametric test whenever one is in doubt about whether the assumptions of a parametric test will hold. But there is a cost. As we shall see, nonparametric tests are usually less powerful than the corresponding parametric test, and so many researchers will use them only when it is absolutely necessary.

As an alternative to the t-test, the Mann-Whitney U test is one of the most powerful nonparametric statistical tests. Using this test, we will reanalyze our data on the effects on problem-solving speed of regular instruction (RI) versus computer-assisted instruction (CAI). We begin by taking the data from Table 17-1 and ordering them as in Table 17-2. The scores have now been arranged in order from lowest to highest and the condition corresponding to each score is noted, along with the score.

We can easily see that the number of RI scores that are higher than CAI scores is far greater than the number of CAIs that are higher than RIs. This observation provides a clue that the CAI group is performing better than the RI group. But the notion needs to be quantified. To do this, we count the number of RI children who scored higher than each CAI child. We first note that all 12 RI children scored higher than the first CAI child (the one whose score was 1). Second, we see that all 12 RI children scored higher than the second CAI child. Similarly, all 12 RI children scored higher than the third, fourth, fifth, sixth, up to the eleventh CAI child. Only 11 RI children scored higher than the last CAI child (the one whose score was 13). The sum of the number of RI children who scored higher than each CAI child equals:

$$12 + 12 + 12 + 12 + 12 + 12 + 12 + 12 + 12 + 12 + 12 + 11 = 143$$

Let's refer to this value as U_1.

We will now calculate the converse statistic, U_2, which is the number of CAI children who scored higher than each RI child. This calculation is easy because only one CAI child scored higher than an RI child (the CAI child whose score was 13 scored higher than the RI child whose score was 12). The other CAI children never scored higher than an RI child. Thus, the sum of the number of CAI children who scored higher than each RI child is equal to 1, that is, $U_2 = 1$.

TABLE 17-2.   Rearrangement of Data from Table 17-1 Needed to Perform a Mann-Whitney U Test

Rank = R_i	1	2	3	4	5	6	7	8	9	10	11	12	13	14	15	16	17	18	19	20	21	22	23	24
Score	1	2	3	4	5	6	7	8	9	10	11	12	13	21	23	27	29	31	33	35	37	39	41	43
Condition	CAI	CAI	CAI	CAI	CAI	CAI	CAI	CAI	CAI	CAI	CAI	RI	CAI	RI	RI	RI	RI	RI	RI	RI	RI	RI	RI	RI

For RIs: $\Sigma R_{i1} = 12 + 14 + 15 + 16 + 17 + 18 + 19 + 20 + 21 + 22 + 23 + 24 = 221$

For CAIs: $\Sigma R_{i2} = 1 + 2 + 3 + 4 + 5 + 6 + 7 + 8 + 9 + 10 + 11 + 13 = 79$

If there were no difference in the speed of solving arithmetic problems between RI and CAI children, we would expect U_1 and U_2 to be about the same, because on the average RI children would score higher than CAI children just about as often as CAI children scored higher than RI children. But a very large value of U_1 accompanying a very small value of U_2 (or vice versa) would suggest that the groups were different. Therefore, when we observe large differences in the value of U_1 versus U_2, we will reject the null hypothesis that the two samples came from the same population. But how large a difference must we observe?

Ordinally we take the smaller value of U_1 and U_2 and refer to this smaller value as U. The sampling distribution of U assuming the null hypothesis to be true is known and depends upon the size of each sample. Tables (like z-tables, t-tables, and the like) have been constructed to provide the criterion values of U for samples of various sizes. As indicated in Appendix F-6, for $n_1 = 12$ and $n_2 = 12$, U must be less than or equal to 37 in order for us to reject the null hypothesis at the 0.05 level. Our observed U of 1.0 is indeed smaller than the critical value of 37; thus, we reject the null hypothesis and conclude that children who learn arithmetic by regular instruction take longer to solve arithmetic problems.

> 17.3 The Mann-Whitney U test is the nonparametric test analogous to a t-test. The Mann-Whitney U, like other nonparametric tests, is primarily concerned with the ordering of scores. It does not use measures such as the mean or variance of the original scores.

It should actually come as little surprise that when we applied the legitimate nonparametric test we came to the same conclusion as we did with the questionable parametric t-test. Recall from our discussion in Chapter 11 that when the obtained score far exceeds a criterion score, we are on fairly safe ground rejecting H_0 even if we have violated one of the assumptions of the parametric test. In the present example the obtained score was so extreme that it was quite unlikely that the legitimate test would produce a different result, and sure enough it didn't. Later in this chapter we will take a look at an example in which the obtained score resulting from a parametric test only slightly exceeds the criterion score. It is this sort of situation that raises problems.

Computational formulas. Although the counting technique was relatively easy in the previous example, you might suspect that this technique can become extremely tedious and time consuming when the number of subjects is large. Your suspicion would be entirely justified, but happily there is an alternate procedure that yields the same results. To calculate U_1 and U_2, we use the formulas

$$U_1 = n_1 n_2 + \frac{n_1(n_1 + 1)}{2} - \sum_{\text{scores}} R_{i1}$$

and

$$U_2 = n_1 n_2 + \frac{n_2(n_2 + 1)}{2} - \sum_{\text{scores}} R_{i2}$$

where R_{i1} is the sum of the *ranks* obtained by Group 1 and R_{i2} is the sum of the ranks assigned to Group 2, as indicated in Table 17-2. (To check your calculations: U_1 plus U_2 should, as it turns out, equal n_1 times n_2. In our example

$$\text{RI: } U_1 = 12(12) + \frac{12(13)}{2} - 221 = 1$$

and

$$\text{CAI: } U_2 = 12(12) + \frac{12(13)}{2} - 79 = 143$$

Thus

$$U_1 + U_2 = 1 + 143 = 144 = (n_1)(n_2)$$

and all is well.)

Ties.   In the example we just considered, every score was different; there were no tied scores. But often there are ties. Consider for example the scores {2, 8, 10, 10, and 14}. In applying the Mann-Whitney U test, the score of 2 would receive a rank of 1, while the score of 8 would receive a rank of 2. The two scores of 10 would be assigned the average of the next two ranks. Because the next two ranks are 3 and 4, their average is 3.5. Consequently the scores of 10 each receive a rank of 3.5. The final score is then given the next rank, 5.

Occasionally three or more scores will be tied. When this occurs, each score receives the average rank that those scores would have received if they were all different. In the set of scores {2, 8, 10, 10, 10, 14}, the scores of 2 and 8 receive ranks of 1 and 2. The three scores of 10 would be assigned the average of the next three ranks. Since the next three ranks are 3, 4, and 5, their average is 4. The final score is then given the rank of 6, the next unused rank. See Table 17-3 for some examples of the appropriate rankings to be used when there are several ties.

TABLE 17-3.   **Examples of Rankings to Be Used When There Are Several Ties**

1.	Score	2	2	8	10	10	14				
	Rank	1.5	1.5	3	4.5	4.5	6				

2.	Score	12	14	14	14	15	16	16	16	16	21
	Rank	1	3	3	3	5	7.5	7.5	7.5	7.5	10

When _n_ is large.   When the number of subjects in the two groups is large (say, greater than 20), the observed value of U approaches a normal distribution with mean

$$\mu = \frac{n_1 n_2}{2}$$

and standard deviation

$$\sigma = \sqrt{\frac{(n_1)(n_2)(n_1 + n_2 + 1)}{12}}$$

This means that we can determine whether U is significant or not by calculating a z-score, using the equation we learned long ago in Chapter 6:

$$z = \frac{x - \mu}{\sigma}$$

or in this case

$$z = \frac{U - (n_1 n_2 / 2)}{\sqrt{(n_1)(n_2)(n_1 + n_2 + 1)/12}}$$

Since our U-tables only go up to an n of 20, this z-approximation is often very handy.

If we were to use the z-approximation in our present example,

$$\mu = \frac{n_1 n_2}{2} = \frac{(12)(12)}{2} = \frac{144}{2} = 72$$

$$\sigma = \sqrt{\frac{(n_1)(n_2)(n_1 + n_2 + 1)}{12}} = \sqrt{\frac{(12)(12)(25)}{12}} = 17.32$$

and

$$U = 1$$

Thus:

$$\text{obtained } z = \frac{1 - 72}{17.32} = -4.10$$

Since our obtained z of -4.10 is much more extreme than our criteria (two-tailed) z of ± 1.96, we reject H_0.

> 17.4 When n is large, it is computationally simpler to use the normal approximation to U instead of calculating U. (Note that this normal approximation has nothing to do with the distributions of the original scores.)

PARAMETRIC VERSUS NONPARAMETRIC TESTS

To reiterate, the difference between a parametric and a nonparametric test is that a parametric test specifies certain conditions about the parameters of the population from which our data have been drawn. To use such tests, certain assumptions about the variance or shape of the distribution must be met. The nonparametric test does not specify conditions about the parameters of the population from which our data were drawn. (These tests do have some assumptions, however. For example the observations must be randomly assigned to the two or more conditions. In the experiment on computer-assisted versus regular instruction, no nonparametric test will save us if we have deliberately put boys in one condition and girls in the other, or 10-year-olds in one condition and 8-year-olds in the other.)

At this point you are probably saying to yourself: "Since the nonparametric test requires so many fewer assumptions, and our conclusions do not depend on things like the shape of the population distribution from which the random sample was drawn, why don't we always use nonparametric tests?" The reason is that nonparametric methods have some disadvantages, which are best illustrated by considering the following example.

Spaced-Out Memory

Two psychologists, whom we will call P and N, together decide to perform an experiment on the effects of various drugs on people's ability to remember emergency instructions. It occurred to them that people who were under the influence of drugs might have a more difficult time remembering a verbal instruction, such as a message to evacuate a building during a fire or a bomb threat. Psychologists P and N began by recruiting 12 individuals via a newspaper advertisement. Four of these subjects were randomly assigned to be given mescaline, four were given LSD, and the remaining four were control subjects who consumed only a harmless sugar pill. All the subjects were given some emergency instructions to remember and were later tested on their memory for these instructions. The number of errors made by each subject were recorded, and these data are presented in Table 17-4.

Looking at the data, we seem to see differences among the three groups. In particular the two drug groups appear to be doing worse than the control group. Given that a drug is taken, LSD appears to be worse than mescaline. The two psychologists decide that they will each analyze the data, and then they will compare their results. This procedure, while requiring them both to do the work, will serve as a check for accuracy.

Psychologist P decides to conduct a one-way analysis of variance, and her calculations are shown in Table 17-5. Using this procedure, she has found an observed F value of 5.96, which is significant beyond the 0.05 level. With 2 and 9 degrees of freedom, the critical value of F is 4.26, and the observed value exceeds this critical value.

TABLE 17-4.   Number of Errors Made by Subjects in Three Different Drug Conditions

Group 1 Sugar Pill	Group 2 Mescaline	Group 3 LSD
4	4	9
7	9	13
1	10	15
1	5	7
$\Sigma X_{i1} = 13$	$\Sigma X_{i2} = 28$	$\Sigma X_{i3} = 44$
$M_1 = 3.25$	$M_2 = 7$	$M_3 = 11$

TABLE 17-5.   Analysis of Data from Table 17-4

Group 1 Sugar Pill	Group 2 Mescaline	Group 3 LSD	
4	4	9	
7	9	13	
1	10	15	
1	5	7	
$\Sigma X_{i1} = 13$	$\Sigma X_{i2} = 28$	$\Sigma X_{i3} = 44$	$T = 85$

$$SSB = \sum_{\text{groups}} \frac{T_j^2}{4} - \frac{T^2}{N} = \frac{13^2}{4} + \frac{28^2}{4} + \frac{44^2}{4} - \frac{85^2}{12} = 722.25 - 602.8 = 119.45$$

$$SSW = \sum_{\text{everything}} X_{ij}^2 - \sum_{\text{groups}} T_j^2/4 = 813 - 722.25 = 90.75$$

Summary Table

Source	SS	df	MS	F
Total	210.92	11		
Between	120.17	2	60.09	5.96
Within	90.75	9	10.08	

Kruskal-Wallis one-way analysis of variance by ranks.   Psychologist N does not want to take a chance that the assumptions of the parametric one-way analysis of variance might not be met, and so he decides to conduct a nonparametric test instead. "I've got nothing to lose," he tells himself. He happily discovers that there is a nonparametric procedure corresponding to the one-way analysis of variance, called the "Kruskal-Wallis one-way analysis of variance by ranks." Since its name is such a mouthful, we'll refer to it as the

TABLE 17-6.   Scores and Ranks for Use in a Kruskal-Wallis Test

Score:	1	1	4	4	5	7	7	9	9	10	13	15
Rank:	1.5	1.5	3.5	3.5	5	6.5	6.5	8.5	8.5	10	11	12

Kruskal-Wallis test for short. This test, he discovers, is extremely useful for deciding whether more than two independent samples are from different populations.

In the computation of the Kruskal-Wallis test, all the observations from all conditions are combined and ranked in a single series. The smallest score is replaced by rank 1, the next smallest by rank 2, and so on. When Psychologist N replaced the observations by ranks, he observed what you see in Table 17-6.

Next the sum of the ranks in each of the original columns is found. If these sums are similar to one another, it is likely that the samples were all drawn from the same population. On the other hand if these sums are highly disparate, it is unlikely that all scores were drawn from the same population.

Psychologist N found the sum of the ranks for each of the three groups shown in Table 17-7.

TABLE 17-7.   Sum of Ranks for each group for a Kruskal-Wallis Test

	Control	Mescaline	LSD
	3.5	3.5	8.5
	6.5	8.5	11.0
	1.5	10.0	12.0
	1.5	5.0	6.5
Sum:	$R_1 = 13.0$	$R_2 = 27.0$	$R_3 = 38.0$

If the samples are actually from the same population or from identical populations (in other words, if H_0 is true), then a statistic referred to as H, defined in the formula which follows, can be used.

$$H = \frac{12}{N(N + 1)} \sum_{groups} \frac{R_j^2}{n_j} - 3(N + 1)$$

where

N = the total number of observations
n_j = the number of observations in sample j
R_j = sum of the ranks in sample j

Psychologist N calculated H to be:

$$H = \left[\frac{12}{12(12 + 1)}\right]\left[\frac{13^2}{4} + \frac{27^2}{4} + \frac{38^2}{4}\right] - 3(12 + 1)$$

$$= (0.076)(585.5) - 39.0$$

$$= 5.498$$

To determine whether the observed value of H exceeds some criterion value, we need a table of criterion values. It turns out that the critical value will depend on both the number of samples and the number of observations within each sample. The present experiment used three samples, with four observations in each. Consulting the H-tables in Appendix F-7 of this book, we see that when the ns are 4, 4, and 4, the critical value of H is approximately equal to 5.69. Since our observed value of 5.498 does not exceed this critical value, we cannot reject the null hypothesis.

17.5 The Kruskal-Wallis test is the nonparametric test that is analogous to a one-way ANOVA.

Digression 17.1

When the *ns* Are Large

When there are more than five observations in each of the various samples, then the statistic H is distributed as a $\chi^2(J - 1)$ where J, as usual, is the number of groups. This makes things easy, for instead of using Appendix F-7, we can determine whether an observed H is significant by reference to Appendix F-5. If the observed value of H is equal to or larger than the value of χ^2 given in Appendix F-5, then H_0 may be rejected. So if we had had 10 observations in each of our samples instead of only 4, we would have required an H [criterion $\chi^2(2)$] of 5.99 or greater in order to reject the null hypothesis at the 0.05 level of significance.

Disagreement between tests. Psychologists P and N finished their analyses and met to compare notes. They soon discovered that one of them (P) had used a parametric test and was able to reject H_0. The other (N) had used a nonparametric procedure and was not able to reject H_0. Thus, if the null hypothesis were really false, the nonparametric test would incorrectly result in a failure to reject, because the nonparametric procedure is less powerful than the parametric procedure.

As you will recall from our discussion of power in Chapter 8, a powerful test allows a researcher to detect the effect of some experimental variable when that variable really does have an effect. To say that parametric tests are more powerful than nonparametric tests means that given exactly the same set of data, a parametric test is more likely to lead to significant results than a nonparametric test. The principal reason for this is that parametric tests make maximum use of all the information inherent in the data. With nonparametric tests, information is lost when scores are converted to some other form—say, ranks.

What should the psychologists do at this point? Should they report the results of the one-way ANOVA or of the Kruskal-Wallis? The answer depends upon whether the assumptions of the ANOVA have been met. If they have, then the investigators should go ahead and report the results of the ANOVA and thereby reject H_0. If, on the other hand, there is reason to suspect that the assumptions have been violated, the researchers cannot in good faith use the ANOVA. They must report the results of the Kruskal-Wallis, and consequently *not* reject H_0.

> 17.6 A nonparametric test is generally less powerful than its corresponding parametric test. Thus it is possible for a nonparametric test to come out nonsignificant while the same data would be significant according to the parametric test.

WHICH NONPARAMETRIC TEST IS APPROPRIATE?

If we have collected some data and wish to conduct a *parametric* test, we now would have little difficulty choosing an appropriate test to apply. Thus, if we had two independent samples, we might apply the *t*-test; with more than two, we might apply a one-way ANOVA, and so on.

When we know that the assumptions of these parametric tests have been violated, and we wish to conduct a nonparametric test instead, we choose an appropriate test in the same way. The test we use will depend upon the data we have collected. For example, the Mann-Whitney *U*-test is appropriate when we have two independent samples, whereas the Kruskal-Wallis test is used when we have more than two independent samples. In Table 17-8 we list the types of data we might have, along with the tests that would be used to analyze these data. Notice several things. First often more than one test is appropriate for a set of data. Second there is no nonparametric test that corresponds to the two-way analysis of variance. This is rather unfortunate, because it means that we cannot evaluate the interaction between two independent variables with a nonparametric procedure.

TABLE 17-8. Parametric and Nonparametric Tests for Analyzing Various Kinds of Data

Data	Purpose of Test	Parametric Test	Nonparametric
Single sample	To determine whether the population from which a particular sample was drawn has a mean or median different from some hypothetical value	z = test or t-test for a single sample	Sign test
Two samples, between subjects	To determine whether the populations from which two independent samples were drawn have the same mean or median	z = test or t-test for independent groups	Mann-Whitney
Two conditions, within subjects	To determine whether the populations from which two samples were drawn have the same mean or median	z = test or within-subjects t-tests	Sign test or Wilcoxon sign ranks test
More than two conditions, between subjects	To determine whether the populations from which more than two independent samples were drawn have the same mean or median	One-way ANOVA (F-test)	Kruskal-Wallis test
More than two conditions, within subjects	To determine whether the populations from which more than two samples were drawn have the same mean or median	Repeated measures one-way ANOVA	Friedman test
Set of items with two measures on each item	To determine whether the two measures or variables are associated	Pearson correlation	Spearman rank-difference correlation (rho)
Single sample	To determine whether the population from which a particular sample was drawn has a certain theoretical form	Also handled by chi-square test (which is nonparametric)	Kolmogorov-Smirnov

Deciding which test to use is sometimes a matter of individual preference. There are literally hundreds of statistical tests that could be presented in this book, each one designed specifically for a particular situation. For a more comprehensive discussion of nonparametric tests, we recommend you consult a reference source devoted solely to this topic. In the meantime we'll present some of the more popular and useful nonparametric procedures to enable you to analyze some kinds of data very quickly.

MORE NONPARAMETRIC TESTS

For the most part a nonparametric test is available as a substitute for a parametric test in any situation. So for example nonparametric tests exist for comparing a set of data with some specified theoretical distribution, for comparing two or more conditions in a within-subjects design, and for evaluating associations in a correlational situation.

Comparing a Set of Data to Some Specified Theoretical Distribution: Kolmogorov-Smirnov One-Sample Test

Suppose you wish to determine whether a population distribution has a certain theoretical form based on some observed data. We saw in Chapter 16 that a χ^2 test could be used for this purpose. Another test that is particularly useful for this purpose, the Kolmogorov-Smirnov one-sample test, specifies the cumulative frequency distribution that would occur under the theoretical distribution and compares that with the observed cumulative frequency distribution. The test involves finding the point at which these two distributions—the theoretical and the observed—show the greatest discrepancy. If a discrepancy is sufficiently large so that it is unlikely to occur if the observations were really a random sample from the theoretical distribution, then the null hypothesis (the assumption that the theoretical distribution holds) can be rejected.

To illustrate the use of the test, suppose you are wondering whether people prefer smiling to nonsmiling faces. To study this problem, you arrange to have five photographs taken of an individual with facial expressions from no smile, through a moderate smile, to a full smile. You get together 10 subjects and ask each which photograph he or she prefers. If smile is not important, you might expect each photograph to be chosen equally often. If smiling is important, then you might expect one of the photographs showing large smiles to be chosen more often than one of the photographs showing little or no smile.

Suppose your observations bring the data shown in Table 17-9. $F_T(S)$ is the theoretical cumulative distribution under H_0, where H_0 is the hypothesis that each of the photographs is chosen equally often, each by one-fifth of the subjects. $F_r(x)$ is the cumulative distribution of the observed choices. The last row in the table gives the absolute deviation of each observed value from its corresponding theoretical value. Thus, the first absolute deviation is obtained by subtracting $\%_{10}$ from one-fifth, to get $\frac{1}{5}$ or $\frac{2}{10}$.

TABLE 17-9. Do People Prefer Smiling Photographs?

	Rank of Photo Chosen, from $1 = $ no smile to $5 = $ full smile				
	1	2	3	4	5
$f = $ Number of subjects choosing each photo	0	1	0	4	5
$F_T(X) = $ Theoretical cumulative distribution under H_0	2/10	4/10	6/10	8/10	10/10
$F_s(X) = $ Cumulative distribution of observed choices	0/10	1/10	1/10	5/10	10/10
Difference: $F_T(X) - F_S(X)$	2/10	3/10	5/10	3/10	0

The important value is the maximum deviation, D, which in this case is five-tenths or 0.50. To determine the significance of the observed D, we need to consult a table of the distribution of D, which can be found in Appendix F-8. From this table it can be seen that the critical value of D, for $N = 10$, at the 0.01 level of significance is 0.49. Since our observed D exceeds this value, we reject H_0, and conclude that our subjects have shown a significant preference for smiling faces.

Comparing Two Conditions (Within-Subjects Design): Wilcoxon Test

Occasionally we are interested in gathering two sets of observations from the same individuals or from related individuals. One way to analyze such data involves the use of a within-subjects t-test, as we saw in Chapter 10. But there are nonparametric counterparts to this test, and one of the most commonly used is the Wilcoxon test for matched samples.

Suppose we are interested in whether a course designed to improve reading skills actually accomplishes that goal. We might take some individuals and measure their reading skills both before and after the course is completed. Suppose we had actually gathered the data shown in Table 17-10, from seven individuals. We could analyze these data using an appropriate t-test, but given the assumptions underlying the Wilcoxon test are less restrictive, we opt for that alternative.

The procedure is quite straightforward. First the matched pairs of scores are subtracted in order to obtain their difference, as is done in Table 17-10. Next the absolute values of the differences are obtained, rather easily. Now the

TABLE 17-10. Does a Course on Reading Skills Actually Improve Those Skills?

Individuals	Before	After	Difference	Absolute Difference	Rank Difference	Signed Rank Difference
A	74	75	−2	2	3	−3
B	81	80	1	1	1.5	1.5
C	85	89	−4	4	6	−6
D	79	88	−9	9	7	−7
E	92	95	−3	3	4.5	−4.5
F	83	80	3	3	4.5	4.5
G	87	86	1	1	1.5	1.5

$$T_+ = 7.5$$
$$T_- = 20.5$$
$$W_{obs} = 7.5$$

absolute values are assigned ranks starting with a rank of 1 for the smallest difference. Finally the appropriate positive or negative signs are restored to the ranks. You can follow this procedure across the columns on Table 17-10.

What do we expect to find? If there is no difference between the reading score obtained before and the reading score obtained after the course, we would expect the sum of the ranks associated with positive differences between the groups to be roughly equal to the sum of the ranks associated with negative differences between groups. If there is a difference between the before and the after scores, then the positive sum of ranks might be quite a bit smaller or larger than the negative sum of ranks. As the before and the after scores become more and more discrepant, either the positive sum of ranks or the negative sum of ranks will get smaller. The Wilcoxon test allows us to determine whether this smaller sum is "significantly" small.

Looking at the last column in Table 17-10, we see the "Signed Rank Difference" for each individual subject. We must compute two quantities: $T+$, which is the total of the ranks having positive signs associated with them, and $T-$, which is the total of the ranks having negative signs. The Wilcoxon test is interested in the smaller of the two, and we denote this value W_{obs} for the observed value of W. In our example $W_{obs} = 7.5$ To determine whether this value is significantly small, we must consult Appendix F-9, which presents the critical values for the sampling distribution of W as a function of the number of individuals, N. If W_{obs} is less than or equal to the critical value, then we reject H_0 and conclude that a real difference in the two sets of scores exists. In the present example we might wish to use a significance level of 0.05, and thus the critical value is 2. Since our W_{obs} is greater than this critical value, we cannot reject H_0.

In the Wilcoxon test, ties are handled in a fairly straightforward way. If there is an even number of zero differences, each one is assigned the average rank for the set. Naturally zero differences rank lowest in terms of absolute size. So with two zero differences, each would receive a rank of 0.5. The next step is to give half these ranks a positive sign and the other half a negative sign. However, if an odd number of zero differences occur, one of the zero differences is randomly removed from the data and the procedure for an even number of zero differences is followed. In this case the total number of pairs, N, must be reduced by 1.

Notice that the table includes values of N between 5 and 50. Actually when N is larger than 25, we can use an approximation to the standard normal curve. The conversion is given by this formula:

$$z = W_{obs} - \frac{\dfrac{N(N + 1)}{4}}{\dfrac{\sqrt{N(N + 1)(2N + 1)}}{24}}$$

Although N is a bit small in our reading skills example, consider what happens when these values are substituted in the formula for z.

$$z = \frac{7.5 - \dfrac{7(7 + 1)}{4}}{\dfrac{\sqrt{7(7 + 1)[2(7) + 1]}}{24}}$$

$$= -1.10$$

As you will recall, when using a 0.05 level of significance and a two-tailed test, the critical values of z are 1.96 and $+1.96$. Our observed value of z does not exceed these values, and so we cannot reject H_0.

Comparing Many Conditions (Within-Subjects Design): Friedman Test

When we wished to compare two independent groups, the Mann-Whitney U test was a possible choice. With more than two independent groups we had the Kruskal-Wallis test available to us. The Friedman test bears a similar relationship to the Wilcoxon test. With two sets of observations on the same or related individuals, we have the Wilcoxon test available to us. With more than two sets of observations, we may wish to turn to the Friedman test.

The procedures involved in conducting a Friedman test are not difficult. Let's say we're interested in whether visual ability changes depending on the time of day it is tested. Suppose we look at the visual ability of $n = 11$ individuals, tested in the morning, noon, afternoon, and evening, in random order. Assume we have obtained the data shown in Table 17-11. So Subject 1 got a score of 1 in the morning, 4 at noon, 8 in the afternoon, and 0 in the evening. The Friedman test begins by first taking each row of scores and assigning each score in the row a rank. The ranks for Subject 1 are: 2 in the morning, 3 at noon,

TABLE 17-11.   **Does Visual Ability Depend on the Time of Day?**

Subject	Morning	Noon	Afternoon	Evening
1	1 (2)	4 (3)	8 (4)	0 (1)
2	3 (2)	2 (1)	4 (3)	13 (4)
3	14 (4)	4 (2)	7 (3)	2 (1)
4	10 (4)	4 (2)	9 (3)	3 (1)
5	10 (4)	4 (2)	5 (3)	3 (1)
6	4 (1)	12 (4)	10 (2)	11 (3)
7	10 (3)	3 (1)	11 (4)	9 (2)
8	1 (2)	3 (3)	10 (4)	0 (1)
9	12 (3)	11 (2)	13 (4)	10 (1)
10	10 (3)	0 (1)	11 (4)	3 (2)
11	2 (2)	3 (3)	13 (4)	1 (1)
ΣR_j	30	24	38	18

4 in the afternoon, and 0 in the evening. This is done for each subject, and the ranks for each column are then summed.

What do we expect to find? Suppose that time of day makes no difference. The column sums of ranks should be identical in the population and similar in the samples, if the null hypothesis of no difference as a function of time of day is true. However, if there tends to be a large number of high ranks in one or two of the columns, we begin to have support needed to reject H_0.

With sufficiently large samples—say, a minimum of $n = 10$ subjects—each receiving $J = 4$ treatments, the test statistic for the Friedman is distributed approximately as χ^2 with $J - 1$ degrees of freedom. The statistic is:

$$\chi_r^2 = \frac{12}{nj(J + 1)} \left[\sum_{\text{groups}} R_j^2 \right] - 3n(J + 1)$$

and in our example this means:

$$\chi_r^2 = \frac{12}{11(4)(5)} [30^2 + 24^2 + 28^2 + 18^2] - 3(11)(5)$$

$$= 11.79$$

With 3 degrees of freedom this value is significant beyond the 0.01 level, so we can handily reject H_0.

The χ' approximation is used only when the number of subjects and treatments is not too small. However, when these numbers are small, exact probabilities can be obtained from tables. For example Appendix F-10 can be consulted when there are three or four treatments, and n is quite small.

Discovering Associations: Spearman's Rank-Difference Correlation (Rho)

Suppose you are interested in whether there is a relationship between the number of visitors permitted in a prison and the quality of prison life. You suspect that the relationship may be positive—that is, with more visitors permitted, the prisoners judge the prison to be of higher quality. You gather data from 10 prisons around the United States, noting for each its rank in terms of number of visitors allowed and its rank in terms of quality determined by a vote of prisoners. Suppose you've gathered the data in Table 17-12. Assume the numbers in the first two columns represent your observed rankings of a sample of 10 prisons on the two variables of interest—visitors and quality. Now you might compute a Pearson correlation coefficient, as described in Chapter 14, but you might also want to consider the Spearman rank-difference correlation (called *rho*) which is a measure of the association of two variables especially useful when the scores are in the form of ranks. *Rho* is merely the Pearson r^2 between the *ranks* of two variables and is quite easy to compute, particularly when n is small.

To calculate *rho*, you first compute the difference in ranks on the two

---

TABLE 17-12.   Does Number of Visitors Permitted in a Prison Affect the Quality of the Prison?

°Prison	Rank-visitors	Rank-quality	d_i	d_i^2
1	1	1.5	0.5	0.25
2	2.5	3.0	0.5	0.25
3	2.5	1.5	−1,0	1.0
4	4	5.0	1.0	1.0
5	5	5.0	0	0
6	6	8.5	2.5	6.25
7	7	5.0	−2.0	4.0
8	8	7.0	−1.0	1.0
9	9	8.5	−0.5	0.25
10	10	10.0	0	0

---

variables for each prison, as is done in the third column of Table 17-12. Then you square these difference scores to obtain:

$$rho = 1 - \frac{6(\Sigma d^2)}{n(n^2 - 1)}$$

Where n is the number of prisons. In this example

$$rho = 1 - \frac{6(290)}{10(10^2 - 1)} \qquad 1 - \frac{6(14)}{1000 - 10}$$

$$= 1 - \frac{1740}{990} \qquad 1 - \frac{84}{990}$$

$$= 1 - 1.76 \qquad 1 - .085$$

$$= 0.76 \qquad 0.915$$

Just as you suspected, there appears to be a positive correlation between the number of prisoners in a prison and its quality, but you still must determine whether this correlation is significantly large to allow you to reject the null hypothesis that the population value of rho is 0.

   To do this, you would consult Appendix F-16. Since you had 10 observations (prisons), the absolute value of your obtained rho must exceed 0.65 in order to reach the 0.05 level of significance. The obtained value of 0.915 does just this, so you can conclude that there is a significant positive correlation between number of visitors and prison quality.

SUMMARY

The statistical tests discussed prior to this chapter (such as the z-test and the t-test) are based upon certain assumptions. The major assumptions are these:

1. The scores constituting the raw data, or the *xs*, are *independent* of one another.
2. The *xs* are drawn from a *normal distribution*.
3. The scores from the different conditions are drawn from populations having *equal variances*.

Violation of the independence assumption has a rather serious effect, but most experiments can be arranged so that this assumption is not violated. Violations of the other two assumptions are less serious. But if one of these assumptions has been strongly violated, it is proper to analyze data using a nonparametric test. Nonparametric tests do not make any assumptions about the parameters of the populations from which samples were drawn. Most parametric tests have nonparametric tests corresponding to them.

PROBLEMS

1. A developmental psychologist is interested in whether kindergarten boys and girls are different from each other in terms of how happy they are. A happiness test is given to five boys and five girls, and the 10 children are ranked from first to tenth in terms of their total happiness score. Are boys significantly happier than girls?

Ranking of happiness	1	2	3	4	5	6	7	8	9	10
Sex of child	G	G	B	G	G	B	B	G	B	B

2. An experimenter is testing the hypothesis that a person's expectations can influence the outcome of an experiment. Nine rats, all from the same strain, are given to psychology students. Some students are told that these rats were bred to be "bright," but others are told that the rats were bred to be "dull." Each student takes a single rat and tests its ability to learn a maze, carefully recording the number of errors made. Do they provide support for the hypothesis that relative to a student who expects the rat to be dull, a student who expects his rat to be bright finds the rat making fewer errors?

Bright	*Dull*
6, 8, 5, 12	10, 2, 14, 9, 4

Errors made by rats that are expected to be

3. Two researchers are interested in the question of whether psychologists or political scientists tend to have larger book collections. They collect some data from nine political scientists and 12 psychologists; they visit the offices of these individuals and simply count the number of books on the shelves. The observations are given below. Determine whether there is a significant difference between the size of collection using a nonparametric test.

Political Scientists	Psychologists
87	131
72	94
65	77
54	88
67	116
76	90
73	87
82	76
104	95
	164
	127
	77

4. Two political researchers are interested in whether the Watergate political scandal had any effect on the voting practices of Democrats versus Republicans. They hypothesize that Republicans would be so discouraged that they would be less likely to vote in local and national elections. The voting behavior of three Democrats and four Republicans is examined. The number of times that each of these people voted in the previous three years is given below. Test the hypothesis that Republicans vote less often than Democrats.

Democrats	Republicans
4	2
3	1
1	0
	0

5. A medical scientist, Dr. Greenwald, investigated the effect of a new drug on patients with insomnia. She took three groups of patients and administered drug A to one group, drug B to a second group, and no drug to the third group. After a week the patients reported the average number of hours they had been able to sleep. A total of 15 patients participated in the experiment, five per group. When the data were in, Dr. Greenwald ranked them from best to worst (1 to 15) in terms of ability to sleep.

   Use a nonparametric procedure to analyze her data:

Drug A	Drug B	No Drug
3	1	2
6	5	4
11	9	7
12	10	8
15	14	13

6. Dr. Decible is interested in whether people can more accurately detect the pitch of a tone when it is presented to the right or left ear rather than to the top of the head. Six individuals hear tones presented to the right ear, six hear tones presented to the left ear, and six hear tones presented over the top of the heads. Does nonparametric statistical analysis support the contention

that the placement of the tone makes a difference? Analyze the following data, which are the number of errors that each subject made.

Right Ear	Left Ear	Top of Head
0	1	3
0	2	5
0	3	2
1	2	1
0	2	4
0	1	5

7. Many Americans are currently seeking another kind of altered state of consciousness through the practice of meditation. Yoga and Zen are two common methods. Both have as their goal the enhanced perceptions of beauty and truth, and these are accomplished in part by relaxation of the muscles.

Dr. Transcendental wishes to study the effects of Zen and Yoga on people's perception of beauty. He devises a test designed to measure the ability to perceive beauty, a test in which people can receive a score between 0 and 40. He finds 36 individuals to participate in his study. One-third are given a week's worth of Yoga, one-third a week of Zen, and one-third are not given a treatment. At the end of the week his Beauty Perception test is administered, and the following scores are obtained. Use a nonparametric procedure to determine whether the groups are different from each other.

Control	Yoga	Zen
6	31	13
11	7	32
12	9	31
20	11	30
24	16	28
21	19	29
18	17	25
15	11	26
14	22	26
10	23	27
8	27	26
14	26	19

8. Many statisticians suggest that when there are ties in the rankings, the value of H should be corrected. The formula to be used for correcting is:

$$C = 1 - \frac{\left[\sum_{i}^{T} (t_i^3 - t_i)\right]}{N^3 - N}$$

where T is the number of sets of tied observations and t_i is the number tied in any set i. In the previous problem there were four sets of two tied observations, one set of three ties, and one set of four tied observations.

Thus:

$$C = 1 - \left(\frac{4[(2)^3 - 2] + [(3)^3 - 3]}{(36)^3 - 36} + [(4)^3 - 4]\right) = 0.997$$

To correct the value of H, we divide it by C.

$$H' = \frac{H}{C} = \frac{13.81}{0.997} = 13.84$$

Now suppose that the first subject score of the Zen condition had received a 7 instead of a 13. What would this do to the value of H? Correct this value for ties.

9. Ten Triumphs and nine Pontiacs drove from Seattle to San Francisco. The total number of hours it took each of the cars to travel the distance is given here:

Triumphs: 16.5, 16.6, 16.7, 16.1, 17.7, 17.4, 16.2, 16.4, 17.1, 16.9
Pontiacs: 16.3, 17.9, 17.8, 16.8, 17.0, 17.2, 17.3, 17.5, 17.6

1. Suppose you had no preconceived notions about whether drivers of Triumphs drive faster than drivers of Pontiacs. Based on a nonparametric test of these data, would you conclude that one type of person was a faster driver?
2. Assume that the ns in this case were large enough so that the normal approximation could be used. Would your conclusion be any different?
3. Compare the above results with those based on a t-test.

10. A student taking a statistics course was required to think up a question to be answered, collect some data, and then analyze the data as part of a course requirement. The question in which she was most interested was whether her favorite Rolling Stones album was cheaper if bought in a record store in California, Oregon, or Washington. She wrote letters to a number of stores in the three states asking what they charged for the album. Responses came from 18 stores in California, 17 in Oregon, and 18 in Washington. The prices each store charged for the album are given here:

Cost for Rolling Stones Album

California		Oregon		Washington	
3.92	4.08	4.70	1.56	2.43	2.52
3.45	3.67	4.15	4.29	2.98	3.10
2.00	2.94	4.55	1.74	3.04	3.53
2.28	5.90	3.31	2.17	4.94	3.06
3.50	2.18	2.13	1.97	3.15	2.57
4.25	5.39	4.68	4.69	2.46	3.48
2.38	2.74	2.68	2.87	3.34	5.94
3.02	3.49	2.36	3.17	2.39	2.51
3.26	2.70	3.93		2.27	1.61

1. Use a nonparametric test to answer the question of whether there is a difference in the prices which stores in the three states charge for the album.
2. Do you think you would get a different answer if you had analyzed these data using a parametric test? Why or why not?

Aptitude Quiz

This brief quiz tests the basic ability that you will need in order to read this book at its intended level of mathematical sophistication. If you do poorly on this quiz, then we suggest that you review your basic algebra.

Solve the following problems for x:

1. $3x = 12$
2. $6x + 4 = 9$
3. $5x - 2 = 3x + 4$
4. $8x + 2x = 1$
5. $2ax - 4b = 2c$

Solve the following pairs of simultaneous equations for x and y:

1. $8y + 3x = 4$
   $x - y = 12$
2. $ay + bx = d$
   $3ay - x = 2d$

Answers to Selected Problems

Chapter 1

1. (verbal descriptions)

3. (verbal descriptions)

5. 1. contingency table—see problem
   2. 0.333
   3. 0.313
   4. 0.593
   5. 0.8810
   6. 0.381
   7. 0.667
   8. 0.615
   9. 0.688

7. 1. contingency table—see problem
   2. $p = 0.05$
   3. $p(J \cap \overline{B}) = 0.20$
   4. $p = 0.15$
   5. $p = 0.40$

9. 1. 0.333
   2. 1.00
   3. 0.50
   4. 0.333
   5. 0.333

11. 1. $11/36 = 0.306$
    2. $1/36 = 0.028$
    3. $1/18 = 0.056$
    4. $1/36 = 0.028$
    5. $1/6 = 0.167$
    6. $5/18 = 0.278$
    7. $5/36 = 0.139$
    8. $4/9 = 0.444$

13. 1. $1/6 = 0.167$
    2. $5/18 = 0.278$
    3. $5/18 = 0.278$
    4. $5/18 = 0.278$

15. 1. $1/36 = 0.028$
    2. $1/6 = 0.167$
    3. $1/9 = 0.111$
    4. $1/48 = 0.021$

17. Though there are more flights between noon and 6:00 p.m., it's easier to see in daylight.

19. (proof)

21. (proof)

Chapter 2

1. 2. $p(1) = 19/77 = 0.247$
      $p(2) = 11/77 = 0.143$
      $p(3) = 17/77 = 0.221$
      $p(4) = 19/77 = 0.247$
      $p(5) = 11/77 = 0.143$

3. 1. $V = \{4, 15, 32, 55, 84, 119\}$
   2. $p(V) = 0.167$

5. 1. $V = \{0,1,2\}$
   2. $p(0) = 0.563; p(1) = 0.35; p(2) = 0.087$

7.

V	$p(V)$	$\Sigma(v)$
A	0.167	216
B	0.278	360
C	0.278	360
D	0.185	240
E	0.092	120

9. 1. $x = \{2,3,4,5,6,7,8,9,10,11,12\}$

x	$p(x)$	x	$p(x)$
2	0.028	8	0.139
3	0.056	9	0.111
4	0.083	10	0.083
5	0.111	11	0.056
6	0.139	12	0.028
7	0.167		

2. $y = \{4,9,16,25,36,49,64,81,98,128,162, 200, 242, 288\}$

y	$p(y)$
4	0.028
9	0.056
16	0.083
25	0.111
36	0.139
49	0.111
64	0.083
81	0.056
98	0.056
100	0.028
128	0.056
162	0.056
200	0.056
242	0.056
288	0.028

11. 1. (graph)
    2. 0.075
    3. 0.125
    4. 0.375
    5. 0.625
    6. 0.50
    7. 1.0

13. 1. $V = \{2,3,4,5,6,7,8,9,10,11,12,13\}$
    2.

v	$p(v)$
2	0.014
3	0.042
4	0.0705
5	0.098
6	0.125
7	0.153
8	0.153
9	0.125
10	0.098
11	0.0705
12	0.042
13	0.014

3. $V = \{3,4,5,6,7,8,9,10,11,12,13\}$
4.

v	$p(v)$
3	0.042
4	0.069
5	0.097
6	0.125
7	0.153
8	0.167
9	0.125
10	0.098
11	0.0705
12	0.042
13	0.014

15. 1. $a = (4,2)$
    $b = \{1,2,3,4,5,6\}$

a	$p(a)$
4	0.50
2	0.50

b	$p(b)$
1	0.167
2	0.167
3	0.167
4	0.167
5	0.167
6	0.167

2. $x = \{-3,-2,-1,0,1,2,3,4\}$

x	$p(x)$
-3	0.083
-2	0.083
-1	0.167
0	0.167
1	0.167
2	0.167
3	0.083
4	0.083

3. $y = \{-2,-1,0,1,2,3,4,5,6,7,8,9,10\}$

y	$p(y)$
-2	0.021
-1	0.042
0	0.063
1	0.083
2	0.104
3	0.125
4	0.125
5	0.125
6	0.104
7	0.083
8	0.063
9	0.042
10	0.021

Chapter 3

1. 1. 7
   2. 7
   3. 5
   4. $s^2 = 3.6 \quad s = 1.90$
   5. range $= 5$ to 10
   6. 7.5

3. 1. 2.75
   2. 7.5

5. 3. $E(v) = 51.603$
      $s^2 = 1596.799$

   5. $E(v) = 0.524$
      $s^2 = 0.423$

   7. $E(v) = 291.672$
      $s^2 = 6757.396$

9. 1. $E(x) = 7.007$
$s^2 = 5.803$

2. $E(y) = 82.354$
$s^2 = 5514.473$

7. $E(a) = 3.0$
$s^2 = 1.0$

$E(b) = 3.507$
$s^2 = 2.898$

$E(x) = 0.50$
$s^2 = 3.906$

$E(y) = 4.004$
$s^2 = 8.188$

9.

v	$p(v)$
0	0.774
0.5	0.200
10.0	0.020
100.0	0.005
600.0	0.001

2. $E(x) = \$1.40$
$s^2 = 410.09$

11. 1. $E(x) = N\ (\$0.50)$
2. $E(x) - \$15.00$

Chapter 4

1. 1. 120
2. assuming no order, 6

3. 1. 120
2. 9,765,625
3. 62,575
4. 9,828,200

5. 1. 7,290,000
2. 864
3. 5,040,000

9. 1. 24
2. 6
3. 12

11. 1. $\binom{6}{1}\binom{8}{1}\binom{2}{1}\binom{10}{1}\binom{3}{1} = 2880$

2. $\binom{6}{1}\binom{8}{1}\binom{2}{1}\binom{7}{1}\binom{3}{1} = 2016$

3. $\binom{6}{1}\binom{16}{2}\binom{2}{1}\binom{10}{1}\binom{3}{1} = 43{,}200$

4. $\binom{5}{1}\binom{7}{1}\binom{1}{1}\binom{9}{1}\binom{2}{1} = 630$

5. $\binom{8}{1}\binom{2}{1}\binom{10}{1}\binom{3}{1} = 480$

15. 1. 0.167
2. 0.833
3. $E(x) = 20,\ \sigma^2 = 16.667$

17. 1. 0.0002
2. 0.349
3. 0.874
4. 0.9998
5. 0.996

19. 1. 0.832
2. 0.360
3. 0.132
4. 0.077

21. $p = 0.0112$

23. 1. 1.0
2. 2.0
3. 0.051
4. 0.027

Chapter 5

1. (list examples)

3. $p = 0.109$, no

5. $p = 0.173$, no

7. $p = 0.113$, no

9. 1. Seven students
2. Seven students

11. $p = 0.018$ supports occutist's hypothesis

Chapter 6

1. 1. $\dfrac{1e^{\frac{-(x-500)^2}{20{,}000}}}{\sqrt{20{,}000\ \pi}}$

2. 0.50
3. 0.159
4. 0.115
5. 0.157
6. 0.354
7. 0.223
8. 0.845
9. 0.816

3. 1. 0.683
2. 0.331
3. 0.331
4. 0.081

5. 1. 0.025
2. 0.094
3. 0.405
4. 0.552
5. 0.001

7. 1. 0.493
2. 0.933

9. 1. 0.999
    2. 0.999

11. 1. 0.535
    2. 0.376
    3. 0.535
    4. 0.119

13. 1. $\mu = 10$
    2. $\mu > 10$
    3. $\mu_s = 10$
        $\sigma_s^2 = 0.04$
        $\sigma_s = 0.2$
    4. $z = 2.5$
    5. yes, greater than 1.64

15. 1. Type I: You decide table is biased but it really is fair. Type II: You decide table is fair but it really is biased.
    2. $\mu = 25$
    3. $\mu < 25$
    4. $\mu = 25$
        $\sigma^2 = 18.75$
        $\sigma = 4.33$
    5. No, $z = -1.16$

17. a. $-5.52*$
    b. $-6.66*$
    c. -1.75 $*p < .05$
    d. $-2.69*$
    e. -1.28
    f. $5.81*$

19. 1. 0.024
        0.382
    3. $z = -1.90$, fail to reject H_0

21. 1. p(string 1 breaks) = 0.23
        p(string 2 breaks) = 0.20
        p(string 3 breaks) = 0.10
        p(string 4 breaks) = 0.10
        p(string 5 breaks) = 0.10
    2. 0.551
    3. 0.000046

Chapter 7

1. 1. 0.281
    2. 0.099
    3. 0.034
    4. 0.50
    5. 0.000
    6. 0.754

3. $z = 0.354$, fail to reject H_0

5. $z = 2.828$, reject H_0

7. 1. $z = 2.0$, reject H_0
    2. $z = 1.0$, fail to reject H_0
    3. $z = 2.121$, reject H_0

9. $z = 3.674$, conclude difference

11. $z = 2.683$, conclude Ultra Brite leads to more sex appeal than Crest ($p < 0.01$)

13. 1. 3.667
    2. 4.383

15. 1. (graph)
    2. 0.023
    3. 0.023

Chapter 8

1. 1. 0.483
    2. 0.999

3. 1. 0.999
    2. 0.261
    3. 0.999
    4. 1. 0.995
        3. 0.999
    5. 1. 0.603
        3. 0.997
    6. (graph)

5. 1. $z = 1.452$, fail to reject H_0
    2. 0.382

7. $y = 5.92$

9. 1. $12.79 \sim 13$
    2. $8.07 \sim 9$

Chapter 9

1. 1. $M = 4.905$
        $S^2 = 5.515$
    2. est $\mu = 4.905$
        est $\sigma^2 = 5.790$
        differ because est $\sigma^2 = S^2 (n/n - 1)$

3. 1. $S^2 = 0.889$
    2. est $\sigma^2 = 1.333$
    3. est $\sigma_m^2 = 0.444$

5. 1. UW: est $\sigma^2 = 5$
            est $\sigma_m^2 = 1$

        Harvard: est $\sigma^2 = 5$
                est $\sigma_m^2 = 1$

    2. *Peanut Butter*   *Jelly*
        est $\sigma^2 = 1.0$ est $\sigma^2 = 5.0$
        est $\sigma_m^2 = 0.333$ est $\sigma_m^2 = 1.0$

        Crest   *Ultra Bright*
        est $\sigma^2 = 4.0$ est $\sigma^2 = 0$
        est $\sigma_m^2 = 1.333$ est $\sigma_m^2 = 0$

Chapter 10

1. 1. $t(19) = 0.459$, not significant (NS)

     95% CI = 42.877 = 61.123
     99% CI = 39.529 = 64.471

   2. $t(11) = 2.487$, significant to $2Q \propto < 0.05$

      95% CI = 171.382 = 192.618
      99% CI = 167.016 = 196.984

   3. $t(63) = 7.937$, significant at all levels
      95% CI = 5′ 8.244″ − 5′ 9.756″
      99% CI = 5′ 7.995″ − 5′ 10.005″

3. 3. $t(7) = 1.57$
      95% CI = −8.793 − 43.543

   4. $t(5) = 2.381$
      95% CI = −0.345 to 9.012

   5. $t(9) = 1.303$
      95% CI = −117.364 to 31.564

   8. $t(11) = −1.654$
      95% CI = −13.206 to 1.872

5. 1. 95% CI 5 ± 1.908
        3 ± 1.908

   2. $t(10) = 2.335$ significant to $2Q \propto 0.05$

7. 1. 95% CI 6.333 + 1.329
        8.0 ± 1.329

   2. $t(10) = 2.796$ significant

11. 1. 95% CI 3.75 ± 0.912
         2.667 ± 1.054

   2. $t(12) = 1.693$ NS

13. 1. $t(5) = 2.317$

   2. 98% CI 125.5 ± 7.441
         112 ± 18.152

15. (verbal descriptions)

17. 1. 2.0

   2. $\sigma^2 m_1 = 0.04$
      $\sigma^2 m_2 = 0.05$
   3. $t(88) = 2.0$ significant $\propto 0.05$
   4. $t(88) = 2.0$

Chapter 11

1. $F = 6.0^*(2, 12)$ df significant $\propto < .05$

3. 6.738 MSW
   7.143 MSB
   $F = 1.060 (1, 12)$ df

5. 1. 3 ± 1.386
        8.5 ± 1.96
        5 ± 1.60

2. $F = 10.097 (2, \infty)$ df
3. $F = 1.143 (6, \infty)$ df
4. $F = 11.540 (2, 6)$ df

7. Squared value of the t-test

9. 1. 0.394

   2. 3 ± 0.404
      8 ± 0.313

   3. $F = 12.784 (1, 6)$ df

   4. 3 = 2.705
      8 ± 2.095

   5. $F = 1.091 (6, \infty)$ df

   6. $F = 5.0 (4, 2)$ $2Q \propto 0.01$

Chapter 12

1. 1. 560 SSB
        3 df

   2. $MSB = 186.667$
         $MSW = 100.00$
         $F = 1.867 (3, 36)$ df
         NS

   3. $SSR = 360$
   4. $MSR = 360$
         $F = 3.6 (1, 36)$ df
         NS
   5. $MSC = 160$
         $F = 1.60 (1, 36)$ df
         NS

   6. $MSI = 40$
         $F = 0.40$
         NS

3. 1. 5.167 ± 1.354
        10.333 ± 1.354
        7.833 ± 1.354
        11.50 ± 1.354

   2. Sex $F = 8.73^* (1, 20)$ df
      Drugs $F = 46.353^* (1, 20)$ df
      Interaction $F = 1.336 (1, 20)$ df

5. 1. 5.9 ± 1.410
        11.4 ± 1.410
        11.6 ± 1.410
        9.7 ± 1.410

   2.
Source	df	SS	MS	F
Between	3	209.3		
Columns	1	136.9	136.9	28.357*
Rows	1	40	40	8.285*
Interaction	1	32.4	32.4	6.711*

   3. (verbal description)

7. 1. Mean $\pm$ 1.048

2.

Source	df	SS	MS	F
Between	8	167.644		
Verb	2	122.978	61.494	46.120*
Adverb	2	1.111	0.556	0.417
Interaction	4	43.555	10.889	8.167*
Within	36	48	1.333	

3. (verbal description)

9. (verbal description)

11. (graph)

Chapter 13

1. 1. $95\% = M \pm 0.978$
2. $F = 24.652$ (2, 6) df, yes

3. 1. $95\% = M \pm 134.493$
2. $F = 8.229$ (2, 8) df, yes
3. $M \pm 0.426$

4.

Source	df	SS	MS	F
Between	19	54.064		
Columns	3	22.733	7.578	11.366*
Rows	4	15.733	3.933	5.90*
Interaction	12	15.601	1.30	1.95 NS
Within	40	26.667	0.667	

Yes, effect of patient and season
No, effect of patient interaction

7. $F = 11.294$ (1, 9) df significant $\alpha < .05$

9. 1. 90
2. $SS_s = 42$
$SS_c = 32$
$SS_I = 16$

3.

Source	df	SS	MS	F
Between	11	90		
Subjects	3	42	14	
Conditions	2	32	16	6.0
Interaction	6	16	2.667	

4. Yes, p < 0.05

6. $SS_s = 0$
$SS_c = 32$
$SS_I = 16$

7.

Source	df	SS	MS	F
Between	11	48		
Subjects	3	0	0	
Conditions	2	32	16	6.0* significant
Interaction	6	16	2.667	*

11. 1.

Source	df	SS	MS	F
Between	7	237.75		
Season	3	162.75	54.25	19.727*
City	1	6.25	6.25	0.273 NS
Interaction	3	68.75	22.917	8.333*
Within	8	22	2.75	

2.

Source	df	SS	MS	F
Between	7	237.75		
Season	3	162.75	54.25	19.727*
City	1	6.25	6.25	2.273 NS
Interaction	3	68.75	22.917	8.333*
Within	8	22	2.75	

Chapter 14

1. (graph)

3. 1. (graph)
2. $y^1 = 0.489x + 97.152$
3. $r = 0.429$ $t(13) = 1.952$ NS
4. women: $y^1 = -0.292x + 171.767$
$r = -0.621$ $t(6) = -1.941$ NS

men: $y^1 = -0.846x + 356.183$
$r = -0.984$ $t(5) = -12.349*$

5. 1. $y^1 = 0.587x + 46.634$
2. $y^1 = 0.659x + 28.828$
3. $r = 0.622$ $r^2 = 0.387$
4. $t(10) = 2.512$ significant $\alpha = 0.05$ 2Q

7. 1. $r = 0.408$ $r^2 = 0.167$
2. $t(8) = 1.265$ NS

Chapter 15

1. 1. $M \pm 1.172$
2. 1.667
3. 2.415
4. $F_L = 86.25$ (1, 36) df, $p < 0.05$,
75.9% accounted for
residual is significant,
$F_R = 9.159$

3. 1. $M \pm 5.946$

2.

Source	df	SS	MS	F
Between	2	12.167	6.083	0.220 NS
Within	9	248.75	27.639	
Total	11	260.917		

3. $\{-1 \quad 2 \quad -1\}$
4. $H = 12.042$ residual = 0.125
5. 99%
6. $F = 0.436$ (1, 9) df NS, no

5. 1. $M \pm 2.372$ $F = 12.667$ (3, 8)*
2. 3.555
3. 4.772
4. $F = 32.4$ (1, 8) df*
5. $F = 2.833$ (2, 8) df NS
$F = 4.0$ (2, 2) df NS

7. 1. Scheffé = 3.404
LSD = 2.693

2. $F = 24.218$ (1, 25) $df*$
    $F_{res} = 1.259$ (1, 25) df NS

9. 1. Scheffé $= 2.221$
        LSD $= 1.745$
        Scheffé $= 0.935$
        LSD $= 0.735$

    2. $F = 0.124$ (1, 15) df NS
        $F = 92.449$ (1, 15) $df*$

    3. $F = 1.036$ (1, 15) df NS
        $F = 0.875$ (1, 15) df NS

11. $F = 17.416$ (1, 12) $df*$

    $F = 7.271$ (1, 12) $df*$

13. 1. LSD $= 1.413$
        cells $= 2.826$

    2. $\{3 \quad 2 \quad 1 \quad -2 \quad -2 \quad -2\}$
    3. $F = 70.673$ (1, 6) $df*$
        79.4%
        $F_{res} = 18.328$ (4, 6) $df*$

15. 1. $\{1 \quad 2 \quad 3 \quad 1 \quad 1 \quad 1 \quad -3 \quad -3 \quad -3\}$
    2. $F = 38.5$ (1, 54) $df*$
        $F_{res} = 43.5$ (7, 54) $df*$

17. $\begin{matrix} 1 & 1 & 1 & -3 \\ 1 & 1 & -2 & 0 \\ 1 & -1 & 0 & 0 \end{matrix}$

21. $t(19,498) \pm 2.776*$
    $\omega^2 = 0.000344$

Chapter 16

1. $F = 1.105$ (6, ∞) $df \propto +0.025$ NS, no
3. $x_1^2 = 4.84*$
5. $x_1^2 = 1.111$ NS
7. $x_3^2 = 33.25*$
9. $x_7^2 = 3799.233$, they do not seem representative!

Chapter 17

1. $p = 0.0754$ Mann-Whitney $U = 5$
    $\qquad\qquad\qquad\qquad\qquad U^1 = 20$

3. U 15 is less extreme than criterion of 26
    U^1 93
    $z = 2.772 \qquad p < 0.03$, difference is significant

5. $H_2 = 0.86$ NS
7. $H_2 = 13.812 \quad p < 0.01$
9. 1. $U = 31$ NS
        $U^1 = 59$
    2. $z = 1.143$ NS
    3. $t(17) = 2.149*$

A Set Theory

The word *set* as we shall use it here refers to "any well-defined collection of objects." Thus for instance, the letters of the alphabet form a set. Likewise, all strings of letters containing a "G" form a set; all persons who went to observe the Seattle Space Needle during the year 1975 form a set; all points on a line form a set.

FINITE AND INFINITE SETS

One of the principal characteristics of a set involves the number of members it contains. If a set contains a finite number of members, it is referred to as a *finite set*. Table A-1 lists some examples of finite sets. Likewise, any set with an infinite number of members is referred to as an infinite set. There are two types of infinite sets, referred to as *countably infinite* and *uncountably infinite* sets. It is somewhat difficult to describe what is meant by countably and uncountably infinite; the distinction is perhaps more easily explained by reference to the examples in Table A-2. A countably infinite set contains an infinite number of members—but each member is "discrete—you could point to each individual member of the set and begin to count the members." So, for example, the set of all possible letter strings is a countably infinite set. Although an *uncountably infinite* set also contains an infinite number of members, the members are not discrete but are "densely packed together." So, for example, the number of points on a line forms an uncountably infinite set, as does the number of points in a plane, or all the possible lengths of time it could take a sprinter to run the 100-yard dash. When in Chapter 2 we use the notion of uncountably infinite sets to describe a continuous probability distribution, we will be able to be a little more mathematically precise about our definitions.

NOTATION FOR SPECIFYING A SET

Typically we use capital letters to denote sets. Hence, we might use the letter I to specify the set of all introductory psychology books or the letter R to specify the

---

TABLE A-1.   Examples of Finite Sets

All women
All Edsels
All books that contain the word "peach"
All trees in Yellowstone National Park
All molecules of water on the earth
All prime ministers of England during the year 1958
All instances during the year 1960 of people uttering the word "fiddlesticks"
All possible four-letter strings

---

set of all real numbers. The most straightforward way of specifying the members of a set is simply to list them—for example

C = {Guatemala, San Salvador, Honduras, Nicaragua, Costa Rica, Panama}

The second (and often easier) way to specify the members of a set is to state a *rule* that defines a set—for example

C = {All the countries in Central America}

Note that the members of an infinite set *must* be specified by a rule (since it would be impossible to list all members of an infinite set). So, we might let

H = {All possible heights that a human being could have}

or

F = {The number of times a person could flip a coin before the coin comes up heads}

Notice that, when denoting a set, we enclose in brackets the listing of members or the rule specifying the set.

TABLE A-2.   Examples of Infinite Sets

I. *Countably Infinite*
   All possible letter strings
   All possible results of die throws prior to getting a 6
   All possible books that could be written
   All possible scores in a baseball game

II. *Uncountably Infinite*
   All real numbers
   All points on a line
   All lengths of time a sprinter could take to run the 100-yard dash
   All possible heights of a tree

SUBSETS

Let's consider any two sets and refer to them as A and B. Then we may define a *subset* as follows: B is a subset of A if and only if all members of B are also members of A. Thus, for example, if

$A = \{$All living things$\}$

and

$B = \{$All reptiles$\}$

then B would be a subset of A since all members of B also must be members of A.

Notationally if B is a subset of A, we represent this relationship as

$B \subseteq A$

Notice the similarity of the symbol "$\subseteq$" to the possibly more familiar $\leq$ which, of course, is used with numbers as opposed to sets, and means "less than or equal to." This similarity makes sense; if B is a subset of A, B contains an equal number of or fewer members than A contains.

Proper Subsets Versus Equal Sets

Suppose that, using ordinary notation and talking about real numbers, we are told that

$b \leq a$

We then know that one of two things must be true: either b is *less than a* or *b equals a*. Likewise, in terms of set notation, if

$B \subseteq A$

then one of two things must be true: either B is a proper subset of (smaller than) A or else B is *equal* to (the same as) A. To be more concrete,

$B \subset A$

which specifies that B is a proper subset of A, means that *at least one member of A is not a member of B*. So if

$M = \{$All movies$\}$

and

$C = \{$All color movies$\}$

then C is a proper subset of M, or

$C \subset M$

since at least one member of M (for instance *Citizen Kane*) is not a member of C.

On the other hand, if

R = {All the members of the Boston Red Sox}

and

B = {All members of an American League baseball team in Boston}

then there is no member of R that is not a member of B; nor is there any member of B that is not also a member of R. Hence

$R = B$

or R and B specify exactly the same set.

In general when we talk about subsets, we will be talking about proper subsets.

INTERACTING SETS: UNIONS AND INTERSECTIONS

Once again, consider any two sets A and B. Suppose, for example, that

A = {1, 2, 3, 4, 5}

and

B = {4, 5, 6, 7}

The intersection of A and B, denoted as

$A \cap B$

consists of a new set that contains all elements common to both A and B—that is, all objects that are *both* in A and in B. In our example,

$A \cap B$ = {4, 5}

since 4 and 5 are the only elements common to both A and B.

The *union* of A and B, denoted as

$A \cup B$

consists of a new set containing all the elements that are in A or B or in both A and B. Thus, in the present example,

$A \cup B$ = {1, 2, 3, 4, 5, 6, 7}

Incidentally, it is the case that

$A \cup B = B \cup A$

and also that

$A \cap B = B \cap A$

UNIVERSAL SETS

When we refer to some set of objects, that set is usually a subset to some larger set in which, for some reason, we are interested. As an example if we were talking about the set of Buicks or the set of Fords, the larger, more general set in which we are interested might be automobiles or vehicles. If we were talking about the set of Harvard students, the larger set might be all students or even all people.

This "larger set" is referred to as a *universal set,* and the letter W is reserved for denoting the universal set. As we can see, once some set, A, is specified, the universal set W that includes A as a subset is *not* automatically specified—rather, the decision of what to include in W must be based on practical considerations. The nature of these practical considerations will become clearer in practical settings, as in Chapter 1 of this book when set theory is used as a foundation for probability theory.

VENN DIAGRAMS: PICTORIAL REPRESENTATIONS OF SETS

Apparently believing in the old adage that "a picture is worth a thousand words," a mathematician named Venn devised the technique that bears his name for visual representations of sets.

Venn Diagrams: Unions and Intersections

To illustrate a Venn diagram, suppose we have some universal set

$W = \{$All cars$\}$

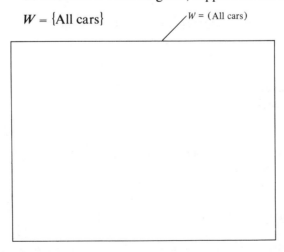

FIGURE A-1.   The start of a Venn diagram. We represent the universal set W (in this case all cars) as a rectangle.

Let's represent W as a box as shown in Figure A-1. Now consider two sets within W—for example

> $A = \{$All General Motors cars$\{$
>
> $B = \{$All blue cars$\}$

We can represent both A and B as circles inside W, as shown in Figure A-2. Notice that the union and intersection of these two sets have specific representations in this Venn diagram. The union of the two sets is represented by the area taken up by both A and B together. Likewise, the intersection of the two sets is represented by the area where A and B overlap.

Venn Diagrams: Subsets

Depiction of a subset using a Venn diagram is very straightforward. Suppose

> $W = \{$All cars$\}$
>
> $A = \{$All General Motors cars$\}$
>
> $B = \{$All Buicks$\}$

"Buick" is of course a subset of "General Motors cars," and the Venn diagram representing this situation is shown in Figure A-3. The subset "Buicks" is simply represented by a circle *within* the superset, General Motors.

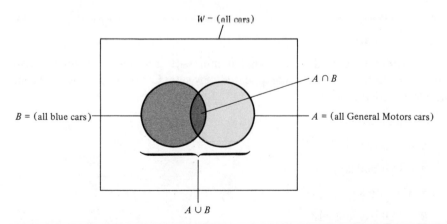

FIGURE A-2. Completion of the Venn diagram. Within the rectangle we have represented two sets within W as circles. The circle labeled B represents all blue cars; the circle labeled A represents all General Motors cars. The intersection of the two sets, $A \cap B$, is represented by the little sliver in the middle; the union of the two sets, $A \cup B$, is represented by the space taken up by the two circles together.

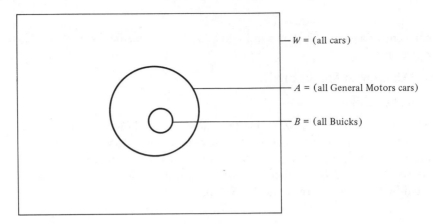

FIGURE A-3.   Representation of a subset in a Venn diagram. Here the subset (Buicks) is represented as a circle *within* the circle representing set A (all General Motors cars).

COMPLEMENTS

Consider some universal set, W, and a set A; for example,

W = {All people}

A = {All females}

The *complement* of A, which we denote $\overline{A}$, (variously called "A-bar" or "not-A") is a new set consisting of *all members of W which are not members of A*. In this example

$\overline{A}$ = {All people who are not female}

or more simply

$\overline{A}$ = {All males}

We can combine the idea of complements with the ideas of unions and intersections to form yet more complicated new sets. Continuing with the above example, once again let

W = {All people}

A = {All females}

$\overline{A}$ = {All males}

and additionally let

B = {All psychologists}

$\overline{B}$ = {All people who are not psychologists}

We have depicted this situation by the Venn diagram shown in Figure A-4. Now

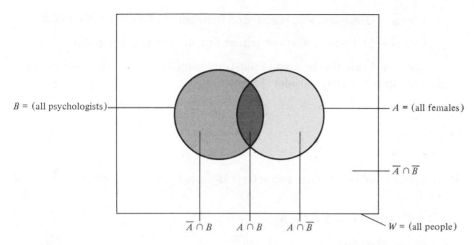

FIGURE A-4. A Venn diagram representing the two sets "all psychologists" and "all females" relative to the universal set "all people." Here four pieces of the universal set are depicted. These four pieces are $\overline{A} \cap B$, $A \cap B$, $A \cap \overline{B}$, and $\overline{A} \cap \overline{B}$. These four pieces form a partition of the universal set.

consider the following four sets, which constitute the four "pieces" of the Venn diagram. For the first piece we already have seen that

$A \cap B$ = {Everyone who is a member of A and also a member of B} or

$A \cap B$ = {All female psychologists} (for example Elizabeth Loftus)

Next, using our complement notation, we can represent the other three sets as

$A \cap \overline{B}$ = {Everyone who is a member of A but not a member of B} or

$A \cap \overline{B}$ = {All females who are not psychologists} (for example Joan of Arc)

$\overline{A} \cap B$ = {Everyone who is not a member of A but a member of B} or

$\overline{A} \cap B$ = {All males who are psychologists} (for example Geoffrey Loftus)

$\overline{A} \cap \overline{B}$ = {Everyone who is neither a member of A nor a member of B} or

$\overline{A} \cap \overline{B}$ = {All males who are not psychologists} (for example George Washington)

We can also use our complement notation in conjunction with union notation to obtain, for example,

$\overline{A} \cup B$ = {Everyone who is either not a member of A or a member of B, or both} or

$\overline{A} \cup B$ = {All people who are either males or psychologists or both}

$\overline{A} \cup \overline{B}$ = {Everyone who is neither a member of A nor a member of B} or

$\overline{A} \cup \overline{B}$ = {All people who are neither females nor psychologists}.

Figure A-4 can also be used to illustrate some important rules for manipulating sets. In particular, consider the two sets

$A \cap B$

and

$A \cap \overline{B}$

If we take the union of these two sets, it is evident from looking at Figure A-4 that

$(A \cap B) \cup (A \cap \overline{B}) = A$

and likewise of course,

$(A \cap B) \cup (\overline{A} \cap B) = B$

THE EMPTY SET

An empty set is simply a set without members. (Since by definition there is only one set containing no members, it is referred to as *the* empty set.) The empty set is denoted by the symbol $\varnothing$.

MUTUAL EXCLUSION AND EXHAUSTION

Mutually Exclusive Sets

Consider some universal set W and two sets, A and B. If

$A \cap B = \varnothing$

that is, if the intersection of A and B contains no members, then A and B are said to be *mutually exclusive*. Suppose, for example,

W = {All people in the United States}

Suppose also that

A = {All registered Democrats}

and

B = {All registered Republicans}

We can easily see that A and B are mutually exclusive since no one is both a registered Democrat and a registered Republican. Notice that another way of looking at mutual exclusion is: If A and B are mutually exclusive and x is a member of A, then x cannot also be a member of B. Conversely, of course, if y is a member of B, then y cannot also be a member of A. That is, membership in A *excludes* membership in B, and vice versa.

Mutually Exhaustive Sets

Two sets, A and B, are said to be *mutually exhaustive* with respect to some universal set W, if

$A \cup B = W$

That is to say, A and B are mutually exhaustive if between them they exhaust all members of W. So suppose that

$W = \{$All book titles sold in Ralph's bookstore$\}$

$A - \{$All book titles sold in hardback$\}$

$B - \{$All book titles sold in softback$\}$

then A and B are mutually exhaustive with respect to W since between them, they exhaust all titles.

Note that A and B are not necessarily mutually exclusive, since some titles are sold in both hardback and softback. However, some pairs of sets are both mutually exclusive and mutually exhaustive. For instance, if

$W = \{$All General Motors cars$\}$

$A = \{$Chevrolets, Pontiacs, Buicks$\}$

$B = \{$Oldsmobiles, Cadillacs$\}$

then A and B are mutually exclusive *and* exhaustive. Whenever A and B are mutually exclusive and exhaustive, then

$B = \overline{A}$

and

$A = \overline{B}$

That is, B contains *every* member of W not in A, and B contains *only* members of W not in A.

Of course sets can also be mutually exclusive but not mutually exhaustive; or mutually exhaustive but not mutually exclusive; or neither mutually exhaustive or mutually exclusive. For example, suppose

$W = \{$All people$\}$

Then if

A = {All people weighing more than 150 pounds}

B = {All people weighing less than 125 pounds}

then A and B are mutually exclusive but not mutually exhaustive. On the other hand, suppose that

A = {All people weighing more than 150 pounds}

B = {All people weighing less than 160 pounds}

then A and B would be mutually exhaustive but not mutually exclusive. Finally, suppose that

A = {All people weighing more than 150 pounds}

B = {All people weighing between 140 and 160 pounds}

Then A and B would be neither mutually exclusive nor mutually exhaustive.

PARTITIONS

Some event, A, and its complement, A, are said to *partition* (or form a partition of) the universal set, W. So, for example, males and females partition the human race.

The notion of a partition is not necessarily restricted to two sets. *Any* collection of mutually exclusive and exhaustive sets are said to partition the universal set. So, for example, R = {red lights}, G = {green lights}, and Y = {yellow lights} form a partition of W = all traffic lights.

Summation and
Subscripts

One thing we will be doing a great deal of in this book is summing numbers. At this point it is therefore appropriate to introduce some notation to vastly simplify our discussions.

Summation Notation: General

Suppose we have a series of sequential integers (for example 0, 1, 2, . . .). We will use the letter i to refer to an interger (just as "x" or "y" is typically used in algebra to refer to any number). To use a specific example, imagine that we are concerned with the set of integers $\{0, 1, 2, 3, 4\}$. Suppose now that we want to *sum up* these integers—for example we want

$$x = 0 + 1 + 2 + 3 + 4$$

A shorthand way of representing this summation is to use the upper-case Greek letter, sigma or "Σ." Sigma means "sum up." Hence in this example,

$$x = \sum_{i=0}^{4} i$$

would mean: "x is equal to the sum (sigma) of the integers (i) from the lower limit of 0 (represented by the "$i = 0$" below the summation sign) to the upper limit of 4 (depicted by the "4") above the summation sign.

Other Ways of Using the
Summation Sign

The summation sign, along with its designation of lower and upper limits, is a very flexible piece of notation. For example, suppose that instead of

$$x = 0 + 1 + 2 + 3 + 4$$

we wanted

$$y = 0^2 + 1^2 + 2^2 + 3^2 + 4^2$$

We could then represent this as

$$y = \sum_{i=0}^{4} i^2$$

Or suppose the integers were abscissa values of some function. Hence, each abscissa value, i, would have some corresponding ordinate (function) value which might be designated as $p(i)$. Imagine now that we want to sum up these ordinate values—that is, we want

$$z = p(0) + p(1) + p(2) + p(3) + p(4)$$

This could be represented, using the summation sign, as

$$z = \sum_{i=0}^{4} p(i)$$

SUMMATION NOTATION AND SUBSCRIPTS

We have already introduced the notion of a summation sign, and now we'll extend its use to *subscripted numbers*.

To see what is meant by the notion of a subscript, imagine that we have five people, and we have listed the *height* for each person. Thus, the heights might be as follows:

Person	Height
1	$x_1 = 70$
2	$x_2 = 69$
3	$x_3 = 71$
4	$x_4 = 73$
5	$x_5 = 68$

Notice that notationally we have referred to the scores of interest—the heights—as xs. However, to refer to a *particular* person's score—say, the second person's score—we add the number corresponding to that person just below the x. This number, 2 in the present example, is referred to as a *subscript*. Thus, by listing a unique subscript to each x, we make all the xs distinct from one another (as they should be, since they refer to different scores).

A "general x" is referred to as x_i, or the score belonging to the ith person. Suppose now that we have n persons, each with a score, and we want to take the *sum* of all n scores. We denote this sum as:

$$\sum_{i=1}^{n} x_i$$

which simply means "sum of all the x_is from the first (lower limit of 1) to the nth

(upper limit of n)." In the example above,

$$\sum_{i=1}^{5} x_i = x_1 + x_2 + x_3 + x_4 + x_5$$

$$= 70 + 69 + 71 + 73 + 68 = 351$$

The summation sign used in conjunction with subscripts is a very flexible and useful piece of notation. For example suppose we wanted to *square* each score and add up these squared scores. We would represent this enterprise as:

$$\sum_{i=1}^{5} x_i^2 = x_1^2 + x_2^2 + x_3^2 + x_4^2 + x_5^2$$

$$= (70)^2 + (69)^2 + (71)^2 + 73^2 + (68)^2$$

$$= 4900 + 4761 + 5041 + 5329 + 4624 = 24,655$$

Likewise, suppose that for some reason we wanted to subtract 10 from each score and add up the results. This would be noted as

$$\sum_{i=1}^{5} (x \quad 10) = (70 \quad 10) + (69 - 10) + (71 - 10) + (73 - 10)$$

$$+ (68 - 10)$$

$$= 60 + 59 + 61 + 63 + 58 = 301$$

A Little Algebra of Summation Signs

We'll go through three algebraic facts about summation signs. We suggest that you play around with some numbers to convince yourself that the facts are valid.

The distributive law. Suppose each of n subjects has *two* scores, x_i and y_i. Imagine that for each subject, we wish to *subtract* y_i from x_i and then take the sum of these difference scores. Such a maneuver would be represented as

$$\sum_{i=1}^{n} (x_i - y_i)$$

The distributive law says that we can *distribute* the summation through this expression or

$$\sum_{i=1}^{n} (x_i - y_i) = \sum_{i=1}^{n} x_i - \sum_{i=1}^{n} y_i$$

That is, the sum of the $(x - y)$s is equal to the sum of the xs minus the sum of the ys.

Multiplication by a constant. Suppose each of n scores is multiplied by a constant (say, 7). Thus, each x_i becomes $7x_i$. Now suppose we want to add these

$7x_i$s. This would be represented as

$$\sum_{i-1}^{n} 7x_i$$

Now we may move the constant (7 in this case) to the left side of the summation sign or

$$\sum_{i-1}^{n} 7x_i = 7 \sum_{i-1}^{n} x_i$$

Thus, taking the sum of the $7x$s is the same as taking the sum of the xs and multiplying this sum by seven.

Summing a constant.   Suppose we had a set of n scores and a constant k that we wanted to subtract from each score. If we were to then sum these $(x - k)$s, we would represent the sum as

$$\sum_{i-1}^{n} (x_i - k)$$

Now using our distributive law, we can see that

$$\sum_{i-1}^{n} (x_i - k) = \sum_{i-1}^{n} x_i - \sum_{i-1}^{n} k$$

The first term of this expression is familiar—it's simply the sum of the xs. However, what about the second term, $\sum_{i-1}^{n} k$? Since k is constant (it does not vary when i is changed as does x_i), this expression simply means "take k and add it to itself n times," or

$$\sum_{i-1}^{n} k = \overbrace{k + k + \cdots + k}^{n \text{ times}}$$

which is equivalent to multiplying k by n—that is

$$\sum_{i-1}^{n} k = nk$$

Derivation of the Raw- C
Score Formula for
Variance

We begin with the deviation-score formula, which is

$$S^2 = \frac{\sum_{i-1}^{n} (x_i - M)^2}{n}$$

Let's work with the numerator, which is referred to as a *sum of squares* (*SS*). Thus,

$$SS = \sum_{i-1}^{n} (x_i - M)^2$$

Expanding out the expression $(x_i - M)^2$, we get

$$SS = \sum_{i-1}^{n} (x_i^2 - 2Mx_i + M^2)$$

$$= \sum_{i-1}^{n} x_i^2 - \sum_{i-1}^{n} 2Mx_i + \sum_{i-1}^{n} M^2$$

Recall (a) that constants may be transferred to the left side of the summation sign and (b) that summing a constant n times is equivalent to multiplying that constant by n. Hence, since 2, M, and M^2 are constants,

$$SS = \sum_{i-1}^{n} x_i^2 - 2M \sum_{i-1}^{n} x_i + nM^2$$

Now recall that $M = \Sigma x / n$. Thus,

$$SS = \sum_{i=1}^{n} x_i^2 - 2 \frac{\sum_{i=1}^{n} x_i}{n} \sum_{i=1}^{n} x_i + n \left(\frac{\sum_{i=1}^{n} x_i}{n} \right)^2$$

$$= \sum_{i=1}^{n} x_i^2 - 2 \frac{\left(\sum_{i=1}^{n} x_i \right)^2}{n} + \frac{\left(\sum_{i=1}^{n} x_i \right)^2}{n}$$

$$= \sum_{i=1}^{n} x_i^2 - \frac{\left(\sum_{i=1}^{n} x_i \right)^2}{n}$$

Finally to get variance from sum of squares, we multiply the sum of squares by $1/n$ to get

$$S^2 = (SS)(1/n) = S \frac{1}{n} \left[\sum_{i=1}^{n} x_i^2 - \frac{\left(\sum_{i=1}^{n} x_i \right)^2}{n} \right]$$

or

$$\boxed{S^2 = \frac{\sum_{i=1}^{n} x_i^2}{n} - \left(\frac{\sum_{i=1}^{n} x_i}{n} \right)^2}$$

One other way of representing S^2 involves the fact that

$$M = \frac{\Sigma x_i}{n}$$

Substituting into the last term,

$$\boxed{S^2 = \frac{\sum_{i=1}^{n} x_i^2}{n} - M^2}$$

A mnemonic for remembering this formula is that the variance is equal to "the mean of the squares minus the square of the mean."

Counting Rules **D**

Suppose you are sitting at a poker game, and for some reason or another it becomes crucial to compute the probability that the person against whom you are betting has a full house. (A full house, in a five-card poker hand, consists of three of one card and two of another. Thus, for example, a hand consisting of two aces and three fives would be a full house.) How are we to calculate this probability?

In Chapter 1 we discussed the most straightforward way of calculating a probability which is to first compute $n(S)$, the total number of elementary events in the sample space of concern and then calculate $n(A)$ where $n(A)$ is the number of elementary events in the set corresponding to the outcome of interest. The probability of A, $p(A)$, is then equal to $n(A)/n(S)$. In the present example one elementary event would consist of a five-card poker hand. So $n(S)$ in this case would consist of the total number of distinct, possible poker hands. Likewise, A consists of the outcome "two of one card and three of another." Thus, $n(A)$ would be the total number of such hands that exist.

How are we to go about computing $n(S)$ and $n(A)$? This task does not, offhand, seem as if it would be particularly easy, and in fact it's not. There are, however, ways of making it easier, and in this appendix we'll be talking about handy techniques that may be used for counting large numbers of complicated things. It may seem somewhat strange to have an entire appendix about counting, since counting is something that's intuitively a straightforward process. It's true that counting is a straightforward process if you are counting the number of people at an intimate dinner party or the number of fish you catch on a fishing excursion, but many other things are not so straightforward to count because there are so many of them and they are generated in a somewhat complicated way. Counting poker hands is an example of such a thing. Other examples are shown in Table D-1.

FIVE COUNTING RULES

In many instances some sample space of concern will be made up of elementary events that consist of *sequences* of things. In our poker hand, for example, the

---

TABLE D-1. Examples of Situations in Which It Is Not Particularly Straightforward to Count Things

---

Situation	*Things to Count*
1. License plates in the state of Washington consisting of three letters followed by three digits.	1a. The number of unique Washington license plates. How many vehicles can exist in the state before a new license-plate system has to be devised?
	1b. The number of license plates with the letter Q.
	1c. The number of license plates with the letter Q but not the digit 5.
2. Out of a 10-person psychology faculty at Bugaboo University, a three-person committee is to be selected.	2a. The number of possible such committees.
	2b. The number of committees that include Gloria Gladstone.
	2c. The number of committees that include both Gloria and her boyfriend Beauregard.

---

things are cards, and each elementary event is a sequence of five of them. In our license plate example (in Table D-1) elementary events are made up of a sequence of three letters followed by another sequence of three digits. Our three-person committee from Bugaboo University consists of a three-person sequence. There are five counting rules that provide some aid in the task of computing the number of sequences that there are in various types of situations. These rules are as follows.

Counting Rule 1: A General Rule for Computing Numbers of Sequences

Let's suppose that an elementary event in a sample space consists of a sequence of N things. Of concern is the number of *states* that each thing can be in (for example, a letter could be in any one of the 26 states, A–Z; a digit could be in any one of the 10 states, 0–9; and so on). In particular let's suppose that the first thing in the sequence can be in any one of k_1 states, the second thing in the sequence can be in any one of k_2 states, and so on until the Nth thing in the sequence can be in any one of k_N states. In this case the total number of existing sequences is obtained by the following formula.

Number of sequences $= (k_1) (k_2) \ldots (k_N)$

To give an example of the use of this type of sequence, suppose that in a small, faraway country, vehicle license plates consist simply of one digit followed by one letter. How many such license plates are there? In this situation, the sequence consists of a series of $N = 2$ things. The first thing—the digit—can take on any one of 10 states; therefore, $k_1 = 10$. Likewise, the second thing in the sequence—the letter—can take on any one of 26 states; therefore, $k_2 = 26$. The total number of such sequences (license plates) is therefore obtained by the formula

Number of license plates $= (k_1) (k_2) = (10) (26) = 260$

Hence, a total of but 260 cars could be assigned unique license plate numbers in this tiny principality.

To give another example of this rule, suppose you're playing a strange game that consists of flipping a coin twice and then throwing a die. How many such two-flip-one-throw sequences are there? In this case $N = 3$ since our sequence consists of three things: k_1 is equal to 2 since the first coin flip can eventuate in either of two states (heads or tails); k_2 is equal to 2 since the second coin flip can also end up in any one of two states; k_3 is equal to 6 since the die throw can be in any one of six states (1–6). Thus, the total number of sequences that we can obtain is obtained by the formula

Number of sequences $= (k_1) (k_2) (k_3) = (2) (2) (6) = 24$

Tree diagrams. Some people have an easier time seeing where these sequences come from through the use of what is called a *tree diagram*. Figure D-1 illustrates the use of a tree diagram to represent the possible sequences in the coin-flip-coin-flip-die-throw example that we have just discussed. In this

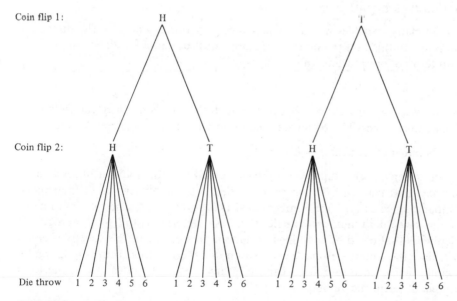

FIGURE D-1. Tree diagram illustrating the 24 sequences obtainable when two coin flips are followed by a die throw.

diagram the top level represents the first thing (in this case the first coin flip); the second level represents the second thing (the second coin flip); and the third level represents the third thing in the sequence (the die throw). Each path down through the tree represents one possible sequence. So for example the far left-hand path represents the sequence head-head-1; the far right-hand path represents the sequence tail-tail-6, and so on. By counting up the number of paths, we can easily see that there are 24 paths and hence 24 separate sequences—the same conclusion that we arrived at by using the formula.

For each . . . there are . . . . Another related way of looking at numbers of sequences is the following. You figure that for each of the k_1 possibilities for the first thing in the sequence, there are k_2 possibilities for the second thing. Hence, there are $(k_1)(k_2)$ possibilities for the first two members of the sequence. Likewise, for each of the $(k_1)(k_2)$ possibilities for the first two members, there are k_3 possibilities for the third, and so on. In the example we have been using, for each of the two outcomes of the first coin toss, there are two possibilities for the second toss. Hence, there are $(2)(2) = 4$ possible sequences for the outcome of the two coin tosses (which are HH, HT, TH, and TT). Analogously for each of these four possibilities there are six possible outcomes of the die throw, giving the $(4)(6) = 24$ total three-thing sequences.

Counting Rule 2: A Specific Case of Counting Rule 1 that Occurs When All ks Are Equal

In many instances we have a sequence of N things where all the things are the same. In this case the number of states must be equal for all N things. That is, in terms of our previous rule,

$$k_1 = k_2 = \cdots = k_N = k$$

In this case the formula we obtained from Rule 1 boils down to the following. The number of possible sequences is equal to k multiplied by itself N times, or

Number of sequences $= k^N$

As an example of this suppose we are concerned with the number of seven-digit numbers that can exist. (Such an issue might be of concern to the telephone company, since a telephone number consists of seven digits. The total number of unique seven-digit numbers would therefore determine the total number of telephones that could be contained within an area code.) In this case N is equal to 7, since the sequence is made up of seven things (the things being digits) and k is equal to 10, since each digit can be in any one of 10 states. Thus, the total number of telephone numbers is

$$k^N = 10^7 = 10,000,000$$

TABLE D-2. Illustrations of Counting Rule 2.

Things to Count	N = Number of Things in the Sequence	k = Number of States Each Thing Can Be in	Number of Sequences
1. Number of sequences that can result from 8 coin flips	8 flips	2 (H or T)	$2^8 = 256$
2. Number of sequences that can result from 8 die throws	8 throws	6 (1 through 6)	$6^8 = 1,679,616$
3. Number of three-letter strings	3 letters	26 (A–Z)	$26^3 = 17,576$

Table D-2 provides some other illustrations of situations that are covered by Counting Rule 2.

Counting Rule 3: Permutations

The issue dealt with by Counting Rule 3 involves the following: Suppose we have N things. In how many ways can these N things be arranged in a sequence? For example suppose we had the three letters A, B, and C. In how many sequential ways could we arrange these three letters? We might have ABC, ACB, CBA, and so on. Each sequence is referred to as a *permutation* of the N things, and the formula for computing the number of permutations is

Number of permutations = $N!$

Here $N!$ does not mean a very excited N. Rather, it is a mathematical expression to symbolize

$(N) (N - 1) (N - 2) \ldots (1)$

and is referred to as "N factorial." Thus,

$3! = 3 \cdot 2 \cdot 1 = 6$

$5! = 5 \cdot 4 \cdot 3 \cdot 2 \cdot 1 = 120$

and so on. By convention,

$0! = 1$

The rationale behind this formula is exemplified in Table D-3 and goes as follows. Let's consider the first member or position of the sequence. There are N possible things that we can put in this position. Now let's consider the second position of the sequence. How many things are there that could go into this position? Since we have already chosen one of the N things to go in the first position, there are only $N - 1$ things left for the second position. Thus, the total number of ways we can fill the first two positions is $N(N - 1)$. Likewise, there

TABLE D-3. Counting Rule 3: Permutations

Question: How many ways are there to arrange the letters A, B, and C? (Note: $N = 3$ in this case.)

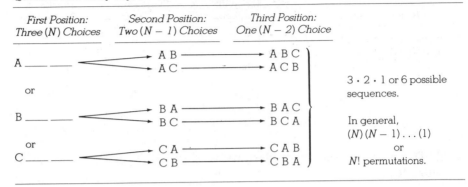

First Position: Three (N) Choices	Second Position: Two ($N-1$) Choices	Third Position: One ($N-2$) Choice	

are ($N - 2$) things that can go into the third position, and so on down the line. Upon arriving at the last position, we are left with only one remaining thing.

Counting Rule 4: Ordered Combinations

The situation covered by this counting rule is as follows. Suppose we have N things, and out of these N things we are going to select and order r of them. How many ways can we do it? For example suppose that we have $N = 4$ letters—a, b, c, and d—and from these four letters we are going to choose and order $r = 3$ of them. How many such three-letter sequences are there?

The logic that we use to deal with this situation is illustrated in Table D-4 and is very similar to the logic we used in the permutation situation described above. We initially consider the first position in our sequence. How many possible things do we have to put in this position? The answer is we could choose any one of our N things. In terms of our example we could choose any 1 of the 4 letters. Now how about the second position? Since we have already used up one of our N things, there are only $N - 1$ things left that can be placed in the second position. Hence, the number of ways of filling the first two positions is $N(N - 1)$. Likewise, there are ($N - 2$) things that can be put in the third position, resulting in $N(N - 1)(N - 2)$ ways of filling the first three positions.

We continue this process until all r things have been used up. The number of ways of filling the r positions is thus

Number of ordered combinations $= N(N - 1)(N - 2) \cdots (N - r + 1)$.

The above expression is somewhat messy and unaesthetic; notationally, there is a better way of representing the same thing, namely,

---

TABLE D-4.   Counting Rule 4: Ordered Combinations

Question: How many ways are there to select and order two (r) letters out of the 4 (N) letters A, B, C, and D?

First Position: *4 (or N)* *Choices*	*Second (r^{th}) Position:* *3 (or N − 1) Choices*	

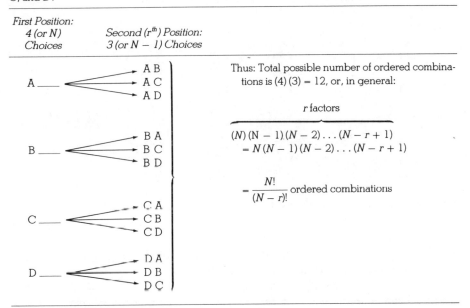

Thus: Total possible number of ordered combinations is (4)(3) = 12, or, in general:

$$\overbrace{(N)(N-1)(N-2)\ldots(N-r+1)}^{r \text{ factors}}$$
$$= N(N-1)(N-2)\ldots(N-r+1)$$

$$= \frac{N!}{(N-r)!} \text{ ordered combinations}$$

$$N(N-1)(N-2)\cdots(N-r+1) =$$

$$\frac{N(N-1)(N-2)\cdots(N-r+1)(N-r)(N-r-1)\cdots(1)}{(N-r)(N-r-1)\cdots(1)}$$

Notice that the last ($N-r$) factors in both numerator and denominator cancel, leaving the number we want. Notice also that the numerator is simply $N!$ and the denominator is simply ($N-r$)! Thus,

$$\text{Number of ordered combinations} = \frac{N!}{(N-r)!}$$

Some examples of permutations and ordered combinations.   Since permutations and ordered combinations are closely related, it is instructive to exemplify them in juxtaposition with one another. A classic example of permutations has to do with seating people in a row of chairs. Imagine that we have $N = 5$ chairs and $N = 5$ people to fill the chairs. How many permutations of the five people are there? We recall from our permutations formula provided in the last section that the number of permutations is equal to N! In this case, $N = 5$, so we

Permutation 1:

Permutation 2:

Permutation 3:

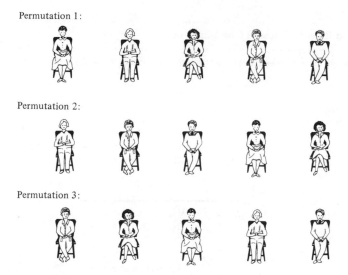

FIGURE D-2.   Three possible permutations of five people sitting in five chairs.

get $5! = 120$ possible arrangements of the five people in the five chairs. Figure D-2 shows examples of some of these permutations.

Suppose that two of the chairs are removed. We are now in sort of a "musical chairs" situation, with five people but only three chairs. Thus, we must *select r* = 3 people out of the five as well as order the lucky three in the chairs once we have selected them. The number of such ordered combinations we can have is thus obtained by

$$\frac{N!}{(N-r)!} = \frac{5!}{(5-3)!} = \frac{5!}{2!} = \frac{5 \cdot 4 \cdot 3 \cdot 2 \cdot 1}{2 \cdot 1} = 5 \cdot 4 \cdot 3 = 6$$

Figure D-3 shows some of these 6 ordered combinations. We emphasize that the ordered combinations may differ from one another either in terms of *which* three out of the five people are selected *or* in terms of how the three selected people are ordered on the chairs. We will discuss this issue further below when we discuss the relationship between ordered and unordered combinations.

Counting Rule 5: Unordered Combinations

Unordered combinations involve a situation in which we have N things and we are interested in the number of ways that we can choose r from these N things *without regard to the order in which the r things are chosen.*

Unordered versus ordered combinations.   As usual our goal is to derive a convenient formula in terms of N and r that will allow us to compute numbers of

Permutation 1:

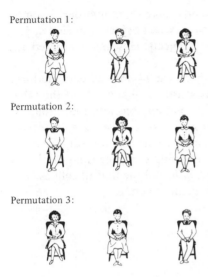

Permutation 2:

Permutation 3:

FIGURE D-3.   Some possible ordered combinations of three out of five people sitting in three chairs.

unordered combinations. To derive such a formula, it is useful to begin by comparing ordered and unordered combinations.

To illustrate this relationship, we will once again consider our 10-person psychology faculty at Bugaboo University. Suppose we're going to choose from the faculty a three-person committee that will have the following characteristics. The first person chosen will be the chair of the committee; the second person will be the vice chair; the third person will be the secretary. In how many ways can such a committee be chosen?

Notice that by this scheme committees can differ from one another in two ways. First, committees can differ in terms of the particular people on them. For example, the Ralph/Shirley/Tom committee will be different from the Ralph/Shirley/Abigail committee or the Daphne/Lance/Gloria committee. Committees may also differ in terms of the order in which the members are chosen. Thus, the Ralph/Shirley/Tom committee will also be different from the Shirley/Ralph/Tom committee (since in the former committee, Ralph will be chair, whereas in the latter, Shirley will be chair). Hence, to find out how many committees there are, we would be dealing with a situation involving *ordered* combinations—order of the members makes a difference.

However, suppose that the order in which committee members were chosen made no difference, that all members were "equals." In this case the Ralph/Shirley/Tom committee would be different from the Ralph/Shirley/Abigail committee (since they involve a different composition of people), but would *not* differ from the Shirley/Ralph/Tom committee (since order is of no import). Thus, in this case, we would be dealing with a situation involving *unordered combinations*.

Table D-5 shows the relationships between ordered and unordered combinations somewhat more explicitly. Here we have $N = 4$ objects (labeled a, b, c, and d) from which we are going to choose $r = 3$ objects. How many ordered and unordered combinations are there?

We can compute via our formula that there are 24 ordered combinations; we have listed these 24 ordered combinations at the left-hand side of the table. Notice that we have grouped these 24 ordered combinations into four groups of 6. And—here is the crucial thing—each group of 6 *involves one unordered combination*. For instance, the six ordered combinations abc, acb, and so on all collapse into the one unordered combination involving the three things a, b, and c. Likewise, the six ordered combinations, abd, adb, and the like all collapse into the one unordered combination involving a, b, d; and so on.

TABLE D-5.   Relationship Between Ordered and Unordered Combinations

---

Consider $N = 4$ things, a, b, c, and d. We will consider both ordered and unordered combinations of $r = 3$ out of these things.

Ordered Combinations	*Unordered Combinations*
abc	
acb	
bac	a, b, c
bca	
cab	
cba	
abd	
adb	
bad	b, a, d
bda	
dab	
dba	
acd	
adc	
cad	a, c, d
cda	
dac	
dca	
bcd	
bdc	
cbd	b, c, d
cdb	
dbc	
dcb	

As we can see, there must be more ordered combinations than unordered combinations. How many more? Well, if each combination involves r things, then each unordered combination (for example involving a, b, and c) can be *permuted r!* ways in order to produce $r!$ ordered combinations. In the example of Table D-5 each unordered combination of three things explodes into $3! = 6$ ordered combinations of the three things.

We now arrive at our formula for calculating the number of unordered combinations. Recall from the last section that the expression for computing the number of ordered combinations was

$$\text{Number of ordered combinations} = \frac{N!}{(N-r)!}$$

But we have just argued that there are $r!$ times as many ordered as unordered combinations; hence, to obtain the number of unordered combinations, we must divide the number of ordered combinations by $r!$. So,

$$\text{Number of unordered combinations} = \frac{\text{number of ordered combinations}}{r!}$$

or

$$\text{Number of unordered combinations} = \frac{N!}{(N-r)!\,r!}.$$

New terminology.   It turns out that we will be using the expression

$$\frac{N!}{(N-r)!\,r!}$$

fairly often. To save space, there is a shorthand way of writing this expression which is $\binom{N}{r}$. The expression $\binom{N}{r}$ should be read as "N choose r" which means "out of N things, I wish to choose r of them." The expression always refers to unordered combinations.

Another way of looking at unordered combinations.   Continuing with the same (a, b, c, d) example, we can also look at unordered combinations as follows. We can line up our N (in this case 4) things and assign either a "$+$" (meaning the thing is "chosen") or a "$-$" (meaning this thing is "not chosen"). Table D-6 shows this scheme. If we are choosing r things, we are simply asking in how many ways may we assign plusses to r (in this case 3) of the objects and minus to the other $N - r$ objects.

Table D-7 shows how this representation may be extended to the question of how many ways there are of obtaining r heads out of N flips of a coin. To maintain the analogy, we have let $N = 4$ flips and $r = 3$ heads. So in this case our "things" are coin flips, and we are "selecting" those to be assigned heads. We

TABLE D-6.   Another Way of Looking at Ordered Combinations

		+: chosen −: not chosen		
a	b	c	d	
+	+	+	−	(corresponds to a, b, c)
+	+	−	+	(corresponds to a, b, d)
+	−	+	+	(corresponds to a, c, d)
−	+	+	+	(corresponds to b, c, d)

can "line up" the four flips (just as we lined up the four things *a*, *b*, *c*, and *d*); we can then "choose" each flip (assign it a head) or not choose the flip (assign it a tail). Thus, the question "how many ways can you get *r* heads out of *N* flips of a coin" is simply a question of unordered combinations. We dwell on this issue because it will be important.

TABLE D-7.   In How Many Ways Can One Obtain $r = 3$ Heads in $N = 4$ Flips?

Flip 1	Flip 2	Flip 3	Flip 4	
H	H	H	T	
H	H	T	H	corresponds to the four unordered
H	T	H	H	combinations from Table D-6
T	H	H	H	

COMBINING COUNTING RULES TO COUNT COMPLICATED THINGS

The five counting rules provide, as we have asserted, convenient aids for counting things. There are, however, many situations that require combining the rules in various ways. Unfortunately, such situations are diverse enough to effectively eliminate a single hard and fast technique that will cover all of them. In this section we'll carry out some examples designed to provide some intuitions and heuristics that we hope will help in complicated situations.

Probability of a Full House

The first example in this appendix involved the probability of being dealt a full house in a poker hand. To compute this probability, we must obtain two numbers: The total number of poker hands and, of these poker hands, the number of them that qualify as a full house.

How many poker hands?   Computing the total number of poker hands is relatively simple. Out of $N = 52$ cards, we are choosing $r = 5$ of them to be in a hand. The order in which the cards are dealt makes no difference; thus, we have a situation involving unordered combinations. We know that the number of such unordered combinations is $\binom{N}{r}$ or in this case

$$\binom{52}{5} = \frac{52!}{(52 - 5)! \, (5!)} = 2{,}598{,}960$$

How many full houses?   Recall that a full house consists of a pair of cards plus three other cards that are the same (a triplet). We will break down the problem into two subproblems: First we will find the number of pairs, and then we will find the number of triplets, assuming the pair has been chosen. The product of these two numbers is the number of full houses. (That is, we use the logic "for each of the x possible pairs, there are y possible triplets." Thus, there is a total of xy hands that qualify as a full house.)

To compute the number of pairs, we begin by pretending that our deck contains only four cards: The four aces, one in each suit. Then the number of pairs would be $\binom{4}{2} = 6$ (since we are choosing two of the four aces to constitute our pair). Figure D-4 shows these six pairs. At this point we take into account that there are in fact 13 cards, not one card, in each suit. Since the pair could be of any one of the 13 cards, we multiply the six ace pairs by 13 (that is, in addition to the six ace-pairs, there are also six king-pairs, six queen-pairs, and so on, down to six deuce-pairs). Therefore the total number of pairs is $(6)(13) = 78$.

Now how many triplets are there? Here we use much the same logic. If there were only one card in each suit, the number of triplets would be $\binom{4}{3} = 4$. It seems now as if we should multiply the four possible triplets per card by the 13

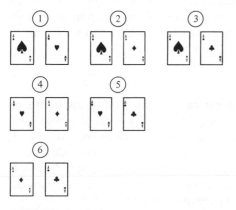

FIGURE D-4.   The six possible pairs of aces.

possible cards per suit to obtain the total number of triplets. But wait! If we have already chosen our pair, we only have 12 cards left to make up the triplet (for example, if our pair had consisted of two aces, then aces would no longer be in contention as the triplet). So we multiply the four triplets per card by 12, not 13, to obtain the total number of triplets.

The total number of full houses may now be computed as

(number of pairs) $\times$ (number of triplets)

$$= \binom{4}{2}(13) \qquad \times \binom{4}{3}(12)$$

$$= (6)(13) \qquad \times (4)(12) = 3744$$

The probability.   The probability of a full house may now be computed as

$$\frac{\text{number of full houses}}{\text{number of poker hands}}$$

$$= \frac{3744}{2,598,960} = 0.00126$$

Common Birthdays

The second example we shall present involves the following situation. Imagine that there are r people at a gathering (assume r is less than 365). What is the probability that at least two of the r people share the same birthday?

How many total patterns of birthdays?   As usual our first task is to compute the total number of elementary events in the sample space—in this case the total number of ways in which the r people can have birthdays. To accomplish this, we invoke Counting Rule 2. Imagine a sequence of r people; each of the people can be in $k = 365$ different states (that is, each person can have a birthday on any one of $k = 365$ different days. We will for purposes of this example arbitrarily eliminate Leap Year Day). The total number of ways that birthdays can be distributed among the r people is therefore computed as:

Number of ways $= 365^r$

How many of these ways involve at least two people having birthdays on the same day?   Let's first invoke a trick. One way we could answer this question is to calculate and sum the number of ways in which two of the people could share a birthday, plus the number of ways three of the people could share a birthday, and so on. Another, easier way is to compute the number of ways in which *no* people share birthdays. We can then subtract this number from the total (365^r) to arrive at the number we are looking for.

Computing the number of ways in which *no* people share the same birthday involves use of Counting Rule 4 (ordered combinations). The reason for this is as follows. Imagine the first of our r people. He or she can have a birthday on any one of 365 days. But the second person is left with a choice of only 364 days (since he or she cannot share a birthday with the first person); the third person has a choice of 363 days, and so on. The number of ways the r people can have no shared birthdays is therefore

$$\frac{365!}{(365-r)!}$$

Now using our trick, we see that the number of ways in which at least two people share a birthday is

$$365^r - \frac{365!}{(365-r)!}$$

The probability. We are now in a position to calculate our probability, which is simply

$$\frac{365^r - \dfrac{365!}{(365-r)!}}{365^r}$$

Figure D-5 shows the probability that at least two people share a birthday as a function of the number of people. As you can see, this probability grows

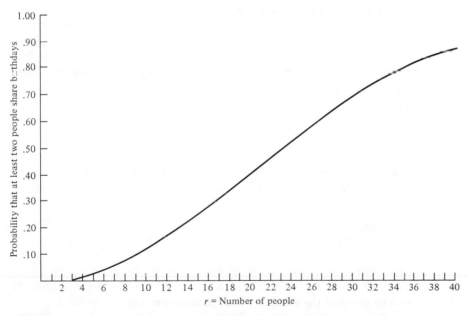

FIGURE D-5. Probability that at least two people share the same birthday as a function of r, the number of assembled people.

surprisingly rapidly. For example, by the time there are 23 people, the probability is over 50% that at least two will have a common birthday. With 32 people, the probability is over 75%, and so on. These are facts that can be used to shock people at cocktail parties and to win bets in bars.

SUMMARY

The purpose of this appendix has been to provide some methods and heuristics to aid in counting up large numbers of complicated things. In particular the following five rules were discussed.

1. The first rule applies when we are counting sequences of N things. We assume that the first thing can be in any one of k_1 states; the second thing can be in any one of k_2 states, and so on until the last thing can be in any one of k_N states. In this case, the total number of sequences of things is computed by the expression

    $$(k_1) (k_2) \cdots (k_N)$$

2. A special case of Counting Rule 1 applies when all the ks are equal—that is, when

    $$k_1 = k_2 = \cdots = k_N = k$$

    In this case, the total number of sequences is

    $$k^N$$

3. *Permutations* apply when we are concerned with the number of ways in which N things can be arranged or sequenced. The number of permutations is

    $$N! = (N) (N - 1) \cdots (1)$$

4. When we must select and order r things out of N possible things, we are dealing with *ordered combinations*. The number of ordered combinations is

    $$\frac{N!}{(N - r)!}$$

    Parenthetically, note that when $r = N$, we are "selecting" N out of N things and ordering them. In such a situation where no selection actually occurs, we are simply dealing with permutations (since all we are doing is ordering the N things). With regard to the formula when $r = N$, the number of "ordered combinations" is simply

$$\frac{N!}{(N-N)!} = \frac{N!}{0!} = \frac{N!}{1} = N!$$

which of course is the expression for the number of permutations of N things.

5. When we are choosing r things from N possible things without regard to the order in which the r things are chosen, we have a situation involving *unordered combinations*. The number of unordered combinations is computed by

$$\frac{N!}{(N-r)!\, r!} = \binom{N}{r}$$

That is to say, there are $r!$ times as many ordered as unordered combinations (since each unordered combination of r things can be ordered in $r!$ different ways).

E Notation for an Analysis of Variance Situation

Since the analysis of variance situations that we will discuss are a little complex, it will be necessary to introduce a bit more notation.

A GENERAL SITUATION

First we note that the number of groups or conditions can be any number two or greater. We shall refer to the total number of groups as J. Furthermore we shall assume that the jth group has n_j subjects. (That is, Group 1 has n_1 subjects, Group 2 has n_2 subjects, and so on.) Therefore, the total number of subjects in the experiment to which we shall refer as N is

$$N = \sum_{j-1}^{J} n_j$$

That is, N is equal to the sum of the number of subjects in each of the J groups.

Now our general situation is depicted in Table E-1. As can be seen, there are several pieces of new notation contained in Table E-1. We have already noted that N refers to the total number of scores in the experiment. Additionally, T is the *grand total*—the *sum* of all the scores in the experiment. Therefore $M = T/N$ is the *grand mean* of all the scores in the experiment.

DOUBLE SUMMATION

Let's take this opportunity to introduce some new tricks that may be performed with summation signs. Suppose we wish to formulate an expression for T, the grand total of all the scores in the experiment. As shown in Table E-1, the simplest way of expressing T is as the sum of all the individual group totals, that

TABLE E-1.   Notation for the General Case; There Are Assumed to Be J Groups and the jth Group Contains n_j Subjects

Group 1 $(n_1$ Ss)	Group 2 $(n_2$ Ss)	. . .	Group j $(n_j$ Ss)	. . .	Group J $(n_J$ Ss)
x_{11}	x_{12}		x_{1j}		x_{1J}
x_{21}	x_{22}		x_{2j}		x_{2J}
x_{31}	x_{32}		x_{3j}		x_{3J}
.	.		.		.
.	.		.		.
x_{i1}	x_{i2}		x_{ij}		x_{iJ}
.	.		.		.
.	.		.		.
$x_{n_1 1}$	$x_{n_2 2}$		$x_{n_j j}$		$x_{n_J J}$
$T_1 = \sum_{i=1}^{n_1} x_{i1}$	$T_2 = \sum_{i=1}^{n_2} x_{i2}$	. . .	$T_j = \sum_{i=1}^{n_j} x_{ij}$	. . .	$T_J = \sum_{i=1}^{n_J} x_{iJ}$
$M_1 = \dfrac{T_1}{n_1}$	$M_2 = \dfrac{T_2}{n_2}$	. . .	$M_j = \dfrac{T_j}{n_j}$	. . .	$M_J = \dfrac{T_J}{n_J}$

Thus:

x_{ij} is the ith subject in the jth group

$T_j = \sum_{i=1}^{n_j} x_{ij}$ is the total of the jth group

$M_j = \dfrac{T_j}{n_j}$ is the mean of the jth group

$T = \sum_{j=1}^{J} T_j$ is the *grand total*

$M = \dfrac{T}{N}$ is the *grand mean*

is

$$T = \sum_{j=1}^{J} T_j \tag{E.1}$$

Notice, however, that we can express T_j in terms of raw scores (the xs). Thus,

$$T_j = \sum_{i=1}^{n_j} \tag{E.2}$$

Substituting Equation E.2 into Equation E.1 yields the somewhat unfamiliar-looking equation

$$T = \sum_{j=1}^{J} \sum_{i=1}^{n_j} \tag{E.3}$$

Equation E.3 makes use of *double summation*. To perform a double summation, the following steps are carried out.

1. Consider the right-hand summation sign. This refers to summation within a group—that is, down a column of the matrix shown in Table E-1. Perform this summation for each column referred to in the left-hand summation sign (from the first column $j = 1$ to the last column $j = J$).

2. Add the results of these J summations as indicated by the left-hand summation sign.

In general double-summing a set of numbers simply means to sum all entries in some matrix or array of numbers—just as single summation means to sum all entries in some column (or row) of numbers.

TABLE F-1.   Binomial Probabilities $\binom{N}{r}p^{r}q^{N-r}$

						p					
N	r	.05	.10	.15	.20	.25	.30	.35	.40	.45	.50
1	0	.9500	.9000	.8500	.8000	.7500	.7000	.6500	.6000	.5500	.5000
	1	.0500	.1000	.1500	.2000	.2500	.3000	.3500	.4000	.4500	.5000
2	0	.9025	.8100	.7225	.6400	.5625	.4900	.4225	.3600	.3025	.2500
	1	.0950	.1800	.2550	.3200	.3750	.4200	.4550	.4800	.4950	.5000
	2	.0025	.0100	.0225	.0400	.0625	.0900	.1225	.1600	.2025	.2500
3	0	.8574	.7290	.6141	.5120	.4219	.3430	.2746	.2160	.1664	.1250
	1	.1354	.2430	.3251	.3840	.4219	.4410	.4436	.4320	.4084	.3750
	2	.0071	.0270	.0574	.0960	.1406	.1890	.2389	.2880	.3341	.3750
	3	.0001	.0010	.0034	.0080	.0156	.0270	.0429	.0640	.0911	.1250
4	0	.8145	.6561	.5220	.4096	.3164	.2401	.1785	.1296	.0915	.0625
	1	.1715	.2916	.3685	.4096	.4219	.4116	.3845	.3456	.2995	.2500
	2	.0135	.0486	.0975	.1536	.2109	.2646	.3105	.3456	.3675	.3750
	3	.0005	.0036	.0115	.0256	.0469	.0756	.1115	.1536	.2005	.2500
	4	.0000	.0001	.0005	.0016	.0039	.0081	.0150	.0256	.0410	.0625
5	0	.7738	.5905	.4437	.3277	.2373	.1681	.1160	.0778	.0503	.0312
	1	.2036	.3280	.3915	.4096	.3955	.3602	.3124	.2592	.2059	.1562
	2	.0214	.0729	.1382	.2048	.2637	.3087	.3364	.3456	.3369	.3125
	3	.0011	.0081	.0244	.0512	.0879	.1323	.1811	.2304	.2757	.3125
	4	.0000	.0004	.0022	.0064	.0146	.0284	.0488	.0768	.1128	.1562
	5	.0000	.0000	.0001	.0003	.0010	.0024	.0053	.0102	.0185	.0312
6	0	.7351	.5314	.3771	.2621	.1780	.1176	.0754	.0467	.0277	.0156
	1	.2321	.3543	.3993	.3932	.3560	.3025	.2437	.1866	.1359	.0938
	2	.0305	.0984	.1762	.2458	.2966	.3241	.3280	.3110	.2780	.2344
	3	.0021	.0146	.0415	.0819	.1318	.1852	.2355	.2765	.3032	.3125
	4	.0001	.0012	.0055	.0154	.0330	.0595	.0951	.1382	.1861	.2344

This table is reproduced by permission from R. S. Burington and D. C. May, *Handbook of Probability and Statistics with Tables*. McGraw-Hill Book Company (ed. 2), 1970.

TABLE F-1.   (continued)

N	r	.05	.10	.15	.20	.25	.30	.35	.40	.45	.50
	5	.0000	.0001	.0004	.0015	.0044	.0102	.0205	.0369	.0609	.0938
	6	.0000	.0000	.0000	.0001	.0002	.0007	.0018	.0041	.0083	.0156
7	0	.6983	.4783	.3206	.2097	.1335	.0824	.0490	.0280	.0152	.0078
	1	.2573	.3720	.3960	.3670	.3115	.2471	.1848	.1306	.0872	.0547
	2	.0406	.1240	.2097	.2753	.3115	.3177	.2985	.2613	.2140	.1641
	3	.0036	.0230	.0617	.1147	.1730	.2269	.2679	.2903	.2918	.2734
	4	.0002	.0026	.0109	.0287	.0577	.0972	.1442	.1935	.2388	.2734
	5	.0000	.0002	.0012	.0043	.0115	.0250	.0466	.0774	.1172	.1641
	6	.0000	.0000	.0001	.0004	.0013	.0036	.0084	.0172	.0320	.0547
	7	.0000	.0000	.0000	.0000	.0001	.0002	.0006	.0016	.0037	.0078
8	0	.6634	.4305	.2725	.1678	.1001	.0576	.0319	.0168	.0084	.0039
	1	.2793	.3826	.3847	.3355	.2760	.1977	.1373	.0896	.0548	.0312
	2	.0515	.1488	.2376	.2936	.3115	.2965	.2587	.2090	.1569	.1094
	3	.0054	.0331	.0839	.1468	.2076	.2541	.2786	.2787	.2568	.2188
	4	.0004	.0046	.0185	.0459	.0865	.1361	.1875	.2322	.2627	.2734
	5	.0000	.0004	.0026	.0092	.0231	.0467	.0808	.1239	.1719	.2188
	6	.0000	.0000	.0002	.0011	.0038	.0100	.0217	.0413	.0703	.1094
	7	.0000	.0000	.0000	.0001	.0004	.0012	.0033	.0079	.0164	.0312
	8	.0000	.0000	.0000	.0000	.0000	.0001	.0002	.0007	.0017	.0039
9	0	.6302	.3874	.2316	.1342	.0751	.0404	.0277	.0101	.0046	.0020
	1	.2985	.3874	.3679	.3020	.2253	.1556	.1004	.0605	.0339	.0176
	2	.0629	.1722	.2597	.3020	.3003	.2668	.2162	.1612	.1110	.0703
	3	.0077	.0446	.1069	.1762	.2336	.2668	.2716	.2508	.2119	.1641
	4	.0006	.0074	.0283	.0661	.1168	.1715	.2194	.2508	.2600	.2461
	5	.0000	.0008	.0050	.0165	.0389	.0735	.1181	.1672	.2128	.2461
	6	.0000	.0001	.0006	.0028	.0087	.0210	.0424	.0743	.1160	.1641
	7	.0000	.0000	.0000	.0003	.0012	.0039	.0098	.0212	.0407	.0703
	8	.0000	.0000	.0000	.0000	.0001	.0004	.0013	.0035	.0083	.0176
	9	.0000	.0000	.0000	.0000	.0000	.0000	.0001	.0003	.0008	.0020
10	0	.5987	.3487	.1969	.1074	.0563	.0282	.0135	.0060	.0025	.0010
	1	.3151	.3874	.3474	.2684	.1877	.1211	.0725	.0403	.0207	.0098
	2	.0746	.1937	.2759	.3020	.2816	.2335	.1757	.1209	.0763	.0439
	3	.0105	.0574	.1298	.2013	.2503	.2668	.2522	.2150	.1665	.1172
	4	.0010	.0112	.0401	.0881	.1460	.2001	.2377	.2508	.2384	.2051
	5	.0001	.0015	.0085	.0264	.0584	.1029	.1536	.2007	.2340	.2461
	6	.0000	.0001	.0012	.0055	.0162	.0368	.0689	.1115	.1596	.2051
	7	.0000	.0000	.0001	.0008	.0031	.0090	.0212	.0425	.0746	.1172
	8	.0000	.0000	.0000	.0001	.0004	.0014	.0043	.0106	.0229	.0439
	9	.0000	.0000	.0000	.0000	.0000	.0001	.0005	.0016	.0042	.0098
	10	.0000	.0000	.0000	.0000	.0000	.0000	.0000	.0001	.0003	.0010
11	0	.5688	.3138	.1673	.0859	.0422	.0198	.0088	.0036	.0014	.0005
	1	.3293	.3835	.3248	.2362	.1549	.0932	.0518	.0266	.0125	.0054
	2	.0867	.2131	.2866	.2953	.2581	.1998	.1395	.0887	.0513	.0269

TABLE F-1. (continued)

N	r	.05	.10	.15	.20	.25	.30	.35	.40	.45	.50
	3	.0137	.0710	.1517	.2215	.2581	.2568	.2254	.1774	.1259	.0806
	4	.0014	.0158	.0536	.1107	.1721	.2201	.2428	.2365	.2060	.1611
	5	.0001	.0025	.0132	.0388	.0803	.1231	.1830	.2207	.2360	.2256
	6	.0000	.0003	.0023	.0097	.0268	.0566	.0985	.1471	.1931	.2256
	7	.0000	.0000	.0003	.0017	.0064	.0173	.0379	.0701	.1128	.1611
	8	.0000	.0000	.0000	.0002	.0011	.0037	.0102	.0234	.0462	.0806
	9	.0000	.0000	.0000	.0000	.0001	.0005	.0018	.0052	.0126	.0269
	10	.0000	.0000	.0000	.0000	.0000	.0000	.0002	.0007	.0021	.0054
	11	.0000	.0000	.0000	.0000	.0000	.0000	.0000	.0000	.0002	.0005
12	0	.5404	.2824	.1422	.0687	.0317	.0138	.0057	.0022	.0008	.0002
	1	.3413	.3766	.3012	.2062	.1267	.0712	.0368	.0174	.0075	.0029
	2	.0988	.2301	.2924	.2835	.2323	.1678	.1088	.0639	.0339	.0161
	3	.0173	.0852	.1720	.2362	.2581	.2397	.1954	.1419	.0923	.0537
	4	.0021	.0213	.0683	.1329	.1936	.2311	.2367	.2128	.1700	.1208
	5	.0002	.0038	.0193	.0532	.1032	.1585	.2039	.2270	.2225	.1934
	6	.0000	.0005	.0040	.0155	.0401	.0792	.1281	.1766	.2124	.2256
	7	.0000	.0000	.0006	.0033	.0115	.0291	.0591	.1009	.1489	.1934
	8	.0000	.0000	.0001	.0005	.0024	.0078	.0199	.0420	.0762	.1208
	9	.0000	.0000	.0000	.0001	.0004	.0015	.0048	.0125	.0277	.0537
	10	.0000	.0000	.0000	.0000	.0000	.0002	.0008	.0025	.0068	.0161
	11	.0000	.0000	.0000	.0000	.0000	.0000	.0001	.0003	.0010	.0029
	12	.0000	.0000	.0000	.0000	.0000	.0000	.0000	.0000	.0001	.0002
13	0	.5133	.2542	.1209	.0550	.0238	.0097	.0037	.0013	.0004	.0001
	1	.3512	.3672	.2774	.1787	.1029	.0540	.0259	.0113	.0045	.0016
	2	.1109	.2448	.2937	.2680	.2059	.1388	.0836	.0453	.0220	.0095
	3	.0214	.0997	.1900	.2457	.2517	.2181	.1651	.1107	.0660	.0349
	4	.0028	.0277	.0838	.1535	.2097	.2337	.2222	.1845	.1350	.0873
	5	.0003	.0055	.0266	.0691	.1258	.1803	.2154	.2214	.1989	.1571
	6	.0000	.0008	.0063	.0230	.0559	.1030	.1546	.1968	.2169	.2095
	7	.0000	.0001	.0011	.0058	.0186	.0442	.0833	.1312	.1775	.2095
	8	.0000	.0000	.0001	.0011	.0047	.0142	.0336	.0656	.1089	.1571
	9	.0000	.0000	.0000	.0001	.0009	.0034	.0101	.0243	.0495	.0873
	10	.0000	.0000	.0000	.0000	.0001	.0006	.0022	.0065	.0162	.0349
	11	.0000	.0000	.0000	.0000	.0000	.0001	.0003	.0012	.0036	.0095
	12	.0000	.0000	.0000	.0000	.0000	.0000	.0000	.0001	.0005	.0016
	13	.0000	.0000	.0000	.0000	.0000	.0000	.0000	.0000	.0000	.0001
14	0	.4877	.2288	.1028	.0440	.0178	.0068	.0024	.0008	.0002	.0001
	1	.3593	.3559	.2539	.1539	.0832	.0407	.0181	.0073	.0027	.0009
	2	.1229	.2570	.2912	.2501	.1802	.1134	.0634	.0317	.0141	.0056
	3	.0259	.1142	.2056	.2501	.2402	.1943	.1366	.0845	.0462	.0222
	4	.0037	.0349	.0998	.1720	.2202	.2290	.2022	.1549	.1040	.0611
	5	.0004	.0078	.0352	.0860	.1468	.1963	.2178	.2066	.1701	.1222
	6	.0000	.0013	.0093	.0322	.0734	.1262	.1759	.2066	.2088	.1833

TABLE F-1.   (continued)

N	r	.05	.10	.15	.20	.25	.30	.35	.40	.45	.50
						p					
	7	.0000	.0002	.0019	.0092	.0280	.0618	.1082	.1574	.1952	.2095
	8	.0000	.0000	.0003	.0020	.0082	.0232	.0510	.0918	.1398	.1833
	9	.0000	.0000	.0000	.0003	.0018	.0066	.0183	.0408	.0762	.1222
	10	.0000	.0000	.0000	.0000	.0003	.0014	.0049	.0136	.0312	.0611
	11	.0000	.0000	.0000	.0000	.0000	.0002	.0010	.0033	.0093	.0222
	12	.0000	.0000	.0000	.0000	.0000	.0000	.0001	.0005	.0019	.0056
	13	.0000	.0000	.0000	.0000	.0000	.0000	.0000	.0001	.0002	.0009
	14	.0000	.0000	.0000	.0000	.0000	.0000	.0000	.0000	.0000	.0001
15	0	.4633	.2059	.0874	.0352	.0134	.0047	.0016	.0005	.0001	.0000
	1	.3658	.3432	.2312	.1319	.0668	.0305	.0126	.0047	.0016	.0005
	2	.1348	.2669	.2856	.2309	.1559	.0916	.0476	.0219	.0090	.0032
	3	.0307	.1285	.2184	.2501	.2252	.1700	.1110	.0634	.0318	.0139
	4	.0049	.0428	.1156	.1876	.2252	.2186	.1792	.1268	.0780	.0417
	5	.0006	.0105	.0449	.1032	.1651	.2061	.2123	.1859	.1404	.0916
	6	.0000	.0019	.0132	.0430	.0917	.1472	.1906	.2066	.1914	.1527
	7	.0000	.0003	.0030	.0138	.0393	.0811	.1319	.1771	.2013	.1964
	8	.0000	.0000	.0005	.0035	.0131	.0348	.0710	.1181	.1647	.1964
	9	.0000	.0000	.0001	.0007	.0034	.0116	.0298	.0612	.1048	.1527
	10	.0000	.0000	.0000	.0001	.0007	.0030	.0096	.0245	.0515	.0916
	11	.0000	.0000	.0000	.0000	.0001	.0006	.0024	.0074	.0191	.0417
	12	.0000	.0000	.0000	.0000	.0000	.0001	.0004	.0016	.0052	.0139
	13	.0000	.0000	.0000	.0000	.0000	.0000	.0001	.0003	.0010	.0032
	14	.0000	.0000	.0000	.0000	.0000	.0000	.0000	.0000	.0001	.0005
	15	.0000	.0000	.0000	.0000	.0000	.0000	.0000	.0000	.0000	.0000
16	0	.4401	.1853	.0743	.0281	.0100	.0033	.0010	.0003	.0001	.0000
	1	.3706	.3294	.2097	.1126	.0535	.0228	.0087	.0030	.0009	.0002
	2	.1463	.2745	.2775	.2111	.1336	.0732	.0353	.0150	.0056	.0018
	3	.0359	.1423	.2285	.2463	.2079	.1465	.0888	.0468	.0215	.0085
	4	.0061	.0514	.1311	.2001	.2252	.2040	.1553	.1014	.0572	.0278
	5	.0008	.0137	.0555	.1201	.1802	.2099	.2008	.1623	.1123	.0667
	6	.0001	.0028	.0180	.0550	.1101	.1649	.1982	.1983	.1684	.1222
	7	.0000	.0004	.0045	.0197	.0524	.1010	.1524	.1889	.1969	.1746
	8	.0000	.0001	.0009	.0055	.0197	.0487	.0923	.1417	.1812	.1964
	9	.0000	.0000	.0001	.0012	.0058	.0185	.0442	.0840	.1318	.1746
	10	.0000	.0000	.0000	.0002	.0014	.0056	.0167	.0392	.0755	.1222
	11	.0000	.0000	.0000	.0000	.0002	.0013	.0049	.0142	.0337	.0667
	12	.0000	.0000	.0000	.0000	.0000	.0002	.0011	.0040	.0115	.0278
	13	.0000	.0000	.0000	.0000	.0000	.0000	.0002	.0008	.0029	.0085
	14	.0000	.0000	.0000	.0000	.0000	.0000	.0000	.0001	.0005	.0018
	15	.0000	.0000	.0000	.0000	.0000	.0000	.0000	.0000	.0001	.0002
	16	.0000	.0000	.0000	.0000	.0000	.0000	.0000	.0000	.0000	.0000
17	0	.4181	.1668	.0631	.0225	.0075	.0023	.0007	.0002	.0000	.0000

TABLE F-1.   (continued)

N	r	.05	.10	.15	.20	.25	.30	.35	.40	.45	.50
								p			
	1	.3741	.3150	.1893	.0957	.0426	.0169	.0060	.0019	.0005	.0001
	2	.1575	.2800	.2673	.1914	.1136	.0581	.0260	.0102	.0035	.0010
	3	.0415	.1556	.2359	.2393	.1893	.1245	.0701	.0341	.0144	.0052
	4	.0076	.0605	.1457	.2093	.2209	.1868	.1320	.0796	.0411	.0182
	5	.0010	.0175	.0668	.1361	.1914	.2081	.1849	.1379	.0875	.0472
	6	.0001	.0039	.0236	.0680	.1276	.1784	.1991	.1839	.1432	.0944
	7	.0000	.0007	.0065	.0267	.0668	.1201	.1685	.1927	.1841	.1484
	8	.0000	.0001	.0014	.0084	.0279	.0644	.1143	.1606	.1883	.1855
	9	.0000	.0000	.0003	.0021	.0093	.0276	.0611	.1070	.1540	.1855
	10	.0000	.0000	.0000	.0004	.0025	.0095	.0263	.0571	.1008	.1484
	11	.0000	.0000	.0000	.0001	.0005	.0026	.0090	.0242	.0525	.0944
	12	.0000	.0000	.0000	.0000	.0001	.0006	.0024	.0081	.0215	.0472
	13	.0000	.0000	.0000	.0000	.0000	.0001	.0005	.0021	.0068	.0182
	14	.0000	.0000	.0000	.0000	.0000	.0000	.0001	.0004	.0016	.0052
	15	.0000	.0000	.0000	.0000	.0000	.0000	.0000	.0001	.0003	.0010
	16	.0000	.0000	.0000	.0000	.0000	.0000	.0000	.0000	.0000	.0001
	17	.0000	.0000	.0000	.0000	.0000	.0000	.0000	.0000	.0000	.0000
18	0	.3972	.1501	.0536	.0180	.0056	.0016	.0004	.0001	.0000	.0000
	1	.3763	.3002	.1704	.0811	.0338	.0126	.0042	.0012	.0003	.0001
	2	.1683	.2835	.2556	.1723	.0958	.0458	.0190	.0069	.0022	.0006
	3	.0473	.1680	.2406	.2297	.1704	.1046	.0547	.0246	.0095	.0031
	4	.0093	.0700	.1592	.2153	.2130	.1681	.1104	.0614	.0291	.0117
	5	.0014	.0218	.0787	.1507	.1988	.2017	.1664	.1146	.0666	.0327
	6	.0002	.0052	.0310	.0816	.1436	.1873	.1941	.1655	.1181	.0708
	7	.0000	.0010	.0091	.0350	.0820	.1376	.1792	.1892	.1657	.1214
	8	.0000	.0002	.0022	.0120	.0376	.0811	.1327	.1734	.1864	.1669
	9	.0000	.0000	.0004	.0033	.0139	.0386	.0794	.1284	.1694	.1855
	10	.0000	.0000	.0001	.0008	.0042	.0149	.0385	.0771	.1248	.1669
	11	.0000	.0000	.0000	.0001	.0010	.0046	.0151	.0374	.0742	.1214
	12	.0000	.0000	.0000	.0000	.0002	.0012	.0047	.0145	.0354	.0708
	13	.0000	.0000	.0000	.0000	.0000	.0002	.0012	.0045	.0134	.0327
	14	.0000	.0000	.0000	.0000	.0000	.0000	.0002	.0011	.0039	.0117
	15	.0000	.0000	.0000	.0000	.0000	.0000	.0000	.0002	.0009	.0031
	16	.0000	.0000	.0000	.0000	.0000	.0000	.0000	.0000	.0001	.0006
	17	.0000	.0000	.0000	.0000	.0000	.0000	.0000	.0000	.0000	.0001
	18	.0000	.0000	.0000	.0000	.0000	.0000	.0000	.0000	.0000	.0000
19	0	.3774	.1351	.0456	.0144	.0042	.0011	.0003	.0001	.0000	.0000
	1	.3774	.2852	.1529	.0685	.0268	.0093	.0029	.0008	.0002	.0000
	2	.1787	.2852	.2428	.1540	.0803	.0358	.0138	.0046	.0013	.0003
	3	.0533	.1796	.2428	.2182	.1517	.0869	.0422	.0175	.0062	.0018
	4	.0012	.0798	.1714	.2182	.2023	.1491	.0909	.0467	.0203	.0074
	5	.0018	.0266	.0907	.1636	.2023	.1916	.1468	.0933	.0497	.0222
	6	.0002	.0069	.0374	.0955	.1574	.1916	.1844	.1451	.0949	.0518

TABLE F-1. (continued)

N	r	.05	.10	.15	.20	.25	.30	.35	.40	.45	.50
							p				
	7	.0000	.0014	.0122	.0443	.0974	.1525	.1844	.1797	.1443	.0961
	8	.0000	.0002	.0032	.0166	.0487	.0981	.1489	.1797	.1771	.1442
	9	.0000	.0000	.0007	.0051	.0198	.0514	.0980	.1464	.1771	.1762
	10	.0000	.0000	.0001	.0013	.0066	.0220	.0528	.0976	.1449	.1762
	11	.0000	.0000	.0000	.0003	.0018	.0077	.0233	.0532	.0970	.1442
	12	.0000	.0000	.0000	.0000	.0004	.0022	.0083	.0237	.0529	.0961
	13	.0000	.0000	.0000	.0000	.0001	.0005	.0024	.0085	.0233	.0518
	14	.0000	.0000	.0000	.0000	.0000	.0001	.0006	.0024	.0082	.0222
	15	.0000	.0000	.0000	.0000	.0000	.0000	.0001	.0005	.0022	.0074
	16	.0000	.0000	.0000	.0000	.0000	.0000	.0000	.0001	.0005	.0018
	17	.0000	.0000	.0000	.0000	.0000	.0000	.0000	.0000	.0001	.0003
	18	.0000	.0000	.0000	.0000	.0000	.0000	.0000	.0000	.0000	.0000
	19	.0000	.0000	.0000	.0000	.0000	.0000	.0000	.0000	.0000	.0000
20	0	.3585	.1216	.0388	.0115	.0032	.0008	.0002	.0000	.0000	.0000
	1	.3774	.2702	.1368	.0576	.0211	.0068	.0020	.0005	.0001	.0000
	2	.1887	.2852	.2293	.1369	.0669	.0278	.0100	.0031	.0008	.0002
	3	.0596	.1901	.2428	.2054	.1339	.0716	.0323	.0123	.0040	.0011
	4	.0133	.0898	.1821	.2182	.1897	.1304	.0738	.0350	.0139	.0046
	5	.0022	.0319	.1028	.1746	.2023	.1789	.1272	.0746	.0365	.0148
	6	.0003	.0089	.0454	.1091	.1686	.1916	.1712	.1244	.0746	.0370
	7	.0000	.0020	.0160	.0545	.1124	.1643	.1844	.1659	.1221	.0739
	8	.0000	.0004	.0046	.0222	.0609	.1144	.1614	.1797	.1623	.1201
	9	.0000	.0001	.0011	.0074	.0271	.0654	.1158	.1597	.1771	.1602
	10	.0000	.0000	.0002	.0020	.0099	.0308	.0686	.1171	.1593	.1762
	11	.0000	.0000	.0000	.0005	.0030	.0120	.0336	.0710	.1185	.1602
	12	.0000	.0000	.0000	.0001	.0008	.0039	.0136	.0355	.0727	.1201
	13	.0000	.0000	.0000	.0000	.0002	.0010	.0045	.0146	.0366	.0739
	14	.0000	.0000	.0000	.0000	.0000	.0002	.0012	.0049	.0150	.0370
	15	.0000	.0000	.0000	.0000	.0000	.0000	.0003	.0013	.0049	.0148
	16	.0000	.0000	.0000	.0000	.0000	.0000	.0000	.0003	.0013	.0046
	17	.0000	.0000	.0000	.0000	.0000	.0000	.0000	.0000	.0002	.0011
	18	.0000	.0000	.0000	.0000	.0000	.0000	.0000	.0000	.0000	.0002
	19	.0000	.0000	.0000	.0000	.0000	.0000	.0000	.0000	.0000	.0000
	20	.0000	.0000	.0000	.0000	.0000	.0000	.0000	.0000	.0000	.0000

TABLE F-2.   Cumulative Normal Probabilities

z	$F(z)$	z	$F(z)$	z	$F(z)$	z	$F(z)$
0.00	.5000000	0.21	.5831662	0.42	.6627573	0.63	.7356527
0.01	.5039894	0.22	.5870604	0.43	.6664022	0.64	.7389137
0.02	.5079783	0.23	.5909541	0.44	.6700314	0.65	.7421539
0.03	.5119665	0.24	.5948349	0.45	.6736448	0.66	.7453731
0.04	.5159534	0.25	.5987063	0.46	.6772419	0.67	.7485711
0.05	.5199388	0.26	.6025681	0.47	.6808225	0.68	.7517478
0.06	.5239222	0.27	.6064199	0.48	.6843863	0.69	.7549029
0.07	.5279032	0.28	.6102612	0.49	.6879331	0.70	.7580363
0.08	.5318814	0.29	.6140919	0.50	.6914625	0.71	.7611479
0.09	.5358564	0.30	.6179114	0.51	.6949743	0.72	.7642375
0.10	.5398278	0.31	.6217195	0.52	.6984682	0.73	.7673049
0.11	.5437953	0.32	.6255158	0.53	.7019440	0.74	.7703500
0.12	.5477584	0.33	.6293000	0.54	.7054015	0.75	.7733726
0.13	.5517168	0.34	.6330717	0.55	.7088403	0.76	.7763727
0.14	.5556700	0.35	.6368307	0.56	.7122603	0.77	.7793501
0.15	.5596177	0.36	.6405764	0.57	.7156612	0.78	.7823046
0.16	.5635595	0.37	.6443088	0.58	.7190427	0.79	.7852361
0.17	.5674949	0.38	.6480273	0.59	.7224047	0.80	.7881446
0.18	.5714237	0.39	.6517317	0.60	.7257469	0.81	.7910299
0.19	.5753454	0.40	.6554217	0.61	.7290691	0.82	.7938919
0.20	.5792597	0.41	.6590970	0.62	.7323711	0.83	.7967306
0.84	.7995458	1.32	.9065825	1.79	.9632730	2.26	.9880894
0.85	.8023375	1.33	.9082409	1.80	.9640697	2.27	.9883962
0.86	.8051055	1.34	.9098773	1.81	.9648521	2.28	.9886962
0.87	.8078498	1.35	.9114920	1.82	.9656205	2.29	.9889893
0.88	.8105703	1.36	.9130850	1.83	.9663750	2.30	.9892759
0.89	.8132671	1.37	.9146565	1.84	.9671159	2.31	.9895559
0.90	.8159399	1.38	.9162067	1.85	.9678432	2.32	.9898296
0.91	.8185887	1.39	.9177356	1.86	.9685572	2.33	.9900969
0.92	.8212136	1.40	.9192433	1.87	.9692581	2.34	.9903581
0.93	.8238145	1.41	.9207302	1.88	.9699460	2.35	.9906133
0.94	.8263912	1.42	.9221962	1.89	.9706210	2.36	.9908625
0.95	.8289439	1.43	.9236415	1.90	.9712834	2.37	.9911060
0.96	.8314724	1.44	.9250663	1.91	.9719334	2.38	.9913437
0.97	.8339768	1.45	.9264707	1.92	.9725711	2.39	.9915758
0.98	.8364569	1.46	.9278550	1.93	.9731966	2.40	.9918025
0.99	.8389129	1.47	.9292191	1.94	.9738102	2.41	.9920237
1.00	.8413447	1.48	.9305634	1.95	.9744119	2.42	.9922397
1.01	.8437524	1.49	.9318879	1.96	.9750021	2.43	.9924506
1.02	.8461358	1.50	.9331928	1.97	.9755808	2.44	.9926564
1.03	.8484950	1.51	.9344783	1.98	.9761482	2.45	.9928572
1.04	.8508300	1.52	.9357445	1.99	.9767045	2.46	.9930531
1.05	.8531409	1.53	.9369916	2.00	.9772499	2.47	.9932443
1.06	.8554277	1.54	.9382198	2.01	.9777844	2.48	.9934309
1.07	.8576903	1.55	.9394292	2.02	.9783083	2.49	.9936128
1.08	.8599289	1.56	.9406201	2.03	.9788217	2.50	.9937903

This table is condensed from Table 1 of the *Biometrika Tables for Statisticians*, Vol. 1 (ed. 3), edited by E. S. Pearson and H. O. Hartley. Reproduced here with the kind permission of E. S. Pearson and the trustees of *Biometrika*.

TABLE F-2.   (continued)

z	F(z)	z	F(z)	z	F(z)	z	F(z)
1.09	.8621434	1.57	.9417924	2.04	.9793248	2.51	.9939634
1.10	.8643339	1.58	.9429466	2.05	.9798178	2.52	.9941323
1.11	.8665005	1.59	.9440826	2.06	.9803007	2.53	.9942969
1.12	.8686431	1.60	.9452007	2.07	.9807738	2.54	.9944574
1.13	.8707619	1.61	.9463011	2.08	.9812372	2.55	.9946139
1.14	.8728568	1.62	.9473839	2.09	.9816911	2.56	.9947664
1.15	.8749281	1.63	.9484493	2.10	.9821356	2.57	.9949151
1.16	.8769756	1.64	.9494974	2.11	.9825708	2.58	.9950600
1.17	.8789995	1.65	.9505285	2.12	.9829970	2.59	.9952012
1.18	.8809999	1.66	.9515428	2.13	.9834142	2.60	.9953388
1.19	.8829768	1.67	.9525403	2.14	.9838226	2.70	.9965330
1.20	.8849303	1.68	.9535213	2.15	.9842224	2.80	.9974449
1.21	.8868606	1.69	.9544860	2.16	.9846137	2.90	.9981342
1.22	.8887676	1.70	.9554345	2.17	.9849966	3.00	.9986501
1.23	.8906514	1.71	.9563671	2.18	.9853713	3.20	.9993129
1.24	.8925123	1.72	.9572838	2.19	.9857379	3.40	.9996631
1.25	.8943502	1.73	.9581849	2.20	.9860966	3.60	.9998409
1.26	.8961653	1.74	.9590705	2.21	.9864474	3.80	.9999277
1.27	.8979577	1.75	.9599408	2.22	.9867906	4.00	.9999683
1.28	.8997274	1.76	.9607961	2.23	.9871263	4.50	.9999966
1.29	.9014747	1.77	.9616364	2.24	.9874545	5.00	.9999997
1.30	.9031995	1.78	.9624620	2.25	.9877755	5.50	.9999999
1.31	.9049021						

TABLE F-3.   Upper Percentage Points of the _t_ Distribution

df	Q = 0.4 2Q = 0.8	0.25 0.5	0.1 0.2	0.05 0.1	0.025 0.05	0.01 0.02	0.005 0.01	0.001 0.002
1	0.325	1.000	3.078	6.314	12.704	31.821	63.657	318.31
2	0.289	0.816	1.886	2.920	4.303	6.965	9.925	22.326
3	0.277	0.765	1.638	2.353	3.182	4.541	5.841	10.213
4	0.271	0.741	1.533	2.132	2.776	3.747	4.604	7.173
5	0.267	0.727	1.476	2.015	2.571	3.365	4.032	5.893
6	0.265	0.718	1.440	1.943	2.447	3.143	3.707	5.208
7	0.263	0.711	1.415	1.895	2.365	2.998	3.499	4.785
8	0.262	0.706	1.397	1.860	2.306	2.896	3.355	4.501
9	0.261	0.703	1.383	1.833	2.262	2.821	3.250	4.297
10	0.260	0.700	1.372	1.812	2.228	2.764	3.169	4.144
11	0.260	0.697	1.363	1.796	2.201	2.718	3.106	4.025
12	0.259	0.695	1.356	1.782	2.179	2.681	3.055	3.930
13	0.259	0.694	1.350	1.771	2.160	2.650	3.012	3.852
14	0.258	0.692	1.345	1.761	2.145	2.624	2.977	3.787
15	0.258	0.691	1.341	1.753	2.131	2.602	2.947	3.733
16	0.258	0.690	1.337	1.746	2.120	2.583	2.921	3.686
17	0.257	0.689	1.333	1.740	2.110	2.567	2.898	3.646
18	0.257	0.688	1.330	1.734	2.101	2.552	2.878	3.610
19	0.257	0.688	1.328	1.729	2.093	2.539	2.861	3.579
20	0.257	0.687	1.325	1.725	2.086	2.528	2.845	3.552
21	0.257	0.686	1.323	1.721	2.080	2.518	2.831	3.527
22	0.256	0.686	1.321	1.717	2.074	2.508	2.819	3.505
23	0.256	0.685	1.319	1.714	2.069	2.500	2.807	3.485
24	0.256	0.685	1.318	1.711	2.064	2.492	2.797	3.467
25	0.256	0.684	1.316	1.708	2.060	2.485	2.787	3.450
26	0.256	0.684	1.315	1.706	2.056	2.479	2.779	3.435
27	0.256	0.684	1.314	1.703	2.052	2.473	2.771	3.421
28	0.256	0.683	1.313	1.701	2.048	2.467	2.763	3.408
29	0.256	0.683	1.311	1.699	2.045	2.462	2.756	3.396
30	0.256	0.683	1.310	1.697	2.042	2.457	2.750	3.385
40	0.255	0.681	1.303	1.684	2.021	2.423	2.704	3.307
60	0.254	0.679	1.296	1.671	2.000	2.390	2.660	3.232
120	0.254	0.677	1.289	1.658	1.980	2.358	2.617	3.160
∞	0.253	0.674	1.282	1.645	1.960	2.326	2.576	3.090

This table is condensed from Table 12 of the _Biometrika Tables for Statisticians_, Vol. 1 (ed. 3), edited by E. S. Pearson and H. O. Hartley. Reproduced here with the kind permission of E. S. Pearson and the trustees of _Biometrika_.

TABLE F-4.   Percentage Points of the F Distribution (upper 5% points)

ν_2 \ ν_1	1	2	3	4	5	6	7	8	9	10	12	15	20	24	30	40	60	120	∞
1	161.4	199.5	215.7	224.6	230.2	234.0	236.8	238.9	240.5	241.9	243.9	245.9	248.0	249.1	250.1	251.1	252.2	253.3	254.3
2	18.51	19.00	19.16	19.25	19.30	19.33	19.35	19.37	19.38	19.40	19.41	19.43	19.45	19.45	19.46	19.47	19.48	19.49	19.50
3	10.13	9.55	9.28	9.12	9.01	8.94	8.89	8.85	8.81	8.79	8.74	8.70	8.66	8.64	8.62	8.59	8.57	8.55	8.53
4	7.71	6.94	6.59	6.39	6.26	6.16	6.09	6.04	6.00	5.96	5.91	5.86	5.80	5.77	5.75	5.72	5.69	5.66	5.63
5	6.61	5.79	5.41	5.19	5.05	4.95	4.88	4.82	4.77	4.74	4.68	4.62	4.56	4.53	4.50	4.46	4.43	4.40	4.36
6	5.99	5.14	4.76	4.53	4.39	4.28	4.21	4.15	4.10	4.06	4.00	3.94	3.87	3.84	3.81	3.77	3.74	3.70	3.67
7	5.59	4.74	4.35	4.12	3.97	3.87	3.79	3.73	3.68	3.64	3.57	3.51	3.44	3.41	3.38	3.34	3.30	3.27	3.23
8	5.32	4.46	4.07	3.84	3.69	3.58	3.50	3.44	3.39	3.35	3.28	3.22	3.15	3.12	3.08	3.04	3.01	2.97	2.93
9	5.12	4.26	3.86	3.63	3.48	3.37	3.29	3.23	3.18	3.14	3.07	3.01	2.94	2.90	2.86	2.83	2.79	2.75	2.71
10	4.96	4.10	3.71	3.48	3.33	3.22	3.14	3.07	3.02	2.98	2.91	2.85	2.77	2.74	2.70	2.66	2.62	2.58	2.54
11	4.84	3.98	3.59	3.36	3.20	3.09	3.01	2.95	2.90	2.85	2.79	2.72	2.65	2.61	2.57	2.53	2.49	2.45	2.40
12	4.75	3.89	3.49	3.26	3.11	3.00	2.91	2.85	2.80	2.75	2.69	2.62	2.54	2.51	2.47	2.43	2.38	2.34	2.30
13	4.67	3.81	3.41	3.18	3.03	2.92	2.83	2.77	2.71	2.67	2.60	2.53	2.46	2.42	2.38	2.34	2.30	2.25	2.21
14	4.60	3.74	3.34	3.11	2.96	2.85	2.76	2.70	2.65	2.60	2.53	2.46	2.39	2.35	2.31	2.27	2.22	2.18	2.13
15	4.54	3.68	3.29	3.06	2.90	2.79	2.71	2.64	2.59	2.54	2.48	2.40	2.33	2.29	2.25	2.20	2.16	2.11	2.07
16	4.49	3.63	3.24	3.01	2.85	2.74	2.66	2.59	2.54	2.49	2.42	2.35	2.28	2.24	2.19	2.15	2.11	2.06	2.01
17	4.45	3.59	3.20	2.96	2.81	2.70	2.61	2.55	2.49	2.45	2.38	2.31	2.23	2.19	2.15	2.10	2.06	2.01	1.96
18	4.41	3.55	3.16	2.93	2.77	2.66	2.58	2.51	2.46	2.41	2.34	2.27	2.19	2.15	2.11	2.06	2.02	1.97	1.92
19	4.38	3.52	3.13	2.90	2.74	2.63	2.54	2.48	2.42	2.38	2.31	2.23	2.16	2.11	2.07	2.03	1.98	1.93	1.88
20	4.35	3.49	3.10	2.87	2.71	2.60	2.51	2.45	2.39	2.35	2.28	2.20	2.12	2.08	2.04	1.99	1.95	1.90	1.84
21	4.32	3.47	3.07	2.84	2.68	2.57	2.49	2.42	2.37	2.32	2.25	2.18	2.10	2.05	2.01	1.96	1.92	1.87	1.81
22	4.30	3.44	3.05	2.82	2.66	2.55	2.46	2.40	2.34	2.30	2.23	2.15	2.07	2.03	1.98	1.94	1.89	1.84	1.78
23	4.28	3.42	3.03	2.80	2.64	2.53	2.44	2.37	2.32	2.27	2.20	2.13	2.05	2.01	1.96	1.91	1.86	1.81	1.76
24	4.26	3.40	3.01	2.78	2.62	2.51	2.42	2.36	2.30	2.25	2.18	2.11	2.03	1.98	1.94	1.89	1.84	1.79	1.73
25	4.24	3.39	2.99	2.76	2.60	2.49	2.40	2.34	2.28	2.24	2.16	2.09	2.01	1.96	1.92	1.87	1.82	1.77	1.71
26	4.23	3.37	2.98	2.74	2.59	2.47	2.39	2.32	2.27	2.22	2.15	2.07	1.99	1.95	1.90	1.85	1.80	1.75	1.69
27	4.21	3.35	2.96	2.73	2.57	2.46	2.37	2.31	2.25	2.20	2.13	2.06	1.97	1.93	1.88	1.84	1.79	1.73	1.67
28	4.20	3.34	2.95	2.71	2.56	2.45	2.36	2.29	2.24	2.19	2.12	2.04	1.96	1.91	1.87	1.82	1.77	1.71	1.65
29	4.18	3.33	2.93	2.70	2.55	2.43	2.35	2.28	2.22	2.18	2.10	2.03	1.94	1.90	1.85	1.81	1.75	1.70	1.64
30	4.17	3.32	2.92	2.69	2.53	2.42	2.33	2.27	2.21	2.16	2.09	2.01	1.93	1.89	1.84	1.79	1.74	1.68	1.62
40	4.08	3.23	2.84	2.61	2.45	2.34	2.25	2.18	2.12	2.08	2.00	1.92	1.84	1.79	1.74	1.69	1.64	1.58	1.51
60	4.00	3.15	2.76	2.53	2.37	2.25	2.17	2.10	2.04	1.99	1.92	1.84	1.75	1.70	1.65	1.59	1.53	1.47	1.39
120	3.92	3.07	2.68	2.45	2.29	2.17	2.09	2.02	1.96	1.91	1.83	1.75	1.66	1.61	1.55	1.50	1.43	1.35	1.25
∞	3.84	3.00	2.60	2.37	2.21	2.10	2.01	1.94	1.88	1.83	1.75	1.67	1.57	1.52	1.46	1.39	1.32	1.22	1.00

TABLE F‑4. (continued) (upper 2.5% points)

ν_2 \ ν_1	1	2	3	4	5	6	7	8	9	10	12	15	20	24	30	40	60	120	∞
1	647.8	799.5	864.2	899.6	921.8	937.1	948.2	956.7	963.3	968.6	976.7	984.9	993.1	997.2	1,001	1,006	1,010	1,014	1,018
2	38.51	39.00	39.17	39.25	39.30	39.33	39.36	39.37	39.39	39.40	39.41	39.43	39.45	39.46	39.46	39.47	39.48	39.49	39.50
3	17.44	16.04	15.44	15.10	14.88	14.73	14.62	14.54	14.47	14.42	14.34	14.25	14.17	14.12	14.08	14.04	13.99	13.95	13.90
4	12.22	10.65	9.98	9.60	9.36	9.20	9.07	8.98	8.90	8.84	8.75	8.66	8.56	8.51	8.46	8.41	8.36	8.31	8.26
5	10.01	8.43	7.76	7.39	7.15	6.98	6.85	6.76	6.68	6.62	6.52	6.43	6.33	6.28	6.23	6.18	6.12	6.07	6.02
6	8.81	7.26	6.60	6.23	5.99	5.82	5.70	5.60	5.52	5.46	5.37	5.27	5.17	5.12	5.07	5.01	4.96	4.90	4.85
7	8.07	6.54	5.89	5.52	5.29	5.12	4.99	4.90	4.82	4.76	4.67	4.57	4.47	4.42	4.36	4.31	4.25	4.20	4.14
8	7.57	6.06	5.42	5.05	4.82	4.65	4.53	4.43	4.36	4.30	4.20	4.10	4.00	3.95	3.89	3.84	3.78	3.73	3.67
9	7.21	5.71	5.08	4.72	4.48	4.32	4.20	4.10	4.03	3.96	3.87	3.77	3.67	3.61	3.56	3.51	3.45	3.39	3.33
10	6.94	5.46	4.83	4.47	4.24	4.07	3.95	3.85	3.78	3.72	3.62	3.52	3.42	3.37	3.31	3.26	3.20	3.14	3.08
11	6.72	5.26	4.63	4.28	4.04	3.88	3.76	3.66	3.59	3.53	3.43	3.33	3.23	3.17	3.12	3.06	3.00	2.94	2.88
12	6.55	5.10	4.47	4.12	3.89	3.73	3.61	3.51	3.44	3.37	3.28	3.18	3.07	3.02	2.96	2.91	2.85	2.79	2.72
13	6.41	4.97	4.35	4.00	3.77	3.60	3.48	3.39	3.31	3.25	3.15	3.05	2.95	2.89	2.84	2.78	2.72	2.66	2.60
14	6.30	4.86	4.24	3.89	3.66	3.50	3.38	3.29	3.21	3.15	3.05	2.95	2.84	2.79	2.73	2.67	2.61	2.55	2.49
15	6.20	4.77	4.15	3.80	3.58	3.41	3.29	3.20	3.12	3.06	2.96	2.86	2.76	2.70	2.64	2.59	2.52	2.46	2.40
16	6.12	4.69	4.08	3.73	3.50	3.34	3.22	3.12	3.05	2.99	2.89	2.79	2.68	2.63	2.57	2.51	2.45	2.38	2.32
17	6.04	4.62	4.01	3.66	3.44	3.28	3.16	3.06	2.98	2.92	2.82	2.72	2.62	2.56	2.50	2.44	2.38	2.32	2.25
18	5.98	4.56	3.95	3.61	3.38	3.22	3.10	3.01	2.93	2.87	2.77	2.67	2.56	2.50	2.44	2.38	2.32	2.26	2.19
19	5.92	4.51	3.90	3.56	3.33	3.17	3.05	2.96	2.88	2.82	2.72	2.62	2.51	2.45	2.39	2.33	2.27	2.20	2.13
20	5.87	4.46	3.86	3.51	3.29	3.13	3.01	2.91	2.84	2.77	2.68	2.57	2.46	2.41	2.35	2.29	2.22	2.16	2.09
21	5.83	4.42	3.82	3.48	3.25	3.09	2.97	2.87	2.80	2.73	2.64	2.53	2.42	2.37	2.31	2.25	2.18	2.11	2.04
22	5.79	4.38	3.78	3.44	3.22	3.05	2.93	2.84	2.76	2.70	2.60	2.50	2.39	2.33	2.27	2.21	2.14	2.08	2.00
23	5.75	4.35	3.75	3.41	3.18	3.02	2.90	2.81	2.73	2.67	2.57	2.47	2.36	2.30	2.24	2.18	2.11	2.04	1.97
24	5.72	4.32	3.72	3.38	3.15	2.99	2.87	2.78	2.70	2.64	2.54	2.44	2.33	2.27	2.21	2.15	2.08	2.01	1.94
25	5.69	4.29	3.69	3.35	3.13	2.97	2.85	2.75	2.68	2.61	2.51	2.41	2.30	2.24	2.18	2.12	2.05	1.98	1.91
26	5.66	4.27	3.67	3.33	3.10	2.94	2.82	2.73	2.65	2.59	2.49	2.39	2.28	2.22	2.16	2.09	2.03	1.95	1.88
27	5.63	4.24	3.65	3.31	3.08	2.92	2.80	2.71	2.63	2.57	2.47	2.36	2.25	2.19	2.13	2.07	2.00	1.93	1.85
28	5.61	4.22	3.63	3.29	3.06	2.90	2.78	2.69	2.61	2.55	2.45	2.34	2.23	2.17	2.11	2.05	1.98	1.91	1.83
29	5.59	4.20	3.61	3.27	3.04	2.88	2.76	2.67	2.59	2.53	2.43	2.32	2.21	2.15	2.09	2.03	1.96	1.89	1.81
30	5.57	4.18	3.59	3.25	3.03	2.87	2.75	2.65	2.57	2.51	2.41	2.31	2.20	2.14	2.07	2.01	1.94	1.87	1.79
40	5.42	4.05	3.46	3.13	2.90	2.74	2.62	2.53	2.45	2.39	2.29	2.18	2.07	2.01	1.94	1.88	1.80	1.72	1.64
60	5.29	3.93	3.34	3.01	2.79	2.63	2.51	2.41	2.33	2.27	2.17	2.06	1.94	1.88	1.82	1.74	1.67	1.58	1.48
120	5.15	3.80	3.23	2.89	2.67	2.52	2.39	2.30	2.22	2.16	2.05	1.94	1.82	1.76	1.69	1.61	1.53	1.43	1.31
∞	5.02	3.69	3.12	2.79	2.57	2.41	2.29	2.19	2.11	2.05	1.94	1.83	1.71	1.64	1.57	1.48	1.39	1.27	1.00

TABLE F-4. (continued) (upper 1 % points)

$\nu_2 \backslash \nu_1$	1	2	3	4	5	6	7	8	9	10	12	15	20	24	30	40	60	120	∞
1	4,052	4,999.5	5,403	5,625	5,764	5,859	5,928	5,982	6,022	6,056	6,106	6,157	6,209	6,235	6,261	6,287	6,313	6,339	6,366
2	98.50	99.00	99.17	99.25	99.30	99.33	99.36	99.37	99.39	99.40	99.42	99.43	99.45	99.46	99.47	99.47	99.48	99.49	99.50
3	34.12	30.82	29.46	28.71	28.24	27.91	27.67	27.49	27.35	27.23	27.05	26.87	26.69	26.60	26.50	26.41	26.32	26.22	26.13
4	21.20	18.00	16.69	15.98	15.52	15.21	14.98	14.80	14.66	14.55	14.37	14.20	14.02	13.93	13.84	13.75	13.65	13.56	13.46
5	16.26	13.27	12.06	11.39	10.97	10.67	10.46	10.29	10.16	10.05	9.89	9.72	9.55	9.47	9.38	9.29	9.20	9.11	9.02
6	13.75	10.92	9.78	9.15	8.75	8.47	8.26	8.10	7.98	7.87	7.72	7.56	7.40	7.31	7.23	7.14	7.06	6.97	6.88
7	12.25	9.55	8.45	7.85	7.46	7.19	6.99	6.84	6.72	6.62	6.47	6.31	6.16	6.07	5.99	5.91	5.82	5.74	5.65
8	11.26	8.65	7.59	7.01	6.63	6.37	6.18	6.03	5.91	5.81	5.67	5.52	5.36	5.28	5.20	5.12	5.03	4.95	4.86
9	10.56	8.02	6.99	6.42	6.06	5.80	5.61	5.47	5.35	5.26	5.11	4.96	4.81	4.73	4.65	4.57	4.48	4.40	4.31
10	10.04	7.56	6.55	5.99	5.64	5.39	5.20	5.06	4.94	4.85	4.71	4.56	4.41	4.33	4.25	4.17	4.08	4.00	3.91
11	9.65	7.21	6.22	5.67	5.32	5.07	4.89	4.74	4.63	4.54	4.40	4.25	4.10	4.02	3.94	3.86	3.78	3.69	3.60
12	9.33	6.93	5.95	5.41	5.06	4.82	4.64	4.50	4.39	4.30	4.16	4.01	3.86	3.78	3.70	3.62	3.54	3.45	3.36
13	9.07	6.70	5.74	5.21	4.86	4.62	4.44	4.30	4.19	4.10	3.96	3.82	3.66	3.59	3.51	3.43	3.34	3.25	3.17
14	8.86	6.51	5.56	5.04	4.69	4.46	4.28	4.14	4.03	3.94	3.80	3.66	3.51	3.43	3.35	3.27	3.18	3.09	3.00
15	8.68	6.36	5.42	4.89	4.56	4.32	4.14	4.00	3.89	3.80	3.67	3.52	3.37	3.29	3.21	3.13	3.05	2.96	2.87
16	8.53	6.23	5.29	4.77	4.44	4.20	4.03	3.89	3.78	3.69	3.55	3.41	3.26	3.18	3.10	3.02	2.93	2.84	2.75
17	8.40	6.11	5.18	4.67	4.34	4.10	3.93	3.79	3.68	3.59	3.46	3.31	3.16	3.08	3.00	2.92	2.83	2.75	2.65
18	8.29	6.01	5.09	4.58	4.25	4.01	3.84	3.71	3.60	3.51	3.37	3.23	3.08	3.00	2.92	2.84	2.75	2.66	2.57
19	8.18	5.93	5.01	4.50	4.17	3.94	3.77	3.63	3.52	3.43	3.30	3.15	3.00	2.92	2.84	2.76	2.67	2.58	2.49
20	8.10	5.85	4.94	4.43	4.10	3.87	3.70	3.56	3.46	3.37	3.23	3.09	2.94	2.86	2.78	2.69	2.61	2.52	2.42
21	8.02	5.78	4.87	4.37	4.04	3.81	3.64	3.51	3.40	3.31	3.17	3.03	2.88	2.80	2.72	2.64	2.55	2.46	2.36
22	7.95	5.72	4.82	4.31	3.99	3.76	3.59	3.45	3.35	3.26	3.12	2.98	2.83	2.75	2.67	2.58	2.50	2.40	2.31
23	7.88	5.66	4.76	4.26	3.94	3.71	3.54	3.41	3.30	3.21	3.07	2.93	2.78	2.70	2.62	2.54	2.45	2.35	2.26
24	7.82	5.61	4.72	4.22	3.90	3.67	3.50	3.36	3.26	3.17	3.03	2.89	2.74	2.66	2.58	2.49	2.40	2.31	2.21
25	7.77	5.57	4.68	4.18	3.85	3.63	3.46	3.32	3.22	3.13	2.99	2.85	2.70	2.62	2.54	2.45	2.36	2.27	2.17
26	7.72	5.53	4.64	4.14	3.82	3.59	3.42	3.29	3.18	3.09	2.96	2.81	2.66	2.58	2.50	2.42	2.33	2.23	2.13
27	7.68	5.49	4.60	4.11	3.78	3.56	3.39	3.26	3.15	3.06	2.93	2.78	2.63	2.55	2.47	2.38	2.29	2.20	2.10
28	7.64	5.45	4.57	4.07	3.75	3.53	3.36	3.23	3.12	3.03	2.90	2.75	2.60	2.52	2.44	2.35	2.26	2.17	2.06
29	7.60	5.42	4.54	4.04	3.73	3.50	3.33	3.20	3.09	3.00	2.87	2.73	2.57	2.49	2.41	2.33	2.23	2.14	2.03
30	7.56	5.39	4.51	4.02	3.70	3.47	3.30	3.17	3.07	2.98	2.84	2.70	2.55	2.47	2.39	2.30	2.21	2.11	2.01
40	7.31	5.18	4.31	3.83	3.51	3.29	3.12	2.99	2.89	2.80	2.66	2.52	2.37	2.29	2.20	2.11	2.02	1.92	1.80
60	7.08	4.98	4.13	3.65	3.34	3.12	2.95	2.82	2.72	2.63	2.50	2.35	2.20	2.12	2.03	1.94	1.84	1.73	1.60
120	6.85	4.79	3.95	3.48	3.17	2.96	2.79	2.66	2.56	2.47	2.34	2.19	2.03	1.95	1.86	1.76	1.66	1.53	1.38
∞	6.63	4.61	3.78	3.32	3.02	2.80	2.64	2.51	2.41	2.32	2.18	2.04	1.88	1.79	1.70	1.59	1.47	1.32	1.00

TABLE F-5.   Upper Percentage Points of the χ^2 Distribution

ν \ Q	0.995	0.990	0.975	0.950	0.900	0.750	0.500
1	$392{,}704 . 10^{-10}$	$157{,}088 . 10^{-9}$	$982{,}069 . 10^{-9}$	$393{,}214 . 10^{-8}$	0.0157908	0.1015308	0.454937
2	0.0100251	0.0201007	0.0506356	0.102587	0.210720	0.575364	1.38629
3	0.0717212	0.114832	0.215795	0.351846	0.584375	1.212534	2.36597
4	0.206990	0.297110	0.484419	0.710721	1.063623	1.92255	3.35670
5	0.411740	0.554300	0.831211	1.145476	1.61031	2.67460	4.35146
6	0.675727	0.872085	1.237347	1.63539	2.20413	3.45460	5.34812
7	0.989265	1.239043	1.68987	2.16735	2.83311	4.25485	6.34581
8	1.344419	1.646482	2.17973	2.73264	3.48954	5.07064	7.34412
9	1.734926	2.087912	2.70039	3.32511	4.16816	5.89883	8.34283
10	2.15585	2.55821	3.24697	3.94030	4.86518	6.73720	9.34182
11	2.60321	3.05347	3.81575	4.57481	5.57779	7.58412	10.3410
12	3.07382	3.57056	4.40379	5.22603	6.30380	8.43842	11.3403
13	3.56503	4.10691	5.00874	5.89186	7.04150	9.29906	12.3398
14	4.07468	4.66043	5.62872	6.57063	7.78953	10.1653	13.3393
15	4.60094	5.22935	6.26214	7.26094	8.54675	11.0365	14.3389
16	5.14224	5.81221	6.90766	7.96164	9.31223	11.9122	15.3385
17	5.69724	6.40776	7.56418	8.67176	10.0852	12.7919	16.3381
18	6.26481	7.01491	8.23075	9.39046	10.8649	13.6753	17.3379
19	6.84398	7.63273	8.90655	10.1170	11.6509	14.5620	18.3376
20	7.43386	8.26040	9.59083	10.8508	12.4426	15.4518	19.3374
21	8.03366	8.89720	10.28293	11.5913	13.2396	16.3444	20.3372
22	8.64272	9.54249	10.9823	12.3380	14.0415	17.2396	21.3370
23	9.26042	10.19567	11.6885	13.0905	14.8479	18.1373	22.3369
24	9.88623	10.8564	12.4011	13.8484	15.6587	19.0372	23.3367
25	10.5197	11.5240	13.1197	14.6114	16.4734	19.9393	24.3366
26	11.1603	12.1981	13.8439	15.3791	17.2919	20.8434	25.3364
27	11.8076	12.8786	14.5733	16.1513	18.1138	21.7494	26.3363
28	12.4613	13.5648	15.3079	16.9279	18.9392	22.6572	27.3363
29	13.1211	14.2565	16.0471	17.7083	19.7677	23.5666	28.3362
30	13.7867	14.9535	16.7908	18.4926	20.5992	24.4776	29.3360
40	20.7065	22.1643	24.4331	26.5093	29.0505	33.6603	39.3354
50	27.9907	29.7067	32.3574	34.7642	37.6886	42.9421	49.3349
60	35.5346	37.4848	40.4817	43.1879	46.4589	52.2938	59.3347
70	43.2752	45.4418	48.7576	51.7393	55.3290	61.6983	69.3344
80	51.1720	53.5400	57.1532	60.3915	64.2778	71.1445	79.3343
90	59.1963	61.7541	65.6466	69.1260	73.2912	80.6247	89.3342
100	67.3276	70.0648	74.2219	77.9295	82.3581	90.1332	99.3341
z_Q	-2.5758	-2.3263	-1.9600	-1.6449	-1.2816	-0.6745	0.0000

This table is taken from Table 8 of the *Biometrika Tables for Statisticians*, Vol. 1 (ed. 3), edited by E. S. Pearson and H. O. Hartley. Reproduced here with the kind permission of E. S. Pearson and the trustees of *Biometrika*.

TABLE F-5.

ν \ Q	0.250	0.100	0.050	0.025	0.010	0.005	0.001
1	1.32330	2.70554	3.84146	5.02389	6.63490	7.87944	10.828
2	2.77259	4.60517	5.99147	7.37776	9.21034	10.5966	13.816
3	4.10835	6.25139	7.81473	9.34840	11.3449	12.8381	16.266
4	5.38527	7.77944	9.48773	11.1433	13.2767	14.8602	18.467
5	6.62568	9.23635	11.0705	12.8325	15.0863	16.7496	20.515
6	7.84080	10.6446	12.5916	14.4494	16.8119	18.5476	22.458
7	9.03715	12.0170	14.0671	16.0128	18.4753	20.2777	24.322
8	10.2188	13.3616	15.5073	17.5346	20.0902	21.9550	26.125
9	11.3887	14.6837	16.9190	19.0228	21.6660	23.5893	27.877
10	12.5489	15.9871	18.3070	20.4831	23.2093	25.1882	29.588
11	13.7007	17.2750	19.6751	21.9200	24.7250	26.7569	31.264
12	14.8454	18.5494	21.0261	23.3367	26.2170	28.2995	32.909
13	15.9839	19.8119	22.3621	24.7356	27.6883	29.8194	34.528
14	17.1170	21.0642	23.6848	26.1190	29.1413	31.3193	36.123
15	18.2451	22.3072	24.9958	27.4884	30.5779	32.8013	37.697
16	19.3688	23.5418	26.2962	28.8454	31.9999	34.2672	39.252
17	20.4887	24.7690	27.5871	30.1910	33.4087	35.7185	40.790
18	21.6049	25.9894	28.8693	31.5264	34.8053	37.1564	42.312
19	22.7178	27.2036	30.1435	32.8523	36.1908	38.5822	43.820
20	23.8277	28.4120	31.4104	34.1696	37.5662	39.9968	45.315
21	24.9348	29.6151	32.6705	35.4789	38.9321	41.4010	46.797
22	26.0393	30.8133	33.9244	36.7807	40.2894	42.7956	48.268
23	27.1413	32.0069	35.1725	38.0757	41.6384	44.1813	49.728
24	28.2412	33.1963	36.4151	39.3641	42.9798	45.5585	51.179
25	29.3389	34.3816	37.6525	40.6465	44.3141	46.9278	52.620
26	30.4345	35.5631	38.8852	41.9232	45.6417	48.2899	54.052
27	31.5284	36.7412	40.1133	43.1944	46.9630	49.6449	55.476
28	32.6205	37.9159	41.3372	44.4607	48.2782	50.9933	56.892
29	33.7109	39.0875	42.5569	45.7222	49.5879	52.3356	58.302
30	34.7998	40.2560	43.7729	46.9792	50.8922	53.6720	59.703
40	45.6160	51.8050	55.7585	59.3417	63.6907	66.7659	73.402
50	56.3336	63.1671	67.5048	71.4202	76.1539	79.4900	86.661
60	66.9814	74.3970	79.0819	83.2976	88.3794	91.9517	99.607
70	77.5766	85.5271	90.5312	95.0231	100.425	104.215	112.317
80	88.1303	96.5782	101.879	106.629	112.329	116.321	124.839
90	98.6499	107.565	113.145	118.136	124.116	128.299	137.208
100	109.141	118.498	124.342	129.561	135.807	140.169	149.449
z_Q	+0.6745	+1.2816	+1.6449	+1.9600	+2.3263	+2.5758	+3.0902

TABLE F-6.   Critical Values of U

Instructions for use: Select the table which corresponds to your alpha level (0.05 or 0.01). Find the N for the smaller of your two samples (N_1) in the left-hand column. Move across the row until you enter the column corresponding to N in the larger of your two samples (N_2). (If the two samples have equal N, this decision is arbitrary.) Your obtained U is significant only if it does *not* exceed the value you have located in the body of the table.

Two-Tailed Test, $\alpha = 0.05$

N_2

N_1	3	4	5	6	7	8	9	10	11	12	13	14	15	16	17	18	19	20
1	—	—	—	—	—	—	—	—	—	—	—	—	—	—	—	—	—	—
2	—	—	—	—	—	0	0	0	0	1	1	1	1	1	2	2	2	2
3	—	—	0	1	1	2	2	3	3	4	4	5	5	6	6	7	7	8
4		0	1	2	3	4	4	5	6	7	8	9	10	11	11	12	13	13
5			2	3	5	6	7	8	9	11	12	13	14	15	17	18	19	20
6				5	6	8	10	11	13	14	16	17	19	21	22	24	25	27
7					8	10	12	14	16	18	20	22	24	26	28	30	32	34
8						13	15	17	19	22	24	26	29	31	34	36	38	41
9							17	20	23	26	28	31	34	37	39	42	45	48
10								23	26	29	33	36	39	42	45	48	52	55
11									30	33	37	40	44	47	51	55	58	62
12										37	41	45	49	53	57	61	65	69
13											45	50	54	59	63	67	72	76
14												55	59	64	67	74	78	83
15													64	70	75	80	85	90
16														75	81	86	92	98
17															87	93	99	105
18																99	106	112
19																	113	119
20																		127

Reprinted by permission from L. C. Freeman, *Elementary Applied Statistics.* New York: John Wiley and Sons, Inc., 1965.

TABLE F-6.   (continued)

Two-Tailed Test, $\alpha = 0.01$

N_2

N_1	3	4	5	6	7	8	9	10	11	12	13	14	15	16	17	18	19	20
1	—	—	—	—	—	—	—	—	—	—	—	—	—	—	—	—	—	—
2	—	—	—	—	—	—	—	—	—	—	—	—	—	—	—	—	0	0
3	—	—	—	—	—	—	0	0	0	1	1	1	2	2	2	2	3	3
4	—	—	—	0	0	1	1	2	2	3	4	4	5	5	6	6	7	8
5			0	1	2	3	3	4	5	6	7	7	8	9	10	11	12	13
6				2	3	4	5	6	7	9	10	11	12	13	15	16	17	18
7					4	6	7	9	10	12	13	15	16	18	19	21	22	24
8						8	10	12	14	16	18	19	21	23	25	27	29	31
9							11	13	16	18	20	22	24	27	29	31	33	36
10								16	18	21	24	26	29	31	34	37	39	42
11									21	24	27	30	33	36	39	42	45	48
12										27	31	34	37	41	44	47	51	54
13											34	38	42	45	49	53	56	60
14												42	46	50	54	58	63	67
15													51	55	60	64	69	73
16														60	65	70	74	79
17															70	75	81	86
18																81	87	92
19																	93	99
20																		105

TABLE F-7.   Table of Probabilities Associated with Values as Large as Observed Values of H in the Kruskal-Wallis One-way Analysis of Variance by Ranks

n_1	n_2	n_3	H	p	n_1	n_2	n_3	H	p
2	1	1	2.7000	0.500	4	3	2	6.4444	0.008
2	2	1	3.6000	0.200				6.3000	0.011
								5.4444	0.046
2	2	2	4.5714	0.067				5.4000	0.051
			3.7143	0.200				4.5111	0.098
3	1	1	3.2000	0.300				4.4444	0.102
3	2	1	4.2857	0.100	4	3	3	6.7455	0.010
			3.8571	0.133				6.7091	0.013
								5.7909	0.046
3	2	2	5.3572	0.029				5.7273	0.050
			4.7143	0.048				4.7091	0.092
			4.5000	0.067				4.7000	0.101
			4.4643	0.105	4	4	1	6.6667	0.010
3	3	1	5.1429	0.043				6.1667	0.022
			4.5714	0.100				4.9667	0.048
			4.0000	0.129				4.8667	0.054
3	3	2	6.2500	0.011				4.1667	0.082
			5.3611	0.032				4.0667	0.102
			5.1389	0.061	4	4	2	7.0364	0.006
			4.5556	0.100				6.8727	0.011
			4.2500	0.121				5.4545	0.046
3	3	3	7.2000	0.004				5.2364	0.052
			6.4889	0.011				4.5545	0.098
			5.6889	0.029				4.4455	0.103
			5.6000	0.050	4	4	3	7.1439	0.010
			5.0667	0.086				7.1364	0.011
			4.6222	0.100				5.5985	0.049
4	1	1	3.5714	0.200				5.5758	0.051
4	2	1	4.8214	0.057				4.5455	0.099
			4.5000	0.076				4.4773	0.102
			4.0179	0.114	4	4	4	7.6538	0.008
4	2	2	6.0000	0.014				7.5385	0.011
			5.3333	0.033				5.6923	0.049
			5.1250	0.052				5.6538	0.054
			4.4583	0.100				4.6539	0.097
			4.1667	0.105				4.5001	0.104
					5	1	1	3.8571	0.143

Adapted and abridged from Kruskal, W. H., and Wallis, W. A. 1952. Use of ranks in one-criterion variance analysis. *J. Amer. Statist. Ass.*, 47, 614–617, with the kind permission of the authors and the publisher. (The corrections to this table given by the authors in Errata, *J. Amer. Statist. Ass.*, 48, 910 have been incorporated.)

TABLE F-7.   (continued)

Sample sizes					Sample sizes				
n_1	n_2	n_3	H	p	n_1	n_2	n_3	H	p
4	3	1	5.8333	0.021	5	2	1	5.2500	0.036
			5.2083	0.050				5.0000	0.048
			5.0000	0.057				4.4500	0.071
			4.0556	0.093				4.2000	0.095
			3.8889	0.129				4.0500	0.119
5	2	2	6.5333	0.008				4.5487	0.099
			6.1333	0.013				4.5231	0.103
			5.1600	0.034	5	4	4	7.7604	0.009
			5.0400	0.056				7.7440	0.011
			4.3733	0.090				5.6571	0.049
			4.2933	0.122				5.6176	0.050
5	3	1	6.4000	0.012				4.6187	0.100
			4.9600	0.048				4.5527	0.102
			4.8711	0.052	5	5	1	7.3091	0.009
			4.0178	0.095				6.8364	0.011
			3.8400	0.123				5.1273	0.046
5	3	2	6.9091	0.009				4.9091	0.053
			6.8218	0.010				4.1091	0.086
			5.2509	0.049				4.0364	0.105
			5.1055	0.052	5	5	2	7.3385	0.010
			4.6509	0.091				7.2692	0.010
			4.4945	0.101				5.3385	0.047
5	3	3	7.0788	0.009				5.2462	0.051
			6.9818	0.011				4.6231	0.097
			5.6485	0.049				4.5077	0.100
			5.5152	0.051	5	5	3	7.5780	0.010
			4.5333	0.097				7.5429	0.010
			4.4121	0.109				5.7055	0.046
5	4	1	6.9545	0.008				5.6264	0.051
			6.8400	0.011				4.5451	0.100
			4.9855	0.044				4.5363	0.102
			4.8600	0.056	5	5	4	7.8229	0.010
			3.9873	0.098				7.7914	0.010
			3.9600	0.102				5.6657	0.049
5	4	2	7.2045	0.009				5.6429	0.050
			7.1182	0.010				4.5229	0.099
			5.2727	0.049				4.5200	0.101
			5.2682	0.050	5	5	5	8.0000	0.009
			4.5409	0.098				7.9800	0.010
			4.5182	0.101				5.7800	0.049
5	4	3	7.4449	0.010				5.6600	0.051
			7.3949	0.011				4.5600	0.100
			5.6564	0.049				4.5000	0.102
			5.6308	0.050					

TABLE F-8.   Table of Critical Values of D in the Kolmogorov-Smirnov One-sample Test

Sample size (N)	Level of significance for $D = maximum \|F_T(X) - F_S(X)\|$				
	0.20	*0.15*	*0.10*	*0.05*	*0.01*
1	0.900	0.925	0.950	0.975	0.995
2	0.684	0.726	0.776	0.842	0.929
3	0.565	0.597	0.642	0.708	0.828
4	0.494	0.525	0.564	0.624	0.733
5	0.446	0.474	0.510	0.565	0.669
6	0.410	0.436	0.470	0.521	0.618
7	0.381	0.405	0.438	0.486	0.577
8	0.358	0.381	0.411	0.457	0.543
9	0.339	0.360	0.388	0.432	0.514
10	0.322	0.342	0.368	0.410	0.490
11	0.307	0.326	0.352	0.391	0.468
12	0.295	0.313	0.338	0.375	0.450
13	0.284	0.302	0.325	0.361	0.433
14	0.274	0.292	0.314	0.349	0.418
15	0.266	0.283	0.304	0.338	0.404
16	0.258	0.274	0.295	0.328	0.392
17	0.250	0.266	0.286	0.318	0.381
18	0.244	0.259	0.278	0.309	0.371
19	0.237	0.252	0.272	0.301	0.363
20	0.231	0.246	0.264	0.294	0.356
25	0.21	0.22	0.24	0.27	0.32
30	0.19	0.20	0.22	0.24	0.29
35	0.18	0.19	0.21	0.23	0.27
Over 35	$\dfrac{1.07}{\sqrt{N}}$	$\dfrac{1.14}{\sqrt{N}}$	$\dfrac{1.22}{\sqrt{N}}$	$\dfrac{1.36}{\sqrt{N}}$	$\dfrac{1.63}{\sqrt{N}}$

Adapted from Massey, F. J., Jr. 1951. The Kolmogorov-Smirnov test for goodness of fit. *J. Amer. Statist. Ass.*, 46, 70, with the kind permission of the author and publisher.

TABLE F-9.   Critical Values of *W* for the Wilcoxon Test

	Level of significance for one-tailed test					Level of significance for one-tailed test			
	0.05	0.025	0.01	0.005		0.05	0.025	0.01	0.005
	Level of significance for two-tailed test					Level of significance for two-tailed test			
N	0.10	0.05	0.02	0.01	*N*	0.10	0.05	0.02	0.01
5	0	—	—	—	28	130	116	101	91
6	2	0	—	—	29	140	126	110	100
7	3	2	0	—	30	151	137	120	109
8	5	3	1	0	31	163	147	130	118
9	8	5	3	1	32	175	159	140	128
10	10	8	5	3	33	187	170	151	138
11	13	10	7	5	34	200	182	162	148
12	17	13	9	7	35	213	195	173	159
13	21	17	12	9	36	227	208	185	171
14	25	21	15	12	37	241	221	198	182
15	30	25	19	15	38	256	235	211	194
16	35	29	23	19	39	271	249	224	207
17	41	34	27	23	40	286	264	238	220
18	47	40	32	27	41	302	279	252	233
19	53	46	37	32	42	319	294	266	247
20	60	42	43	37	43	336	310	281	261
21	67	58	49	42	44	353	327	296	276
22	75	65	55	48	45	371	343	312	291
23	83	73	62	54	46	389	361	328	307
24	91	81	69	61	47	407	378	345	322
25	100	89	76	68	48	426	396	362	339
26	110	98	84	75	49	446	415	379	355
27	119	107	92	83	50	466	434	397	373

Source: From F. Wilcoxon, S. Katte, and R. A. Wilcox, *Critical Values and Probability Levels for the Wilcoxon Rank Sum Test and the Wilcoxon Signed Rank Test*, New York. Reproduced with the permission of American Cyanamid Co., 1963, and F. Wilcoxon and R. A. Wilcox, *Some Rapid Approximate Statistical Procedures*, New York, Lederle Laboratories, 1964 as used in Runyon and Haber, *Fundamentals of Behavioral Statistics*, 1967, Addison-Wesley, Reading, Mass.

For a given *N* (the number of pairs of scores), if the observed value is *less than or equal to* the value in the table for the appropriate level of significance, then reject H_0.

TABLE F-10. Table of Probabilities Associated with Values as Large as Observed Values of χr^3 in the Friedman Two-way Analysis of Variance by Ranks*

χr^3	p	χr^2	p	χr^3	p	χr^2	p
N = 2		**N = 3**		**N = 4**		**N = 5**	
0	1.000	0.000	1.000	0.0	1.000	0.0	1.000
1	0.833	0.667	0.944	0.5	0.931	0.4	0.954
3	0.500	2.000	0.528	1.5	0.653	1.2	0.691
4	0.167	2.667	0.361	2.0	0.431	1.6	0.522
		4.667	0.194	3.5	0.273	2.8	0.367
		6.000	0.028	4.5	0.125	3.6	0.182
				6.0	0.069	4.8	0.124
				6.5	0.042	5.2	0.093
				8.0	0.0046	6.4	0.039
						7.6	0.024
						8.4	0.0085
						10.0	0.00077

χr^2	p	χr^2	p	χr^2	p	χr^2	p
N = 6		**N = 7**		**N = 8**		**N = 9**	
0.00	1.000	0.000	1.000	0.00	1.000	0.000	1.000
0.33	0.956	0.286	0.964	0.25	0.967	0.222	0.971
1.00	0.740	0.857	0.768	0.75	0.794	0.667	0.814
1.33	0.570	1.143	0.620	1.00	0.654	0.889	0.865
2.33	0.430	2.000	0.486	1.75	0.531	1.556	0.569
3.00	0.252	2.571	0.305	2.25	0.355	2.000	0.398
4.00	0.184	3.429	0.237	3.00	0.285	2.667	0.328
4.33	0.142	3.714	0.192	3.25	0.236	2.889	0.278
5.33	0.072	4.571	0.112	4.00	0.149	3.556	0.187
6.33	0.052	5.429	0.085	4.75	0.120	4.222	0.154
7.00	0.029	6.000	0.052	5.25	0.079	4.667	0.107
8.33	0.012	7.143	0.027	6.25	0.047	5.556	0.069
9.00	0.0081	7.714	0.021	6.75	0.038	6.000	0.057
9.33	0.0055	8.000	0.016	7.00	0.030	6.222	0.048
10.33	0.0017	8.857	0.0084	7.75	0.018	6.889	0.031
12.00	0.00013	10.286	0.0036	9.00	0.0099	8.000	0.019
		10.571	0.0027	9.25	0.0080	8.222	0.016
		11.143	0.0012	9.75	0.0048	8.667	0.010
		12.286	0.00032	10.75	0.0024	9.556	0.0060
		14.000	0.000021	12.00	0.0011	10.667	0.0035
				12.25	0.00086	10.889	0.0029
				13.00	0.00026	11.556	0.0013
				14.25	0.000061	12.667	0.00066
				16.00	0.0000036	13.556	0.00035
						14.000	0.00020
						14.222	0.000097
						14.889	0.000034
						16.222	0.000011
						18.000	0.0000006

Adapted from Friedman, M. 1937. The use of ranks to avoid the assumption of normality implicit in the analysis of variance. *J. Amer. Statist. Ass.*, 32, 688–689, with the kind permission of the author and the publisher.

TABLE F-10.   (continued)

χr^2	p	χr^2	p	χr^2	p	χr^2	p
	$N = 2$		$N = 3$			$N = 4$	
0.0	1.000	0.2	1.000	0.0	1.000	5.7	0.141
0.6	0.958	0.6	0.958	0.3	0.992	6.0	0.105
1.2	0.834	1.0	0.910	0.6	0.928	6.3	0.094
1.8	0.792	1.8	0.727	0.9	0.900	6.6	0.077
2.4	0.625	2.2	0.608	1.2	0.800	6.9	0.068
3.0	0.542	2.6	0.524	1.5	0.754	7.2	0.054
3.6	0.458	3.4	0.446	1.8	0.677	7.5	0.052
4.2	0.375	3.8	0.342	2.1	0.649	7.8	0.036
4.8	0.208	4.2	0.300	2.4	0.524	8.1	0.033
5.4	0.167	5.0	0.207	2.7	0.508	8.4	0.019
6.0	0.042	5.4	0.175	3.0	0.432	8.7	0.014
		5.8	0.148	3.3	0.389	9.3	0.012
		6.6	0.075	3.6	0.355	9.6	0.0069
		7.0	0.054	3.9	0.324	9.9	0.0062
		7.4	0.033	4.5	0.242	10.2	0.0027
		8.2	0.017	4.8	0.200	10.8	0.0016
		9.0	0.0017	5.1	0.190	11.1	0.00094
				5.4	0.158	12.0	0.000072

TABLE F-11.   Critical Values of Rho

Instructions for use: Locate the number of observations (difference scores) appropriate to your study in the left-hand column (*N*). Move across this row until you enter the column that corresponds to your selected alpha level. If the value in the body of the table is exceeded by your obtained value of rho, it is significant at the alpha level you have selected.

| N | Level of Signficance α | | | | |
	0.20	0.10	0.05	0.02	0.01
4		1.00			
5	0.80	0.90		1.00	
6	0.66	0.83	0.89	0.94	1.00
7	0.57	0.71	0.79	0.89	0.93
8	0.52	0.64	0.74	0.83	0.88
9	0.48	0.60	0.68	0.78	0.83
10	0.45	0.56	0.65	0.73	0.79
11	0.41	0.52	0.61	0.71	0.77
12	0.39	0.50	0.59	0.68	0.75
13	0.37	0.47	0.56	0.65	0.71
14	0.36	0.46	0.54	0.63	0.69
15	0.34	0.44	0.52	0.60	0.66
16	0.33	0.42	0.51	0.58	0.64
17	0.32	0.41	0.49	0.57	0.62
18	0.31	0.40	0.48	0.55	0.61
19	0.30	0.39	0.46	0.54	0.60
20	0.29	0.38	0.45	0.53	0.58
21	0.29	0.37	0.44	0.51	0.56
22	0.28	0.36	0.43	0.50	0.55
23	0.27	0.35	0.42	0.49	0.54
24	0.27	0.34	0.41	0.48	0.53
25	0.26	0.34	0.40	0.47	0.52
26	0.26	0.33	0.39	0.46	0.51
27	0.25	0.32	0.38	0.45	0.50
28	0.25	0.32	0.38	0.44	0.49
29	0.24	0.31	0.37	0.44	0.48
30	0.24	0.31	0.36	0.43	0.47

Reproduced by permission from E. G. Old, Distribution of sums of squares of rank differences, and The 5% significance levels for sums of squares of rank differences and a correction, *Annals of Mathematical Statistics,* 9:133–148, 1938 and 20:117–118, 1949.

Bibliography

We provide here a brief list of references to supplement the material in this book.

Measurement

Baker, B. O., Hardyck, C. D., and Petrinovich, L. F. Weak measurements vs. strong statistics: an empirical critique of S. S. Stevens' proscriptions on statistics. *Educational and Psychological Measurement,* 1966, *26,* 291–309.

Coombs, C. H., Dawes, R. M., and Tversky, A. *Mathematical psychology, an elementary introduction.* Englewood Cliffs, NJ: Prentice-Hall, 1970.

Loftus, G. R. On interpretation of interactions. *Memory and Cognition,* 1978, *6,* 312–319.

Torgerson, W. *Theory and methods of scaling.* New York: Wiley, 1958.

Probability Theory

Feller, W. *An introduction to probability theory and its applications,* Vol. I (3rd ed.). New York: Wiley, 1968.

Jeffreys, H. *Theory of probability.* Oxford: Clarendon Press, 1961.

Mosteller, F., Rourke, R.E.K., and Thomas, G. B. *Probability with statistical applications.* Reading, MA: Addison-Wesley, 1961.

Parzen, E. *Modern probability theory and its applications.* New York: Wiley, 1960.

Descriptive Statistics

Freedman, D., Pisani, R., and Purves, R. *Statistics.* New York: Norton, 1978.

Tukey, J. W. *Exploratory data analysis.* Reading, MA: Addison-Wesley, 1976.

General Statistics

Alexander, H. W. *Elements of mathematical statistics.* New York: Wiley, 1961.

Bakan, D. The test of significance in psychological research. *Psychological Bulletin,* 1966, *66,* 423–437.

Dixon, W., and Massey, F. *Introduction to statistical analysis* (2nd ed.). New York: McGraw-Hill, 1957.

Freund, J. E. *Mathematical statistics.* Englewood Cliffs, NJ: Prentice-Hall, 1962.

Grant, D. A. Testing the null hypothesis and the strategy and tactics of investigating theoretical models. *Psychological Review,* 1962, *69,* 54–61.

624

Green, B. F., and Tukey, J. W. Complex analyses of variance: General problems. *Psychometrika,* 1960, *25,* 127–152.

Hays, W. L. *Statistics for the social sciences* (2nd ed.). New York: Holt, Rinehart & Winston, 1973.

Huff, D. *How to lie with statistics.* New York: Norton, 1954.

Rosenthal, R., and Gaito, J. The interpretation of levels of significance by psychological researchers. *Journal of Psychology,* 1963, *55,* 33–38.

Scheffé, H. *The analysis of variance.* New York: Wiley, 1959.

Sampling

Cochran, W. G. *Sampling techniques* (2nd ed.). New York: Wiley, 1966.

Power

Cohen, J. *Statistical power analysis for the behavioral sciences.* New York: Academic Press, 1969.

Post-Hoc Tests

Dunnett, C. W. A multiple comparison procedure for comparing several treatments with a control. *Journal of the American Statistical Association,* 1955, *50,* 1096–1121.

Dunnett, C. W. New tables for multiple comparisons with a control. *Biometrics,* 1964, *20,* 482–491.

Gaito, J. Unequal intervals and unequal *n* in trend analyses. *Psychological Bulletin,* 1969, *72,* 273–276.

Harter, H. L. Error rates and sample sizes for range tests in multiple comparisons. *Biometrics,* 1957, *13,* 511–536.

Petrinovich, L. F., and Hardyck, C. D. Error rates for multiple comparison methods: Some evidence concerning the frequency of erroneous conclusions. *Psychological Bulletin,* 1969, *71,* 43–54.

Strength of Association

Fleiss, J. L. Estimating the magnitude of experimental effects. *Psychological Bulletin,* 1969, *72,* 273–276.

Vaughan, G. M., and Corballis, M. C. Beyond tests of significance: Estimating strength of effects in selected ANOVA designs. *Psychological Bulletin,* 1969, *72,* 204–213.

Nonparametric Statistics

Bradley, J. V. *Distribution-free statistical tests.* Englewood Cliffs, NJ: Prentice-Hall, 1968.

Kendall, M. G. *Rank correlation methods* (3rd ed.). London: Griffin, 1963.

Kruskal, W. H. Ordinal measures of association. *Journal of the American Statistical Association,* 1958, *53,* 814–861.

Siegel, S. *Nonparametric statistics for the behavioral sciences.* New York: McGraw-Hill, 1956.

Index